Quantitative Structure-Activity Relationships (QSAR) for Pesticide Regulatory Purposes

Quantitative Structure-Activity Relationships (QSAR) for Pesticide Regulatory Purposes

Edited by

EMILIO BENFENATI

Laboratory of Environmental Chemistry and Toxicology,
Istituto di Richerche Farmacologiche "Mario Negri",
Milano, Italy

ELSEVIER

Amsterdam • Boston • Heidelberg • London • New York • Oxford
Paris • San Diego • San Francisco • Singapore • Sydney • Tokyo

Elsevier
Radarweg 29, PO Box 211, 1000 AE Amsterdam, The Netherlands
Linacre House, Jordan Hill, Oxford OX2 8DP, UK

First edition 2007

Library of Congress Cataloging-in-Publication Data
A catalog record for this book is available from the Library of Congress

British Library Cataloguing in Publication Data
A catalogue record for this book is available from the British Library

ISBN-13: 978-0-444-52710-3

Printed and bound in The Netherlands

07 08 09 10 11 10 9 8 7 6 5 4 3 2 1

Contents

CHAPTER 5
Hybrid systems 149

Nicolas Amaury, Emilio Benfenati, Severin Bumbaru, Antonio Chana, Marian Craciun, Jacques R. Chrétien, Giuseppina Gini, Gongde Guo, Frank Lemke, Viorel Minzu, Johann-Adolf Müller, Daniel Neagu, Marco Pintore, Silviu Augustin Stroia, Paul Trundle

CHAPTER 6
Validation of the models 185

Emilio Benfenati, Jacques R. Chrétien, Giuseppina Gini, Nadège Piclin, Marco Pintore, Alessandra Roncaglioni

Forewords

THE DEMETRA PROJECT: AN INNOVATIVE CONTRIBUTION TO REGULATORY QSAR

Within DEMETRA, we wanted to produce software for a specific application: the prediction of ecotoxicological properties of pesticides. This application-driven approach is very different from most of the thousands of QSAR methods so far published, in its very premises. We wanted to develop software to be used by industries and regulators, so the first action was to identify their *needs* and the related *constrains*, to be applied to the software. As a consequence, two major differences compared to all the other models resulted:

1) the targets organisms to be modelled (i.e. the toxicity endpoints) have been decided *not* by the modellers (as typically done in most of the QSAR published studies) but by the users, according to a series of criteria that have been deeply studied and clearly defined by users. The criteria have been applied to identify where there was more utility for a predictive model, to reduce the costs, the use of animals and to gain a maximum benefit of the QSAR model. This activity is, of course, very specific, but in our opinion should be done in all the cases where a QSAR model is developed to a specific purpose and not as a general theoretical tool.

2) The QSAR models have been developed and optimized according to the specific criteria defined with the users and not only according to generic mathematical features, i.e. the so obtained QSAR models encode the desired features. This refers for instance to the quality and source of the data and to the careful check for the presence of the so-called false negatives: indeed, what it has primarily to be avoided is to define not toxic a compound which *vice versa* is toxic, because this may result in serious environmental problems.

In terms of the model, this careful definition of the constrains mainly involves the input and the output of the model, because the users have to define what is important as output and what is reliable as input. We have to remember that the final target of the QSAR model for toxicity prediction is the risk assessment of chemical compounds and that, if the related issues are not fully addressed, the model can be perfect, but it will not be used. In other words, the QSAR model is only a segment in a more complex process, and if this is not taken into account, the necessary links with the real world are lost.

The activity to satisfy all users' needs requested a long discussion, which involved not only partners of the consortium but also users outside. However, for a QSAR model to be used, it is not sufficient that it addressed an endpoint useful for regulatory purposes: the model has to be recognised as valid, reliable and reproducible. Thus, we also need to put very solid basis for the model, i.e. the toxicity data have to be of high quality. It means that these data have to be produced *only* according to official protocols. Indeed, in the real word, to assess the toxicity of a pesticide, only experiments done according to official protocols are valid. This point refers to the characterization of the inputs, in terms of quality.

However, we added, as far a possible, further quality criteria, comparing data between three high-quality databases, in the modelling phase, and we further checked the quality of the used data with five other high-quality databases in the validation phase. The use and comparison of so many high-quality data is unique in the QSAR modelling.

The comparison of the experimental data allowed us also to have a much better characterisation of the variability of the experimental data. This is another unique feature of our project, because in practically all QSAR models, only a single value for toxicity is used per chemical, without any knowledge on the related variability of the toxicity data. But of course the accuracy of any model is related to the variability of the input data.

We also mention here that we applied restrictive rules, for instance using for our models only pesticides with a relatively small variability, to have even more reliable data, eliminating pesticides whose toxicity values have been accepted for regulatory purposes. In this, our model relies on data of higher quality than those used for regulatory purposes, which do not define an acceptability level for the variability of the experimental data.

The quality of the input data is not only related to the toxicity data. The data on the chemical compounds have to be of high quality too, of course. This issue is not often mentioned in QSAR studies, because it is assumed that the chemical information is correct. Actually, there are possibilities of many mistakes starting from the very simple chemical formula, chemical identification, chemical structure, etc. We checked all these elements (and we also found some mistakes in the original used databases), and all chemical structures have been independently checked by at least two researchers, using different sources and methods.

For the chemical structures, we also used crystallographic data and *ab initio* calculations, even to identify which tautomer use. This is a further procedure to achieve high-quality data, even if not strictly necessary, because less valid approaches can be used. In this way, we addressed the issue of the validity and reproducibility of the model.

All these steps in the quality assessment of the input testify the great attention given to put the best basements to our models. This took a long effort and time-consuming activities that are very seldom done because expensive in terms of human resources.

In the specific case of QSAR for pesticide ecotoxicity, many different mechanisms exist, producing the final toxicity. We wanted to model the heterogeneous classes of pesticides, because developing some simpler models for specific classes, such as triazines, would be surely much easier, but the utility much reduced, or maybe null, because nobody would use them, for the low interest in developing other triazines. In future development, more focused models can be added, but we gave priority to the development of a general tool.

We verified that, in this heterogeneous complex situation, no simple model can produce acceptable results for ecotoxicological properties. A recent document from the Danish EPA reached the same conclusions (Hansen, 2004). The approach we used was to develop advanced models, taking advantage of different innovative methodologies, both to describe the chemical compounds and to produce sophisticated algorithms. *A priori* it is difficult to choose a suitable combination of chemical descriptors and algorithms. Many attempts have been done, within a good collaboration between partners with different skills.

We are convinced that for a heterogeneous data set, as in our case, no unique model exist. Several possibilities exist, which provide more or less valid models. Our decision was to combine different models, into an integrated, hybrid system. In this way, positive features of different methods can be added.

The reader may feel uncomfortable for the use of exoteric techniques. Actually, all our modelling studies, all the mathematical algorithms, have been used to identify a list of best models. These models have been combined, but the final model is relatively simple, such as a linear equation with a series of coefficients and chemical parameters. What has been difficult was to achieve these coefficients and to identify the most important chemical features. And for this advanced methods were necessary, because simple tools are not sufficient, as also reported by the Danish EPA.

Nowadays advanced information technology tools are becoming part of our life. For instance, artificial neural networks, which have been tested in our project, are continuously used by everybody for common electronic tools.

Actually we put some efforts in reducing the complexity of the models, for instance preferring chemical descriptors which can be calculated by users without buying many expensive packages or avoiding complex three-dimensional descriptors which require time and experience.

As we said, the final models are relatively simple, and no *ab initio* calculation, no complex mathematical knowledge is required to run the model. All the theory and the techniques we used are presented in this book, but this knowledge is not a requirement for the use of the model. It is presented for correctness and for interested scientists and users.

We think it is important to clarify the criteria that have to be used in the evaluation of the models we propose. These criteria have to be the same used for the general assessment of pesticides, applied to the specific case. We said that the model should be useful, reliable and reproducible. Furthermore, for the use of the model, we have to provide information on some other points, to achieve a reliable model: the validation procedure and the applicability domain.

For the validation, we applied a battery of tools, including internal and external validation. The applicability domain refers to pesticides, of course, but specific boundaries have been evaluated and characterized.

Finally, the issue of reproducibility. We already introduced some considerations on the reproducibility of chemical structures. More important, we produced models that will give the same result once applied by the different users. Some models, especially those involving optimization of the three-dimensional structure, can provide different chemical descriptors, depending on the manual procedure to optimize the three-dimensional structure. We avoided this risk. Parameters for the final model are fixed. We also make available some of the general modelling tools for scientists who want to develop their own models, also for other purposes, but these tools should not be confused with the models produced for QSAR for regulatory purposes, which passed all the quality criteria we introduced.

Now the word is given to the users. These models represent the state-of-the-art in modelling properties of pesticides. They are not the ultimate models, and improvements are possible. The main source of improvement is the extension of the toxicity experimental data used to build up the model. We solicit industry to make available experimental data of their toxicity studies for more chemicals, to improve the knowledge basis of future models.

The present book explains in detail the activities done within DEMETRA. We believe that this experience represents an useful example not only for the case of pesticides but also for the prediction of ecotoxicity and toxicity in general, for the innovative approach and methodologies developed. For this reason, the lessons here given apply to a much broader field, when QSAR wants to contribute solving real world problems.

Emilio Benfenati
Coordinator of DEMETRA

REFERENCE

Hansen O.C. (2004) Quantitative Structure-Activity Relationships (QSAR) and Pesticides. Danish Environmental Protection Agency, Pesticides Research No. 94.

Preface

Emilio Benfenati, Mosè Casalegno

Laboratory of Environmental Chemistry and Toxicology, Istituto di Ricerche Farmacologiche “Mario Negri”, Milano, Italy

THE PESTICIDES AND THEIR ECOTOXICOLOGICAL PROPERTIES

Since before 500 BC, humans have used pesticides to prevent damage to their crops. The first known use of pesticide was sulphur. By the fifteenth century, toxic chemicals such as arsenic, mercury and lead were being applied to crops to kill pests. In the seventeenth century, nicotine sulphate was extracted from tobacco leaves for use as an insecticide. The nineteenth century saw the introduction of two more natural pesticides, pyrethrum, which came from chrysanthemums, and rotenone, from the roots of tropical legumes. In 1939, Paul Müller discovered that DDT was a very effective insecticide. It quickly became the most widely used pesticide in the world. However, in the 1960s, it was discovered that DDT was preventing many fish-eating birds from reproducing, which was a huge threat to biodiversity. DDT was also found to cause birth defects in animals and humans. DDT is now banned in at least 86 countries but is still used in some developing nations to prevent malaria and other tropical diseases by killing mosquitoes and other disease-carrying insects. DDT represented the main precursor of modern pesticides.

Nowadays, the term pesticide indicates different chemicals intended for preventing, destroying, repelling or mitigating any pest. These include algicides, antifouling agents, antimicrobials, biopesticides, biocides, disinfectants, fungicides, fumigants, herbicides, insecticides and many others (Tomlin, 1997; Hurst *et al.*, 1991). Pesticides are used worldwide to reduce the damage to plants by insects and other pests, to control overgrowth of undesirable plant species and to protect public health from disease vectors such as mosquitoes, ticks, cockroaches, rats and disease-causing organisms. Accepting all the benefits coming from the use of pesticides means also accepting the related risks. Agricultural and domestic use of pesticides inevitably leads to exposure of non-target organisms, including humans. For this reason, pesticides’ toxic potential has to be carefully evaluated before marketing and distribution, taking into account that each compound might be harmful to humans, as well as to other animal species, and the environment. Risk assessment procedures aimed at evaluating the impact of pesticides on the environment are extremely demanding in terms of money and time.

MOVING FORWARDS THE USE OF QSAR TO PREDICT TOXICOLOGICAL PROPERTIES

The multitude of different compounds, non-target organisms (pets, birds, fishes and mammalians) and adverse effects (eye and skin irritation, neurotoxicity, cancer and birth defects) to be tested require huge efforts in terms of testing animals and money. To reduce direct costs, a number of possible alternatives have been proposed and evaluated by regulatory authorities. Among them, the use of (Quantitative) Structure-Activity Relationships [(Q)SAR] is strongly encouraged by both EU and USA regulators as a tool for supporting and optimizing risk assessment strategies. In Chapter 1, we will present many cases were QSAR is used for regulatory purposes in the world.

(Q)SARs are estimation methods developed and used to predict certain properties of chemicals which are primarily based on the structure of the substance. Use of (Q)SAR techniques will allow potential savings of million test animals and billion euros, boosting cost- and time-effectiveness of risk assessment procedure. Within the EU risk assessment framework, the role of QSAR has been clearly stated in the technical guidance document in support to the EU directive 93/67/EEC (EEC, 1996). In the document, a general framework in which (Q)SARs can be used within the risk assessment process is presented. Use of (Q)SARs is proposed for the following purposes:

1) Assisting data evaluation.

2) Contributing to the decision-making process on whether further testing is necessary to clarify an endpoint of concern.

3) Establishing input parameters necessary to conduct exposure assessment.

4) Identifying effects which may be of potential concern on which test data are not available.

All the four above listed purposes testify the importance of (Q)SAR in supporting the risk assessment procedures.

More recently, the REACH legislation clearly mentioned QSAR as tool to reduce the use of animals and resources for the purpose of assessing industrial chemicals. This also suggests that (Q)SARs are ideal tools for addressing regulatory tasks, and indeed, they are widely used by USEPA for these purposes. In Europe, their use will be strongly encouraged within current and forthcoming EU chemical policies. At present, however, several concerns about the validity and applicability of the (Q)SARs have not yet been solved.

The lack of standardised, reproducible, and reliable (Q)SAR protocols has raised serious concerns about the reliability of current *in silico* predictions. To date, more than 20 000 (Q)SARs have been developed and published, each adopting different combinations of human hand-feeding actions and computational resources. Despite their effectiveness, none of them could individually face the challenge posed by the current EU chemical legislation.

THE DEMETRA PROJECT

To be used for optimally exploit each model's potential and target regulatory objectives, the most reliable solution would be to combine several models, integrating them into a decision support system. The DEMETRA project has been developed following this principle, with the aim to assist regulators in taking decisions during the risk assessment process. The project's heart is the decision to refer in all steps to the target, which means to refer to the intended use of the models according to the latest EU regulatory directives. With its innovative applications and services, DEMETRA places the current (Q)SAR dimension closer to the regulatory one.

The main objective of DEMETRA was to develop tools for pesticides and related compounds (such as their metabolites) toxicity prediction against five endpoints. Regulatory bodies, industries, non-governmental organizations and researchers are major potential users that might be interested in exploiting the software. In addition, regulatory evaluators might greatly benefit by using DEMETRA in the data evaluation process for approval applications. The intrinsic complexity of the project planning and development has made mandatory the splitting of the main goal in several sub goals, as defined in the project and here reported:

- To select at least five relevant toxicity endpoints for modelling in this project and to compile the quality-controlled data sets required for the project.
- To calculate chemical descriptors of the chemicals selected in the five data sets as above defined.
- To analyse, develop and propose algorithms for toxicity of pesticides.
- To integrate the knowledge acquired through different approaches in a homogeneous manner, within a hybrid system, for each endpoint.
- To validate the hybrid systems for the five selected endpoints.
- To develop the web site as user-oriented portal to access the hybrid systems, the online documentation and user manuals.
- To exploit and to disseminate results of the project, notably to wider number of regulatory bodies and users and to organize a European workshop to present results.

The list above depicts a clear picture of all different aspects covered by DEMETRA during its development. Each objective represents a single project step to be addressed before passing to the next one.

Contractors and subcontractors involved in the project are listed below.

Participant	**Participant Name**	**Participant Short Name**	**Team Leader**	**Town, Country**
P1 (CO)	"Mario Negri" Institute	IRFMN	Emilio Benfe-nati	Milano, Italy
P2	BioChemics Consulting SAS	BCX	Marco Pintore	Orléans, France
P3	Central Science Laboratory	CSL	Qasim Chaudhry	York, UK
P4	University "Dunarea de Jos" of Galati	UGAL	Viorel Minzu	Galati, Romania
P5	Politecnico di Milano	POLIMI	Giuseppina Gini	Milano, Italy
P6	University of Patras	UNIPATRAS	Nikolaos Avouris	Patras, Greece
P7	Syngenta Croop Protection AG	SYNGENTA	Bruno Lefevbre	Basel, Switzerland

Participant	**Participant Name**	**Participant Short Name**	**Team Leader**	**Town, Country**
S1	The Pesticide Safety Directorate	PSD	Mark Clook	York, UK
S2	BASF Agricultural Centre	BASF	Peter Dolmen	Limburgerhof, Germany

Participant	Participant Name	Participant Short Name	Team Leader	Town, Country
S3	KnowledgeMiner Software Frank Lemke	KNOWLEDGE MINER	Frank Lemke	Panketal, Germany
S4	Technology for Growth	TfG	Nick Price	York, UK
S5	Bradford University	UBRAD	Daniel Neagu	Bradford, UK

THE BOOK CHAPTERS

The following chapters will present the main theoretical issues, the methodologies and the results. Chapter 1 gives details of the legislative requirements according to the EU regulation. On the basis of these requirements, the main used endpoints for pesticides are presented, and criteria established to identify the most useful QSAR models, to reduce the number of animals, the cost of the experiment, the frequency of the tests, their severity, etc. QSAR requirements are also discussed on the basis of opinions from users.

Chapter 2 describes the sources of the toxicity data we choose. Only high-quality data have been considered. The reproducibility of the data is presented. We also compared the values in the different databases. In order to achieve a standardized format for data representation, a module was produced for the XML format.

Chapter 2 also describes how data have been selected, from the source identified above. It is common that for the same pesticide more then one toxicity value is reported, because of the variability and uncertainty of the experimental procedure. We defined a protocol to select the most useful and reliable values. We report the five datasets for the five endpoints that have been finally considered – two aquatic endpoints: trout and daphnia; quail (dietary and oral exposure); finally bee.

Chapter 3 explains how to define and process chemical structures, and what kind of chemical descriptors to calculate. Two- and three-dimensional descriptors are introduced. Another possibility is the use of chemical fragments. Several thousands of chemical parameters can be obtained.

Chapter 4 describes the algorithms for *in silico* modelling. We discuss the methods to select chemical descriptors, which is useful considering the presence of a very high number of parameters. Methods to obtain continuous or categorical toxicity values are discussed.

Chapter 5 explains the possible ways to integrate different models into a combined, hybrid model. Taking advantage of the positive features of the separate individual models within an intelligent strategy improves the overall final results.

Chapter 6 deals with the validation of QSAR models. Internal and external (with a separate test set) validation methods are presented. Validation in case of regression methods and classifiers are described. Besides mathematical methods, we discuss the specific features requested by the intended application of the models, describing the false-negative issue.

Chapter 7 presents the results of the DEMETRA models. Thousands of models have been obtained, and here, we report the most successful. Results are discussed relatively to classification methods regression models and hybrid systems. Five separate hybrid models have been obtained, one for each selected toxicity endpoint.

Chapter 8 discusses the innovative aspects of the DEMETRA models. The OECD guidelines for validation of QSAR are introduced and DEMETRA models commented in relation to them. Future perspectives are also presented.

Chapter 9 describes the public use of the obtained models. The five final models are public available, for non-commercial use, through the Internet. The final models have been optimized for the five endpoints, as described in the book. In addition, a general, flexible toolbox is available, to develop further models, for other purposes.

ACKNOWLEDGEMENT

The editor gratefully acknowledges financial support from the Commission of the European Communities, under the European Union's Fifth Framework for Research and Technological Development Programme, for the project "Development of Environmental Modules for Evaluation of Toxicity of pesticide Residues in Agriculture" QLK5-CT-2002-00691.

DISCLAIMER

This publication does not necessarily reflect the European Commission's views and in no way anticipates the Commission's future policy in this area. Its content is the sole responsibility of the authors.

REFERENCES

EEC (1996) Technical Guidance Document in support of Commission directive 93/67/EEC for new notified substances and Commission regulation (EC) No 1488/94 on risk assessment for existing substances. Brussels, Office for Official publications of the European Communities, 1996, ISBN 92-827-8013-9.

Hurst P., Hay A., Dudley N. (1991) The Pesticides Handbook. Journeyman.

Tomlin C. (1997) The Pesticide Manual (11th edition). British Crop Protection Council, Farnham, UK.

Quantitative Structure-Activity Relationships (QSAR) for Pesticide Regulatory Purposes
Edited by Emilio Benfenati

Chapter 1

QSARs for regulatory purposes: the case for pesticide authorization

Emilio Benfenati[1], Mark Clook[2], Steven Fryday[3], Andy Hart[3]

[1]Istituto di Ricerche Farmacologiche "Mario Negri", Milano, Italy

[2]Pesticides Safety Directorate, Department for Environment, Food and Rural Affairs, Mallard House, York, UK

[3]Central Science Laboratory, Department for Food and Rural Affairs, Sand Hutton, York, UK

1. OVERVIEW OF THE CURRENT PESTICIDE AUTHORIZATION PROCEDURE

1.1. Description of the current pesticide legislation (EU Directive 91/414/EEC)

Currently in the EU, pesticides are assessed through the EU Directive 91/414/EEC (EEC, 1991). This Directive and the associated Annexes cover the risk to the operator, consumer and the environment. Annex II outlines what data are required on the active substance (AS), whilst Annex III indicates the data required for the associated product (AP). Annex VI, or the Uniform Principles, outlines, amongst other issues, the decision-making criteria that need to be considered prior to an AS being placed on Annex I and the associated plant protection product being authorized.

The risk to the environment covers both the fate and behaviour of an AS (i.e. exposure) as well as its possible effects to non-target organisms. Non-target organisms considered under 91/414/EEC include the following: birds, mammals, aquatic life (including fish, aquatic invertebrates, algae and aquatic plants), non-target arthropods, honeybees, earthworms, soil macro-invertebrates, soil microbial processes and terrestrial non-target plants.

The risk assessment carried out for non-target organisms currently takes a single point estimate of toxicity as well as exposure. This results in either a 'toxicity-exposure ratio' (i.e. TER) or 'hazard quotient' (HQ) which is then compared to a regulatory trigger value in the Uniform Principles of 91/414/EEC (Council Directive 94/43/EC). If the relevant trigger value is breached then no authorization can be granted 'unless it is clearly established through an appropriate risk assessment that under field conditions no unacceptable impact occurs after use of the plant protection product according to the proposed conditions of use'. This 'appropriate risk assessment' usually takes the form of further information on either the toxicity of the compound or the exposure of non-target organisms to the compound. Examples of how this tiered approach works are outlined below.

When the risk to birds or mammals is assessed, the process as outlined in SANCO/4145/2000 is followed (EEC, 2002a). For the first tier assessment, the look-up tables (Tables 4, 6 and 7 in EEC, 2002a) are used to produce the 'estimated theoretical exposure' for the acute, short-term and long-term assessment respectively. These end-points are then compared to the appropriate toxicity end-points. If any of the resulting TERs are less than the appropriate Annex VI trigger values, then the assessment may be refined using the various steps outlined in the document. For example, the amount of food obtained from the treated area as well as the proportion of different food types in the diet can be factored into the assessment (Section 5.6 of EEC, 2002a). It is also possible, but not recommended, to refine the acute risk assessment using additional acute toxicity data (Section 5.1 in EEC, 2002a). Sometimes, a weight of evidence approach is used where information from laboratory studies, field trials and wildlife incidents are used to determine the impact of a plant protection product.

When the risk to aquatic life is assessed, the risk to fish, aquatic invertebrates (i.e. always a free-swimming invertebrate – *Daphnia magna* and if triggered a sediment dwelling invertebrate – *Chironomus riparius*), algae and higher aquatic plant (if the compound is a herbicide) is considered. Currently, the predicted environmental concentration (or PEC) in surface water is determined via the use of standard assessment models (see FOCUS website at http://viso.ei.jrc.it\focus\sw\index.html). The resulting predicted environmental concentration (PEC) is compared to the appropriate toxicity end-point. If the resulting TER is less than the appropriate Annex VI trigger value, then the assessment should be refined. Generally, refinement takes the form of higher tier toxicity studies (EEC, 2002b – Section 5 of SANCO 3268rev4). These studies fall into two different types: one assesses the effects of the compound on test species under more realistic exposure scenarios, whilst another type of tests assesses the toxicity of the compound on a wide range of species [see Section 5.3 in EEC (2002b)]. Finally, complex field or mesocosm studies cover not only realistic fate and large number of species, but also other relevant parameters such as interactions, secondary effects and recovery. The output from these studies is then used to produce a regulatory end-point. This end-point may or may not include additional uncertainty factors depending upon the type of study used. This end-point is then compared to the PEC and an assessment made of its 'acceptability'. In addition, it may be possible to refine the risk by reassessing the exposure issue; however, this has yet to be fully developed within the European context.

When the risk to non-target arthropods is determined, laboratory 'worst case' toxicity studies with two indicator species (*Typhlodromus pyri* and *Aphidius rhopalosiphi*) are performed and LR50 end-points produced (i.e. the rate causing

50% mortality of the test organisms). These end-points are then used together with the in- and off-field exposure estimates to determine the HQ, i.e. in-field application exposure scenario/LR50. If the resulting HQ is greater than 2, then it has to be demonstrated by the use of further information, for example, higher tier toxicity studies or risk management measures, that the risk is acceptable (Candolfi *et al.*, 2001).

From the above, it can be seen that the initial first tier assessment uses standard toxicity studies. It can also be seen that if concern is raised, i.e. a regulatory trigger is breached, then higher tier data are requested. These may either address the effects or exposure elements of the risk assessment.

1.2. Outline of the ecotoxicology tests required for pesticide authorization under 91/414/EEC

Presented in Table 1 is a list of all the ecotoxicological data that may be requested or submitted when an AS (and where appropriate metabolites or degradation products) and its AP are assessed under 91/414/EEC. The table includes a brief description of the type of study that is often submitted (column 3). Column 4 gives an indication of when these data may be required. This information is based on information provided in the relevant Annexes, guidance documents and associated risk assessments.

It should be noted that whilst the list of annex points and data requirements are taken from the appropriate Annexes, the views regarding what data are submitted and when required are based on the relevant guidance documents (EEC, 2002a,b,c) as well as the experience of the authors.

One area not covered in the following table is that of mammalian toxicity. Currently, when the risk to non-target mammals are assessed under 91/414/EEC, two toxicity studies are used – the acute oral LD50 study and the rat multi-generation study. According to Annex II point 5.2.1 of 91/414/EEC, an acute oral study is always required. There are several studies that can be used to address this Annex point: initially, the point could have been addressed by the procedure defined in the OECD guideline 401; however, due to animal welfare issues this has now been withdrawn; therefore, registrants are required to carry out either OECD 420, 423 or 425. Annex II point 5.6.1 states that a multi-generation study is always required and hence this Annex point is addressed by conducting a study according to OECD 416.

1.3. How frequently are certain studies submitted and how many studies are submitted to address an Annex point?

Table 1 provides an indication of all the toxicity studies that may be requested when an AS (and where appropriate, metabolites or degradation products) and its AP/formulation is considered under 91/414/EEC. Outlined below is an indication

Table 1
Ecotoxicological data required under 91/414/EEC

Annex point	Data requirement	What is required[a]?	When is it required[b]?
Annex II – Data requirements from studies with the active substance			
8.1	Effects on birds		
8.1.1	Acute oral toxicity	One study is required which may be carried out to the US EPA guideline (see http://www.epa.gov/opptsfrs/publications/OPPTS_Harmonized/850_Ecological_Effects_Test_Guidelines/Drafts/850-2100.pdf). These studies use either mallard duck, Japanese quail or bobwhite quail	Always required – except for active substances/products used indoors
8.1.2	Short-term dietary toxicity	One study is required (except where a reproduction study is submitted). Study usually carried out to either the OECD guideline 205 (OECD, 1984a) or the US EPA guideline 71-2 (US EPA, 1985a). Test species are usually either the bobwhite quail or the mallard duck (see http://www.epa.gov/opptsfrs/publications/OPPTS_Harmonized/850_Ecological_Effects_Test_Guidelines/Drafts/850-2200.pdf)	Always required unless reproduction study submitted – except for active substances/products used indoors
8.1.3	Sub-chronic toxicity and reproduction quail or the mallard duck	Study usually carried out to the OECD guideline 206 (OECD 1984b), draft OECD guideline (see http://www.epa.gov/opptsfrs/publications/OPPTS_Harmonized/850_Ecological_Effects_Test_Guidelines/Drafts/850-2300.pdf) or the US EPA guideline 71-3 (US	A reproduction study is required unless it can be justified that continued or repeated exposure of adults or exposure of nest sites during the breeding season is unlikely to occur. This has been

		EPA, 1985b). Test species are usually either the bobwhite quail, Japanese quail or the mallard duck	interpreted in the EU terrestrial guidance document as always being required unless the plant protection products are used indoors or if a product with a short half life of <14 days on food items is applied in autumn. It should be noted that low acute and dietary avian toxicity are not sufficient to indicate a low reproductive toxicity (see EEC, 2002a). In reality this means that a reproduction study is submitted for nearly all uses
8.2	Effects on aquatic life		
8.2.1	Acute toxicity to fish	A warm and cold water fish species must be tested. These are usually the bluegill sunfish and the rainbow trout, although studies have been conducted with carp or fathead minnows. Protocol usually followed is the OECD 203 guideline (OECD, 1984c), however US EPA guideline 72-1 may be used (US EPA, 1996)	These two studies are always required regardless of use as they are used for risk assessment and classification purposes
8.2.2	Chronic toxicity to fish		According to Annex II, a chronic toxicity study must be carried out unless it can be demonstrated that continued or repeated exposure does not occur. This has been interpreted as if the DT50 of the active substance in the water phase is greater than 2 days, then chronic data are required.

(Continued)

Table 1
(*Continued*)

Annex point	Data requirement	What is required[a]?	When is it required[b]?
			As to what study is submitted is dependant upon the log K_{ow} of the compound etc (see EEC, 2002b)
8.2.2.1	Chronic toxicity test on juvenile fish	Generally a study conducted to OECD 204/215 guideline is submitted (OECD, 1984d, 2000a)	As above
8.2.2.2	Fish early life stage toxicity test	Generally a study conducted to OECD 210 guideline is submitted (OECD, 1992). Test species are usually fathead minnow or rainbow trout	As above
8.2.2.3	Fish life cycle test	Generally a study submitted to US EPA protocol 72-5 is submitted (see http://www.epa.gov/opptsfrs/publications/OPPTS_Harmonized/850_Ecological_Effects_Test_Guidelines/Drafts/850-1400.pdf)	Guidance is provided in EEC (2002b), as to the when this study may be requested/submitted
8.2.3	Bioconcentration in fish	Generally a study conducted to OECD 305 guideline is submitted (OECD, 1981)	If the log K_{ow} for the active substance is greater than 3, then a study is required
8.2.4	Acute toxicity to aquatic invertebrates	Generally a study conducted to OECD 202 guideline is submitted (OECD, 2004a)	This study is always required as it is used in risk assessment and for labelling of the active substance
8.2.5	Chronic toxicity to aquatic invertebrates	Generally a study conducted to OECD 211 guideline is submitted (OECD, 1998a)	This study is always required where the DT50 of the active substance in the water phase is greater than 2 days

8.2.6	Effects on algal growth	Generally a study is conducted to OECD 201 guideline (OECD, 1984e)	Data on one species is always required; however, if the compound under consideration is a herbicide, then an additional species is required
8.2.7	Effects on sediment dwelling organisms	Usually a study conducted along the lines of the draft OECD 218 and/or 219 guideline (OECD, 2004b,c)	The need for this study is dependant upon the potential for exposure through the sediment and the chronic toxicity of the parent to *Daphnia magna* (see EEC, 2002b)
8.2.8	Aquatic plants	Usually a study conducted along the lines of the draft OECD guideline 221	Study is only required for herbicides and plant growth regulators
8.3	Effect on arthropods		
8.3.1	Bees		
8.3.1.1	Acute toxicity	Study covers the contact and oral toxicity of the active substance. Studies submitted are based on OECD 213 and 214 (OECD, 1998b,c), and the EPPO guideline 170 (EPPO, 1992)	Required whenever bees may be exposed. A list of possible exceptions is provided in Annex II Annex point 8.3.1
8.3.1.2	Bee brood feeding test	Studies conducted according to Oomen *et al.* (1992)	Only submitted if the active substance is an insect growth regulator
8.3.2	Other arthropods	Data may be submitted on two standard sensitive species as well as two crop relevant species. The testing procedure is outlined in Barrett *et al.* (1994). However, due to recent developments a new procedure has been adopted (Candolfi *et al.*, 2001). This latter procedure will be considered here. Rate response data on two standard species is conducted. The protocols used are as presented in Blümel *et al.* (2000) and	The two rate response studies are always required unless the proposed use of the product is to be used in food storage, is a wound sealant or a rodenticidal bait

(*Continued*)

Table 1
(*Continued*)

Annex point	Data requirement	What is required[a]?	When is it required[b]?
		Mead-Briggs *et al.* (2000). Studies are normally conducted with the product instead of AS.	
8.4	Effects on earthworms		
8.4.1	Acute toxicity	Usually a study conducted to OECD 207 is submitted (OECD, 1984f)	This study is always required where the active substance is applied to or can contaminate soil
8.4.2	Sublethal effects	Usually a study conducted along the lines of the OECD 222 guideline (OECD, 2004d)	According to Annex II expert, judgement is required in determining the need for this study. The EU guidance document on terrestrial ecotoxicology provides the following: The test is not required when both the $DT90_f$ is less than 100 days and the number of applications is less than 3 The test is always required if the $DT90_f$ is above 365 days (regardless of the number of applications) The test is always required if the number of applications is greater than 6 (regardless of persistence) If the $DT90_f$ is between 100 and 365 days and/or the number of applications is between 3 and 6, a case by case decision is made
8.5	Effects on soil non-target micro-organisms		
		Usually studies are submitted which indicate the likely effects on nitrogen transformation and carbon	Studies are required whenever the active substance is applied or can contaminate soil

		mineralization. Possible protocols include the OECD 216/217 guidelines (OECD, 2000b,c)	
8.6	Effects on other non-target organisms (flora and fauna) believed to be at risk		
		Usually a summary of screening data is submitted. However, more recently, dose response studies conducted according to draft OECD 208 (OECD, 2003) may be submitted	
8.7	Effects on biological methods for sewage treatments		
		Usually a study conducted to the OECD 209 guideline is submitted (OECD, 1984g)	Study is required whenever exposure of sewage treatment works is considered likely
Annex III – Data required from studies with formulated products			
10.1	Effects on birds		
10.1.1	Acute oral toxicity	A standard acute oral study using the formulation is required (see Annex II 8.1.1.)	According to Annex III of 91/414/EEC, the study is required where the TERa or TERst for the active substance in birds are between 10 and 100 or where results from mammal testing give evidence of a significantly higher toxicity of the preparation compared to the active substance
10.1.2	Supervised cage or field trials	No standard protocol, studies are designed on a case-by-case basis depending upon the concern highlighted	Study only required depending upon the results of the first tier risk assessment

(Continued)

Table 1
(*Continued*)

Annex point	Data requirement	What is required[a]?	When is it required[b]?
10.1.3	Acceptance of bait, granules or treated seeds by birds	No standard protocol, studies are designed on a case-by-case basis depending upon the concern highlighted	Study only required depending upon the results of the first tier risk assessment
10.1.4	Effects of secondary poisoning	No standard protocol, studies are designed on a case-by-case basis depending upon the concern highlighted	Study only required depending upon the results of the first tier risk assessment
10.2	Effects on aquatic organisms		
10.2.1	Acute toxicity to fish, aquatic invertebrates or effects on algal growth	Studies are as noted in Annex II point 8.2.1, 8.2.4 and 8.2.6	Studies on the toxicity of the formulation are always required for both classification and risk assessment purposes. However, tests may be restricted to the most sensitive group providing that it is at least 100 times more sensitive than the next sensitive group (EEC, 2002b)
10.2.2	Microcosm or mesocosm study	Study only required depending upon the results of the first tier risk assessment (Campbell *et al.*, 1999, Giddings *et al.*, 2002)	Study only required depending upon the results of the first tier risk assessment
10.2.3	Residue data in fish	Results from Annex II point 8.2.3 usually address this issue	When log K_{ow} of the active substance is greater than 3
10.2.4	Additional studies	Studies are dependent upon the results of the first tier risk assessment	The need for further studies depends upon the results of the first tier risk assessment
10.3	Effects on terrestrial vertebrates other than birds		
		Studies are dependent upon the results of the first tier risk assessment	The need for further studies depends upon the results of the first tier risk assessment

10.4	Effects on bees		
10.4.1	Acute oral and contact toxicity	See Annex II 8.3.1.1	Required when product contains more than one active substance or the toxicity cannot be predicted from the study on active substance
10.4.2	Residue test	No standard protocol, studies are designed on a case-by-case basis	Study is only required depending upon the results of the first tier risk assessment
10.4.3	Cage tests	Study usually conducted according to the EPPO 170 guideline (EPPO, 1992)	Study is only required depending upon the results of the first tier risk assessment
10.4.4.	Field tests	Study usually conducted according to the EPPO 170 guideline (EPPO, 1992)	Study is only required depending upon the results of the first tier risk assessment
10.4.5	Tunnel tests	Study usually conducted according to the EPPO 170 guideline (EPPO, 1992)	Study is only required depending upon the results of the first tier risk assessment
10.5	Effects on arthropods other than bees		
10.5.1	Laboratory, extended laboratory and semi-field tests	Standard studies according to Candolfi *et al.* (2001) – see 8.3.2 above	Depends upon the results of the first tier risk assessment
10.5.2	Field tests		
10.6.1	Effects on earthworms		
10.6.1.1	Acute toxicity tests	Standard OECD 207 study (see above 8.4.1)	Required where the product contains more than one active substance or the toxicity of the formulation can not be reliably predicted from the formulation tested

(Continued)

Table 1
(*Continued*)

Annex point	Data requirement	What is required[a]?	When is it required[b]?
10.6.1.2	Tests for sublethal effects	Data submitted to address 8.4.2 address this annex point; however, study is usually carried out with the formulation	See Annex II point 8.4.2 for details
10.6.1.3	Field studies	No standard test protocol, however methodology described by ISO (11268-3: 1999), further details in Greig-Smith *et al.* (1992), and Sheppard *et al.* (1997)	The need for such a study depends upon the results of the risk assessment and persistence of the compound [see EU Terrestrial guidance document and EPPO (2002)]
10.6.2	Effects on other soil non-target macro-organisms	Test methodologies used are ISO 11267:1999, Løkke and van Gestel (1998) and Bakker *et al.* (2003). No agreed protocol for the litter bag study	Studies are required depending upon the persistence of the active substance
10.7	Effects on soil non-target micro-organisms		
10.7.1	Laboratory testing	See Annex II point 8.5 above	See Annex II point 8.5 above
10.7.2	Additional testing	No agreed protocols	
10.8	Available data from biological primary screening in summary form		
		See Annex II point 8.6 above	See Annex II point 8.6 above

[a] The information presented in this column is based on experience of evaluation of data packages submitted for authorization under 91/414/EEC. The protocol stated may not correspond directly with that in the relevant Annexes.

[b] The statement in this column provides an indication of when the study is required. This is based on the information provided in the directive and associated guidance documents.

of the approximate proportion of pesticides for which each end-point is required and the typical number of studies submitted by registrants. In order to try and determine this, 16 ASs were examined. These ASs were under evaluation in the UK and consisted of seven new and nine existing ASs. There were two herbicides, one plant growth regulator, one nematicide, two mollusicides, seven insecticides and three fungicides. Uses included two indoor-only products, one granule, two pellets, two seed treatments and twelve sprays.

In trying to determine the approximate proportion of pesticides for which each end-point is required and the typical number of studies submitted by Registrants, only first tier laboratory studies were considered. The reason for this is that higher tier studies, for example, honeybee field trials, non-target arthropod extended laboratory or semi-field or field studies and avian avoidance studies, tend to be designed on a case-by-case basis to address a specific issue. It is also possible that if a risk assessment highlights concern in one specific area, for example, non-target arthropods, a registrant will conduct several studies to try and address this particular concern. It should be noted that in determining the number of studies submitted, no differentiation has been made regarding whether the study was conducted on the AS, metabolite or the formulation. Hence, the aim of this exercise was to determine an indication of the number of times that data were required to address a specific Annex point as well as the actual number of studies submitted. Table 2 presents the results from the above assessment.

As regards the two mammalian toxicity studies outlined in Section 1.1.2, it is usual that one acute oral toxicity study is submitted for each AS. It is also usual that an acute oral toxicity study is submitted on the formulation. However, it is possible that an appropriate case, based on calculated toxicity of the formulation, may be used to address this issue. As regards the multi-generation study, it is usual that one study, using the AS, is submitted for each AS.

In the above assessment, there was consideration of the number of formulation studies conducted and hence submitted. It is not intended that quantitative structure-activity relationships (QSARs) should be able to predict the toxicity of a formulation. The information on formulations was included to indicate the number of studies that may be submitted to address a specific Annex point through the use of toxicity studies. If a QSAR is developed that predicts the toxicity of the AS, then it may be possible to use this information to determine the toxicity of simple formulations. It is not proposed that formulation studies are used in the development of a QSAR, unless it can be clearly demonstrated that the co-formulants have not affected the toxicity of the AS.

The above analysis is crude and is obviously limited to a small range of compounds. However, it does broadly reflect the frequency that some studies are submitted as well as the number of studies submitted to address a particular Annex point. Outlined below is a brief explanation of why the number of studies submitted differs from what would be expected based on the outline in Table 1.

Table 2
Summary table of the assessment of 16 compounds

Annex point	Data requirement	Number of times at least one study was submitted out of 16 compounds (%)	Total number of studies submitted to address Annex point	Average number of studies submitted per Annex point
8.1	Effects on birds			
8.1.1/10.1.1	Acute oral toxicity	15 (94)	28	1.8
8.1.2	Short-term dietary toxicity	15 (94)	36	2.4
8.1.3	Sub-chronic toxicity and reproduction quail or the mallard duck	15 (94)	31	1.9
8.2	Effects on aquatic life			
8.2.1/10.2.1	Acute toxicity to fish	16 (100)	76	4.8
8.2.2	Chronic toxicity to fish			
8.2.2.1	Chronic toxicity test on juvenile fish	7 (44)	8	0.5
8.2.2.2	Fish early life stage toxicity test	12 (75)	14	0.9
8.2.2.3	Fish life cycle test	1 (6)	1	0.06
8.2.3	Bioconcentration in fish	10 (62)	13	0.8
8.2.4/10.2.1	Acute toxicity to aquatic invertebrates	16 (100)	60	3.8

8.2.5	Chronic toxicity to aquatic invertebrates	13 (87)	22	1.4
8.2.6/10.2.1	Effects on algal growth	16 (100)	51	3.2
8.2.7	Effects on sediment dwelling organisms	8 (50)	15	0.9
8.2.8	Aquatic plants	4 (25)	6	0.4
8.3	Effects on arthropods			
8.3.1	Bees			
8.3.1.1/10.4.1	Acute toxicity (oral/contact)	14 (87)	28	1.8
8.3.2/10.5	Other arthropods (NB: only glass plate style studies, not extended lab studies)	12 (75)	25	1.6
8.4	Effects on earthworms			
8.4.1/10.6.1.1	Acute toxicity	15 (94)	41	2.6
8.4.2/10.6.1.2	Sublethal effects	10 (62)	17	1.1
8.5	Effects on soil non-target micro-organisms			
8.5/10.7	Effects on soil non-target micro-organisms	14 (87)	44	2.8
8.7	Effects on biological methods for sewage treatments	10 (62)	15	0.9

1.3.1. Avian toxicity studies

According to Table 1, it could be concluded that one acute toxicity study, possibly a reproduction or dietary study, would normally be submitted. However, from Table 2, it can be seen that approximately two studies are submitted to address each of these points. It is assumed that the main reason for this is that two acute, two dietary and two reproduction studies are required for authorization in the USA.

1.3.2. Aquatic toxicity studies

According to Table 1, two acute fish studies with the AS should always be submitted along with possibly one on the formulation. However, it is usual that the two with the AS are submitted along with two or possibly more with the formulation. In addition to these studies, it is a common practice to submit toxicity data on certain metabolites as well. Hence, the reason why the number submitted is greater than initially expected or predicted.

As regards aquatic invertebrates and algae, it is customary to submit one study on the AS as well as one with the formulation and any metabolites of concern.

It is likely that each time approval is requested for a specific formulation, then at least one formulation study on the most sensitive species will be conducted and submitted for product authorization. This is in line with Annex III point 10.2.1.

1.3.3. Honeybees

It is customary to conduct an acute contact and oral study with the AS and hence submit this as one study. Likewise, if one study on the formulation is submitted then this usually covers both contact and oral routes of exposure. Therefore, the actual number of studies noted above probably underestimate the actual number of tests carried out.

1.3.4. Non-target arthropods

In the above assessment, only first tier dose response studies on *Aphidius* spp. and *T. pyri* have been considered. For some of the uses, for example, pellets and granules, alternative species that are more relevant to the proposed uses have been tested.

1.3.5. Earthworms

According to Table 1, it would be expected that there would only be one or possibly two acute earthworm studies, i.e. one with the AS and one with the formulation. However, frequently studies with soil metabolites are submitted and hence this increases the number of studies submitted to address this Annex point.

1.4. What changes are likely to occur that could alter the frequency and number of toxicity studies submitted?

It is possible that there may be changes in what studies are requested or submitted in future. Outlined below are some views on how the data set may change over the short-term.

1.4.1. Avian studies

As regards the acute oral study, a draft OECD guideline has been written that reduces the number of birds used whilst maintaining the same level of precision. It is understood that this guideline is nearing completion and will, hopefully, be adopted in the near future. The OECD has also produced a revised avian reproduction guideline that reduces the duration of the study from 21 weeks down to 6 weeks. Whilst the duration of the study has decreased and the scientific robustness increased, it is unclear whether the number of test animals will significantly change. As regards the dietary study, a new guideline has been drafted; however due to technical issues, it has not progressed as far as either the acute oral or reproduction study. Overall, these changes could alter the type of study conducted; however, it is unclear whether this will lead to a change in the number of studies conducted.

Recently, the EU has noted a new risk assessment approach for birds (EEC, 2002a); this risk assessment uses the concept of daily dose for the acute, short-term and long-term risk assessment. This means that the end-point from the short-term dietary study and reproduction study have to be converted from ppm diet (or mg AS/kg diet) to mg AS/kg bw/day. This change is relatively straightforward for the reproduction study, as adult birds are used and doses used in the study should not be repellent to birds. However, conversion of the end-point from the dietary study is somewhat problematic as it is not possible to determine individual doses and hence determine an appropriate end-point. The end-point may, in certain circumstances, be the result of both toxicity and repellency. This change in risk assessment procedure may lead to repeat dietary studies being conducted.

As part of the ongoing discussions regarding endocrine disruptors, it is possible that a two-generation reproduction study may be a requirement for compounds that are suspected of being an endocrine disruptor.

Legislation exists in many countries to limit the severity of toxicity testing and to reduce pain and suffering experienced by animals. Therefore, whatever testing is carried out it must be fully justified for both scientific and public scrutiny. It should be noted that the OECD 401 mammalian acute oral toxicity test has been withdrawn due to animal welfare concerns, and it is possible that similar pressure may come to bear on similar lethality style studies involving vertebrates.

1.4.2. Aquatic studies

The need for acute toxicity studies on both the AS and the formulation is unlikely to change in the short term. As regards chronic studies, due to the trigger value used for these studies the number of chronic studies is likely to increase. Of all the chronic studies highlighted in Table 1, it is considered that due to its sensitivity the early life stage study will become more frequently submitted.

As part of the ongoing discussions regarding endocrine disruptors, it is likely that a range of new studies may be developed, for example, a modified early life stage may be developed which assesses effects on the pre-mating procedure. It is also a common practice for certain regulators to request a full fish life cycle study for compounds that are considered to be potential endocrine disruptors.

1.4.3. Non-target arthropods

In the above assessment, consideration has only been made of the standard dose response studies submitted according to ESCORT 2 (Candolfi *et al.*, 2001). It is unlikely that the number of these types of studies will increase, however, the number of higher tier studies may vary depending upon the interpretation of ESCORT 2 and the associated risk assessment.

1.4.4. Soil invertebrates

In the above assessment, only earthworms and microbial processes have been considered. According to Annex III point 10.6.2, if an AS (or metabolite) has a $DT90_f$ of less than 100 days, then no further data are required. If, however, the $DT90_f$ is greater than 365 days then a study designed to examine organic matter breakdown is required. This issue is usually addressed by a litter bag study. For compounds with a $DT90_f$ greater than 100 days but less than 365 days, a stepwise risk assessment is used (Section 6 of EEC, 2002c). In this stepwise assessment, data on the toxicity of the AS/product to *Hypoaspis aculeifer* or *Folsomia candida* are submitted. It is likely that these studies will increase in future due to increasing concern regarding persistent compounds.

1.4.5. Mammalian toxicity studies

Of the two mammalian studies outlined above, the acute oral toxicity study OECD 401 has been withdrawn due to animal welfare issues and replaced with three possible alternatives – OECD 420, 423 and 425. It is, therefore, unlikely that there will be any further major changes to the study protocols in the short to medium term. As regards the need for the study, this is unlikely to change in the short to medium term due to the fact that these studies are used to classify the AS as well as the formulation or product.

As regards the multi-generation study, this has only recently been modified and hence it is unlikely to change significantly in the short term. It may in the medium term increase in complexity to incorporate issues from, for example, developmental neurotoxicity studies. It is considered that the need for this study is unlikely to change, as this is the only mammalian study to provide an indication of reproductive performance.

1.4.6. General issue

Currently, within the EU, metabolites receive much attention (EEC, 2002b,c; 2003), and it is likely that, in the short term at least, this will increase and hence the number of studies conducted will also increase. According to Maund (personal communication), the estimated cost for assessing one metabolite is approximately €633000, of which €146000 is for the ecotoxicological testing. Therefore, there is a high probability that a QSAR approach that could deal with metabolites would gain favour within the EU regulatory forum.

1.4.7. Revision of 91/414/EEC

It should be noted that 91/414/EEC and the associated technical annexes are currently under revision.

1.4.8. Conclusion

It is difficult to predict changes in the type or number of studies submitted. However based on the above, it is likely that on the one hand there could be pressure to reduce the number of avian acute and dietary avian studies whilst there could be pressure to increase the number of avian reproduction studies due to concerns regarding endocrine disruption. Also linked to the issue of endocrine disruption, there could be an increase in the number of chronic fish studies. As regards invertebrate studies, it is likely that there will be an increase in those involving certain soil dwelling invertebrates, due to concerns related to persistent compounds.

2. INTRODUCTION ON QSARS FOR PESTICIDES

Studies on QSARs for toxicological end-points are numerous (Benfenati and Gini, 1997; Cronin and Livingstone 2004) and tend to give satisfactory results for industrial chemicals, although studies on pesticides give more variable results. The main reasons for this are that the studies on industrial chemicals are more straightforward as the toxic mechanisms are simpler. For instance, models on aquatic toxicity, such as fish or *Daphnia*, are numerous and in many cases dominated by the use of relatively simple equations involving $\log P$ as descriptor. $\log P$, also called $\log K_{ow}$, is the coefficient of partition between octanol and water, and as octanol mimics the biological membrane of fish cells, it is a good experimental model for the phenomenon called narcosis. $\log P$, whether measured or calculated, can, therefore, be used to predict toxicity. As most of the industrial chemicals act through narcosis in the case of aquatic toxicity (Russom *et al.*, 1997), a simple model based on $\log P$ can account for most of the toxicity phenomena. However, the toxicity processes involved in the case of pesticides are much more complex.

Studies on the relationship between the structure and the activity of pesticides have been used by industry during the screening of potential new compounds.

These studies are quite different from the studies done within DEMETRA for the following reasons:

1) Studies to produce new pesticides are aimed at achieving the maximum activity towards the target organism.

2) Related to point (1) is the fact that industry wants to optimize the pesticide effect and thus to produce one or a limited number of compounds with a given activity. Indeed, it is expensive to evaluate the activity of a large number of candidate pesticides in the laboratory. Thus, industry wants to avoid false positive (which means compounds that are predicted active, but are not active) in its modelling studies. Vice versa, we wanted to avoid as much as possible false negatives, as this may cause adverse environmental effects.

3) We intended to develop general models addressing pesticides in a general meaning, while industry, in many cases, focuses its attention to a very restricted class of compounds, to optimize the effect towards pests.

4) Our models are publicly available, while the industrial ones are commercial and therefore restricted.

Recently, the Danish Environmental Protection Agency (EPA) evaluated QSAR models for pesticides (Hansen, 2004). Reliable models have been reported for physicochemical properties, such as solubility and $\log P$, whereas models for other physicochemical properties, such as melting point, estimations were found to be inaccurate. Models for fish, *Daphnia* and algae were also assessed; however, the results were unsatisfactory due to the fact that they were simple and based on classical physicochemical parameters, such as $\log P$. Results are reported in Table 3. The 'fitting' of the models (i.e. the performances on the used compounds, without an attempt to predict values on new compounds) for fish was quite low: R^2, the correlation coefficient between experimental and estimated values for models that considered specific modes of action, was 0.3, which means that $\log P$ is not sufficient to explain toxicity for pesticides, as mentioned above.

As noticed by Hansen, most of the QSARs developed and those used as general tools in the evaluation of available QSAR models are mainly based on industrial chemicals which generally do not have a potent biological activity (Hansen, 2004). This is in contrast to pesticides that have been developed to exert a specific activity towards certain organisms.

The same report shows results after dividing pesticides into classes based on their activity: fungicides, herbicides, herbicides with a growth inhibition mode of action, herbicides with a photosynthesis mode of action, insecticides, insecticides with a cholinesterase inhibition mode of action, insecticides with a contact and systemic mode of action, plant growth regulators and rodenticides. Correlations

Table 3
Correlation between log K_{ow} and toxicity expressed as R^2

Pesticide class	Fish	Daphnia acute	Daphnia reproduction	Algae
Pesticides	0.32 (300)	0.31 (255)	0.32 (96)	0.15 (183)
Fungicides	0.20 (59)	0.24 (57)	0.11 (28)	0.12 (48)
Herbicides	0.50 (118)	0.38 (97)	0.29 (37)	0.07 (82)
Herbicides with a growth inhibition mode of action	0.52 (69)	0.35 (58)		0.10 (51)
Herbicides with a photosynthesis mode of action	0.47 (40)	0.50 (39)		0.00 (29)
Insecticides	0.40 (81)	0.26 (79)	0.62 (22)	
Insecticides with a cholinesterase inhibition mode of action	0.17 (53)	0.16 (43)		
Insecticides with a contact and systemic mode of action	0.38 (24)	0.37 (26)		
Plant growth regulators	0.69 (14)	0.07 (11)		
Rodenticides	0.62 (5)			

The number of pesticides are given in parenthesis.

with log *P* for these classes are shown in Table 3. Some interesting relationships between log *P* and some pesticide classes appear, as in the case of plant growth regulators and rodenticides for fish toxicity or insecticides for *Daphnia* reproduction.

Some other models have been developed in the past and we briefly describe some representative studies not done within DEMETRA below. We will report on DEMETRA models and papers in the following chapters, along with comments on the models.

- Devillers *et al.* (2002) describe a model that estimates the acute toxicity of pesticides to honeybees using data on 100 pesticides. While this produced good results, the data sources are somewhat heterogeneous, because while a main source was used, additional data were taken from a variety of other sources. This reduced the data quality.

- Another model of pesticides in general has been developed to classify the lethal dose for the rat (Pintore *et al.*, 2003). A classifier may be considered a simpler approach compared to the previously discussed models, because the prediction does not need to be as precise as it only has to predict a toxicity class, such as high, medium and low. Results from this model using a fuzzy logic approach tend to be quit accurate. More details on this technique will be given in Chapter 5 (Amaury *et al.*, this volume).

- In another study on the classification of toxicity of pesticides, several classifiers were combined, which improved the overall results (Benfenati *et al.*, 2002). Here, rat toxicity data for 57 organophosphorus pesticides were used. Good results were obtained, but the source of data was the Pesticide Manual.
- There has been an interest in understanding and explaining some specific activities, for instance, of compounds with anticholinesterase activity (Maxwell and Brecht, 1992; Lin *et al.*, 1999; Ali *et al.*, 2005). Of course, in this case only some pesticides are of interest, and, therefore, the model takes advantage of the direct or indirect knowledge of the existence of a specific receptor, with a given structure.
- Vighi *et al.* (1991) studied *Daphnia* and bee toxicity of organophosphorus pesticides (22 chemicals). For *Daphnia*, $\log P$ was one of the descriptors, as in most of the models. However, other descriptors were introduced to explain other features, such as electronic characteristics. The limited number of compounds may limit the possibility to extend the models.

3. REGULATORY PERSPECTIVES IN THE USE OF QSARS

3.1. Current use of QSARs in regulation

Outlined below is a brief overview of how QSARs (and SARs) are used in various regulatory bodies.

3.1.1. USA

QSARs are used in the USA as part of the US Toxic Substances Control Act (TSCA) for new chemicals in order to complement the information provided by a manufacturer in the frame of a pre-manufacture notice (TSCA, 1976). TSCA Section 5 requires a manufacturer or importer of a new chemical substance to submit a Premanufacture Notice (PMN) to the EPA 90 days before commencing manufacture or import of a new chemical. Regulatory decisions often have to be made in the absence of any experimental data on the chemical. The US EPA evaluates approximately 1500–2000 PMN cases a year. The agency has completed a total of about 40,000 reviews since 1978. A series of SAR and QSAR models is in use: EPI Suite, ECOSAR, OncoLogic, ChemSTEER and E-FAST. (Q)SAR predictions are currently accepted under the US HPV Program to fill data gaps for certain end-points (US EPA, 1999).

The 2003 ‘Framework for a Computational Toxicology Program in ORD’ (US EPA, 2003) addresses use of computational toxicology to improve the Agency’s prioritization of data requirements and risk assessments. US Office of

Water uses OncoLogic to predict cancer potential of chemicals for prioritization and uses AQUATOX to predict the environmental fate of chemicals.

The US Food and Drug Administration has developed, validated and used (Q)SAR for decision support information for end-points required for chemical marketing.

In the case of pesticides, US Office of Pesticide Programs uses ECOSAR to establish food tolerance exemptions under the Food Quality Protection Act and utilizes (Q)SAR to determine biodegradation potential.

The US EPA Endocrine Disruptor Screening program is conducting QSAR analysis on most pesticide ASs, pesticide formulation ingredients and selected commodity chemicals as part of a priority setting process (see www.epa.gov/scipoly/oscpendo/index.htm).

3.1.2. Canada

The Healthy Environments & Consumer Safety Branch (HECSB) addresses responsibilities for Existing Industrial Chemicals under the Canadian Environmental Protection Act (CEPA). CEPA is the primary authority to assess and manage the risks posed by toxic substances (health and environment) in Canada. CEPA within the Existing Substances Program develops and refines predictive tools for effects, including QSARs. One particular model is ComHaz that adopts a hierarchical approach for multiple end-points and uses data from QSARs as well as a weight of evidence approach. Sources of information considered hierarchically are

- comprehensive literature searching (electronic and hardcopy resources),
- reviews or secondary accounts of toxicological or epidemiological studies,
- original toxicological and epidemiological studies,
- (Q)SAR models (TOPKAT and CASETOX) and
- chemical structures of concern, SAR models (DEREK) and surrogate/analogue approaches.

(Q)SARs introduced positive features in the overall process of systematic consideration of all Existing Substances.

Environment Canada, New Substances Branch NWRI developed a system based on an artificial neural network. It can be used to predict a variety of end-points. It identifies the 10 closest analogues with experimental data.

3.1.3. Japan

In Japan, chemical substances are controlled by various laws. Among these laws, the Chemical Substance Control Law (CSCL) is based on pre-manufacturing examination of chemicals, under which the impacts on human health of chemicals through environment are controlled.

Three Ministries are involved in the data management under CSCL: METI, Ministry of Economy, Trade and Industry; MHLW, Ministry of Health, Labor and Welfare and MOE, Ministry of Environment. The National Institute of Technology and Evaluation (NITE) has a major role in the integrated management of the chemicals, collaborating with these ministries in an integrated way. NITE's program for the utilization of QSAR in the examination of Existing Chemicals includes (Shigekura, 2005) the following activities and features:

- QSAR prediction: important pieces of information for the evaluation of chemicals in the regulatory decision,
- Target under CSCL: biodegradation, log *P* and bioaccumulation,
- Organization of the 'NITE's QSAR Committee' of which members are experts of QSAR and hazard of chemicals and
- QSAR reports on individual chemicals to METI and the Council for their decision-making.

For the new chemicals examination in the Chemical Substance Council, QSAR prediction data of biodegradation for reference started in March 2005.

3.1.4. Denmark

The Danish EPA developed a QSAR database for chemicals in general – not only pesticides as mentioned above (Danish EPA, 2001). Very recently, this database became publicly available through the European Chemical Bureau [European Chemicals Bureau (ECB), 2006]. This QSAR database contains predictions on 166,000 chemicals (around 47,000 discrete EINECS chemicals). Predictions originate from over 70 models including TOPKAT, MultiCASE, LeadScope, and CATABOL. This database contains predictors for physico-chemical properties, ecotoxicological/environmental fate properties and health end-points.

Ecotoxicological and environmental fate end-points include (the software used is indicated in parentheses): bioconcentration factor (Syracuse and Bintein), Biodegradation (BIOWIN1-6 and Mcase), Steady-state fish (T95), Fish LC50s (ECOSAR and DK-EPA), *Daphnia* LC50 (ECOSAR and DK-EPA), Algae EC50 (ECOSAR and DK-EPA) and Tetrahymena EC50 (DK-EPA).

Health end-points include LD50 rat (TOPKAT), NOAEL rat (28 d) (TOPKAT), skin irritation (Mcase), skin sensitization (TOPKAT and Mcase), reproductive toxicity (TOPKAT), developmental toxicity – human (Mcase), oestrogen binding (three models, DK EPA), anti-androgen reporter gene (DK EPA) and Ames mutagenicity (three models).

Information from the DK EPA QSAR database is aimed to help producers and importers assess hazard of untested substances in relation to self-classification.

3.1.5. European Commission

In the technical guidance document related to the EU directive concerned with existing substances [i.e. EEC/1488/94, EEC (1995)], the use of QSARs is encouraged as a supporting tool in the process of risk assessment of substances. This use is restricted to (1) data evaluation, (2) decision support for further testing, (3) selection of input parameters/data that are needed in the risk assessment and (4) identification of (other) potential concerns.

When the Scientific Committee on Toxicity, Ecotoxicity and the Environment commented on the 'Technical guidance document in support of the Directive 98/8/EC concerning the placing of biocidal products on the market: Guidance on data requirements for ASs and biocidal products', they noted the following regarding the use of QSARs:

> In order to minimise experimental testing, the use of QSAR may also be suggested. However, when using QSAR data it must be considered that these products are intentionally designed to produce effects on living organisms, and therefore similarity in the chemical structure alone cannot be accepted, and care must be taken to use QSAR information obtained from chemicals with a similar mode of action. Moreover, in order to take into account the higher level of uncertainty of predicted data in comparison to experimental data, different rules (i.e. stricter application factors and triggers) should be applied to QSAR information (EEC, 2000).

In recently published guidance documents related to the assessment of pesticides under 91/414/EEC reference is made to SARs. In the guidance document on the assessment of the relevance of metabolites in groundwater of substance regulated under Council Directive 91/414/EEC (EEC, 2003), the following reference is made regarding determining biological activity:

> Structure-activity relationships may be considered on the basis of the mode of activity of the parent molecule (i.e. usually the active substance). In many cases for compounds belonging to a well defined group of active substances (e.g. sulfonyl thiourea herbicides) this may already provide useful and sufficient information for the assessment of this question in the absence of experimental data.

Likewise in the aquatic guidance document on aquatic ecotoxicology (EEC, 2002b), reference is made to an SAR approach for determining the toxicity of minor metabolites. This approach is based on work done by Sinclair and Boxall (2002).

It should be noted that both QSAR and SAR approaches have been considered on a case-by-case basis as part of the human health and environmental assessment of pesticides carried out under 91/414/EEC.

3.1.6. OECD

According to the OECD (OECD, 2001), QSARs can be relied upon to provide prediction of acute toxicity for fish, *Daphnia* and algae for non-electrolyte, non-electrophilic and otherwise non-reactive substances. The report highlights potential problems regarding substances such as organophosphates that operate by means of special mechanisms such as functional groups that interact with biological receptors.

More recently, OECD proposed guidelines for the validation of (Q)SAR for regulatory purposes. Five principles are identified (OECD, 2005):

'To facilitate the consideration of a (Q)SAR model for regulatory purposes, it should be associated with the following information:

1) a defined end-point
2) an unambiguous algorithm
3) a defined domain of applicability
4) appropriate measures of goodness-of-fit, robustness and predictivity
5) a mechanistic interpretation, if possible'.

The intent of principle 1 is to ensure that the end-point is clearly defined because it can be determined in different ways. However, further discussion is ongoing about some complex end-points, such as no-observed-effect level. The intent of principle 2 is to ensure transparency in the model algorithm, particularly in the case of commercial models. The idea behind this is to be able to independently assess a given model, which is likely to represent a barrier for regulatory acceptance. This is linked to the issue of reproducibility of the predictions.

The defined domain of applicability (principle 3) refers to the need to assess the boundaries of the validity of a given model, which is linked to the training set. Thus, a given model is linked, for instance, to the types of chemical structures, physicochemical properties and mechanisms of action used to build up the model. Principle 4 aims to ensure appropriate measures of the validity of a given model. The simple goodness-of-fit, done internally, is not sufficient.

It is recognized that it is not always possible, from a scientific viewpoint, to provide a mechanistic interpretation of a given (Q)SAR (principle 5) or that there even be multiple mechanistic interpretations of a given model. The absence of a mechanistic interpretation for a model does not mean that a model is not potentially useful in the regulatory context.

3.1.7. Industrial initiatives

ECETOC promoted a Workshop on Regulatory Acceptance of (Q)SARs for Environmental and Human Health End-points, March 2002, Setubal. This workshop is at the basis of the OECD principles (see Section 3.1.6).

CEFIC promotes initiatives in the area, for instance, through the Long Range Initiative (LRI) work programme. Some projects are active, one for the identification of SAR alerts for repeated dose toxicity and one for the prediction of dermal permeation rate. Furthermore, within another LRI project, a public tool box for QSAR has been developed, called AMBIT (AMBIT, 2005).

From the above, it can be seen that from a regulatory perspective in the EU, Canada, Japan and USA as well as the OECD, at least some use of QSARs has been accepted.

3.2. Potential barriers for using QSARs in the pesticide authorization procedure

Any QSAR produced resulting from DEMETRA would be primarily for use in risk assessments carried out under 91/414/EEC. Compared to the above examples, this would be a new use for QSARs and could put a different emphasis on what is ultimately required, and hence, identification of any policy or practical problems is essential.

In discussion with other regulators, it was generally considered that in order for a QSAR approach to be accepted there would need to be a demonstration that they were as reliable as, and hence equivalent to, what is currently used. It was also stressed that the scientific community would have to be satisfied that it was a credible, reliable and appropriate approach.

3.2.1. Practical barriers

As regards practical barriers or potential problems, these could be divided into the following: validity, reliability and reproducibility, lack of an agreed approach, overall usefulness for a potentially wide range of compounds and types of studies QSARs could replace. These issues were discussed with a group of regulators and their views are discussed below.

3.2.2. Validity

When a group of regulators were asked what they considered to be the most important factor regarding the acceptance of a QSAR approach, the overriding concern was regarding validity, that is, would the QSAR model give an end-point that was equivalent to what is currently used. They stated that in order to gain acceptance, there needs to be a clear demonstration that the QSAR approach can predict the toxicity end-point as accurately as an actual toxicity study. If this is not possible, the degree of accuracy should be known and hence the end-point adjusted so that the final figure used in risk assessments is as accurate as the toxicity study.

Linked to the above was the issue regarding the quality and quantity of data used to construct the model. Regulators felt that the model should be based on sound reliable data as this would help to increase the overall validity of both the concept and the output.

3.2.3. Reliability and reproducibility

Toxicity tests are usually conducted to international guidelines, for example, OECD or ISO test guidelines. These guidelines have usually been developed and tested to ensure that they are both reliable and reproducible. Regulators considered that providing the end-point was valid, the QSAR model would also need to be as reliable and reproducible as the current toxicity studies (i.e. if toxicity studies were to be replaced by a QSAR approach, then they would have to be as reliable and reproducible as the status quo).

Whilst the EU has its own legislation dealing with the registration and regulation of pesticides, it is appreciated that notifiers and registrants operate in a global environment. Therefore, the data submitted to address one set of legislation are often used to address several. For example, the data set produced to obtain registration in the USA is often used to obtain registration in the EU. An illustration of this is the fact that in the EU according to 91/414/EEC, only one acute oral, one dietary and one reproduction study need be generated (Table 1). However, it is normal that two studies are conducted and hence submitted for all three areas as this is what is required by the US EPA. Therefore, if QSARs are to be of use in the production of regulatory data packages, they need to be acceptable to a wide range of potential users especially regulatory organizations.

3.2.4. Lack of an agreed approach

Regulatory risk assessments have to be consistent to ensure equity between assessments. In order to ensure this, they tend to be very standardized and carried out to an approach agreed between regulators and notifiers. Therefore, if a QSAR approach were to be incorporated into regulatory risk assessments, its interpretation and subsequent use would also have to be agreed between notifiers and regulators. Due to the lack of an agreed approach, QSARs have not been incorporated into routine risk assessments.

3.2.5. Overall usefulness for a potentially wide range of compounds

Pesticides represent a wide range of compounds that in turn have a wide range of modes of action, for example, anticholinesterase, sterol biosynthesis inhibitors and amino acid synthesis inhibitors. Therefore, to be of use in pesticide regulation, a QSAR approach would need to be able to be used to assess the toxicity of any pesticide regardless of mode of action.

Classification and labelling of the AS and APs are required according to Annex IV and Annex V of 91/414/EEC. It is required under the Chemical Hazard Information and Packaging for Supply Regulations 2002 (EEC, 2002d). Both the AS and formulation may need to be classified. Therefore, if data were available from a QSAR this could reduce the need for animal testing. Data are currently required for classification purposes on the acute toxicity to fish, *D. magna* and algae. For further information see EEC (2002d).

3.2.6. Types of studies QSARs could replace

The take up of a QSAR approach, by both regulators and notifiers, would be aided if the studies they were replacing were costly either in terms of number of animals used, time taken or financial implications.

To conclude there are potentially many practical barriers for QSARs; however, most of these can be distilled down to the two following points:

1. Validity – do they give the same result as the status quo?
2. Reliability – are they as reliable and reproducible as the current toxicity test(s).

If the answer to both these issues can be satisfactorily addressed, then there should not be any policy or practical barriers preventing their use in regulatory risk assessments.

3.3. End-user criteria for the use of QSARs in regulatory assessment

The above assessment of potential barriers was carried out by discussing the QSARs with regulators. Further work was carried out through the use of an end-user questionnaire. Details of this are presented below.

As part of the exercise to determine key regulatory issues regarding the use of QSARs in the regulatory assessment, quality criteria of the data set used to determine the QSARs has been determined along with the degree of precision (see Chapter 2, Section 3). In addition to these activities, an assessment of criteria required by potential end-users was determined. This was carried out in the following way.

A short abstract describing the DEMETRA projects and proposed questions to be included in the questionnaire for end-users was prepared and sent to partners for comment. The comments received were then used to produce a final version of the survey that contained the following six questions:

1. What types of end-points would you like to be addressed within a QSAR model (e.g. acute, chronic, which species)?
2. How accurate are current registration data for these end-points (e.g. ± a factor of 2, 5, 10 etc.)?
3. How accurately should QSARs predict actual toxicity (e.g. ± a factor of 2, 5, 10 etc.)?
4. What reassurance would you require about the quality and quantity of the data used to produce the QSAR?
5. What type of validation would you expect to have been conducted?
6. Are there any other issues you would require to be addressed before you could accept the use of QSARs for regulatory purposes?

3.3.1. Distribution

The abstract and questionnaire were sent to over 40 of the end-users listed in Annex 1 of the DEMETRA contract in May 2003 with a request that they be returned during June 2003. Given the limited response, the questionnaire was also distributed to all member states and accession states (25 in total) at the ECCO Working Group evaluation meeting held in York in September 2003. No further responses were forthcoming.

3.3.2. Responses

There were only five replies that responded to the specific questions and a further one containing more general comments on the project. These are summarized in Table 4 and a full list of responses is provided in Appendix A.

4. QUALITY CRITERIA FOR MODELLING ECOTOXICITY DATA

4.1. Data quality and precision required

This section defines the quality criteria that need to be applied during the course of the project. Firstly, it defines the quality criteria that need to be applied to the data sets used. Secondly, it describes an analysis that was conducted to quantify the degree of precision required from QSARs to make them useful in the regulatory assessment, particularly in terms of potential for reduced animal testing.

4.2. Quality criteria to be applied to ecotoxicity data used in a QSAR

In order to determine what quality criteria should be applied to data used for modelling, it is proposed to initially consider the quality criteria applied to regulatory data sets and then discuss the possible quality criteria that could be applied to data sets used in any QSAR model.

It should be noted that this section is restricted to a consideration of the quality of the ecotoxicity data considered in constructing a QSAR. It does not cover issues such as range of potency or chemical structure.

4.2.1. Quality criteria currently applied to regulatory data sets

When notifiers make regulatory submissions, the data included with that submission must be conducted to 'Good Laboratory Practice' (GLP). This is to ensure the work is standardized and of a sufficiently high quality to produce reliable results which can with confidence be compared with those of others carrying out the same work and applying the same general principles. Further details can be found at http://www.pesticides.gov.uk/uploadedfiles/glp.pdf.

In addition to the need for studies to be conducted to GLP, it is usual that first tier studies are carried out to standardized study protocols using standard test organisms (Table 1). The majority of the first tier studies highlighted in Tables 1

Table 4
Summary of main responses to the survey questions

Survey question	Responses
1) What types of end-points would you like?	Those required by pesticide/biocide regulations
	Long-term toxicity, because tests are more expensive
2) How accurate are current test data?	Factors of 2–5, higher for chronic end-points
3) How accurate should QSARs be?	Factor of 5
	As accurate as input data (!)
	Accurate enough for pesticide to pass regulatory trigger
4) What reassurance would you require about the quality and quantity of input data?	Homogeneous – same experimental protocol
	At least 10 data points per descriptor
	Full details of training set
	Clear definition of domain of applicability
	Indication of experimental variability
	Existing guidelines/literature/GLP
5) What type of validation do you expect?	Cross-validation, ideally with 'independent' data (JRC could organize this?)
6) Any other requirements?	Accepted by regulators at EU level
General comments	Find a way of using the good quality study data held by industry
	Use existing QSAR programs as a starting point
	Reliability of QSAR results must be indicated
	Keep it simple and transparent

and 2, for example, OECD 203, have been developed and tested to ensure that they are both reliable and reproducible. This ensures reliability and consistency in the end-point used in the first tier risk assessments. For examples of reproducibility for *Daphnia* see Section 4.3, see also discussion on reproducibility in Chapter 2, Section 3.

4.2.2. Quality criteria to be applied to modelling data set

It is hoped that QSARs produced as a result of this project will be used in pesticide regulation and hence replace certain toxicity studies. If these are going to gain widespread acceptance and hence use, then it could be argued that the data upon which the models are based should be equivalent to the regulatory data they are replacing. This would go someway to addressing some of the concerns highlighted in Section 3.2. This approach would certainly address the issues raised by Schultz and Cronin (2003) who state that 'acute and chronic ecotoxicological data used in formulating QSARs should be reliable (and) of high quality'.

It should be considered that whilst the regulatory quality criteria are used to ensure reliability of data and hence consistency of approach, it is not necessarily *totally* appropriate to apply *exactly* the same criteria to large prospective data sets. It is however appropriate to apply some quality criteria to ensure that the QSAR models are reliable and hence likely to be adopted by regulators. Outlined below are suitable quality criteria that could be used when screening data sets that are to be used in the production of QSARs.

4.2.3. Consistency of study design

Strict adherence to an OECD or US EPA guideline is usually necessary from a regulatory perspective. It is, therefore, proposed that any study considered for use in constructing a QSAR should have been based on and hence similar to a recognized protocol. It is considered acceptable that studies conducted to draft protocols are acceptable, providing information regarding the study protocol or design is available.

Outlined below are some other factors regarding the design of the study that should be considered.

As regards aquatic toxicity studies, the points made above regarding protocols are appropriate; however, there are three issues that warrant further detailed consideration – namely conditions of test (i.e. static, semi-static or flow-through), analytical measurements and 'difficult substances'. These are dealt with below:

- The first issue is the use of static, semi-static or flow-through aquatic studies. Schultz and Cronin (2003) indicate that 'if toxicity values from different protocols (e.g. . . . static vs flow-through protocol) are used to develop QSARs, the results will be less accurate'. From a regulatory perspective, providing that the concentrations were maintained appropriately any test, that is, static, semi-static or flow-through, is appropriate providing that some form of justification is provided regarding the choice of test. It should further be noted that providing test concentrations were maintained appropriately (see below), there is rarely any major difference in the toxicity end-point regardless of how it was conducted. This issue, usually, only applies to the acute fish study submitted to address Annex point 8.2.1.

- Regarding analytical measurements, it is usual, although not essential, for aquatic toxicity studies to be accompanied by some form of chemical analysis. This analysis usually indicates whether the nominal concentrations were initially achieved and if so whether they were maintained. This information is used to determine if the study was acceptable and hence whether it can be used either for risk assessment or classification purposes. Ideally, analytical information should be available for studies used to construct QSARs. If analytical information is not present, then information regarding whether concentrations would have been maintained should be available.

- Some ASs may be classed as 'difficult substances' due to their low water solubility. Often, solvents are used to try and overcome this issue, and hence, when a study is conducted using a solvent, the end-point is usually the result of the solvent ensuring the AS is in solution. Data generated using this approach may be inappropriate to be used in developing a QSAR, and hence, this issue will need careful consideration when data sets are considered.

4.2.4. Consistency in the choice of species

For birds, fish and algae, it is possible to carry out a standard OECD or US EPA toxicity study on one of several species. For example, according to OECD 203, it is possible to test zebra fish, fathead minnow, common carp, ricefish, guppy, bluegill sunfish or the rainbow trout. Likewise, for algae the following species can be tested: *Selenastrum capricornutum* (now *Pseudokirchnereilla subcapitata), Scenedesmus subspicatus* or *Chlorella vulgaris*, and for birds, the bobwhite quail, Japanese quail or the mallard duck are acceptable. When the end-point from these studies is used for either classification, labelling or risk assessment purposes, the species tested is not the key issue, that is, the end-point is used to indicate the acute or chronic toxicity to birds, fish and algae. For example, the bobwhite quail is used to represent all birds, not just galliforms; likewise, the rainbow trout is a representative cold water fish species. From a regulatory point of view, therefore, data are used to indicate toxicity to 'birds', 'fish' and so on; however, this may not be compatible with the development of a QSAR. This is due to the fact that a QSAR may have to focus on data on one specific species. That is to say, a QSAR will predict the toxicity to rainbow trout and not fish generally. Therefore, in building the QSAR for rainbow trout, only data on rainbow trout should be used.

As regards studies on *D. magna* (see Annex points 8.2.4, 8.2.5 and 10.2.1), sediment dwelling invertebrates (see Annex point 8.2.7), aquatic plants (see Annex point 8.2.8), earthworms (see Annex points 8.4.1, 8.4.2, 10.6.1.1 and 10.6.1.2) and honeybees (see Annex points 8.3.1.1 and 10.4.1), there is only one test species specified. Therefore, only studies conducted using the specified organism should be accepted. This is in line with the statement made by Schultz and Cronin (2003) that 'if toxicity values from different protocols (e.g. *D. magna* or *D. pulex*) are used to develop QSARs, the results will be less accurate'.

4.2.5. End-point

The exact end-point needs to be clearly stated, along with the experimental error.

4.2.6. Identification of the test material

Identification and purity of the test material is essential.

4.2.7. Good laboratory practice

As was stated in Section 4.2.1, regulatory studies have to be conducted to GLP (for further information, see http://www.pesticides.gov.uk/uploadedfiles/glp.pdf). It could be argued that studies used to construct a regulatory QSAR should also be conducted to GLP. However, it is considered that while GLP status is desirable, providing that the studies are scientifically valid (using the above criteria and the summary in Table 5), then it is not necessary for studies to be conducted to GLP.

In addition to the above criteria, consideration should be made to two more points that are not related directly to the issue of quality criteria but are important in determining the overall acceptance of any QSAR model.

Table 5
Summary of the key quality criteria that should be considered when assessing data

Criteria	Definition
Consistency of study design	Studies used to construct a QSAR model should be based on a recognized protocol As regards aquatic studies, consideration should be made regarding whether initial nominal concentrations were attained and maintained appropriately and hence whether the study can be considered reliable
Consistency in the choice of species	Studies only on the relevant standard species should be considered
Identity of the test material	Information on the test material and its purity is essential
Good laboratory practice	This is desirable but not essential
Clearly defined and continuous toxicity end-point	The final choice of selected end-points (Section 5.1) should have a clearly defined toxicity end-point
Availability of all supporting data	All data used in constructing the QSAR should be made available to users

4.2.8. Clearly defined and continuous toxicity end-point

Schultz and Cronin (2003) state that acute and chronic toxicity data used in formulating a well-defined QSAR should reflect a well-defined and continuous toxic end-point. It should be noted that the term 'continuous' is not specifically defined by the authors of this paper; however, it is assumed to mean that the end-point is clearly defined and unambiguous, for example, LD/LC/EC50. Most of the studies highlighted in Table 2 have either a LD/LC/LR/EC50 or NOEC/NOEL as an end-point. It is considered that LD/LC/LR/EC50 end-points are clearly defined and unambiguous. With those studies that produce either a NOEC or NOEL as an end-point, it is agreed that the end-point can be clearly defined; however, it is not always unambiguous.

The aim of some studies is only to provide sufficient detail so that the AS can be classified. The classification categories may be broadly banded. An example of the possible broad banding is that used in the Chemicals Hazard Information and Packaging for Supply Regulations 2002 or 'Chips' classification (EEC, 2002d). Compounds that have a 96-h LC50 for fish or a 48-h EC50 for *D. magna* or a 72-h EC50 for algae of less than 1 mg/l are classified as 'very toxic to the aquatic organisms'. Those with an end-point between greater than 1 mg/l and less than or equal to 10 mg/l are classified as 'toxic to the aquatic organisms'. Those with an end-point greater than 10 mg/l but less than or equal to 100 mg/l are classified as 'harmful to aquatic organisms'. Therefore, it should be noted that for the purposes of producing an appropriate QSAR model, the classification alone is not considered sufficient.

In some studies, therefore, before deciding upon the priority end-points (see Section 5.5), consideration should be made to ensure that the end-point is appropriate to be modelled.

4.2.9. Availability of all supporting data

In order to address potential problems highlighted in Section 3.2, it is strongly recommended that wherever possible, all data used to construct a QSAR should be available to those scientists involved in developing the QSAR. It is considered extremely important that the data used to produce a QSAR should be independently checked by the scientist developing the QSAR and that the quality criteria outlined above applied.

Presented in Table 5 is a summary of the key quality criteria that should be considered when assessing data.

4.2.10. European Chemicals Bureau Quality Criteria for developing a QSAR

The ECB has developed 'principles' that should be considered when developing a QSAR. These are presented at http://ecb.jrc.it/QSAR. These 'principles' cover issues such as purpose, proposed application of the QSAR and domain of applicability. In Chapter 8 (Benfenati, this volume), we will specifically analyse the OECD principles for the DEMETRA models. As regards quality criteria that should be applied to the data set, 'full details of the QSAR training set should be provided. We give all details (chemical structure, and data for all descriptor and response variables) in the web site (http://www.demetra-tox.net). These views are in line with what is presented above.

4.2.11. Conclusion

The above criteria are general and could theoretically be applied to any data set. Outlined in Table 6 are quality criteria that should be applied to the toxicity data set to ensure that the data used to construct the QSAR is appropriate.

Table 6
Validity criteria for the end-points with high potential for developing in to a QSAR

Test	Preferred guideline(s)[a]	Validity criteria[b]
Acute toxicity test for the mallard duck	US EPA 71-1	Bird should be at least 16 weeks old Control mortality should not exceed 10% If a 'vehicle' is used to administer the test material, then it should be specified The LD50 as well as the 95% confidence limits should be specified Study should ideally be 14 days long
Acute toxicity test for the bobwhite quail	US EPA 71-1	Bird should be at least 16 weeks old Control mortality should not exceed 10% If a 'vehicle' is used to administer the test material, then it should be specified Study should ideally be 14 days long The LD50 as well as the 95% confidence limits should be specified
Dietary toxicity test for the mallard duck[c]	US EPA 71-2 OECD 206	Birds should be 5–17 days old There must be evidence that the concentration of the test substance was maintained during the study, if it was not maintained then measured concentrations should be used Control mortality should not exceed 10% The LC50 as well as the 95% confidence limits should be specified
Dietary toxicity test for the bobwhite quail[c]	US EPA 71-2 OECD 206	Birds should be 10–17 days old Control mortality should not exceed 10% If a 'vehicle' is used to administer the test material, then it should be specified

Table 6
(*Continued*)

Test	Preferred guideline(s)[a]	Validity criteria[b]
		The LC50 as well as the 95% confidence limits should be specified
		Study should ideally be 14 days long
Acute toxicity test for the rainbow trout	OECD 203 US EPA 72-1	Control mortality should not exceed 10%
		Water temperature should be between 13 and 17 °C
		Fish should be 5.0 cm
		Fish loading should be 1.0 g fish/l water
		Test should be run for 96 h
		Concentrations should be at least 80% of the nominal concentrations throughout the test. If concentrations are not maintained, then the end point should be quoted as measured concentration
		Test substance and any carrier should be clearly stated
		Dissolved oxygen concentration should be at least 60% of the air saturation value and pH should be between 6.0 and 8.5
		The LC50 as well as the 95% confidence limits should be specified
Acute toxicity test for the bluegill sunfish	OECD 203	Control mortality should not exceed 10%
		Water temperature should be between 21–25°C
		Fish should be 2.0 ± 1.0 cm
		Fish loading should be 1.0 g fish/l water
		Test should be run for 96 h
		Concentrations should be at least 80% of the nominal concentrations throughout the test. If concentrations are not maintained, then the end-point should be quoted as measured concentration
		Test substance and any carrier should be clearly stated
		Dissolved oxygen concentration should be at least 60% of the air saturation value and pH should be between 6.0 and 8.5
		The LC50 as well as the 95% confidence limits should be specified

(*Continued*)

Table 6
(*Continued*)

Test	Preferred guideline(s)[a]	Validity criteria[b]
Acute toxicity study for *Daphnia magna*	OECD 202	Control mortality should not exceed 10% Dissolved oxygen concentration should be at least 60% of the air saturation value Test substance and any carrier should be clearly stated *Daphnia magna* should not be more than 24 h old Water temperature should be between 18–22°C The EC50 as well as the 95% confidence limits should be specified
Acute toxicity to bees	OECD 213, 214, and EPPO 170	Need to ensure that the same end-point is considered, that is, contact or oral Control mortality should be quoted Test substance and any carrier should be clearly stated The LD50 for both oral and contact end-points as well as the 95% confidence limits should be specified
Acute toxicity test for earthworms	OECD 207	Control mortality should not exceed 10% Test species should be *Eisenia foetida* Artificial soil should be used which consists of 10% sphagnum peat, 20% kaolin clay and 70% sand
Algal growth	OECD 201	Need to ensure that the same end-point is considered, that is, EbC50 or ErC50 Test substance and any carrier should be clearly stated The EC50 as well as the 95% confidence limits should be specified

[a] If a study has been carried out according to the appropriate guideline, then it can be assumed to have satisfied the validation criteria and can, therefore, be used to build a QSAR.

[b] The criteria outlined are one that are considered to be crucial in determining the validity of a study. It is acknowledged that these may not always be in line with the validation criteria specified in the test guideline; however, it is considered that if these criteria are met then the study would be of use in a regulatory context.

[c] The end-point from the avian dietary study is currently ppm diet (or mg a.s./kg diet); however, the current risk assessment (EEC, 2002a) uses a daily dose approach, that is mg a.s./kg bw/kg day. In order to convert from one end-point to another, information on daily food consumption is required for the control as well as the treated birds. It is not possible, due to issues of repellency and food avoidance, to use a generic conversion factor; therefore, the feasibility of using data to build a QSAR to address the avian dietary toxicity is considered to be low.

In addition to the information outlined in Table 6, information should always be available on the purity of the test substance.

It is felt that when available data and databases are scrutinized, the above criteria are used, if a study fails one or more of the criteria then it should be rejected. If this is rigorously followed, potential users of the QSARs developed by this project should have confidence in the data used to build and, subsequently, test the QSARs.

4.3. Degree of precision required of QSARs for pesticide assessments

For QSARs to be used in the pesticide authorization process, an indication of the degree of precision is required to provide an indication of their reliability and hence usefulness in reducing animal testing. In order to achieve this, standard risk indices (e.g. TERs) were computed for a range of representative pesticide uses and compared to the regulatory triggers in the Annexes to Directive 91/414/EEC. We then determined the proportion of pesticides for which a QSAR would be sufficient to 'pass' the regulatory trigger, in which case it is assumed an animal study would not be required. The answer depends on the precision of the QSAR (and, of course, this applies to the precision of experimental data, if the same exercise would be done with these), which was expressed as the percentage of variance explained by the QSAR, so the calculation was repeated for a range of precision levels (0–100% variance explained). The answer also depends on the degree of confidence that is required, which was expressed as the width of the confidence interval that assessors would wish to place on the QSAR prediction when substituting it for an animal test, for example, 80, 90 or 95% confidence limits. The calculation was therefore repeated for a range of confidence intervals up to 99%.

This analysis was carried out for three types of toxicity end-points. Priority was given to conducting this analysis for birds (avian acute oral LD50) and fish (acute LC50), because of the priority given to reducing testing of vertebrates. In addition, an analysis was conducted for *Daphnia* (acute EC50), as an example of an invertebrate end-point.

4.3.1. Methods

The basic method was the same for each end-point examined, as follows:

1) End-point data were obtained for as many pesticides as possible. For bobwhite quail acute oral LD50, the data were provided by the BVL (Federal Office for Consumer Protection and Food Safety, Germany). For fish and *Daphnia* acute LC/EC50s, the data were provided by Partner 1 (Mario Negri Institute) and derived mainly from public databases of the US EPA. All of the data used therefore derived from standard studies accepted by government authorities for regulatory assessments. In addition, the data for fish and *Daphnia* were subjected to a quality check by P1.

2) For each pesticide in the data set the following was done.
   a) Estimated exposure using standard assumptions used in EU regulatory assessments. For each pesticide, we used a 'worst case' use scenario, that is, the combination of crop, application rate and application date that gave maximum exposure for the end-point concerned. As there was no readily accessible information on the scenarios considered at EU level, we used information on crops and application rates currently used in the UK (based on information provided by the UK Pesticide Usage Survey, Miles Thomas, personal communication). As a consequence, the analysis was limited to pesticides used in the UK. For birds, the exposure assumptions were taken from the EU guidance document for risk assessments for birds and mammals (EEC, 2002a). For fish and *Daphnia*, exposure was estimated for spray drift only, and assumptions were taken from the EU guidance document for aquatic ecotoxicology (EEC, 2002a). In all cases, a single application at the maximum permissible rate was considered, and day 0 predicted exposure values were used to calculate TERs.
   b) Estimated the lower x% one-tailed confidence bound for toxicity of that pesticide assuming QSAR explains y% of variance (more details of this calculation are given below). This calculation was repeated for values of x between 50 and 99% and for values of y between 0 and 100%.
   c) Calculated the TER and compared it to the relevant EU trigger value (10 for birds, 100 for fish and *Daphnia*).

3) We calculated the percentage of pesticides that passed the regulatory trigger (i.e. where the pesticide would not require further testing, if the QSAR were accepted for regulatory assessment). We plotted the results to show how the proportion of pesticides passing the regulatory trigger with a QSAR depends on the % variance explained and degree of confidence required. The lower confidence bound for the QSAR toxicity prediction of each pesticide was estimated using the formula for the confidence limits on a single new observation from a linear regression relationship (Draper and Smith, 1981):

$$\widehat{Y}_0 \pm t(v, P)\left[1+\frac{1}{n}+\frac{(X_0-\bar{X})^2}{\Sigma(X_i-\bar{X})^2}\right]^{1/2} \cdot s \qquad (1)$$

where, $\widehat{Y}_0$ is the expected value of Y for X_0; $t(v, P)$ is Students t statistic with one-tailed probability P and v degrees of freedom; n is the sample size for the regression (number of pesticides used to estimate the QSAR); X_0 is the value of X for which a predicted toxicity is required; $\bar{X}$ is the average value of X from the QSAR regression; X_i are the values of X for the pesticides on which the QSAR was based ($i = 1 \ldots n$); and s is the standard error of the residuals for the fitted QSAR.

A one-tailed t was used because it is assumed the regulator's interest is in the certainty that the true toxicity value is greater than the specified confidence limit.

Because this analysis was conducted before developing the QSAR (so as to assess the potential benefit of developing it), it was not known what particular type of QSAR model would be most successful, nor the number or type of structural parameters it would include. It was therefore necessary to make a number of assumptions in the analysis, as listed below.

It was assumed that

- The QSAR was based on $\log_{10}$ molar toxicity.
- The QSAR is a simple linear regression with only one predictor variable.
- The residuals of the QSAR regression follow the normal distribution.
- The expected value of the QSAR prediction ($\widehat{Y}_0$) is equal to the reported toxicity. This assumption was necessary because the predictor variable (X_0) is unknown. It results in over-estimation of the confidence limits for some pesticides and under-estimation for others, but over a reasonable number of pesticides, these errors should approximately cancel out.

In order to calculate the last term in brackets within the Eq. 1, it was necessary to estimate the predictor values (X_0). As we already assumed that $\widehat{Y}_0$ is equal to the observed toxicity (see above), we estimated X_0 as standard normal deviate (i.e. value from a distribution with a mean of 0 and standard deviation of 1) for the same percentile as that of the observed toxicity value within a normal distribution based on the mean and standard deviation of all the toxicity values. Using the standard normal deviate for X is equivalent to assuming that the predictor variable for the QSAR had been transformed to a standard normal deviate before the QSAR analysis was conducted (i.e. transforming each value of X by subtracting the mean of X and dividing by the standard deviation of X). For a perfect regression relationship (100% of variance explained), this estimate would be precisely correct. As the percentage of variance explained decreases, the estimate of X_0 becomes increasingly uncertain. However, over a number of pesticides, the errors should approximately cancel out.

In calculating Eq. 1, it was assumed that estimation of the QSAR regression model excluded toxicity data reported as 'greater than' or 'less than' values, because this was the intended approach for DEMETRA. These values were also omitted when calculating the proportion of pesticides for which QSAR estimates would pass the regulatory triggers. However, for birds only, a second analysis was conducted in which 'greater than' values were included when calculating the proportion of pesticides for which QSAR estimates would pass the regulatory triggers. The implications of this difference are considered when discussing the results (below).

4.3.2. Results and discussion

4.3.2.1. Bobwhite quail – acute oral LD50

A total of 70 pesticide toxicities were available for analysis, of which 47 were reported as 'greater than' values. The results when TERs were calculated excluding and including the 'greater than' values are shown in Figs. 1 and 2 respectively.

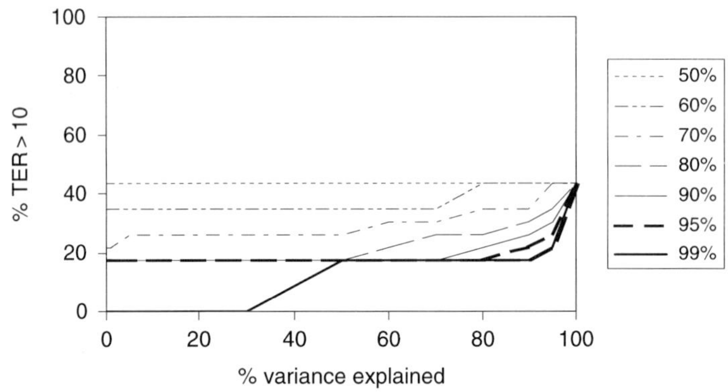

Figure 1: Results of precision analysis for bobwhite quail acute oral LD50. In this case, pesticides with 'greater than' toxicity values were omitted both when calculating the quantitative structure-activity relationship confidence limits and from the calculation of toxicity-exposure ratios (TERs). The different lines show results for different levels of certainty (one-sided confidence limits between 50 and 99%).

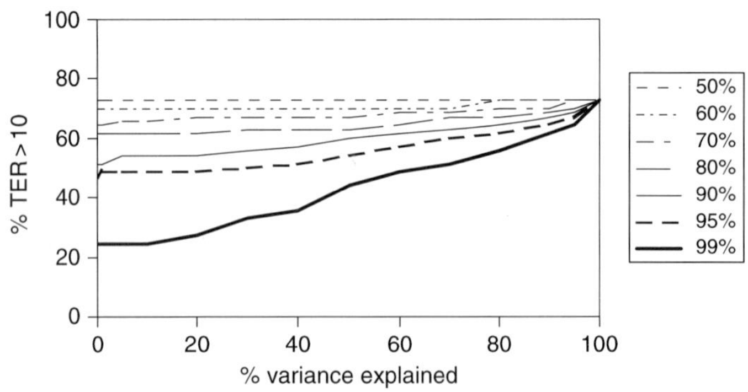

Figure 2: Results of precision analysis for bobwhite quail acute oral LD50. In this case, pesticides with 'greater than' toxicity values were omitted when calculating the quantitative structure-activity relationship confidence limits but not from the calculation of toxicity-exposure ratios (TERs). The different lines show results for different levels of certainty (one-sided confidence limits between 50 and 99%).

Both Figs. 1 and 2 show a convergence of the plotted lines at the right-hand side of the graph. This is because, when 100% of the variance is explained, all confidence intervals (99%, 95% and so on) are zero. Therefore, the toxicity estimates (the expected values) used for calculating all the TERs are the same, and the proportion exceeding 10 is the same. This proportion is simply equal to the proportion of TERs exceeding 10 when the observed toxicity data are used (because we assumed the expected toxicity is equal to the observed toxicity, as explained earlier).

Both Figs. 1 and 2 include a line for a 50% one-tailed confidence interval. The 50% confidence lines are always horizontal, because they imply using the expected value from the QSAR (which is assumed equal to the observed toxicity value) in every case regardless of the % of variance explained. In this case again, therefore, the proportion of TERs exceeding 10 is equal to the proportion obtained when the observed toxicity data are used.

Comparison of Figs. 1 and 2 shows that when pesticides with 'greater than' toxicity data are included, a higher proportion of studies could potentially be replaced by QSAR estimates. This is because compounds with high LD50s are very unlikely to have TER < 10 unless they are applied at unusually high rates, or unless the formulation is consumed directly by birds (e.g. treated seeds). It is therefore important to develop QSARs that are valid even when the predicted toxicity is above the maximum level usually tested in toxicity studies (2000 mg/kg body weight).

Provided that QSARs can be developed that are valid for pesticides with high LD50s, Fig. 2 shows considerable potential for replacing avian LD50 testing. For example, if regulators would be content with 95% certainty that the QSAR is below the true LD50, then even a poorly predictive QSAR (explaining only 10–20% of the variance) could reduce the need for testing by half. This slightly counter-intuitive result occurs because a large proportion of pesticides have low toxicity to birds and TERs well above the EU trigger value of 10: therefore, even if the QSAR has wide confidence limits, the TER still remains above the regulatory trigger for further testing.

The corollary of this is that QSAR is unlikely to be useful for end-points that frequently fail EU trigger (e.g. the avian reproduction study), because a QSAR can only be equally or more likely to fail.

4.3.2.2. Trout – acute LC50

A total of 59 pesticide toxicities were available for trout. 'Greater than' toxicity values had already been excluded, and it is estimated that about 10–20% of pesticides were omitted because of this. This proportion is much lower than that for the acute avian LD50, where 67% had 'greater than' values (see Section 4.3.2.1).

Results for trout are shown in Fig. 3. There seems to be less potential for replacing acute fish testing than acute bird testing. For example, if regulators would be content with 95% certainty that the QSAR is below the true LD50, then

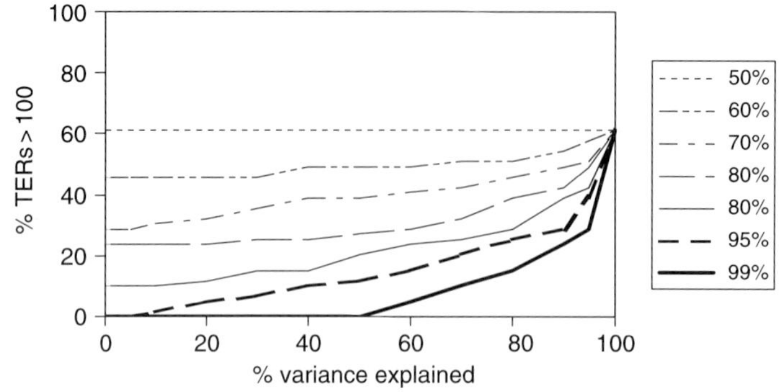

Figure 3: Results of precision analysis for rainbow trout acute LC50. Pesticides with 'greater than' toxicity values were omitted both when calculating the quantitative structure-activity relationship confidence limits and from the calculation of toxicity-exposure ratios (TERs). The different lines show results for different levels of certainty (one-sided confidence limits between 50 and 99%).

the QSAR will need to explain over 90% of the variance in order to reduce acute fish testing by more than 20%.

4.3.2.3. Daphnia – acute EC50

A total of 61 pesticide toxicities were available for *Daphnia*. 'Greater than' toxicity values had already been excluded, and as for trout, it is estimated that about 10–20% of pesticides were omitted because of this. This proportion is again much lower than that for the acute avian LD50 (see Section 4.3.2.1).

Results for *Daphnia* are shown in Fig. 4. There seems to be slightly more potential for replacing acute *Daphnia* testing than acute fish testing. For example, if regulators would be content with 95% certainty that the QSAR is below the true LD50, then the QSAR will need to explain over 60% of the variance in order to reduce acute *Daphnia* testing by more than 20%.

4.3.2.4. Conclusions

This section has examined the relative potential for replacing different end-points with QSAR predictions, depending on the proportion of variance explained by the QSAR and the degree of certainty required by regulators. The results suggest that there is substantial potential for replacing acute testing for birds, and some (though less) potential for replacing acute testing of fish or *Daphnia*. The potential for replacement is heavily dependent on the degree of certainty required by regulators, as can be seen by examining the lower lines in Figs. 1–4. Finally, it is concluded that there is very little potential for replacing end-points that frequently fail the EU regulatory triggers, such as the avian reproduction study.

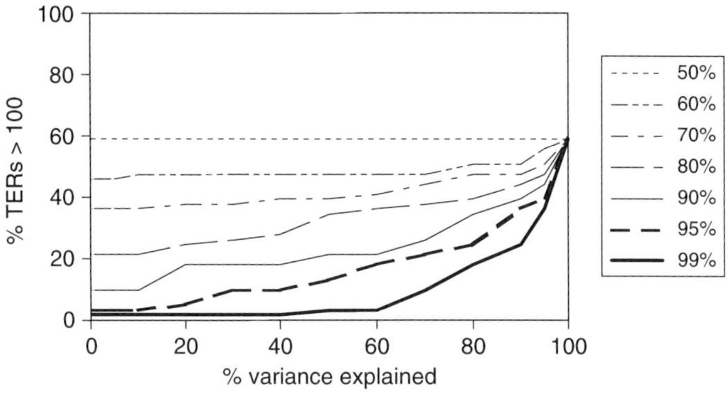

Figure 4: Results of precision analysis for *Daphnia* acute EC50. Pesticides with 'greater than' toxicity values were omitted both when calculating the quantitative structure-activity relationship confidence limits and from the calculation of toxicity-exposure ratios (TERs). The different lines show results for different levels of certainty (one-sided confidence limits between 50 and 99%).

4.3.2.5. Regulatory perspective

From the above section, it is suggested that there is substantial potential for replacing certain toxicity studies with an appropriate QSAR. However, the above assessment indicates that the potential for replacement is heavily dependent on the degree of certainty required by regulators. The following section will try and expand upon this issue and provide a regulatory view as to the degree of certainty required. It will also build on some of the key issues highlighted in the above assessment.

From a regulatory perspective, it is important that if a QSAR approach is adopted then the QSAR should be as reliable as the method it replaces (see Section 3.2). This means that it should predict the level of toxicity with the same degree of reliability and reproducibility as standard regulatory data sets.

When a regulatory data set is submitted for evaluation, a range of studies are submitted depending upon the proposed use of the plant protection product as well as characteristics of the AS (see Section 1.1). Usually only a limited number of studies are submitted to address a particular point, for example, in Table 2, it is indicated that 1.8 acute avian toxicity studies are submitted. These studies will usually be on the bobwhite quail and/or the mallard duck. The key issue is that usually there is only one study provided on the toxicity of an AS to a certain species. Due to the limited number of studies available, regulators do not know how accurate or representative a particular end-point is, that is, we do not know its 'true LD50' as referred to above. Based on the authors experience, it is possible that there could be sufficient variation that when one toxicity test is used in the regulatory risk assessment, concern is raised (i.e. the regulatory TER is breached) whilst another study conducted to exactly the same protocol did not.

A further illustration of the above is presented in the OECD report of the Final ring test of the *D. magna* reproduction test (OECD, 1997). As part of the validation procedure, variability between laboratories was assessed. Of all the tests that provided estimates of NOEC, EC50, EC25 and EC10:

- for DCA, approximately 50% of the NOECs, EC50s, EC20s and EC10s lay within a factor of 2, over 75% lay within a factor of 4 and over 90% lay within a factor of 8;
- for cadmium, 38% of the NOECs lay within a factor of 2 and 62% within a factor of 8; the figures for the EC50s and EC20s were 27 and 45% respectively, and for EC10s 27 and 55% and
- for phenol, 45% of the NOECs lay within a factor of 3.2 and 82% within a factor of 10; the figures for the EC50s were 70 and 100% respectively, for the EC20s 50 and 90% and for the EC10s 50 and 80%.

From the above, it can be concluded that it is possible that the end-point used from a study submitted for regulatory purposes could vary by a factor of 10. However, as was stated above, in regulatory risk assessment it is usually the case that only one of two studies are submitted on one species. Therefore, these end-points are used in the risk assessment together with an appropriate uncertainty factor [see EEC (1991) for more details]. Bearing this in mind, it is difficult to state what degree of precision is required, except that it should be as 'accurate' as the status quo. Or put another way, if a compound failed the regulatory trigger with 'real' data, then it should do likewise with an end-point from a QSAR.

To try and put the above regulatory view in terms of precision and confidence is extremely difficult, if not impossible to do. When end-users were consulted (see Section 3.3), a question was posed regarding the accuracy of QSARs, as can be seen from Table 4 the responses indicated that they should be within a factor of 5 of the 'true end-point'. Taking on board the above work on precision, it is considered more appropriate to work in terms of levels of confidence. However, it is felt that as an appropriate starting point, it is considered feasible that 95% certainty, or confidence, should be sufficient, this does mean that 1 in 20 times, the QSAR could be 'incorrect'. However, from a risk assessment point of view, this is considered acceptable as a QSAR can only be equally or more likely to fail a TER than 'real' data.

In the above work on acute avian toxicity, it is noted that a QSAR approach could be of benefit if it could be used to predict the toxicity of low toxicity compounds, that is, those where the toxicity is usually above the highest dose tested. This would be of great benefit in terms of animal testing, as currently a notifier or applicant would have to submit data on a low toxicity compound, even though the risk is likely to be low.

An additional issue linked to the development and selection of QSARs is the potential saving of test animals. As can be seen from the avian example, there is potential to reduce the number of toxicity tests carried out and hence reduce the number of test animals. This is considered a potential benefit as it would help address both individual Member State legislation and concerns regarding animal welfare (e.g. UK Home Office requirement to replace animal testing with other methods where possible and to reduce the number of animals used in tests) as well as other international drivers (e.g. the OECD has a policy of reducing the use of animals in testing). The important issue here is that the QSAR is sufficient to explain an appropriate degree of variance. For example, referring to the avian examples, it is clear that a poorly predictive QSAR may have the ability to reduce animal testing even when it only explains between 10 and 20% of the variance.

To conclude, it is difficult, if not impossible to state the level of precision required from a QSAR as the 'true LD50' is not known due to inter- and intra-laboratory variation. However, it is proposed that the issue of precision is considered fully in both the development of a QSAR and the end-user testing phase of DEMETRA. This will ensure that any output from DEMETRA is not only of practical use to regulators and applicants but also likely to be used by regulators. It is also concluded that at the level of certainty proposed above, there could be a significant reduction in the number of animals tested, depending upon the final choice of studies and the level of precision that can be achieved in the models that are developed.

5. TOXICITY END-POINTS WITH A HIGH POTENTIAL TO BE REPLACED WITH A QSAR APPROACH

In order to determine those end-points with a high potential to be replaced with a QSAR approach, information on the availability of data, numbers of animals used in tests and study costs were considered. Outlined below is a summary of this assessment.

5.1. Data availability

A survey was conducted amongst DEMETRA partners and subcontractors to gather information about databases that contain pesticide toxicity data to determine the availability of sufficient good quality data.

The main conclusion is that good data are not readily available. The most suitable sources of those listed above were the EPA-OPP, SEEM and BVL databases.

5.2. Number of animals tested

One of the main benefits of the use of QSARs in risk assessment is the reduction in the number of animals tested. Currently, the main issue is with the reduced use of vertebrate species in testing (birds and fish in this case). Numbers of animals used in each vertebrate guideline were collated from the appropriate guidance documents and these are summarized in Table 7.

Table 7
Numbers of animals used in each test guideline

End-point	Guideline	Treated group	Control group	Notes
Avian acute toxicity	Draft OECD 223	24	–	Using full three-stage design (12 in two-stage design)
	EPA 71-1	50	20	Assuming 10 birds per group, 5 concentrations and two control groups
Avian dietary toxicity	OECD 205	50	20	Assuming 10 birds per group, 5 concentrations and two control groups
	EPA 71-2	50	20	Assuming 10 birds per group, 5 concentrations and two control groups
Avian reproduction	Draft OECD	120	40	Assuming 3 concentrations and 20 pairs per group with 1 control group (Note: 20 pairs per group to ensure at least 16 are layers)
	OECD 206	108	36	For quail assuming, 3 concentrations, 12 pens per concentration and 1 control group
	EPA 71-4	72	36	For BWQ assuming 1 male to 2 females, 2 test groups, replication of 12 pens per group

Table 7
(*Continued*)

Endpoint	Guideline	Treated group	Control group	Notes
Acute toxicity to fish	OECD 203	50	10	Assuming only five concentrations and one control group
	EPA 72-1	50	10	Assuming only five concentrations and one control group
Chronic toxicity test on juvenile fish	OECD 204	50	10	Assuming only five concentrations (number not given) as above for OECD 203
Fish early life stage toxicity test	OECD 210	300	60	Embryos initially exposed, assumes no vehicle control
Fish life cycle test	EPA 72-5	–	–	Multi-stage test at five concentrations
Bioconcentration in fish	OECD 305A 'Sequential static test'	95	?	For catfish (may be more if smaller fishes are used)
	OECD 305B 'Semi-static test'	98	14	Assuming two concentrations
	OECD 305C 'Degree of bio-concentration'	40	20	Using carp and for accumulation part only – no details for initial acute toxicity test with orange red killifish

In most cases, these are the minimum number for a full test required by the guideline (e.g. minimum recommended number of concentrations) and do not include initial range finding tests. Where the species tested affects numbers, tested, the species selected is the recommended one that requires the greatest number.

Of the birds studies listed, the acute and dietary toxicity tests are the ones that are most severe and would carry a substantial severity limit under current UK Home Office licensing. Avian reproduction testing would carry a moderate severity limit under the same system, but uses more animals especially when compared to the draft new avian acute study. All three tests are almost always requested with around two studies per Annex point (Table 2). Thus, those tests that may appear to offer the greatest benefits in terms of welfare if replaced (e.g. avian acute and dietary toxicity) are not necessarily those for which most animals would be saved.

Similar issues arise with fish studies with the acute toxicity studies being the most severe but not necessarily involving the largest numbers. However, if we consider the number of studies requested then some studies such as fish acute toxicity (70 animals per test) requested with a mean of 4.75 studies per Annex point (Table 2), the potential number of animals saved would be substantially increased (over 300 animals).

On the basis of numbers and severity, this would suggest that the avian acute, avian dietary and fish acute studies show themselves to be among the preferred end-points for modelling.

5.3. Study costs

Approximate study costs for ecotoxicological end-points were obtained from commercial contacts. These included avian toxicity studies, fish toxicity studies (except fish life cycle test as this is rarely used) and invertebrate studies (including aquatic invertebrates, bees, other arthropods and earthworms). These are summarized in Table 8.

Table 8
Approximate costs in euros for a range of studies most used in ecotoxicological risk assessment

Data requirement	Approximate costs (euros)	Notes
Acute oral toxicity to birds	7000	–
Short-term dietary toxicity to birds	13,000 to 17,000	–
Sub-chronic toxicity and reproduction in birds	71,500 to 114,500	Japanese quail cheapest, mallard most expensive
Acute toxicity to fish	8500 to 14,500	Includes analysis
Chronic toxicity test on juvenile fish	32,000	For an OECD215
Fish early life stage toxicity test	20,000 to 55,500	Depending on country
Fish life cycle test	–	Rarely used
Bioconcentration in fish	28,500 to 57,000	Typical price (can range from 20,000 to 137,000 depending on study/country)
Acute toxicity to aquatic invertebrates	2000 to 3500 (up to 10,000)	Not including analysis (including analysis)

Table 8
(*Continued*)

Data requirement	Approximate costs (euros)	Notes
Chronic toxicity to aquatic invertebrates	5500 to 10,000 (up to 28,500)	Not including analysis (including analysis)
Effects on algal growth	2000 to 3000 (up to 11,500)	Not including analysis (including analysis)
Acute toxicity to bees	5000 to 6000	–
Acute toxicity to other arthropods	4500 to 32,000	Depending on species/study
Acute toxicity to earthworms	2500 to 7000	–
Sublethal effects in earthworms	2000	–
Effects on soil non-target micro-organisms	8500 to 17,000	–

Again, it is clear that some of the studies that might produce the most benefits in terms of welfare if replaced (e.g. avian and fish acute toxicity studies) are not those that would provide the most financial benefits due to their relatively low cost although the number of fish studies per end-point again increases the benefits of modelling this end-point. For invertebrates, the main benefit will be financial savings, but many of these are relatively low in cost unless analysis is required.

5.4. End-points with high potential for replacement with a QSAR

In trying to determine the end-points with high potential for replacement with a QSAR, data were collected on:

- the frequency each end-point is requested,
- the number of animal used in the study,
- the severity of the test (vertebrates),
- the proportion of TERs that do not trigger further testing,
- the availability of data and
- an assessment of feasibility (e.g. a single clear end-point).

On the basis of this discussion, a list of end-points considered to have high potential was selected and is presented in Table 9.

Table 9
Information used in discussion to select the list of priority end-points for the remainder of the project

End-point	% requested	Number of animals	Severity	Cost	% not trigger	Data availability	Feasibility	Select
Acute oral toxicity to birds (BWQ and Mallard)	High	High	High	Low	High	Good	High	Yes (two spp.)
Short-term dietary toxicity to birds (BWQ and Mallard)	High	High	High	Medium	High	Good	?	Yes (two spp.)
Sub-chronic toxicity and reproduction quail or the mallard duck.	High	High	Medium	High	Low	Good	?/Poor	No
Acute toxicity to fish (Rainbow trout and Bluegill sunfish)	High	(High)	(High)	Medium	High	Good	High	Yes (two spp.)
Chronic toxicity test on juvenile fish	High	(High)	Low	High	?	?/poor	Poor (multiple end-points)	No

Acute toxicity to aquatic invertebrates (Daphnia)	High	Inverts	Low	Low	Low	Good	High	Yes
Chronic toxicity to aquatic invertebrates (Daphnia)	High	Inverts	Low	(High)	Low	?	Poor (NOEC+ multiple end-points)	No
Effects on other non-target arthropods	High	Inverts	Low	(High)	?	Poor	High	No
Acute toxicity to bees (two studies)	High	Inverts	Low	Low	High	Poor	High	Yes
Acute toxicity to earthworms	High	Inverts	Low	Low	High (chronic)	Poor (in US)	High	Yes
Effects on algal growth	High	None	Low	?	High	Poor	two End-points	Yes

The final column indicates the decision of the group.

5.5. Priority end-points

On the basis of the information presented in Table 9, the end-points considered to be the most suitable and important for modelling are:

- Mallard acute toxicity
- Bobwhite quail acute toxicity
- Mallard dietary toxicity
- Bobwhite quail dietary toxicity
- Rainbow trout acute toxicity
- Bluegill sunfish acute toxicity
- *Daphnia* acute toxicity
- Acute toxicity to bees
- Acute toxicity to earthworms
- Effects on algal growth

It is considered that these have well defined end-points and continuous and hence satisfy the criteria specified by Shultz and Cronin (2003) and discussed in Section 4.

REFERENCES

Ali H.M., Sharaf E.H.A., Hikal M.S. (2005) Selectivity, acetylcholinesterase inhibition kinetics and quantitative structure–activity relationships of a series of N-(2-oxido-1,3,2-benzodioxa-phosphol-2-yl) amino acid ethyl or diethyl esters. *Pestic. Biochem. Physiol.*, 83, 58–65.

AMBIT (2005). Building blocks for a future (Q)SAR decision support system: databases, applicability domain and structure conversions (codename AMBIT). http://ambit.acad.bg/.

Bakker F., Feije R., Grove A.J., Hoogendoorn G., Jacobs G., Loose E.D., van Stratum P. (2003). A laboratory test protocol to evaluate effects of plant protection products on mortality and reproduction of the predatory mite *Hypoaspis aculeifer* Canestrini (Acari: Laelapidae) in standard soil. *J soils Sediment* 3:73–77.

Barrett K.L., Grandy N., Harrison E.G., Hassan S., Oomen P. (1994) Guidance document on regulatory testing procedures for pesticides and non-target arthropods. From the ESCORT Workshop. SETAC-Europe, 51 p.

Benfenati E., Gini G. (1997) Computational predictive programs (expert systems) in toxicology. *Toxicology*, 119, 213–225.

Benfenati E., Mazzatorta P., Neagu D., Gini G. (2002) Combining classifiers of pesticides toxicity through a Neuro-fuzzy approach. Multiple classifier systems. MCS2002. In: *Lecture Notes in Computer Science*. Roli F., Kittler J., Eds. Springler-Verlag, Berlin, Germany, pp. 293–303.

Blümel S., Bakker F., Baier B., Brown K., Candolfi M.P., Goßmann A., Grimm C., Jäckel B., Nienstedt K., Schirra K.J., Ufer A., Waltersdorfer A. (2000) Laboratory residual contact test with the predatory mite *Typhlodromus pyri* Scheuten (Acari: Phytoseiidae) for regulatory testing of plant protection products. In: *Guidelines to Evaluate Side-Effects of Plant Protection Products to Non-Target Arthropods*. IOBC, BART and EPPO Joint Initiative. Candolfi M.P., Blümel S., Forster R., Eds. IOBC/wprs, Gent, 121–144.

Campbell P.J., Arnold D.J.S., Brock T.C.M., Grandy N.J., Heger, W., Heimbach F., Maund S.J., Streloke M. (1999) Guidance document on higher tier aquatic risk assessment for pesticides (HARAP). From the SETAC-Europe/OECD/EC workshop, Lacanau Ocean, France.

Candolfi M.P., Barrett K.L., Campbell P.J., Forster R., Grandy N., Huet M.C., Lewis G., Oomen P.A., Schmuck R., Vogt H. (2001) Guidance document on regulatory testing and risk assessment procedures for plant pretection products with non-target arthropods. ESCORT 2 workshop. SETAC, Pensacola, FL, USA.

Cronin M.T.D., Livingstone D.J. (2004) *Predicting Chemical Toxicity and Fate*. CRC Press, Boca Raton, USA.

Danish EPA (2001) The advisory for self classification of dangerous substances. http://www.mst.dk/chemi/01050000.htm.

Devillers J., Pham-Delegue M.H., Decourtye A., Budzinski H., Cluzeau S., Maurin G. (2002) Structure-toxicity modeling of pesticides to honey bees. *SAR QSAR Environ. Res.*, 13, 641–648.

Draper N.R., Smith H. (1981) *Applied Regression Analysis*. John Wiley and Sons Inc, New York, USA.

ECB (2006) Danish (Q)SAR Database. http://ecb.jrc.it/QSAR/.

EEC (1995) Technical guidance documents in support of the Commission Regulation of laying down the principles for the assessment of risks to man and the environment of existing substances. (EEC/1488/94).

EEC (1991) Council Directive concerning the placing of plant protection products on the market (91/414/EEC). http://europa.eu.int/comm/food/fs/ph_ps/pro/legal/dir91-414-eec_en.pdf.

EEC (2000) Technical guidance document in support of the Directive 98/8/EC concerning the placing of biocidal products on the market: guidance on data requirements for active substances and biocidal products. Scientific Committee on Toxicity, Ecotoxicity and the Environment (CSTEE). http://europa.eu.int/comm/health/ph_risk/committees/sct/docshtml/sct_out53_en.htm.

EEC (2002a) Working document: guidance document on risk assessment for birds and mammals. Council Directive 91/414/EEC SANCO/4145/2002. http://europa.eu.int/comm/food/fs/ph_ps/pro/wrkdoc/wrkdoc19_en.pdf.

EEC (2002b) Working document guidance document on aquatic ecotoxicology under Council Directive 91/414/EEC SANCO/3268/2002. http://europa.eu.int/comm/food/fs/ph_ps/pro/wrkdoc/wrkdoc10_en.pdf.

EEC (2002c) Working document Guidance document on terrestrial ecotoxicology under Council Directive 91/414/EEC SANCO/10329/2002. http://europa.eu.int/comm/food/fs/ph_ps/pro/wrkdoc/wrkdoc09_en.pdf.

EEC (2002d) Approved classification and labelling guide (5th edition) Chemicals (Hazard Information and Packaging for Supply) Regulations 2002. Guidance on regulations. L131 HSE books.

EEC (2003) Working document guidance document on the assessment of the relevance of metabolites in groundwater of substance regulated under Council Directive 91/414/EEC SANCO/221/2000 – rev 10.

EPPO (1992) Guideline on test methods for evaluating the side effects of plant protection products on honeybees, Method 170, Bulletin OEPP/EPPO, 22, 203–215.

EPPO (2002) Environmental risk assessment scheme for plant protection products. Chapter 8: Soil organisms and functions. EPPO Bulletin.

Giddings J.M., Brock T.C.M. Heger W., Heimbach F., Maund S.J., Norman S.M., Ratte H.T., Schafers C., Streloke M. (2002) *Community-Level Aquatic Systems Studies – Interpretation Criteria*. SETAC, Pensacola, FL, USA.

Greig-Smith P.W., Becker H., Edwards P.J., Heimbach F., Eds. (1992) *Ecotoxicology of Earthworms*. Intercept, Andover, UK.

Hansen O.C. (2004) Quantitative Structure-Activity Relationships (QSAR) and Pesticides. Danish Environmental Protection Agency, Pesticides Research No. 94.

Lin G., Lai C.Y., Liao W.C. (1999) Molecular recognition by acetylcholinesterase at the peripheral anionic site: structure–activity relationships for inhibitions by aryl carbamates. *Bioorg. Med. Chem.*, 7, 2683–2689.

Løkke H., Van Gestel C.A.M. (1998) *Handbook of Soil Invertebrate Toxicity Tests*. Ecological and Environmental Toxicology Series, Wiley and Sons, Chichester, Sussex, England.

Maxwell D.M., Brecht K.M. (1992) Quantitative structure–activity analysis of acetylcholinesterase inhibition by oxono and thiono analogues of organophosphorus compounds. *Chem. Res. Toxicol.*, 5, 66–71.

Mead-Briggs M.A., Brown K., Candolfi M.P., Coulson M.J.M., Miles M., Moll M., Nienstedt K., Schuld M., Ufer A., McIndoe E., (2000) A laboratory test for evaluating the effects of plant protection products on the parasitic wasp, *Aphidius rhopalosiphi* (DeStephani-Perez) (Hymenoptera: Braconidae). In: *Guidelines to Evaluate Side-Effects of Plant Protection Products to Non-Target Arthropods*. IOBC, BART and EPPO Joint Initiative. Candolfi M.P., Blümel S., Forster R. Eds. IOBC/wprs, Gent.

OECD (1981) OECD Guidelines for testing of chemicals, Method 305, Bioconcentration: flow-through fish test, Paris, France.

OECD (1984a) OECD Guidelines for testing of chemicals, Method 205, Avian dietary toxicity test, Paris, France.

OECD (1984b) OECD Guidelines for testing of chemicals, Method 206, Avian reproduction test, Paris, France.

OECD (1984c) OECD Guidelines for testing chemicals, Method 203, Fish, Acute toxicity test, Paris, France.

OECD (1984d) OECD Guidelines for testing chemicals, Method 204, Fish, Prolonged toxicity test: 14-day study, Paris, France.

OECD (1984e) OECD Guidelines for testing of chemicals. Method 201. Alga, growth inhibition test. Paris, France.

OECD (1984f) OECD Guideline for testing chemicals, Method 207, Earthworms, Acute toxicity test. Paris, France.

OECD (1984g) OECD Guideline for testing chemicals, Method 209, Activated sludge, respiration inhibition test, Paris, France.

OECD (1992) Guideline for testing of chemicals, Method 210 Fish, Early-Life Stage Toxicity Test, Paris, France.

OECD (1997) Report of the final ring test of the *Daphnia magna* reproduction test. (OCDE/GD(97)19, Paris, France.

OECD (1998a) OECD Guidelines for testing of chemicals. Method 211. Dapnia magna Reproduction test. Paris, France.

OECD (1998b) Guideline for testing of chemicals, Method 213 Honeybees, Acute Oral Toxicity Test, Paris, France.

OECD (1998c) Guideline for testing of chemicals, Method 214 Honeybees, Acute Contact Toxicity Test, Paris, France.

OECD (2000a) Guideline for testing of chemicals, Method 215, Fish, Juvenile Growth test, Paris, France.

OECD (2000b) Guideline for testing of chemicals, Method 216, Soil micro-organisms: Nitrogen transformation test, Paris, France.

OECD (2000c) Guideline for testing of chemicals, Method 217, Soil micro-organisms: Carbon transformation test, Paris, France.

OECD (2001) Guidance document on the use of the harmonised system for the classification of chemicals which are hazardous for the aquatic environment. OECD series on testing and assessment No. 25. ENV/JM/MONO(2001)8.

OECD (2003) OECD guideline for the testing of chemicals proposal for updating guideline 208 terrestrial plant test: seedling emergence and seedling growth test. http://www.oecd.org/dataoecd/11/31/33653757.pdf.

OECD (2004a) OECD Guidelines for testing of chemicals. Method 202. Daphnia sp., Acute immobilisation test. Paris, France.

OECD (2004b) Guideline for testing chemicals, Method 218, Guideline for sediment-water chironomid toxicity test using spiked sediment, Paris, France.

OECD (2004c) Guideline for testing chemicals, Method 219, Guideline for sediment-water chironomid toxicity test using spiked water, Paris, France.

OECD (2004d) OECD Guidelines for testing of chemicals, Method 222, Earthworm reproduction test (*Eisenia fetida/Eisenia andrei*). Paris, France.

OECD (2005). OECD Principles for the Validation, for Regulatory Purposes, of (Quantitative) Structure-Activity Relationship Models. Paris, France. http://www.oecd.org/document/23/0,2340,en_2649_34365_33957015_1_1_1_1,00.html.

Oomen P.A., De Ruijter A., Van der Steen J. (1992) Method for honeybee brood feeding test with insect growth-regulating insecticides. *EPPO Bull.*, 22, 613–616.

Pintore M., Piclin N., Benfenati E., Gini G., Chrétien J.R. (2003) Database mining with adaptive fuzzy partition (AFP): application to the prediction of pesticide toxicity on rats. *Environ. Toxicol. Chem.*, 22, 983–991.

Russom C.L., Bradbury S.P., Broderius S.J., Hammermeister D.E., Drummond A. (1997) Predicting modes of toxic action from chemical structure: Acute toxicity in the Fathead minnow (*Pimephales promelas*). *Environ. Toxicol. Chem.*, 16, 948–967.

Schultz T.W., Cronin M.T.D. (2003) Essential and desirable characteristics of ecotoxicity quantitative structure-activity relationships. *Environ. Toxicol. Chem.*, 22(3), 599–603.

Sheppard S.C., Bembridge J.D., Holmstrup M., Posthuma L. Eds. (1997) *Advances in Earthworm Ecotoxicology*. SETAC Press, Pensacola FL.

Shigekura M. (2005) Toward the regulatory use of QSAR, present status in Japan. Presented at the ECB Consultation Meeting on the Development of an International QSAR Decision Support System, Milan, Italy.

Sinclair C.J., Boxall A.B.A. (2002) Assessment of the environmental properties and effects of pesticide transformation products. Report to DEFRA, Project No. PN 0930.

TSCA (1976) Toxic Substances Control Act. United States Publ. Law 94–469, 90 Stat 2003, USA.

US EPA (1985a) Hazard evaluation division standard evaluation procedure avian dietary LC50. US EPA Office of Pesticide Programs, Washington D.C. 20460, EPA-540/9-85-008, USA.

US EPA (1985b) Hazard evaluation division standard evaluation procedure avian reproduction test. US EPA Office of Pesticide Programs, Washington D.C. 20460, USA.

US EPA (1996) Ecological effects test guidelines OPPTS 850.1075 Fish Acute Toxicity Test, Freshwater and Marine. http://www.epa.gov/opptsfrs/ publications/OPPTS_Harmonized/850_Ecological_Effects_Test_Guidelines/Drafts/850-1075.pdf.

US EPA (1999) Development of Chemical Categories in the HPV Challenge Program, http://www.epa.gov/HPV/pubs/general/catdoc29.pdf.

US EPA (2003) A framework for a computational toxicology research program in ORD. US Environmental Protection Agency Office of Research and Development. Washington, DC 20460, USA. http://www.epa.gov/comptox/publications/comptoxframework06_02_04.pdf.

Vighi, M., Masoero, M., Calamari G., Calamari D. (1991) QSARs for toxicity of organophosphorous pesticides to Daphnia and honeybees. *Sci. Total Environ.*, 109–110, 605–622.

Edited by Emilio Benfenati

Chapter 2

Databases for pesticide ecotoxicity

Emilio Benfenati[1], Elena Boriani[1], Marian Craciun[2], Ladan Malazizi[3], Daniel Neagu[3], Alessandra Roncaglioni[1]

[1]Laboratory of Environmental Chemistry and Toxicology, Istituto di Ricerche Farmacologiche "Mario Negri", Milano, Italy

[2]Department of Computer Science and Engineering, University "Dunarea de Jos", Galati, Romania

[3]Department of Computing, School of Informatics, University of Bradford, Bradford, UK

1. INTRODUCTION

To develop successful quantitative structure-activity relationship (QSAR) models, the availability of biological data for a large number of compounds is essential. In this chapter, we describe the databases of ecotoxicological data scrutinized for selecting appropriate sources of data. As we discussed in Chapter 1 (Benfenati *et al.*, this volume), it is very important that the quality of data is high. This is a general requirement for any QSAR model especially for maintaining a reasonable variability within the data to be modeled, avoiding an introduction of noisy data. However, in the case of models designed for their use within a regulatory perspective, it becomes important to strictly apply the same rules adopted for the production of experimental in vivo data. Thus, if the in vivo data have to be produced according to a given guideline, this rule should be adopted for the choice of data.

A second requirement, which is more typical of QSAR models, is that the body of data should be as large and representative as possible, to have a good basis for models to be built up.

Thus, one criterion is more qualitative and the other one is more quantitative, and typically they go in opposite directions: indeed, it is easier to find a large amount of data if collected in an uncritical way, while it is more difficult to gather a sufficient number of data of high quality. Within DEMETRA, we placed a higher priority on the qualitative criterion, as explained. Thus, only high-quality data were used.

2. DATA AVAILABILITY

A survey was conducted among the DEMETRA consortium and on the Web to gather information about databases that contain pesticide toxicity data to determine the availability of sufficient good quality data. An overview of the database identified from this search is provided in Table 1.

A review of toxicity data for pesticides has been published (Felsot, 2002). The availability of data is problematic. The quality of the data is difficult to assess. Data relative to old compounds are more easily found. If a pesticide is on the market, its properties have to be publicly available, and they can be found on a number of Web sites. However, much data are considered confidential, and the industry is

Table 1
Summary list of available toxicity databases supplied by DEMETRA consortium

Database	Source	Web
COMET project database	Mario Negri Institute	No
ECOTOX database	US EPA	http://www.epa.gov/ecotox
US EPA-OPP database	US EPA-OPP evaluated dossier for pesticide-active ingredients	No
AGRITOX	French officials (SSM)	http://www.inra.fr/agritox/
EXTOXNET	University of California-Davis, Oregon State University, Michigan State University, Cornell University, and the University of Idaho	http://extoxnet.orst.edu/
ARS PPD	USDA, ARS, Alternate Crops & Systems Lab	http://www.ars.usda.gov/Services/docs.htm?docid=6433
TDMS	US National Toxicology Program (NTP) established by the Secretary, Department of Health, Education and Welfare	http://ntp-server.niehs.nih.gov/
TOXNET	US National Library of Medicine	http://toxnet.nlm.nih.gov/
CERC acute toxicity database	USGS CERC	http://www.cerc.usgs.gov/data/acute/acute.html
BVL ecotoxicology database	BBA	No

against the dissemination of data for compounds that are under development, even for compounds that will not go on the market or for metabolites. This attitude is not beneficial to a wider, more mature assessment of pesticides in general, because a large part of data is precluded. General policy tendencies go toward the sharing approach, which requires a distribution of toxicity data, to eliminate duplication and reduce the costs and the use of animals (see for instance the OECD Existing Chemicals Programme and the European REACH legislative initiative).

The most suitable sources were the US EPA-OPP, SEEM, and BBA databases, due to the fact that in this case the quality of the data was reported and was evaluated as appropriate. To increase the number of compounds, we had to consider confidential data from these sources. Indeed, besides the data publicly available on the Web, there are other databases that include confidential data. Thus, all the used data have kindly been provided by the developing institutes as confidential data. We acknowledge the collaborative attitude of the involved institutes.

As explained in Section 3.1, in practice we used US EPA-OPP data. During the project, OECD published criteria for the validation of QSAR, as introduced in Chapter 1 (Benfenati *et al.*, this volume). One of these criteria states that components of the model have to be given, which includes the toxicity data. For this reason, we asked US EPA-OPP to allow public access of the toxicity data only for the effective chemicals and values we used within DEMETRA, which represents a minor part of their data. US EPA-OPP kindly accepted to disclose their data, which are reported in Appendix B (this volume).

We describe below the three databases (US EPA-OPP, SEEM, and BBA) used to develop the data sets and other relevant databases that have been used subsequent to the modeling step for a further check of the toxicity data used, although these does not represent an exhaustive list of available databases for pesticide ecotoxicity.

2.1. The EPA-OPP database

The database developed by the US EPA-OPP contains reviewed and categorized ecological toxicity data produced for the fulfillment of pesticide registration requirements. It was developed to make accessible an updated summary of the EPA reviewed data corresponding to the ecotoxicological effects of all pesticide-active substances (ASs) presently registered or previously manufactured in the USA.

Toxicity data are collected from several sources:

- ecotoxicological studies conducted by audited commercial laboratories and submitted by pesticide companies in support of the registration of their products,
- studies conducted by US EPA, USDA, and USFWS laboratories from 70's to 90's, and

- published data considered to meet guideline criteria for acceptable data (Hill *et al.*, 1975; Johnson and Finley, 1980; Hudson *et al.*, 1984; Mayer, 1986; Mayer and Ellersieck, 1986).

Data have been produced according to guidelines, and checks were performed to ensure compliance with the guidelines and data quality. Thus, some criteria were used to reject studies that were not scientifically sound.

Quality-assurance procedures have been adopted to ensure the quality of the data entered. Data quality is ensured by the review process for pesticide regulatory assessment, and additional procedures have been used to avoid typographical errors.

The database was provided in a database file accessible with MS Excel, containing several fields. A brief description of the most relevant fields is given in Table 2.

The database contains a collection of 15,000 toxicological data for approximately 800 pesticide ASs.

Data cover many species such as amphibians, earthworms, algae and aquatic plants, birds, daphnids, fishes, and insects, and some data cover mammals, mollusks, and several terrestrial plants.

2.2. The SEEM database

International Centre for Pesticide and Health Risk Prevention (ICPS) produced a database of ecotoxicological endpoints for pesticides and their metabolites within the SEEM project (Statistical evaluation of available ecotoxicology data on plant protection products and their metabolites, sponsored by the EC, DG Health and Consumer Protection, contract no. B1-3330/2001216).

The SEEM database contains data produced in accordance with the Council Directive 91/414/EEC, concerning the marketing of plant protection products, for ASs approved at the EU level through the ECCO peer-review process. Additional data coming from national review process have been collected from the Italian Ministry of Health and were used for the project, as these two sources contain available validated data.

Literature was searched as a complementary source. The on-line databases EXTOXNET (2006), FADINAP (2006), and EPA Fact Sheets (2006) were investigated. The Pesticide Manual (Tomlin, 1997) was used to fill physicochemical data gaps (for log P and pKa).

Data referring to the ASs were always considered; data on the product were used when data on the ASs themselves were not available.

The database was created using an MS Excel spreadsheet. It contains data for fishes, daphnids, sediment dwellers, algae and aquatic plants, earthworms, log P, and pKa.

A total number of 200 ASs and 130 metabolites were included in the database from ECCO and national review process.

Table 2
Relevant data entry fields for the EPA-OPP database

Data field	Description
Chemical name	The common name associated with this particular active substance
Chemical Abstract Service (CAS) number	CAS number associated with this particular active substance
Use pattern	The major use pattern generally associated with this active substance
Taxa – common name – scientific name	These fields permit to identify in detail the tested organism by genus and species
Age	Some indications of the age of the organism is given
Guideline number	Guideline codes for the test conditions
Test type	Further definition of the method for administering the dose (e.g., oral, dietary, contact, static, and flow through)
% active ingredient	The percent of active ingredient or substance contained in the test material
Study length	Study period of duration for the endpoint
Dose type	Definition of the selected endpoint (e.g., LD_{50}, LC_{50}, and Lowest Observed Effect Level (LOEL))
Toxic data Greater or Less (TGL)	This field contain separately the info for Toxic data Greater or Less
Toxicity	The numerical data for the effect
Tox level	Measure units for the activity
95% confidence levels	Confidence limits for the statistical analysis
Curve slope	Probit slope for the statistical analysis
Study date	Year of completition for the study
Category	The three study categories are used: core, supplemental, and invalid Core: All essential information was reported, and the study was performed according to recommended methodology. Studies within this category fulfill the requirements of guidelines and are acceptable for use in risk assessment. Supplemental: Studies in this category are scientifically sound but performed under conditions that deviated from recommended protocols. However, the information may be useful in risk assessment. Invalid: These studies provide no useful information in risk-assessment procedure.
Laboratory code	Identification for the laboratory that submitted the study

About 65% of AS data and 90% of metabolite data were obtained directly from ECCO peer-review process, the remaining from the Italian authorization process.

Additional 30 ASs were included with data coming from literature.

2.3. The BBA database

These data are a collection of endpoints from studies conducted for regulatory purposes by the Federal Biological Research Centre for Agriculture and Forestry (BBA) in Germany. It is based on two separate collection of data: one for aquatic species and one for terrestrial ones.

The aquatic database contains valid studies only for aquatic laboratory toxicity tests for technical ASs (no formulations). It include referenced data for invertebrates, algae and aquatic plants, amphibians, fishes, and sediment dwellers.

The terrestrial database is constituted by valid studies coming from referenced data for acute, dietary, and reproductive avian endpoints for technical ASs (no formulations).

Nearly 200 and approximately 170 pesticides are included in the aquatic and terrestrial database, respectively.

2.4. Other databases

Other databases have been used in a second phase of the project to check some compounds for the correctness of their ecotoxicological values. All these databases except the RIVM are publicly accessible sources of ecotoxicological data. The RIVM database is a confidential database built up within the EC funded project HAIR (project no. SSPE-CT-2003-501997). It will become publicly available at the end of the project HAIR (2006).

The ECOTOXicology database (ECOTOX, 2006) is a source of toxicity data for aquatic life, terrestrial plants, and wildlife. Peer-reviewed literature is the primary source of information encoded in the database. Another source of test results is independently compiled data files provided by various US and International government agencies.

Toxicity data are derived from three US EPA ecological effect databases: AQUIRE, TERRETOX, and PHYTOTOX. The AQUIRE database includes toxic effects data on all aquatic species including plants and animals, freshwater and saltwater species. TERRETOX is the terrestrial animal database, primarily focused on wildlife species. PHYTOTOX is a terrestrial plant database.

Pertinent information on the species, chemicals, and test methods and results presented by the authors are abstracted and entered into the database.

Some independently compiled data sets have been transferred into ECOTOX from external sources (including EPA-OPP database). The external data sets must meet the ECOTOX data parameter and quality-assurance guidelines.

The AGRITOX (INRA AGRITOX, 2006) database was prepared by the INRA (Franch National Institute for Agricultural Research). It is a database containing physicochemical properties, toxicity and ecotoxicity, environmental fate, and regulatory requirements of plant protection product ASs.

About 80% of the data come from the dossier for the pesticide authorization process at a French or European level, whereas the remaining 20% of the data are from bibliographic sources. Totally, it contains more than 400 pesticides and for each of them a summary of their assessment.

The EXTOXNET (2006) database provides various information about pesticides. In particular, pesticide information profiles (PIPs) are documents that provide information on a list of 184 specific pesticides relating to health and environmental effects. PIPs are not based on an exhaustive literature search but contains fully referenced data.

The PAN (2006) pesticides database is a resource of Pesticide Action Network North America, locating toxicity and regulatory information for pesticides. The PAN pesticide database brings together a diverse array of information on pesticides from many different sources, providing human toxicity (chronic and acute), ecotoxicity, and regulatory information for about 6400 pesticide-active ingredients and their transformation products. Most of the data are taken from official sources such as the US EPA, WHO, NTP, NIH, and EU. When official lists do not exist, various original data sources that refer to the peer-reviewed scientific literature are presented.

3. SELECTION OF THE DATA

From the list of suitable endpoints listed in Chapter 1 (Benfenati *et al.*, this volume), the five endpoints to be addressed within the DEMETRA project have been identified based on the amount of available data for each endpoint in several databases. The selected endpoints were:

- acute toxicity for rainbow trout (*Oncorhynchus mykiss*): LC_{50} after 96 h of exposure,
- acute toxicity for water flea (*Daphnia magna*): EC_{50} after 48 h of exposure,
- acute oral toxicity for bobwhite quail (*Colinus virginianus*): LD_{50} after 14 days of exposure,
- dietary toxicity for bobwhite quail (*C. virginianus*): LC_{50} after 8 days of exposure, and
- acute contact toxicity for honeybee (*Apis mellifera*): LD_{50} after 48 h of exposure.

The data sources have been identified in Section 2, this chapter. Still, this is not enough, because further steps have to be made, to specifically identify the data points for each chemical. To construct a reliable data set to develop a QSAR model for pesticide toxicity, we have derived a protocol to critically evaluate the quality of the underlying data. Many sources of error can influence the reliability of database collections of ecotoxicological data (Benfenati *et al.*, 2001), and we characterize the protocol here described, with the intention of keeping only reliable data for modeling.

Below the characteristics that have driven the choice of the US EPA-OPP database as source of ecotoxicological data as well as the entire procedure for pruning the data set under construction from any identified noisy data are explained.

3.1. Key features in the choice of the database

For selecting the preferred database, several factors had to be balanced and taken into account at the same time. It was decided to use as starting point data contained in a single database instead of merging data from different sources and use other databases for comparing and verifying the data.

A first essential characteristic concerns the number of chemicals with toxicity data available. Among the available database, the US EPA-OPP was the one with more data and was selected as primary source for the data.

Other criteria have been considered in this choice, focusing the attention on the quality criteria addressed in Chapter 1 (Benfenati *et al.*, this volume):

- The compliance to the quality criteria applied to regulatory data sets (EEC, 1991) is referred as 'good laboratory practice' (GLP), and standardized study protocols are ensured by the selected database, because for all the selected endpoint the exact protocol is stated. In particular, the guidelines for the five selected endpoints were the following: OPP 71–2, OPP 71–1, OPP 72–1, OPP 72–2 (US EPA Report, 1982a), and OPP 141–1 (US EPA Report, 1982b).

- Data in the EPA-OPP database included data derived from public authorities in USA as well as data provided by industries during regulatory submission within dossier preparation, so the intrinsic quality of these data have been analyzed by a reviewer and a label was assigned to those data considered of enough quality to be used for risk assessment.

- For the identification and purity of the test material, this database was preferred, because for each data point, the percentage of the tested AS is specified. This allowed us to ensure the use of data points referred only to pure compounds.

- Other ancillary data were stored in this database, and their use permitted a broader view of the characteristics of these ecotoxicological data such as year of the experiment, uncertainty of the experimental result, the confidence level of each measure, and details about some experimental conditions.

3.2. Comparison of the data internally to the database

For some of the compounds, more than one ecotoxicological data point was available in the US EPA-OPP database. For this reason, a two-step approach was adopted to derive a single data point. At the first stage, a pruning strategy was used to exclude unreliable data. After this process, a specific strategy was settled to select a single data point among multiple data for each remaining compound.

3.2.1. Pruning

The main goal of this step was to exclude those compounds lacking of a reliable data point. This was assessed by using for each endpoint separately the same criteria:

- Eliminating studies conducted on compounds with an AS purity lower than 85%. This criterion was essential for adopting a QSAR approach because the relation between the chemical structure and the activity is studied. The chemical structure of the tested substance should be identified and it should be a pure compound. For this reason, we have to consider only ecotoxicological data referred to pure compounds. Activity data ranged from low purity (10%) to very high one. We used 85% as a reasonable purity grade for all substances and for activity data for all endpoints.

- A second criterion was to exclude studies identified as invalid by EPA-OPP, where invalid studies are defined as studies that may not be scientifically sound, or they were performed under conditions that deviated from the recommended protocols so that the results would not be useful in a risk assessment. As our final goal was to provide a support for risk assessment, we eliminated those studies judged not useful from this point of view.

- A third criterion was to avoid studies where a precise numerical value to the experimental endpoint was not assigned, so only studies with actual values were kept discarding data given as higher or lower than a certain value. For these studies, we have no idea about the precise target value for the modeling task. Having a model trained on wrong values or very approximated ones will increase the noise and reduce its precision.

At this stage, all data that are not of sufficient quality to be used to construct a QSAR have been eliminated.

3.2.2. Selection of the toxicity data

For many pesticides after the pruning, only one ecotoxicological value was kept for a certain endpoints. In this case, we used this value. For other pesticides, more than one value was still present because several studies, acceptable in accordance with our criteria, were available. In some cases, the spread of toxicity values was quite high. In these cases, we introduced a further evaluation step of the multiple values. We considered as acceptable a range of variability for the toxicity data less than a factor of 4 between the minimal and the maximal values. It was considered that a factor of 4 could be considered acceptable or even quite optimistic for the reproducibility of ecotoxicity data. After the decision of this conservative threshold for variability (a factor of 4), an EU guidance document (EEC, 2005) has been published that provides guidance on the assessment of the equivalence of technical materials of substance regulated under Council Directive 91/414/EEC. The aim of this document is to try and determine whether two sources of the same AS, which may include different quantities of impurities, are equivalent in terms of their toxicity. This guidance document states that

> Where data are available for the new source, the ecotoxicological profile will be considered equivalent to that of the reference profile where the ecotoxicological data provided on the technical AS do not differ by more than a factor of 5 compared with the reference profile (or by a factor more than that of the appropriate dosage increments, if greater than 5), when determined using the same species.

The above approach can be interpreted as saying that endpoints from the same study protocol for the same compound can vary by a factor of 5 and still be equivalent, that is, inter-study variability can be up to a factor of 5.

Thus, our threshold is in line, or actually more conservative, with the above-mentioned EU guidance document and with work by the OECD on the ring testing of the chronic *D. magna* study which indicated that the endpoint could vary by a factor of 8 (OECD, 1997).

If we found that the spread was higher, we eliminated the pesticide. With this step, we introduced in the data set only robust data and excluded compounds with contradictory experimental outcomes.

The entire procedure is described in more detail in Fig. 1. After the pruning, the variability, when more that one data point is present, was assessed and compounds associated with too variable toxicity were discarded. Of course, this step only involved chemicals presenting more than one data point.

At this stage, the following step in the procedure was how to choose a single data point for each compound retained after the variability check.

To address this point, we developed the scheme reported in Fig. 1. This scheme allowed us to choose a unique value if more than one data point exists, in an objective and reproducible way.

If more than one value existed, within a factor of 4, we used the minimum, choosing among the studies defined by the US EPA as core studies.

With this procedure, we were able to identify a single data point for each chemical starting from the large number of available data but preserving the quality of the selected values.

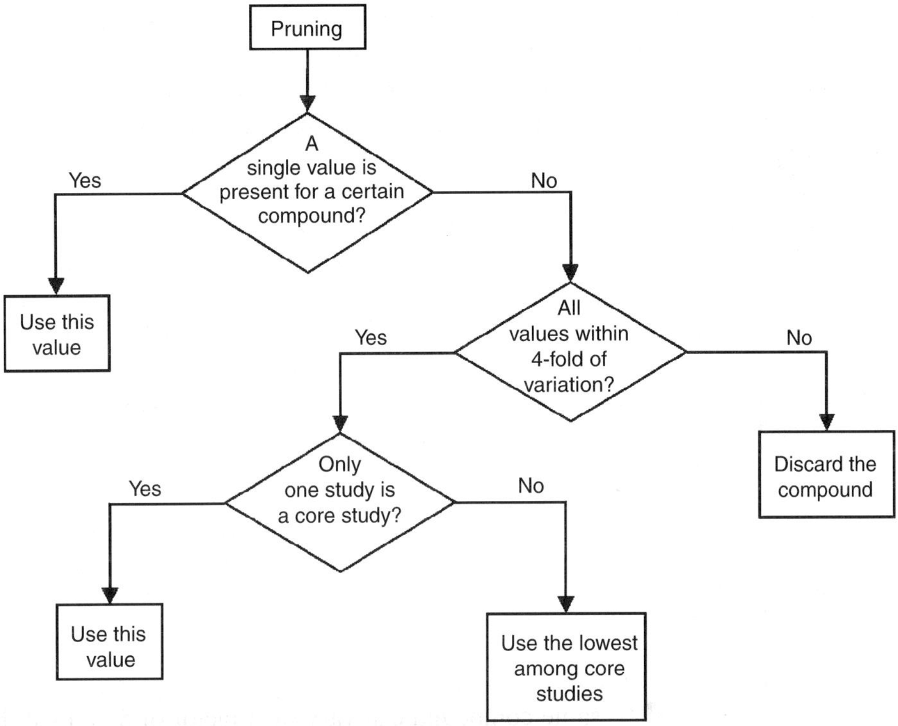

Figure 1: The scheme adopted to select a single data point for each chemicals form the US EPA-OPP database. Copyright 2006. From Roncaglioni *et al.* (2004). Reproduced with permission by Taylor & Francis Group, LLC., http://www.taylorandfrancis.com.

The effect of the application of this procedure to the selected endpoints is shown in Fig. 2, where for each endpoint the starting number of ASs contained in the US EPA-OPP database is shown with the number of chemicals discharged at each stage of the selection.

Trout and daphnia have a similar proportion of data (about the half) discarded during the pruning, whereas another 40 or so compounds were excluded from the variability check. A considerable proportion of data was retained, which had at least two toxicity values that were in agreement. The other three terrestrial data sets differ in the relative proportion of chemicals belonging to these categories, in particular the number of compounds excluded during the pruning is much higher due to the presence of many values 'greater than' a specific threshold, whereas only a few compounds were present with multiple values.

3.2.3. External check for the toxicity data

Finally, we crosschecked the experimental value so obtained from the US EPA-OPP database with the SEEM and BBA database. If again from this comparison

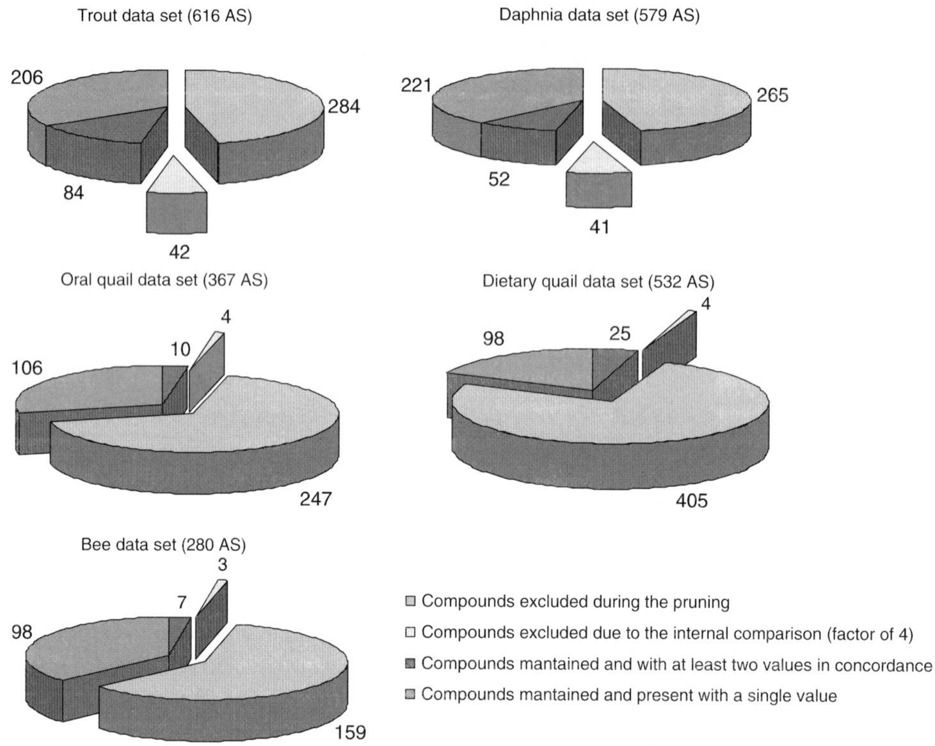

Figure 2: Effects of the comparison of the data internally to the US EPA-OPP database.

questionable values appeared, we further excluded the chemicals, increasing the robustness of the data sets. For the aquatic endpoints, both SEEM and BBA database contained data on these species, whereas for the oral and dietary quail, data were available only in the BBA database. For acute contact studies against bee, no further data were identified in both databases.

Only for relative few compounds was it possible to identify other data from these two sources as shown in Fig. 3. The portion of the data verified with the other databases is relatively small especially for the terrestrial endpoints, but most data were found in agreement with those contained in the US EPA-OPP database.

4. DATA REPRESENTATION FOR PREDICTIVE TOXICOLOGY

Public toxicity databases are a valuable source of information of available toxicity data from diverse sources. These databases offer banks of chemicals and chosen endpoints that are in place for use by the public, scientists, government, and industry. The generality, quality, and usability of toxicity databases highlight the

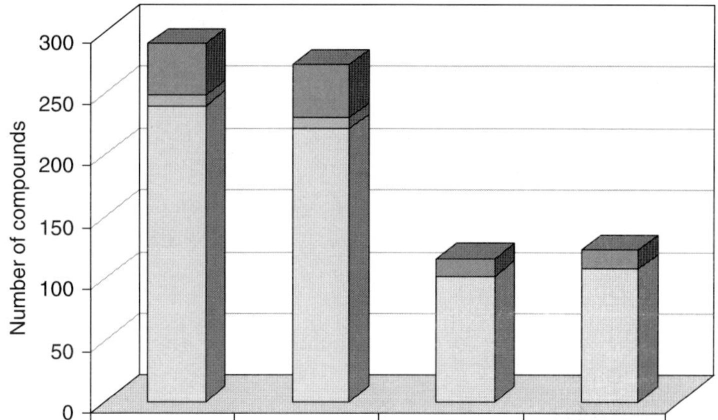

	1	2	3	4
Verified and accepted data	42	43	14	15
Verified and not accepted data	8	9	0	0
Not verified data	240	221	102	108

Figure 3: Effects of the comparison of the data derived from the US EPA-OPP database with two other databases (SEEM and BBA).

importance of the data representation from various points of view. Their data quality, structure and format, data availability, and accessibility are the issues that need proper attention to produce reliable projects to mine information related to chemical toxicity.

These databases are central repository of chemical toxicity information. They represent results of experimental work on toxic compounds. On the contrary, because of their limitations such as diverse file format, data quality, and linkage to compound chemical structure, many of them are not recognized as trusted sources. The effort of environmental agencies to organize and manage toxicity databases rests on standardization of the elements of these data to improve their integrity and reliability. One of these organizations is National Institute of Standards and Technology, which focuses on producing a common vocabulary and standardization of weights, measures, names, and symbols to scientific enterprises and agreement of a data file terminologies (Warr, 2003). Another issue that raises the importance of the matter further is the use of this data for QSAR modeling or relating some aspect of compound structure to biological activities to predict chemical toxicity of new compounds.

Data analysis and integration for producing models using data mining/machine-learning techniques also rely on quality of data. The idea of developing artificial intelligence (AI) in silico modeling for toxicity prediction is also a main interest to regulatory bodies and environmental protection agencies that also encourages a non-animal alternative to toxicity testing.

4.1. A public database example: DSSTox

DSSTox public database has been chosen as an example of a useful data source for the study because it is considered as one of the reliable resources implementing new elements to overcome some of the deficiencies of public toxicity databases. DSSTox (2006) is a free publicly accessible site, which has been launched as an US Environmental Protection Agency project (Richard and Williams, 2002). This web site is connected to a number of databases with all the associated documentation from collaborating organizations from different areas of toxicology and chemistry.

The following are the main elements of this database (Richard and Williams, 2002):

- Adopt and encourage the use of a common standard structure data file (SDF) format for public toxicity databases that include chemical structure, text, and property information.
- Implement a distributed source approach that will enable decentralized, free public access to toxicity data files and that will effectively link toxicity data sources with potential users and modelers of these data from other disciplines.
- Engage public/commercial/academic/industry groups in contributing to and expanding the DSSTox public database network.
- Connect through a single interface various data sources [such as Carcinogenic Potency Database (CPDBAS), EPA Water Disinfection By-Products with Carcinogenicity Estimates (DBPCAN), EPA Fathead Minnow Acute Toxicity (EPAFHM), FDA Maximum Daily Dose Database (FDAMDD), and National Centre for Toxicological Research Estrogen Receptor Binding Database (NCTRER)].

Naumann and Roker (2000) also have highlighted some other issues for data presentation such as consistency, completeness, comprehensiveness, and conciseness that impact the quality of the data as well.

4.2. Current toxicity database limitations

There are a large number of publicly available toxicology databases, which reflect the information and experimental investigations on biological systems from broad fields of toxicology. These databases have been scattered across public and private sources. Most do not have a standard format and contain different types of descriptive information. A major problem with many of them is that they also do not contain chemical structure information. Some of them are searchable and indexed by common chemical names and/or Chemical Abstract Service (CAS) registry number. Although CAS identifiers are unique, they are subject to transcription, typing, and formatting errors (Helma *et al.*, 2000). In contrast, chemical structures have universally understood scientific content. Linkage of chemical structures with chemical toxicity information is a very important issue in designing QSAR models for further chemical compound toxicity predictions.

4.3. XML-based standards in chemistry and toxicology

The standardization of data and knowledge representation is an important issue in all domains, not only in toxicology and related fields. There are many initiatives in this direction both as database and as XML applications such as that of the US EPA in the DSSTox project involving many research institutes, private companies, governmental entities, and regulatory bodies. The idea of downloadable, standardized toxicity data files that include chemical structures is not new but only in the last ten years has the first results materialized.

In the chemistry and toxicology areas, there are already a few proposed XML-based standards such as

- Chem eStandards™: an open, platform-independent, uniform, and free of charge set of standards developed for buying, selling, and delivery of chemicals, based on the XML standard for electronic data exchange (CStandard, 2006);
- Chemical markup language (CML, 2006): a new approach in managing molecular information, covering macromolecular sequences to inorganic molecules and quantum chemistry; and
- ToxML (2006): the scientists at Leadscope® are currently leading a public initiative to promote and develop the 'first toxicology-controlled vocabulary for data mining'. ToxML is intended to evaluate and integrate data from diverse sources to enable the possibility of assessment of toxicity by computational models.

Until this year, when the ToxML schema has become publicly available and researchers in the field have started to use it, none of the earlier mentioned XML proposal fits the needs of a free predictive modeling toxicological data description. Thus, in the framework of DEMETRA project, we proposed PToxML as an application of XML to describe chemical information related to predictive toxicology tasks.

4.4. PToxML – a simple XML-based description in predictive toxicology

To gather and evaluate the information for the development of computational models capable of assessing the toxic effect of chemicals using structural descriptors is a difficult task, given distributed and heterogeneous data sources. An important step is the initiative of predictive toxicology data and vocabulary standardization, where XML provides a powerful way to describe objects (chemicals in our case). Because descriptor-generating software platforms and modeling techniques vary considerably, XML promises to be the most likely language for the representation of toxicology data and predictive models.

The simplest way to describe a data collection such as used by predictive toxicology is based on the entity attribute value (EAV) representation. Unlike ToxML, a more complex and almost complete language that describes the toxicological information from the biologist, chemist, and toxicologist, PToxML simply presents an enriched view of the data set (Neagu *et al.*, 2005). The underlying idea is to offer semantic knowledge for the numbers in the data matrix, for rows and columns.

Chemical compounds can be seen and identified in a number of ways: using a CAS number, a chemical formula, one or more chemical names, and a list of different sorts of descriptors to characterize various facets of chemical compounds (e.g., 2D or 3D structure and physicochemical or electrostatic properties).

On the contrary, toxicologists associate a chemical compound with its effect against life forms (plants, animals, and human beings), and they are mainly interested in the negative effects the chemical compounds could have and their mode of action. Data used in laboratory experiments represent chemical substances associated with a certain mechanism of action and a certain dose that have toxic effects in certain condition against some target species.

Therefore, PToxML presents the list of chemical compounds, the list of the calculated descriptors (attributes), and the matrix of values corresponding to each descriptor for every chemical compound (Fig. 4, Table 3). PToxML documents consist of three main sections: the header, the identity, and the data.

The first section (Table 4) defines authoring information [version of the data set, date, author/owner, initial source(s), and comments], whereas the second section relates properties to the entities (chemical compounds list and descriptors list). The third section presents the data rows.

The chemical compounds list is a simple unordered list of zero or more chemical compounds characterized by the required attribute 'length'. The concordance between this attribute and the number of chemicals presented in the list must be verified by the application using this data. Practically, chemical identification section (chemical compounds list and descriptors list) provides the meta-structure of quantitative data (chemical compounds as rows and descriptors as columns): the matrix of values in PToxML data section. Such organization of PToxML documents permits further correlation checks, that is, the number of chemical compounds is equal to the number of matrix rows, whereas the number of descriptors gives the size of matrix columns.

Previously, it was possible to identify a chemical compound unequivocally using the CAS number, but this is becoming inadequate. So, to be more specific

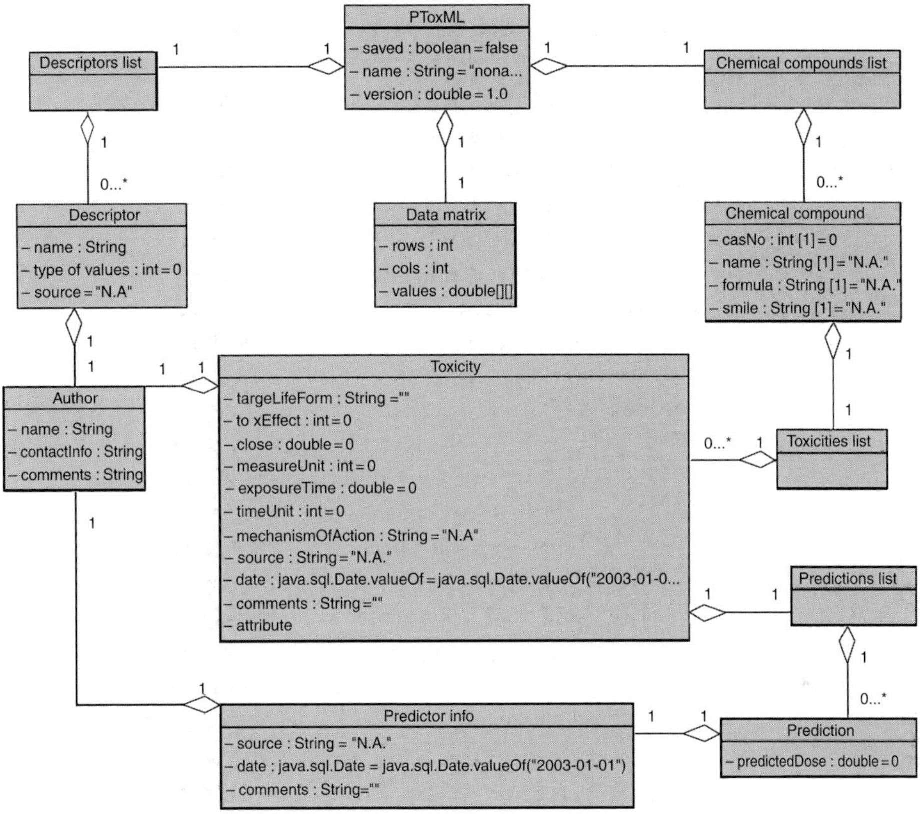

Figure 4: The PToxML structure.

Table 3
The main section of PToxML documents

```
<!– Root element –>
<xsd:element name=‘PToxML’>
    <xsd:complexType>
        <xsd:sequence>
            <xsd:element ref=‘header’/>
            <xsd:element ref=‘chemicalCompoundsList’/>
            <xsd:element ref=‘descriptorsList’/>
            <xsd:element ref=‘dataMatrix’/>
        </xsd:sequence>
        <xsd:attribute name=‘version’ type=‘xsd:string’ use=‘required’/>
    </xsd:complexType>
</xsd:element>
```

Table 4
The header section of PToxML documents

```
<!—The description of header ->
<xsd:element name='header' id='el.header'>
    <xsd:complexType>
        <xsd:sequence>
            <xsd:element ref='version'/>
            <xsd:element ref='date'/>
            <xsd:element ref='author'/>
            <xsd:element ref='source'/>
            <xsd:element ref='comments' minOccurs='0'/>
        </xsd:sequence>
    </xsd:complexType>
</xsd:element>
```

Table 5
The structure of a chemical compound

```
<!- The description of a chemical compound ->
<xsd:element name='chemicalCompound' id='el.chemicalCompound'>
    <xsd:complexType>
        <xsd:sequence>
            <xsd:element ref='CASNo'/>
            <xsd:element ref='Name'/>
            <xsd:element ref='Formula' minOccurs='0'/>
            <xsd:element ref='SMILE' minOccurs='0'/>
            <xsd:element ref='toxicity' minOccurs='0'/>
        </xsd:sequence>
    </xsd:complexType>
</xsd:element>
```

in the process of identifying a compound, both 'CAS number' and 'name' (one or more names) are required. However, there are also chemical 'formula', 'simplified molecular input line entry specification (SMILES)' code, and 'toxicity' (all optional, Table 5) that can be used. (SMILES widely used as a general purpose chemical nomenclature and data exchange format.)

The toxicity structure is one of the main features of a PToxML document (Table 6). The information regarding toxicity (possibly unavailable for some chemicals) contains a list with one or more 'toxInfo' elements to describe: the species (target life form), the measured toxic effects (e.g., LC_{50} – lethal concentration for 50% of a population) – a compulsory field, and the mechanism of action (if available), the dose (if available) and the measure unit (mg/l, mmol/l, etc.),

Table 6
The toxicity and prediction structures

```
<!– The toxicity related information –>
<xsd:element name='toxInfo' id='el.toxInfo'>
    <xsd:complexType>
        <xsd:sequence>
            <xsd:element ref='targetLifeForm'/>
            <xsd:element ref='toxEffect'/>
            <xsd:element ref='dose' minOccurs='0'/>
            <xsd:element ref='measureUnit'/>
            <xsd:element ref='exposureTime'/>
            <xsd:element ref='timeUnits'/>
            <xsd:element ref='mechanismOfAction' minOccurs='0'/>
            <xsd:element ref='source'/>
            <xsd:element ref='date'/>
            <xsd:element ref='comments' minOccurs='0'/>
            <xsd:element ref='author' minOccurs='0'/>
            <xsd:element ref='prediction' minOccurs='0'
                    maxOccurs='unbounded'/>
        </xsd:sequence>
    </xsd:complexType>
</xsd:element>

<!– The prediction related information –>
<xsd:element name='prediction' id='el.prediction'>
    <xsd:complexType>
        <xsd:sequence>
            <xsd:element ref='predictedDose'/>
            <xsd:element ref='predictorInfo'/>
        </xsd:sequence>
    </xsd:complexType>
</xsd:element>
```

the exposure time against the life form and the time units (seconds, minutes, hours, etc.). It is also required to supply information about the source and the date when this data was published, together with some comments and the name of the author(s) (Neagu *et al.*, 2005).

Prediction information might be also required, as a list with zero or more 'prediction' fields: predicted values and the predictor-related information [the model used to obtain the value, the source of the model, date, comments, and author(s)].

Finally, the descriptors – another way to describe a chemical compound – are characterized by their name, type of values (i.e., numerical or categorical), source, and some authoring data (Table 7).

Table 7
The structure of a descriptor

```
<!—Descriptors related information –>
  <xsd:element name='descriptor'>
    <xsd:complexType>
      <xsd:sequence>
        <xsd:element ref='name'/>
        <xsd:element ref='typeOfValues'/>
        <xsd:element ref='source'/>
        <xsd:element ref='author'/>
      </xsd:sequence>
    </xsd:complexType>
  </xsd:element>
```

From one point of view, the measured dose showing toxic effects against the life forms could be interpreted as a special class of descriptors and included in this section too. However, in PToxML, this information is strongly connected with the chemical compounds because sometimes the software calculating the descriptors values are proprietary and the values obtained or even the descriptors themselves are copyrighted. On the contrary, the chemical compounds and the information in relation to their toxicity should be publicly available. In this way, it is easier to export this information attached by the chemical compounds using just parts of the PToxML structure.

5. THE CHARACTERISTICS OF THE FINAL DATA SETS

At the end of this process, the data sets for each endpoint were generated. For the aquatic endpoints, we could rely on wider data sets (more than 250 compounds), whereas for the terrestrial ones, only about 100 original compounds were selected. An overview on these data is reported in Table 8 where the number of chemicals in each data set is shown along with the Pearson correlation coefficient (r) for those compounds that are in common between two data sets. Overall, each data set is quite diverse from all the others sharing a relatively low similarity. As expected, a higher correlation has been identified for aquatic endpoints that share the exposure medium and for the oral and dietary quail endpoints that refer to the same species even if associated with a different route of exposure.

6. CONCLUSIONS

Several sources of ecotoxicological data have been identified and taken into account to produce the DEMETRA data sets. Many different sources of data have

Table 8
Pearson correlation coefficients for the DEMETRA data sets

	Daphnia, –log (LC_{50}) mmol/l	Oral quail, –log (LD_{50}) mmol/kg b.w.	Dietary quail, –log (LC_{50}) mmol/kg diet	Bee, –log (LD_{50}) μmol/bee
Trout (282), –log (LC_{50}) mmol/l	0.74 (190)	0.30 (73)	0.30 (90)	0.40 (70)
Daphnia (264), –log (LC_{50}) mmol/l	–	0.65 (80)	0.44 (81)	0.62 (62)
Oral quail (116), –log (LD_{50}) mmol/kg b.w.		–	0.79 (50)	0.51 (30)
Dietary quail (123), –log (LC_{50}) mmol/kg diet			–	0.51 (46)
Bee (105), –log (LD_{50}) μmol/bee				–

been evaluated and compared. Only high-quality data have been used and compared between databases, as a further check. We defined a procedure to identify a single toxicity value in a reproducible way when multiple values were available. After all these quality checks the reliability of the data sets is very high. Toxicity values can be eventually chosen on the basis of different principles. We adopted a conservative approach using the lowest value because this choice is typically preferred in the EU (Rudén and Hansson, 2003). Another choice would be to use the average or the median value; this approach has been used (Russom *et al.*, 1997) for QSAR studies of aquatic toxicity. It is also possible to use all available information, which can be useful in the case of probabilistic risk assessment. On a practical point of view, in our case, the differences were not high, for the threshold on variability that has been introduced. Indeed, other approaches have been already evaluated for trout species as an example (Roncaglioni *et al.*, 2004), but no major differences have been noticed.

An XML schema has also been proposed for further standardization of toxicology data representation: PToxML is one among very few international proposals to consider together chemical data, toxicology data, and author and source information. This scheme is also used by the software NIKE as the way to manage the toxicity data for use during neural, fuzzy, neuro-fuzzy, and hybrid models development.

We conclude that the selected values are reliable, with low internal variability, as we eliminated compounds with a toxicity range exceeding a factor of 4.

ACKNOWLEDGMENTS

Authors thank ICPS, US EPA-OPP, BBA, and RIVM for kindly providing their databases and their use within DEMETRA project.

REFERENCES

Benfenati E., Piclin N., Roncaglioni A., Varì M.R. (2001) Factors influencing predictive models for toxicology. *SAR QSAR Environ. Res.*, 12, 593–603.

CML (2006) http://xml-cml.org.

CStandard (2006) http://www.cidx.org/ChemeStandards.

DSSTox (2006) *Distributed Structure-Searchable Toxicity*. http://www.epa.gov/nheerl/dsstox/.

ECOTOX (2006) http://www.epa.gov/ecotox.

EEC (1991) *91/414/EEC: Council Directive of July 1991 Concerning the Placing of Plant Protection Products on the Market*. http://europa.eu.int/comm/food/fs/ph_ps/pro/legal/dir91-414-eec_en.pdf.

EEC (2005) *Guidance Document on the Assessment of the Equivalence of Technical Materials of Substances Regulated Under Council Directive 91/414/EEC*. Sanco/10597/2003 – rev 7 final 2, 14th December 2005. http://europa.eu.int/comm/food/plant/protection/evaluation/guidance/wrkdoc23_en.pdf.

EPA Fact Sheets (2006) *Chemicals in the Environment: OPPT Chemical Fact Sheets*. http://www.epa.gov/chemfact/.

EXTOXNET (2006) *The EXtension TOXicology NETwork*. http://extoxnet.orst.edu/.

FADINAP (2006) *Rural Development Section*. http://www.fadinap.org/index.html.

Felsot A.S. (2002) WEB resources for pesticide toxicology, environmental chemistry, and policy: a utilitarian perspective. *Toxicology*, 173, 153–166.

HAIR (2006) *HArmonised Environmental Indicators for Pesticide Risk*. http://www.rivm.nl/stoffen-risico/NL/hair.htm.

Helma C., Kramer S., Pfahringer B., Gottmann E. (2000) Data quality in predictive toxicology: identification of chemical structures and calculation of chemical properties. *Environ. Health Perspect.*, 108, 1029–1033.

Hill E.F., Heath R.G., Spann J.W., Williams J.D. (1975) *Lethal Dietary Toxicities of Environmental Pollutants to Birds*. USFWS Special Scientific Report Wildlife No. 191, EPA MRID 00022923. US Fish and Wildlife Service, Washington, D.C. 20460, USA.

Hudson R.H., Tucker R.K., Haegle M.A. (1984) *Handbook of Toxicity of Pesticides to Wildlife*. USFWS Publication No. 153, EPA MRID 00160000, US Fish and Wildlife Service, Washington, D.C. 20460, USA.

INRA AGRITOX (2006) http://www.inra.fr/agritox/.

Johnson W.W., Finley M.T. (1980) *Handbook of Acute Toxicity of Chemicals to Fish and Aquatic Invertebrates*. USFWS publication No. 137, EPA MRID 40094602 US Fish and Wildlife Service, Washington, D.C. 20460, USA.

Mayer F.L. (1986) *Acute Toxicity Handbook of Chemicals to Estuarine Organisms*. USEPA Environmental Research Laboratory, Gulfbreeze, FL. EPA Publication 600/x-86/231, EPA MRID 40228401.

Mayer F.L., Ellersieck M.R. (1986) *Manual of Acute Toxicity: Interpretation and Database for 410 Chemicals and 66 Species of Freshwater Animals*. USFWS Publication No. 160, EPA MRID 40098401 US Fish and Wildlife Service, Washington, D.C. 20460, USA.

Naumann F., Roker C. (2000) *Assessment Methods for Information Quality Criteria*. Proceedings of the International Conference on Information Quality (IQ2000), October 20–22. Cambridge, MA.

Neagu D., Craciun M.V., Stroia S.A., Bumbaru S. (2005) *Hybrid Intelligent Systems for Predictive Toxicology – a Distributed Approach*, 5th International Conference on Intelligent Systems Design and Applications (ISDA2005), September 8–10 IEEE Computer Society Press, 26–31, Wroclaw, Poland.

OECD (1997) *Report of the Final Ring Test of the Daphnia magna Reproduction Test.* (OCDE/GD(97)19, Organisation for Economic Co-operation and Development, Paris, France.

PAN (2006) Pesticides Database. http://www.pesticideinfo.org/Index.html.

Richard A.M., Williams C.R. (2002) Distributed Structure-Searchable Toxicity (DSSTox) public database network: a proposal. *Mutat. Res.*, 499, 27–52.

Roncaglioni A., Benfenati E., Boriani E., Clook M. (2004) A protocol to select high quality datasets of ecotoxicity values for pesticides. *J. Environ. Sci. Health, Part B*, 39, 641–652.

Rudén C., Hansson S.O. (2003) How accurate are the European Union's classifications of chemical substances. *Toxicology Lett.*, 144, 159–172.

Russom C.L., Bradbury S.P., Broderius S.J., Hammermeister D.E., Drummond R.A. (1997) Predicting modes of action from chemical structure: acute toxicity in the fathead minnow (Pimephales promelas). *Environ. Toxicol. Chem.*, 16, 948–957.

Tomlin C. (1997) *The Pesticide Manual.* 11th Edition. British Crop Protection Council, Farnham, UK.

ToxML (2006) http://www.leadscope.com/news/.

US EPA Report (1982a) *540/09-82-024, Pesticide Assessment Guidelines, Subdivision E-Hazard Evaluation; Wildlife and Aquatic Organisms, OPP 71–2 Avian Dietary LC50 Test, OPP 71–1 Avian Single-Dose LD50 Test, OPP 72–1 Acute Toxicity Test for Freshwater Fish, OPP 72–2 Acute Toxicity Test for Freshwater Aquatic Invertebrates.* US EPA Office of Pesticide Programs, Washington, D.C. 20460, USA.

US EPA Report (1982b) *540/09-82-019, Pesticide Assessment Guidelines, Subdivision L-Hazard Evaluation; Nontarget Insects, OPP 141–1 Honey Bee Acute Contact LD50.* US EPA Office of Pesticide Programs, Washington, D.C. 20460, USA.

Warr W.A. (2003) *IUPAC Project Meeting: Extensible Markup Language (XML) Data Dictionaries and Chemical Identifier.* National Institute of Standards and Technology, Gaithersburg, Maryland, USA, November 12–14.

Quantitative Structure-Activity Relationships (QSAR) for Pesticide Regulatory Purposes
Edited by Emilio Benfenati

Chapter 3

Characterization of chemical structures

Emilio Benfenati[1], Mosè Casalegno[1], Jane Cotterill[2], Nick Price[3], Morena Spreafico[1], Andrey Toropov[1]

[1]Laboratory of Environmental Chemistry and Toxicology, Istituto di Ricerche Farmacologiche "Mario Negri," Milano, Italy

[2]Central Science Laboratory (CSL), Department for Food and Rural Affairs, Sand Hutton, York, UK

[3]CSL Fellow, Technology for Growth, York, UK

1. INTRODUCTION

As the basic principle of quantitative structure-activity relationship (QSAR) is the relationship between physico-chemical properties of compounds and their biological activity, it is evident that it is fundamental to the development of any QSAR that the structures of the compounds must be fully and accurately represented. These days, it is most usual for the compounds in the chosen data set to be modelled on computers for which a range of molecular modelling and computational chemistry software is available (CCR, 2006).

In order for this highly labour-intensive facet of a QSAR study to be successful, it clearly must result in a set of accurate chemical structures.

The subsequent calculation of a range of molecular descriptors on which the QSAR will depend ideally will only be valid if the chemical structures are as close as possible to their actual characteristics in the aqueous phase and at ambient temperatures and pressures. Although it is not possible unambiguously to determine this by experiment, a combination of data from X-ray crystallography, NMR and other spectroscopic techniques can give good clues as to the structural features of chemicals, with which to validate computer molecular models.

When commencing a QSAR study, it is unlikely that the detailed three-dimensional (3D) features of the compounds in the data set will be known. Sources of structural information on the compounds in the data set may impart one-dimensional (1D), two-dimensional (2D) or 3D information (Fig. 1).

The systematic name and molecular formula of a chemical contains information on the relative proportions of the elements in the compound and some information on connectivity but incomplete information on how they are positioned with respect to each other (2D) or in space (3D). Coded 1D systems such as Simplified Molecular Input Line Entry System (SMILES) (SMILES, 2006)

Information type	Information	Dimensionality encoded
Common Name	Dicrotophos	NONE
CAS Number	141662	NONE
Molecular Formula	$(CH_3O)_2P(=O)OC(CH_3)=CHC=ON(CH_3)$	
Systematic Name	(E)-3-(Dimethylamino)-1-methyl-3-oxo-1-propenyl dimethyl phosphate	2D + some 3D
IUPAC Name	(E)-2-dimethylcarbamoyl-1-methylvinyl dimethyl phosphate	2D + some 3D
SMILES string	O=P(O/C(=C/C(=O)N(C)C)C)(OC)OC	2D + some 3D
INCHI	1/C8H16NO5P/c1-7(6-8(10)9(2)3)14-15(11,12-4)13-5/h6H,1-5H3/b7-6+	3D
Structural Diagram EG SKC file	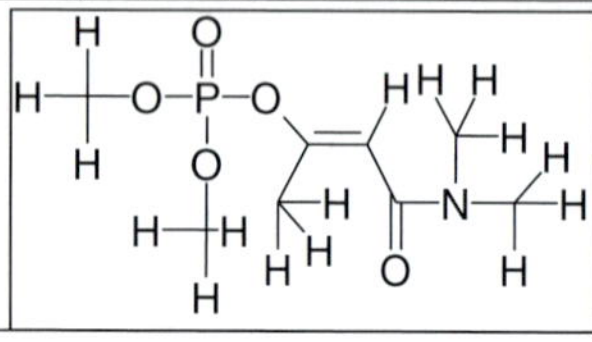	2D + some 3D
Computer model (EG MOL2 FILE)	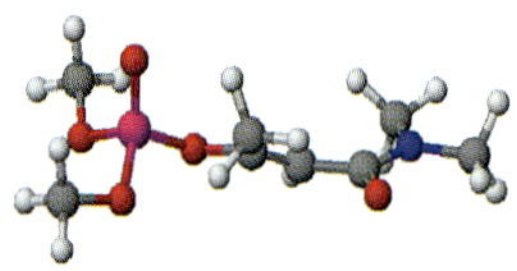	3D

Figure 1: Chemical information and dimensionality of the organophosphate insecticide dicrotophos.

or connectivity tables can convey 2D information, and the emerging InChI system (InChI, 2006) is aimed at encoding even 3D information in a 1D string. Information encoded in 2D representations can unambiguously assign characteristics such as tautomerism (Section 2.3) and isomerism, but conformational information can only be encoded by 3D models (Fig. 1).

For QSAR, it is usually necessary to have 3D structure models. 3D data contains a full description of the position of all the atoms of the structure in space. This is important as many of the descriptors we may wish to use in the QSAR carry information about 3D aspects of the molecules or are quantum descriptors which rely on having run semi-empirical or ab initio molecular orbital (MO) calculations on a structure, and such algorithms require coordinate data. Depending on the format, 3D data files can carry additional information about the molecule, such as atomic partial charges.

For accurate 3D models, it is necessary to have at least good quality 2D data from which to build a de novo computer molecular model. Such 1D and 2D information is available for many compounds from either manufacturers data sheets or many chemical databases available in the literature or online. Having collected the 1D and/or 2D data, there are two ways to construct 3D models; either the data can be converted to a set of 3D coordinates with an appropriate piece of software or else the model can be built de novo with molecular modelling software. File conversion software such as the open source program Babel (2006) can carry out file conversions and can, for example, convert 1D SMILES strings to a range of 3D structural files. In addition, a number of chemical utility programs such as Batchman (Batchman, 1999) and CHEMPAK (Chempak, 2006), incorporate Babel into a batch process for handling multiple files. CHEMSketch (Chemsk, 2006) from ACD Labs Inc can convert the widely used 2D Isis sketch files into 3D molfiles, as can a number of other chemical utilities such as the Sage program ChemTK (Chemtkx, 2006). Many molecular modelling suites can read 1D or 2D files and then export to 3D files for further refinement. However it is important that, if using file conversions, the resulting 3D structures are examined closely to make sure that sensible conversions have been achieved.

Building models de novo using molecular modelling software is possibly a better route especially if access to the Crystallographic Database Service (CDS) is available (Section 3.1). However whatever approach is taken, it is essential that rigorous checks are put in place to ensure that all resulting structures are the best that can be achieved for the required use. In the remaining sections of this chapter, we will discuss how the structures were developed and validated for the DEMETRA project, including the removal of compounds with ambiguous features or whose composition or structure were not amenable to the methods available, sources of starting data, modelling procedures, conformational analysis, molecular mechanics (MM) and quantum mechanics choices. Throughout these procedures, three DEMETRA partners were involved in discussions and experimentation to establish the most appropriate methods for the purposes of DEMETRA.

The origins of the DEMETRA data set are outlined in Chapter 2 (Benfenati *et al.*, this volume). The compounds in the data sets represent a wide range of chemical types and pesticide properties, including herbicides, insecticides, fungicides, rodenticides, fumigants and repellents. By the nature of this diversity, it was not possible to adopt a 'template' approach to the building of 3D models. Where no 3D information was available, 2D images were obtained from The Pesticide Manual (Tomlin, 1997), The Pesticide HandBook (Hurst *et al.*, 1991) and a range of internet sources, primarily Chemfinder (Cfinder, 2006). However, where possible, similar (or identical if available) compounds from the X-ray crystallographic database [Chemical Database Service (CDS), Daresbury, UK] were used as guidelines for the building of structures.

2. CHARACTERIZATION OF BI-DIMENSIONAL STRUCTURES

There are a number of ways to optimize the 3D model of a chemical compound, and many programs have been written for the purpose. Similarly, for the calculation of chemical descriptors and fragments there are several tools available, most of them are commercial. We do not intend here to support specific software. Likewise, it was not our intention to make a comparison of different programs. We describe below the specific packages that were used in DEMETRA, only for the purposes of clarity, quality control and the requirement of some of the packages to make specific reference in published work.

2.1. Preprocessing of compounds in the data set

In any QSAR study, it is important that the biological data and the structure modelled are precisely related. So, items in the data set may be discarded if it is judged that the biological activity (toxicity in the case of DEMETRA) is not of high quality, for example, from studies carried out under GLP guidelines. Considering the chemicals, we adopted the following criteria:

- Inorganic compounds, for example, copper sulphate, were discarded. Inorganic compounds have not traditionally been used in QSAR models; as modelling software is parameterized for organic compounds, it would not be possible to calculate accurate chemical descriptors for inorganic structures.
- Mixtures of more than one active substance were discarded.
- Mixtures of chemicals with different molecular weights were discarded, for example, abamectin (CAS 71751-41-2), which is a mixture of avermectin B1a and avermectin B1b.
- Impure compounds were discarded when the percentage of active substance was lower than 85% and also where a formulation rather than the pure compound was used in the toxicity test.
- We excluded complexes, such as those with iodine.

These criteria also define a first level of boundaries for the applicability of the developed QSAR models. Many of these criteria have been adopted following recommendations of other EC projects on QSAR modelling, such as COMET (ENV4-CT97-0508) and IMAGETOX (HPRN-CT-1999-00015).

In DEMETRA, data points were also discarded if it was expressed as greater than or less than a value, or if the variability of multiple values was high.

It is also important to mention that hydrates were modelled in the anhydrous form, and organometallics were retained. Further mixtures of chemicals with the same molecular weight were considered as follows:

- Mixtures of chemicals with the same molecular weight but with different connectivities were discarded; however, tautomers were kept. In other words, if the mixture involves a shift of one or more atoms different from hydrogen, from one position to another, the mixture was excluded. If the shift is only

Table 1
Number of compounds in the data set for each of the five endpoints

Endpoint	Total number of compounds in data set (for 2D modelling)	Number of compounds in data set suitable for 3D modelling
Trout	282	275
Daphnia	264	258
Quail (oral exposure)	116	114
Quail (dietary exposure)	123	118
Bee	105	102

2D, two-dimensional; 3D, three-dimensional.

due to a hydrogen atom (tautomers), the mixture was kept. Mixture of stereochemicals was kept for general QSAR models (Sections 2.2 and 3.3).

- Salts were modelled without the inorganic ion. For example, in the case of a HCl salt, it was used in its basic form, and in the case of a K salt, the K atom was replaced with a H atom. Nitrogen linked to four carbons was modelled without the anion.

- Following the pruning of the data set, the final number of compounds for each endpoint is summarized in Table 1. Because a few compounds in the data set contained atoms that are not parameterized in 3D descriptor generating software (e.g. tin, arsenic and boron), a small proportion of compounds for each endpoint were only suitable for modelling using 2D descriptors.

- The data set contained a diverse range of chemicals, including pesticides (insecticides, herbicides, fungicides, nematicides, acaracides, molluscicides, rodenticides, plant growth regulators and bioirritant) and also microbiocides and repellents. As previously mentioned, it is critical to the modelling to ensure that the correct 2D structure was obtained for each chemical. The 2D structure, CAS number and chemical name were checked for consistency through at least three different sources – The Pesticide Manual (Tomlin, 1997), Chemfinder (Cfinder, 2006) and The Compendium of Pesticide Common Names (Hclrss, 2006). Further useful freely available sources of 2D structures, the latter of which has recently become available, are ChemIDPlus (Chemid, 2004) and PubChem (PubChem, 2006). Although these were not used in the DEMETRA project, it is possible to download 2D structures from these on-line programs, although as always it is imperative to check the structures are correct using others such as the three sources described earlier. For the DEMETRA project, the chemical structures were 'sketched' in 2D using Isis/Draw (Isis/Draw, 2006) and exported as 2D MDL (Molecular Design Ltd) files.

2.2. Geometrical isomers

Geometrical isomerism arises because of either restriction about a double bond or restriction of rotation in certain ring systems. For example, in the case of the C=C bond or a carbon ring, the isomers are distinguished by the prefixes *cis*- and *trans*- (or E- and Z- when there is no pair of identical or similar groups) according to whether two given groups or atoms lie on the same or opposite sides of the plane of the double bond or ring. The problem in using such compounds in QSAR modelling is that in some cases the substance tested may be a mixture of isomers or it is not specified which isomer was used. When specified in trade literature of pesticide databases, the correct isomers were modelled for DEMETRA. When the isomeric form was not specified, the compound was discarded from the DEMETRA data set.

2.3. Tautomers

Tautomerism or dynamic isomerism is a special case of structural isomerism. Two tautomers are directly interconvertible because of mobility of a hydrogen atom, which can move from one position to another in the molecule with rearrangement of a double bond. A common example of this is keto-enol tautomerism, whereby a hydrogen atom attached to the carbon atom of a carbonyl compound moves to the carbonyl oxygen atom. Tautomers exist in equilibrium in the liquid state or in solution, but not in the solid state, and the position of equilibrium for tautomerism is dependent on molecular structure.

The challenge in building QSAR models is to determine which structure is most likely the correct in vivo tautomer. For the purpose of building the high-quality data set in the DEMETRA project, we independently optimized the two isomers with ab initio methods and chose the more stable isomer. Although it is impossible to confirm that the most stable tautomer is the biological conformation, this was considered to be the best systematic approach to the identification of the most likely tautomeric structure. An alternative approach, particularly in more complex structures where there may be a number of possible permutations, and hence the systematic optimization of all tautomers would require a lot of time and effort, would be to search for the crystallographic structure or other literature to identify which tautomer is dominant.

For common use no great differences should appear using one or the other isomer; indeed, for instance, in the case of pesticides in the trout data set (the largest one) for the seven compounds with tautomery present in our data set, the differences in final geometry were small.

If a shift was due to a hydrogen (tautomers), the mixture was retained in the DEMETRA data set, as shown in Fig. 2.

Fig. 3 shows two tautomers with different stability.

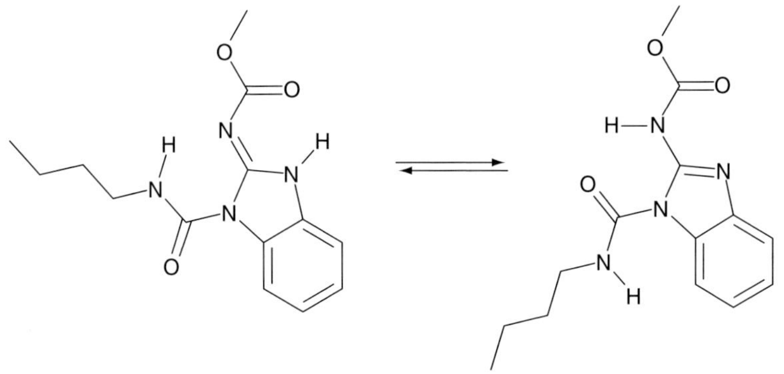

Figure 2: Tautomers of compound benomyl, CAS Nr. 17804352.

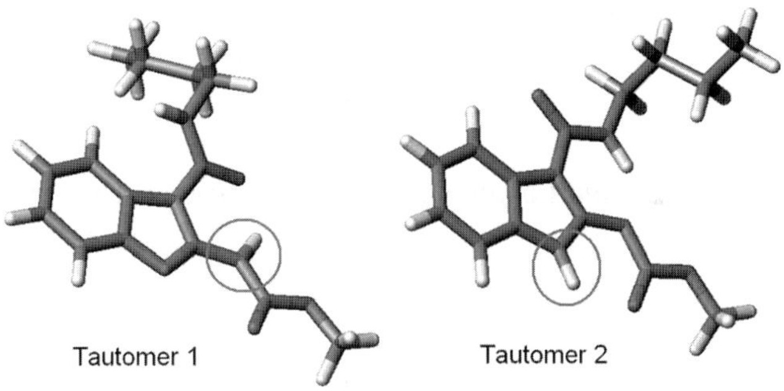

Figure 3: Tautomer 2 is by 45.41 kJ mol^{-1} more stable than tautomer 1.

3. CHARACTERIZATION OF TRI-DIMENSIONAL STRUCTURES

3.1. Crystallographic data

The CDS (CDSweb, 2006) is a UK national centre, with access free to UK University staff and postgraduate students. One of the databases available from CDS is the Cambridge Structural Database (CSD), which contains crystal structure data for over 322,000 organic and organometallic compounds. Most of the records in CSD include coordinate data held as Cambridge Structure Search and Retrieval (CSSR) files, a format compatible with many molecular modelling suites.

The use of CDS is strictly non-commercial and only for registered users, who must be affiliated to a UK Higher Educational establishment. Other non-fee Internet sources include ChemFinder (Cfinder, 2006), and The USA National Library of Medicine's TOXNET (Toxnet, 2006), which employs the ChemIDplus system for access to structure files in:

- Databases such as the Developmental and Reproductive Toxicology/ Environmental Teratology;
- Information Center (DART/ETIC) Database, Chemical Carcinogenesis Research. Information System (CCRIS), EPA's GENE-TOX mutagenicity database, IRIS, and EPA's Toxic Chemical Release Inventory (TRI) and
- The Distributed Structure-Searchable Toxicity (DSSTox) (DSSTox, 2006) Database Network is part of the USA EPA Computational Toxicology Program (EPACTP, 2006), aimed at building a public data foundation for improved structure-activity and predictive toxicology capabilities. The website provides a public forum for publishing downloadable, standardized toxicity data files that include chemical structures. (For a review of structural databases and access to them, see Richard *et al.*, 2002).

3.2. Conformational searching and energy minimization

The modelling strategy adopted by DEMETRA was as follows listed according to order of preference for the starting point for modelling:

- Crystal structure of data set entry,
- Crystal structure of related compound,
- 2D file converted to 3D model and
- Compound built de novo by computer molecular modelling.

Two DEMETRA project partners worked on comparison of structures modelled from full or partial crystal structure data. In the case of compounds with full crystal data, energy minimization techniques were applied and basic conformational analysis performed to confirm that the crystal structure did represent a low energy conformer. Data from the two partners were compared, and where results were not the same or very similar (less than 10% of cases), the final structure was decided by discussion and debate.

If crystal structure coordinates were not available for compounds in the data set, then

- 2D 'sketches' (.skc files) were compiled from structural diagrams from the sources previously described using MDL Isis Draw. In some cases, such as when structures were retrieved from ChemFinde, .skc files could be directly imported into Isis Draw;
- there is an option in Isis Draw to export a .skc file as a Sybyl (Tripos, 2006) .mol2 file, a widely used 3D file type. However, using a range of

DEMETRA structures exported as .mol2 files resulted in failure to read the subsequent .mol2 files by four different molecular modelling suites; CAChE (Fujitsu, 2006), CHEMX (Chemx, 2006), ChemDraw (CambridgeSoft, 2006) and Molecular Modelling Pro (MMP) (Chemistry Software, 2006);

- although .skc files can be exported from Isis Draw as the industry standard .mol file, there are several issues involved in the conversion of 2D .mol files to 3D.mol files or other 3D files. The main problem is the 3D interpretation of stereochemistry, which is shown in 2D diagrams by dotted and wedged or thickened bonds. Problems can vary from complete loss of stereochemical information to the subtle and unpredictable changes in the configuration. These errors occur apparently at random, with no warnings and make accurate modelling of large numbers of compounds virtually impossible.

Sketch files exported as MDL molfiles were treated in different ways by the four molecular modelling suites described above.

CAChE reads the converted mol files as 2D only minus hydrogen atoms, but gives no error messages. The command 'Beautify comprehensive' only adds hydrogens but does not introduce a 3D.

MMP read the .skc converted .mol file as a 2D file. MMP performs a 2D to 3D conversion, which can then be minimized. This is very powerful, but CHEMX appears initially more 'intelligent' in its 2D to 3D conversion.

CHEM 3D DRAW also reads the .skc to .mol files as 2D files but will run MM2 optimization to give 3D structures results.

The 2D to 3D conversion in CHEMX gives a 3D result, but as with all conversions, the issue of atom type remains (i.e. hydridization, valence or charge type). If the 2D to 3D algorithm produces the wrong atom type, then the final structure will be incorrect and generate erroneous property descriptors. Most molecular modelling programs have built in fragment libraries in which common chemical substructures found in many classes of organic chemicals are prebuilt using the correct atom types and geometry. The use of such fragment libraries is recommended when building compounds de novo as this ensures a valid starting structure.

As well as the issue of atom types, the position of flexible groups in the structure is also of importance. Single bonds allow relatively free rotation of attached groups, and it is the orientation of such flexible groups, which characterizes the conformation of the final model. In cases involving particularly flexible molecules, the number of possible conformations can be very large (Fig. 4), and so it is important that a conformational analysis is carried out on all the compounds in the data set. Some programs allow the calculation of an almost unlimited number of conformations resulting from rotation of any number of flexible bonds. However to make the numbers of conformations realistic to analyse, some form of filtering of results is necessary. Typically, the conformation more stable is used, as in most cases nothing is known about the active conformation.

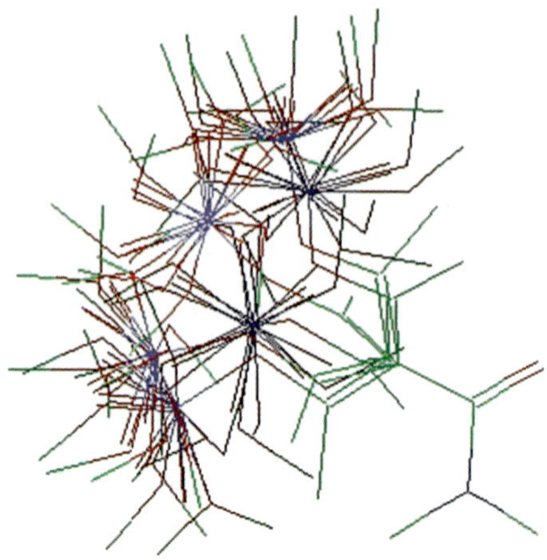

Figure 4: Some of the many possible conformations of dicrotophos.

Rejecting all but the lowest energy conformations is one approach, whilst using a molecular dynamics simulation to find stable low-energy conformations is another. Carrying out conformational analysis will find low-energy conformations, but it is then necessary to carry out energy minimization or 'fine tuning' of the resulting selected low-energy conformer.

Once the chosen low-energy conformer has been obtained, the structure is optimized using one of the many MM, or MO, algorithms. Which is chosen depends on a several criteria such as the number of molecules to be processed, presence of heavy elements, the need to generate extra information, such as atomic partial charges and other electronic parameters and most particularly the type of compounds being modelled. It is not appropriate to explain the fundamentals of MM and MO methods here. However, the use of a modern semi-empirical MO approach was chosen in DEMETRA as a good compromise with the mixed data set. MOPAC AM1 was used with keywords to generate partial charges on the atoms.

Following extensive exchanges of data and debate among the three chemical modelling partners, the following strategy was validated as suitable for the purpose of DEMETRA project:

- Structures 'sketched' in 2D using Isis Draw if crystal coordinates were not available;
- 2D files exported as 2D MDL files;
- 2D MDL files read into molecular modelling packages: Macromodel, Tripos Sybyl, CAChE, CHEMX and MMP;

- MM optimization of the structures carried out using MM94 or MM3;
- Full conformational analysis carried out;
- Structures 'fine tuned' using MOPAC-AM1 optimization;
- AM1 partial charges to each molecule assigned and
- Files distributed in Sybyl mol2 format to all three partners for quality assurance.

Using this approach, initially 30 compounds selected from different chemical classes were assigned to all three modelling partners as a trial set. Then, one partner carried out quality assurance on the compounds and reported back. In this way, we can be confident that our modelling strategy produces:

- Correctly assigned chemical structures,
- Good representation of the 3D structure and
- Reasonable partial charges.

All of this is essential, as the quality of the 3D descriptor-based models will rely on molecular descriptors calculated from the 3D structures. Using the above approach, a further 258 compounds, representing the rest of the whole data set, were modelled by two partners and quality assessment carried out by a third one. The type and origin of many descriptors was agreed. 2D descriptor files were exchanged and calculation of the 3D descriptors conducted.

3.3. Stereoisomers

Stereoisomers differ in their spatial orientation of atoms in the molecule and include geometric isomers (Section 2.2) and also optical isomers. Optical isomers are asymmetric molecules that rotate the plane of polarized light differently depending on the sequence of atoms around the chiral centre. Terminologies include the following: D/L, R/S or +/− isomers. Diastereoisomers are stereoisomers which are not related as mirror images, and enantiomers are molecular structures which are mirror images of each other and non-superimposable. As the specific configuration of stereoisomers differs, their ability to bind to receptors may also vary, which in turn may affect toxicity.

In relation to QSAR modelling, it is possible to build models that do not distinguish between stereoisomers, for example, using mono-dimensional, by-dimensional descriptors and connectivity indices. However, diastereoisomers can be distinguished with tri-dimensional descriptors. Thus, in the case of use of these descriptors, the correct stereochemistry of diastereoisomers has to be known.

Enantiomers can be distinguished by methods such as CoMFA and Quantum Chemical Similarity; thus, in the case of use of these methods, the correct stereochemistry of enantiomers has to be known.

Decisions on the inclusion of compounds with stereoisomerism were taken on the same basis as for geometric isomerism. Some pesticides that have isomeric forms are known to be biologically active, largely or entirely because of one specific isomer. Often, the economics of the pesticide market determine that a mixture of active and inactive isomers is more commercially viable than to incur the extra costs associated with purification of the active isomer. Thus, it was important to establish the exact chemical identity of the compounds used to generate the biological data. In some cases, it was uncertain whether pure pesticide or formulated material had been used to generate the toxicity data, and in such cases, the compound had to be discarded from the data set.

3.4. Procedure for the quality control of the chemicals and chemical structures

CAS numbers, chemical names and 2D chemical structures were checked for their consistency through at least three different sources:

1) The Pesticide Manual, 11th edition (Tomlin, 1997),

2) Chemfinder (Cfinder, 2006) and

3) Compendium of Pesticide Common Names (Hclrss, 2006).

If any further elaboration was done, a check on each structure was performed using a separate procedure. For instance, structures that were imported with DRAGON were checked by a second software package comparing appropriate descriptors. Where these comparisons showed no difference, the structure was retained. In case of differences in structures, four sources were checked to verify consistency with the IUPAC name. All compounds were checked for the presence of tautomers and stereoisomers. If identified, the criteria defined in Sections 2.3 and 3.3 were applied.

4. CHEMICAL STRUCTURE FILE FORMATS

The problem of 1D, 2D file types and their interconversion has already been touched upon. However, even when the final 3D structure has been agreed, there is the question of which of the many coordinate file types will be used. In DEMETRA, the three chemical modelling partners all used different computational chemistry software, and the issue of file transfer was discussed early in the project. Although crystallographic files are of the .cssr type (or .xr when downloaded from CDS), it was found that some modelling programs did not handle these files accurately, especially with respect to atom types. MDL mol files were handled by all our programs but do not carry additional data such as atomic partial charges. We chose Tripos Sybyl .mol2 files as being a comprehensive file format common to a wide range of modelling software from high end UNIX workstation suites to low end PC freeware. All the optimization and conformational calculations can be carried out on .mol2 files, and parameters generated by AM1 calculations are held within the file structure.

When swapping files between platforms and modelling suites, one should always be aware of issues concerning atom type. Atom types are specific to each software package and are handled differently in different modelling packages; no one method is right or wrong. Molecular modelling is a relatively simplistic attempt to reproduce chemical structures by using rules which do not really exist in nature but, by and large, give approximately correct results for those ensembles of atoms used in the parameterization process. For example, benzene rings may be represented as a structure with aromatic carbons and aromatic bonds or alternatively with sp2 carbons, three single bonds and three double bonds. Both representations may be correct, but it is important to keep coherence for all molecules with benzene rings in the database.

This applies to chemical forms that may have more than one structural interpretation, such as nitro groups, sulphones and suphoxides amongst others. When swapping these files between modelling suites, changes in atom type may occur; the important issue is whether this will affect any descriptors subsequently calculated. In DEMETRA, significant effort was put into ensuring consistency of atom types in the final files used for descriptor calculation. One additional problem was encountered which was a function of file structure. When using the program DRAGON (DRAGON, 2003) to calculate descriptors, it was found that .mol2 files generated by some modelling software were not accepted by DRAGON whilst .mol2 files from other packages were. Like most 3D structure files, .mol2 files are simply text files. It was found that .mol2 files from UNIX platforms generate encoded text files, and DRAGON requires plain text files. Using a batch converter within Microsoft Word© to change the file to plain text cured this problem.

4.1. Bi-dimensional descriptors

In traditional QSAR practice, the term ‘bi-dimensional descriptors’ refers to molecular descriptors whose numerical value is insensitive to a change in the molecular conformation. A change in the molecular conformation occurs when the spatial arrangements of atoms in the molecule are modified through the free rotation of the atoms about single chemical bonds. Bi-dimensional descriptors are the easiest molecular descriptors one might compute, as only information about the constituent atoms and their reciprocal connectivity patterns are required. Owing to their simplicity, bi-dimensional descriptors have been widely used in QSAR.

Table 2
Classes of 2D descriptors used for modelling purposes with the software package used for computing them

2D Descriptor Class	Software
Constitutional	Codessa, Dragon
Topological	Codessa, Dragon
Physico-chemical	ACD, Pallas
Correlation weights	OCWLGI
Fragments	See section 4.3

To date, a number of software programs have been developed for computing them. Within the current project, five main classes of bi-dimensional descriptors have been exploited. Those classes, together with the software package used for computing them, are summarized in the Table 2. Details about the procedures adopted for computing 2D descriptors are given in Appendix C (this volume). Software versions are referenced (Codpro, 1996; DRAGON, 2003).

For sake of clarity, fragment-based descriptors have been included in this table; however, their description and application towards pesticide toxicity prediction is given in Section 4.3.

Constitutional descriptors can be defined as descriptors reflecting the molecular composition of a compound without connectivity and geometry information (Sfqsar, 2006). The following categories embrace a large majority of constitutional descriptors (Descod, 2006):

- Total number of atoms in the molecule,
- Absolute and relative numbers of atoms of certain chemical identity (C, H, O, N, F, etc.) in the molecule,
- Absolute and relative numbers of certain chemical groups and functionalities in the molecule,
- Total number of bonds in the molecule,
- Absolute and relative numbers of single, double, triple, aromatic or other bonds in the molecule,
- Total number of rings and number of rings divided by the total number of atoms,
- Total and relative number of six-atom aromatic rings and
- Molecular weight and average atomic weight.

Despite their conceptual simplicity, constitutional descriptors introduce a key argument in building QSAR models, the definition of molecular aromatic character. In contrast to constitutional descriptors, topological ones are descriptors reflecting the molecular connectivity without geometry information. The principal typologies of topological descriptors are

- molecular connectivity indices,
- molecular valence indices,
- information content indices,
- molecular shape indices,
- topological electronic indices,
- weighted distance matrix indices,
- sum of distances between different elements and
- indices based on distance/detour paths

Physico-chemical descriptors refer to physico-chemical molecular properties such as the boiling point, melting point, water solubility, lipophilicity and solvation free energy. Among them, the octanol–water partition coefficient, log *P* has been the attribute used by most investigators to correlate structure and toxic effects. A partition coefficient is a measure of differential solubility of a compound in two solvents. The log ratio of the concentrations of the solute in the solvent is called log *P*. The octanol–water partition coefficient is a measure of the hydrophobicity and hydrophilicity of a substance. The more lipophilic a molecule is the more soluble it is in lipophilic organic phase. Cell membranes are composed of phospholipids, which have hydrophobic tails that produce a very hydrophobic environment in the middle of the membrane bilayer. In the absence of active membrane transport, more hydrophobic compounds have an easier time getting through a membrane. Compounds' hydrophobicity plays a fundamental role in the modulation of many key ADME (Absorption, Distribution, Metabolism and Excretion) processes, such as drug-membrane interactions, drug transport, biotransformation, distribution, accumulation, protein and receptor binding (Huuskonen *et al*., 2000; Sfqsar, 2006). For this reason, log *P* has been regarded as one of the most important properties in QSAR studies. Despite its efficacy, however, it is important to mention that log *P* cannot describe chemical features such as the compound reactivity, meaning the possibility for a substance to dissociate or react, giving a number of by-products, usually referred as metabolites. The study of metabolites is of special importance in pesticide risk assessment, as in many cases, metabolites have to be inspected for pesticide activity, as well as their parent compounds. For this reason, in this study, log P is used in combination with other descriptors accounting for chemical reactivity, such as HOMO-LUMO ones.

Descriptors based on correlation weights come from a series of topological investigations. Milestones in topological studies have been established in the pioneer researches of H. Wiener (Wiener, 1947a,b; Wiener, 1948a,b). The main idea of these studies is the use of a molecular graph or more exactly a matrix of topological distances calculating special coefficients (descriptors), which can be correlated with the properties of organic compounds.

From the beginning of 1980s, a number of different descriptors conceptually analogical to Wiener number started to increase (Hosoya, 1972; Amidon and Anik, 1976; Bonchev *et al.*, 1980; Randic, 2001; Randic and Basak, 2001; Randic *et al.*, 2001; Jalbout and Li, 2003; Roy and Leonard, 2004). Most of these descriptors or indices were based on two special matrixes, named as the adjacency matrix and the above-mentioned matrix of topological distances in a molecular graph. Considering, for instance, the molecular graph of 2-methyl butane, with numbering of vertices as in Fig. 5, the adjacency matrix A(G) and matrix of topological distances D(G) are shown in Fig. 6.

In spite of the existence of a large number of descriptors (Randic, 2001), the main idea of their calculation can be illustrated with the Wiener number (W) and connectivity indices of zero-order ($^0\chi$) and first-order ($^1\chi$), the latter is also known as Randic (Randic, 2001; Randic and Basak, 2001; Randic *et al.*, 2001; Roy and Leonard, 2004) index. Some descriptors have been based on the hydrogen-suppressed molecular graph. In other cases, the descriptors are based on hydrogen-filled molecular graph.

Within DEMETRA, in order to increase the amount on independent sources of chemical information, we considered innovative descriptors based on the SMILES (Weininger, 1988, 1990; Weininger and Weininger, 1989) notation. SMILES-based models have been tested in QSPR modelling of solubility (Vidal *et al.*, 2005) and normal boiling points (Toropov *et al.*, 2005).

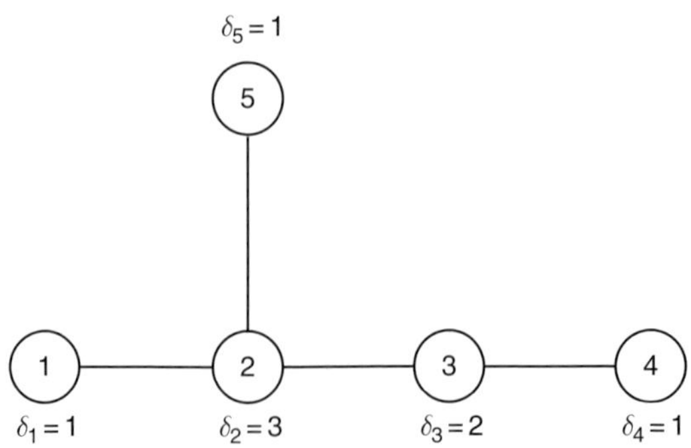

Figure 5: The molecular graph of 2-methyl butane.

	1	2	3	4	5
1	0	1	0	0	0
2	1	0	1	0	1
3	0	1	0	1	0
4	0	0	1	0	0
5	0	1	0	0	0

A(G)

	1	2	3	4	5
1	0	1	2	3	2
2	1	0	1	2	1
3	2	1	0	1	2
4	3	2	1	0	3
5	2	1	2	3	0

D(G)

Figure 6: Example of adjacency and distance matrices.

SMILES code contains a string of characters. For instance, the pesticide bromacyl can be described with the following SMILES code: O=P(SCCS(=O)CC)(OC)OC. These characters reflect the molecular structure, or at least some part of attributes of the molecular structure. Under such circumstances, we defined descriptors that are a mathematical function of the SMILES characters. Inputs are the characters in the SMILES code and their combination (for instance the group C=C), which can be considered as descriptors. We define LI_k a local SMILES invariant (e.g., C, c, N, n, . . . , =, #, . . . , C=C and [N+]). A simple way to deal with these SMILES-based descriptors is to count the presence of some given characters, listed in a defined order. Thus, we can easily obtain a sequence of 0 and 1, depending on if the character is present. This sequence will depend on the chemical, and it is defined as GI_j that is the global SMILES invariant for the *j*th compound. These SMILES based descriptors are used within the model, and appropriate correlation weights are obtained. We defined the optimal descriptor to model toxicity as:

$$DCW = DCW\left(SMILES_j\right) = CW\left(GI_j\right)^x \prod_{k=1,N} CW\left(LI_k\right)$$

The Monte Carlo method has been used to optimize the correlation weights that produce the largest value of correlation of the optimal descriptors with the toxicity on the training set.

4.2. Tri-dimensional descriptors

In contrast to bi-dimensional descriptors (Section 4.1), tri-dimensional descriptors rely on the 3D molecular structure. That is, the numerical value of those descriptors

Table 3
3D descriptors classes and software packages used for their calculation

3D descriptor class	Software
Geometrical	Codessa, Dragon
Electrostatic	Codessa, Dragon
Quantum-mechanic	Codessa, Dragon
Thermodynamic	Codessa, Dragon

3D, three-dimensional.

might considerably vary changing the molecular conformation (Benfenati *et al.*, 2000). For this reason, 3D descriptors can be reasonably compared only when referred to the same molecular conformation. Thus, in order to have a set of comparable 3D descriptors, a standard, or reference, conformation should be set for any molecule. Details about conformational search and energy minimization are available at Section 3.2. In the Table 3, 3D descriptors classes have been reported, together with the software package used to compute them. The CODESSA software package accounts for 745 3D descriptors, while DRAGON 735.

Geometrical descriptors are conformationally dependent descriptors based on the molecular geometry. They are, so far, the simplest 3D descriptors one might compute. Here, we report the main classes of geometrical descriptors (Sfqsar, 2006).

- Distance to atom.
- Gravitational index.
- Gravitational index (Square and Cube Roots).
- Moments of inertia.
- RDF Proton descriptor.

Electrostatic descriptors are used to quantify the molecular charge distribution or the molecular electron density. Both these quantities can be related to the presence of reactive (charged) sites that might be responsible for enhanced toxicological properties. Electrostatic descriptors can be divided into three sub-classes:

1) Atomic-related, such as partial atomic charges, maximum and minimum atomic charges. In some cases, this group also includes atomic polarizabilities, hardness and softness.

2) Molecule-related, such as topographic electronic index, average ionization energy, polarizability and hyperpolarizability.

3) Charged partial surface area (CPSA) descriptors. These descriptors were originally designed for use in structure-physical relationship studies to capture information about the features of molecules responsible for polar intermolecular interactions. Most recently, the CPSA descriptors have been found to be practically useful in the study of acute aquatic toxicity where they appear to provide an alternative to LUMO energy level measures for describing global and local electrophilicity in cases of non-covalent molecular interactions. For this reason, they have become popular among QSAR practitioners. Most of CPSA descriptors are calculated summing up atomic contributions.

Quantum-mechanic descriptors rely on molecular properties computed at quantum mechanical level and directly deal with molecular electronic density and electrons distributions. They are, indeed, the most expensive descriptors to be computed from the computational point of view. The possibility to compute descriptors at the quantum mechanical level offers the possibility to gather meaningful information. For example, the HOMO/LUMO energy difference can be used to quantify the stability of the hypothetical transition state formed during a chemical reaction. Therefore, it can be used as indicator of the molecular reactivity. Nucleophilic and electrophilic atomic indexes are used to recognize possible reactive centres. Despite the class of quantum mechanical descriptors accounts for more than 500 descriptors, only some classes are needed to cover all of them:

- HOMO/LUMO energies;
- Nucleophilic reactivity atomic indices;
- Electrophilic reactivity atomic indices;
- Maximum/minimum atomic charges;
- PPSA, partial positive surface area;
- PNSA, partial negative surface area;
- Hydrogen donor/acceptor surface area;
- Total molecular surface area;
- Maximum/minimum atomic bond orders;
- Maximum/minimum electron–electron repulsion for each element;

- Maximum/minimum electron–nucleus attraction for each element;
- Maximum/minimum bond resonance/exchange energies and
- Maximum/minimum coulomb bond interaction.

Finally, thermodynamic descriptors rely on well-known thermodynamic concepts, such as enthalpy, entropy and heat capacity. As far as descriptors are regarded, the aforesaid quantities are computed with respect to different internal energy contribution, namely translational, vibrational and rotational energies. This leads to the following descriptors:

- Vibrational enthalpy of the molecule,
- Translational enthalpy of the molecule,
- Vibrational entropy of the molecule,
- Rotational entropy of the molecule,
- Translational entropy of the molecule and
- Vibrational heat capacity of the molecule

The calculation of 3D descriptors in DEMETRA is described in Appendix C (this volume), where details about the software parameters specifications are given for the software macromodel and MOPAC.

4.3. Fragments and Residues in DEMETRA

The DEMETRA project provided a unique opportunity for testing and validating fragment-based QSAR models for use in pesticide risk assessment. Although fragment-based methods have been successfully exploited in correlating toxicity endpoints to several terrestrial and aquatic species (Gao *et al.*, 1992; Russom *et al.*, 1997; Martin and Young, 2001), their application towards pesticides has been only marginally investigated (Perez *et al.*, 2003; Toropov and Benfenati, 2006).

The main goal of fragment-based methods is to recognize substructures that may be associated with an observed toxicity endpoint. In contrast to holistic approaches, which involve a range of descriptors referring to the entire molecular structure, fragment-based methods investigate the structure-activity relationships at a sub structural level. This allows identification of local chemical features that would not be represented by global, holistic descriptors. A meaningful example is considering well-known reactive substructures such as allylic, vinylic or propalgylic ones. The presence of any of such substructure makes the substances more toxic than an analogous of equal hydrophobicity. Thus, the possibility of successfully recognizing these local indicators of toxicity would noticeably improve both accuracy and predictive power of QSARs. Furthermore, the recognition of reactive structural patterns might be exploited for improving the current knowledge about the chemical bioactivity.

Fragment-based approaches can be also exploited for prioritization purposes. In this case, biologically active substructures previously identified are searched within a database of new pesticides. The compounds containing substructures of potential concern are isolated. The process is similar to virtual screening, where among thousands of candidates, only few are picked up and considered valuable for further screening. In this way, testing priorities can be assigned to the pesticides showing structural alerts.

Those considerations suggest fragment-based approaches to be ideal candidates for developing QSAR models aimed at pesticides impact evaluation. Whilst the successful implementation of fragment-based techniques is fairly straightforward, several factors may affect model performance.

The very first step in fragment-based QSAR development is the definition of fragments size and chemical composition. No predefined rules for determining the optimal fragment size and chemical composition for addressing a specific structure-activity problem exist. There are many ways for defining fragments, depending on the structural features one would emphasize and on the representation of the molecular structure. Different definitions of the aromatic character, for example, yield different fragments. For this reason, in common QSAR practice, this choice is left to the user. Often, most important chemical features are defined on purpose, by visually inspecting the molecules under investigation. In order for a model to be suitable for such a purpose, fragments need to be generated in an automated way. Use of automated algorithms would guarantee the possibility to handle any data set and the opportunity to standardize the fragment generation procedure. Those advantages would ensure high reproducibility, that is a mandatory requirement for in silico tools to be adopted at a regulatory level. Two kinds of automated algorithms have so far been developed. The first searches predefined fragments within all molecules under investigation. In this case, the total number of fragments is fixed, like in MDL fingerprints (MDL, 2005). The second, conversely, breaks down all molecules in fragments up to a predefined length (three to seven atoms, for example), or extending atomic connectivity to their neighbours, like in atomic-centred fragments (Adamson *et al.*, 1971). In this case, the final number of fragments strictly depends on the size and the chemical variability of the data set. Indeed, the total number of fragments obtained provides a measure of the data set chemical diversity. In both cases, the overall number of fragments ranges from hundreds to thousands. Fragment-based descriptors are usually coded in fingerprints accounting for their number. In case of simple presence(1)/absence(0), the fingerprint is simply a key bit. Otherwise, it accounts for the number of

Mol A	0	2	0	3	0	6	2	3	1	...	...	...

Figure 7: Example of fingerprint.

occurrences of each fragment in a molecule. A typical fingerprint accounting for occurrence numbers looks like the one shown in Fig. 7 (Mol A indicating a generic molecule).

For QSAR modelling purposes, all fingerprints can be conveniently collected giving a matrix-like representation of a data set. The matrix-like representation of any data set can be easily constructed by indexing each fragment with increasing integer numbers. Then, for each molecule, a fingerprint array is filled with integer accounting for fragment occurrence. In this way, a matrix listing all molecules with their corresponding fingerprints is created. Figure 8 gives an example of such a representation.

Observing the matrix in Fig. 8, one may note that several cells are filled with 0s, simply indicating that the fragment is absent in that molecule. Matrix-like representation can be directly submitted to model building algorithms.

As mentioned above, the number of fragments to be handled can be very large. Once all molecules have been codified in the corresponding fingerprints, one needs a procedure for descriptors selection, in order to select the most meaningful chemical features, whilst preventing over-fitting which occurs when too many descriptors are simultaneously considered as independent variables. An absurd and false model may fit perfectly if the model has enough complexity by comparison to the amount of data available (Wiki1, 2006). Over-fitting is generally

ID	FG 1	FG 2	FG 3	FG 4	FG 5	FG ...
100027	1	4	2	2	0	...
101053	0	2	0	3	1	...
101213	0	4	0	2	0	...
10222012	0	0	0	0	0	...
10336109	0	4	0	0	2	...
10453868	0	5	0	4	0	...
10465334	0	8	0	10	2	...
10551206	0	7	0	2	1	...
10572667	0	0	0	0	0	...
10582778	0	3	2	1	0	...
ID ...	...	...	...	...	...	...

Figure 8: Excerpt of matrix listing molecular fingerprints.

recognized to be a violation of Occam's razor. In order to prevent over-fitting, certain algorithms for selecting variables are usually adopted. The main expectation is to end up with a pool of fragments that are not just statistically but also chemically relevant. In other words, one would be able to identify the chemical features responsible for the observed toxicity. As mentioned above, this is the ultimate goal of fragment-based method. Unfortunately, the limited data usually available often hamper an objective investigation. Having a small number of examples at disposal might even lead to erroneous conclusions. From the statistical point of view, fragments occurring in highly toxic molecules are likely to be targeted as toxicophores, even when those are known to be biologically inactive. An expert chemist might distinguish between potentially active substructures and inactive ones. Conversely, an algorithm alone cannot make such a distinction unless more information has been supplied.

Having briefly introduced the advantages and drawbacks of fragment-based algorithms, we discuss the application of such techniques within DEMETRA. Following is the results for Trout data set (Chapter 7, Amaury *et al.*, this volume), where three different approaches were evaluated with the aim to extract preliminary indications about the possibility of successfully applying fragment-based methods in DEMETRA:

1) Top priority fragment approach (Casalegno *et al.*, 2006): The top priority fragment (TPF) approach relies on a prioritization scheme, developed to take into account the fact that a fragment contribution to toxicity may vary, depending on the surrounding chemical environment. In order to prioritize a fragment contribution, a simple numerical criterion was implemented, based on the observed toxicity data of the training set molecules. According to the original project strategy, 236 molecules were selected for training. This criterion was used to discover winner/loser relationships among fragments and retrieve them in a priority matrix. The information collected in the priority matrix was then exploited with the aim to quantitatively predict a toxicity endpoint. The final scheme was applied to 46 test molecules, in order to validate the model.

2) MDL Keys (MDL, 2005): Nine hundred and sixty structural keys, provided by MDL information System. The selection of most relevant fingerprints was performed through a HSA, using cross-validation. Afterwards, the molecules were classified according to the current EC regulation (Piclin *et al.*, 2006), using an adaptive fuzzy partition (AFP) algorithm. Model performances were evaluated in terms of correct classification percentages. The training set used accounted for 235 molecules, whereas the test set for 46 molecules, according to the project strategy.

3) LEADSCOPE (Leadscope Personal, version 3.2) (LFG, 2005): LEADSCOPE was used here for clustering compounds on the basis of similarity

of structural features among the compounds in DEMETRA Trout data set. The program produces structural fragments for each molecular structure, compares them with a database of 34,653 structural fragments and assigns the molecule to a particular cluster on the basis of structural similarity. This fragment-based classification was initially performed on the entire data set with the aim to classify all compounds. After the classification, three major classes were obtained, accounting for the larger fraction of compounds. The remaining compounds, belonging to small different classes, were finally retrieved in a fourth class. A genetic algorithm (GA) was used to select most relevant descriptors among a pool provided with the same software package. Four to five descriptors resulted after the selection. Each class was then randomly split into training (60 %)/test (40 %) sets. At this point, a multiple linear regression (MLR) was performed on the selected descriptors with the aim to assess, for each class, model performances on training and test sets. Table 4 reports the results obtained with different fragments. Use of the TPF method yielded satisfactory results within this data set, as testified by the test set correlation coefficient of 0.69. For this reason, this method was afterwards combined with others to generate hybrid architectures. The resulting hybrid system was not chosen as the final one; nevertheless, these results have been reported in Chapter 7 (Amaury *et al.*, this volume).

As indicated in the Table 4, MDL Key results account for the percentage of correctly classified compounds. This percentage was computed following the toxicological classification provided by the EU legislative documents (EEC, 1992). The training and test performances are comparable, but poor in comparison with other models developed within the same study (Piclin *et al.*, 2006). For this reason, use of MDL Keys was not further encouraged with the aim of developing hybrid systems.

Results obtained by LEADSCOPE software were, instead, provided within a range of correlation coefficients. The worst and the best correlation coefficients were reported for training sets and test ones, obtained by random splitting. Good results were obtained for the training sets

Table 4
Models performances towards trout data set

Method	NDes[a]	R2_TR	R2_TS
TPF	135	0.85	0.69
MDL-Keys	8	0.64**	0.57[b]
Leadscope	4–5	0.80–0.98***	0.15–0.55[c]

[a]Number of descriptors.
[b]Percentage of compounds correctly classified (Piclin *et al.*, 2006).
[c]Worst/best regression coefficients obtained within different classes.

(0.80–0.98), but results were not satisfactory for the test set. One of the main difficulties with this approach was that the number of compounds in most of the clusters was not large enough to allow building of a good sub-model for every cluster. The high number of chemical classes prevented the possibility to adequately populate each class, finally leading to low accurate predictions.

A detailed analysis of the results so far obtained with all three strategies may indicate that fragment-based methods need improvements in modelling pesticide activity. This conclusion is, however, only partially true. The present study found no evidence for methodological failures. What hampered the successful application of fragment-based model was the difficulty of dealing with a wide variety of structurally diverse compounds. More than 20 pesticide classes were represented in the Trout data set. These included organotins, organochlorines, organophosphates, carbamates, formamidines, terpenes, pyrethroids, phenols, spinosyns, pyrroles, pyridazinones, benzoylureas and so on. It is important to note that some of those classes accounted for only one or two compounds. This clearly contrasts with a method that demands for structurally similar compounds to work properly. As mentioned above, missing adequate data cannot result in the development of reliable models. At the same time, fragment-based techniques provide a direct and intuitive way to relate structure and activity. For this reason, improving current approaches and developing new ones should be encouraged in the area of pesticides risk assessment.

REFERENCES

Adamson G.W., Lynch M.F., Town W.G. (1971) Analysis of structural characteristics of chemical compounds in a large computer-based file, II. Atom-centred fragments. *J. Chem. Soc.* C, 3702–3706.

Amidon G.L., Anik S.T. (1976) Comparison of several molecular topological indexes with molecular surface area in aqueous solubility estimation. *J. Pharm. Sci.*, 65, 801–808.

Babel (2006) http://sourceforge.net/projects/openbabel/.

Batchman (1999) www.univorleans.fr/SCIENCES/ICOA/batchman/.

Benfenati E., Piclin N., Roncaglioni A., Varì M.R. (2000) Factors influencing predictive models for toxicology. *SAR QSAR Environ. Res.*, 12, 593–603.

Bonchev D., Balaban A.T., Mekenyan O. (1980) Generalization of the Graph Center Concept, and Derived Topological Centric Indexes. *J. Chem. Inf. Comput. Sci.*, 20, 106–113.

CambridgeSoft (2006) http://www.cambridgesoft.com/.

Casalegno M., Sello G., Benfenati E. (2006) Top priority fragment QSAR approach in predicting pesticide aquatic toxicity. *Chem. Res. Toxicol.*, 19, 1533–1539.

CCR (2006) WWW Computational Chemistry Resources, http://www.chem.swin.edu.au/chem_ref.html#Software.

CDSweb (2006) Chemical Database Service, http://cds.dl.ac.uk/cds/cds.shtml.

Cfinder (2006) www.chemfinder.com.

Chemid (2004) http://chem.sis.nlm.nih.gov/chemidplus/.

Chemistry Software (2006) www.chemistry-software.com.

ChemPak (2006) http://www.madisonsoftware.com/chempak.html.

Chemsk (2006) http://www.acdlabs.com/download/chemsk.html.

Chemtkx (2006) http://www.sageinformatics.com/chemtkx.html.

Chemx (2006) CHEMX, originally Chemical Design Ltd, then Oxford Molecular. Currently unsupported.

Codpro (1996) CODESSA: COmprehensive DEscriptors for Structural and Statistical Analysis Copyright © 1994–96, Version 2.20, University of Florida.

Descod (2006) http://www.codessa-pro.com/.

DRAGON (2003) Dragon Web Version 3.0. http://www.talete.mi.it/products/dragon_description.htm.

DSSTox (2006) http://www.epa.gov/nheerl/dsstox/.

EEC (1992) Directive 92/32/EEC, the seventh amendment to Directive 67/548/ECC, OJL 154 of 5.VI.92, p1.

EPACTP (2006) http://www.epa.gov/comptox/.

Fujitsu (2006) http://www.fujitsu.com/.

Gao C., Govind R., Tabak H.H. (1992) Application of group contribution method for predicting the toxicity of organic compounds. *Environ. Toxicol. Chem.*, 11, 631–636.

Hclrss (2006) Compendium of Pesticides Common Names. www.hclrss.demon.co.uk/index.html.

Hosoya H. (1972) Topological index as a sorting device for coding chemical structures. *J. Chem. Doc.*, 12, 181–183.

Hurst P., Hay A., Dudley N. (1991) *The Pesticides Handbook*. Journeyman, London.

Huuskonen J.J., Livingstone D.J., Tetko I.V. (2000) Neural network modeling for estimation of partition coefficient based on atom-type electrotopological state indices. *J. Chem. Inf. Comput. Sci.*, 40, 947–955.

InChi(2006) http://www.iupac.org/inchi/.

Isis/Draw (2006) Isis/Draw Version 2.5, MDL Information Systems Inc.

Jalbout A.F., Li X. (2003) Anti-HIV-1 inhibitors of various molecules using principles of connectivity. *J. Mol. Struct.*, 663, 19–23.

LFG (2005) LEADSCOPE. LeadScope Personal (version 3.2.0), www.leadscope.com/lsp_product.php.

Martin T.M., Young D. M. (2001) Prediction of the acute toxicity (96-h LC_{50}) of organic compounds to the fathead minnow (Pimephales promelas) using a group contribution method. *Chem. Res. Toxicol.*, 14, 1378–1385.

MDL (2005) *MDL-Isis Host 5.0*, Enterprise Edition. MDL Information Systems Inc., San Leandro, California, USA.

Piclin N., Pintore M., Wechman C., Roncaglioni A., Benfenati E., Chrétien J.R. (2006) Ecotoxicity prediction by adaptive fuzzy partitioning. Comparing descriptors computed on 2D and 3D structures. *SAR QSAR Environ. Res.*, 17, 225–251.

Pubchem (2006) http://pubchem.ncbi.nlm.nih.gov/.

Randic M. (2001) Graph valence shells as molecular descriptors. *J. Chem. Inf. Comput. Sci.*, 41, 627–630.

Randic M., Basak S. C. (2001) On use of the variable connectivity index $^1\chi^f$ in QSAR: toxicity of aliphatic ethers. *J. Chem. Inf. Comput. Sci.*, 41, 614–618.

Randic M., Plavsic D., Lers N. (2001) Variable connectivity index for cycle-containing structures. *J. Chem. Inf. Comput. Sci.*, 41, 657–662.

Richard A.M., Williams, C.R., Cariello N.F. (2002) Improving structure-linked access to publicly available chemical toxicity information. *Curr. Opin. Drug Discov. Devel.*, 5, 136–143.

Roy K., Leonard T.J. (2004) QSAR modeling of HIV-1 reverse transcriptase inhibitor 2-amino-6-arylsulfonylbenzonitriles and congeners using molecular connectivity and E-state parameters. *Bioorg. Med. Chem.*,12, 745–754.

Russom C.L., Bradbury S.P., Broderius S.J., Hammermeister D.E., Drummond R.A. (1997) Predicting modes of toxic action from chemical structure: acute toxicity in the fathead minnow (Pimephales Promelas). *Environ. Toxicol. Chem.*, 16, 948–967.

Sfqsar (2006) http://qsar.sourceforge.net/dicts/qsar-descriptors-metadata/index.xhtml.

SMILES (2006) http://www.daylight.com/smiles/f_smiles.html.

Tomlin C. (1997) *The Pesticide Manual, (11th Edition).* British Crop Protection Council, Farnham, UK.

Toropov A.A., Benfenati E. (2006) QSAR models for Daphnia toxicity of pesticides based on combinations of topological parameters of molecular structures. *Bioorg. Med. Chem.*, 14, 2779–2788.

Toropov A.A., Toropova A.P., Mukhamedzhanova D.V., Gutman I. (2005) Simplified molecular input line entry system (SMILES) as an alternative for constructing quantitative structure-property relationships (QSPR). *Indian J. Chem.*, 44A, 1545–1552.

Toxnet (2006) Toxicology Data Network. http://toxnet.nlm.nih.gov/.

Tripos (2006) http://www.tripos.com/.

Vidal D., Thormann M., Pons M., Lingo M. (2005) An efficient holographic text based method to calculate biophysical properties and intermolecular similarities. *J. Chem. Inf. Model.* 2005, 45, 386–393.

Weininger D. (1988) SMILES, a chemical language and information system. 1. Introduction to methodology and encoding rules. *J. Chem. Inf. Comput. Sci.*, 28, 31–36.

Weininger D. (1990) SMILES. 3. DEPICT. Graphical depiction of chemical structures. *J. Chem. Inf. Comput. Sci.*, 30, 237–243.

Weininger D., Weininger J.L. (1989) SMILES. 2. Algorithm for generation of unique SMILES notation. *J. Chem. Inf. Comput. Sci.*, 29, 97–101.

Wiener H. (1947a) Structural determination of paraffin boiling points. *J. Am. Chem. Soc.*, 69, 17–20.

Wiener H. (1947b) Correlation of heats of isomerization, and differences in heats of vaporization of isomers, among the paraffin hydrocarbons. *J. Am. Chem. Soc.*, 69, 2636–2638.

Wiener H. (1948a) Relation of the physical properties of the isomeric alkanes to molecular structure. Surface tension, specific dispersion, and critical solution temperature in aniline. *J. Phys. Chem.*, 52, 1082–1089.

Wiener H. (1948b) Vapor pressure-temperature relationships among the branched paraffin hydrocarbons. *J. Phys. Chem.*, 52, 425–430.

Wiki1 (2006) Overfitting. http://en.wikipedia.org/wiki/overfitting.

Edited by Emilio Benfenati

Chapter 4

Algorithms for (Q)SAR model building

Qasim Chaudhry[1], Jacques Chrétien[2], Marian Craciun[3], Gongde Guo[4], Frank Lemke[5], Johann-Adolf Müller[5], Daniel Neagu[4], Nadège Piclin[2], Marco Pintore[2], Paul Trundle[4]

[1]Central Science Laboratory, Sand Hutton, UK

[2]BioChemics Consulting SAS, Orléans, France

[3]Department of Computer Science and Engineering, University "Dunarea de Jos", Galati, Romania

[4]Department of Computing, School of Informatics, University of Bradford, Bradford, UK

[5]KnowledgeMiner Software, Berlin, Germany

1. INTRODUCTION

Predictive computational models based on structure-activity relationship (SAR), or quantitative structure-activity relationship (QSAR), are among the very few available alternatives to testing chemical toxicity in animals (Price and Watkins, 2003; Cronin and Livingstone, 2004). The basic concept behind (Q)SAR approach is that a specific biological activity of a series of related compounds is proportional to one or more physicochemical properties of the molecules, that is,

$$\text{Biological activity} = fn(\text{property}_1 + \text{property}_2 + \ldots \text{property}_n)$$

The concept of mathematically relating biological activity with physicochemical properties of related chemical compounds emerged in the 1960s (Hansch and Fujita, 1964). Early (Q)SARs were based on simple principles, such as substituent parameters (Hansch and Leo, 1979), and linear mathematics. It was, however, gradually realized that (Q)SAR models based on such simplistic properties and statistical algorithms only worked well in certain well-defined situations, for example, where compounds were closely related or congeneric, varied only at a single substituent position, shared the same mode of action, and where biological properties were measured close to the point of action, for example, receptor binding or enzyme assays. (Q)SAR models for relatively simple sets of molecular data are still based on linear algorithms, but this approach has only a limited usefulness in finding multi-dimensional relational patterns in complex data sets. Linear models are also often hard to generalize across chemical classes and/or test species. This has led to the use of non-linear algorithms and soft-computing techniques, such as fuzzy systems, probabilistic methods, and artificial neural networks (ANNs) to decipher relational patterns in large, imprecise, and complex data sets (Agrafiotis *et al.*, 2002; Kaiser, 2003; Pintore *et al.*, 2003a). This shift in (Q)SAR paradigm has made it possible to predict biological properties of a wide range of chemicals, which otherwise would be difficult, or impossible, to determine experimentally.

In DEMETRA (Development of Environmental Modules for Evaluation of Toxicity of Pesticide Residues in Agriculture) project, several generalized models have been developed to predict toxicity of a range of pesticides and derivatives, which belong to different chemical classes, and have different modes of action. As part of the modeling, both traditional and newer statistical algorithms were evaluated for suitability to model large and complex data sets. This chapter provides an essential overview of the different algorithms that can be used in developing (Q)SAR models.

2. METHODS FOR DATA PRE-PROCESSING AND SELECTING DESCRIPTORS

The chemical structure of a compound contains a lot of embedded information on physicochemical characteristics that determine biological activity. For (Q)SAR modeling, this information is first converted from structural features into a numerical representation of physicochemical properties, termed as molecular descriptors. The first step in modeling thus involves in silico assembling of precise chemical structures in one-dimensional (1D), 2D, and 3D forms using various specialist software. Where available, a chemical structure may be downloaded from online chemical databases, such as ChemIDplus (http://chem.sis.nlm.nih.gov/chemidplus) or PubChem (http://pubchem.ncbi.nlm.nih.gov/).

Once assembled, the molecular structures need optimizing in 2D and 3D space to obtain a molecular configuration that resembles its ‘real-life’ form. This is a very important step, as the use of wrong configuration can jeopardize quality and performance of the resulting (Q)SAR model. As determining the ‘real-life’ conformation of a molecule with any degree of certainty is not easy, it is usually the lowest energy conformation that is used in modeling. In DEMETRA, the optimized 3D configurations of molecules were compared with their crystal structures (where available) to ascertain that the right conformation was used in modeling. Several other factors were considered during structure optimization; for example, whether a molecule could exist in different interchangeable tautomeric forms and whether that the geometrical or stereo-isomeric form used in modeling was the right one.

Once optimized, the molecular structures were used in calculating a wide range of descriptors, such as constitutional, hydrophobic, steric, electronic, electrotopographical, topological, number of functional groups, certain molecular fragments, and so on. Indeed, several thousand descriptors for each chemical compound were calculated using a suite of specialist software (see Chapter 3, Benfenati *et al.*, this volume). Modeling with such a large number of calculated descriptors, however, carried the risk of leading to a false 'chance' correlation between the descriptors and biological activity. To avoid this problem, various methods were employed in DEMETRA to reduce the 'noise' data (termed as feature selection). The methods used for feature selection were based on the following:

1. Cross correlation matrix: Cross correlation involves building a table of correlations of each descriptor with all others. In this way, by pairwise comparison, non-orthogonal descriptors can be eliminated from the data set from which models are to be built.

2. Stepwise regression: In stepwise regression, the user is able to set a threshold value of significance of variables in the model. Descriptors whose inclusion in a multi-linear regression (MLR) model do not meet this threshold are excluded.

3. Genetic algorithm (GA): The powerful techniques based on GA concept are able to thoroughly explore the molecular-descriptor hyperspace. GA operates by optimizing solutions based on the generation of random combinations of descriptors and then 'spawning' new models based on repeated improvement in performance.

4. Principal component analysis (PCA): PCA derives linear combinations of the descriptors to produce composite variables describing various proportions of the variance in the observed data. Thus, PCA can indicate the best combinations of the original variables for use in model construction.

5. *k*-nearest neighbor (*k*NN): *k*NN method classifies a new object on the basis of attributes and training samples.

In addition to these, a hybrid selection algorithm (HSA), which combines the GA concepts and a stepwise regression, was also used to select the best descriptors for classifying DEMETRA data sets (Ros *et al.*, 2002). This is because the use of GA alone was not particularly suitable for local search, and the incorporation of a stepwise approach in HSA method to reach local convergence provided a quick and adaptable solution to the identification of 'promising' areas. A specific index was derived from the fuzzy clustering method to evaluate the fitness function. This index gave the advantage of quick calculations for the estimation of descriptor relevance by analysing complex molecular distributions in which finding separating edges between different categories would otherwise be difficult. A cross validation procedure was also included in the algorithm during the selection procedure to prevent over-fitting and a poor generalization. This procedure worked by randomly dividing the database into training and validation sets, and the fitness score for each chromosome was derived from the combination of scores for the training and validation sets. Other statistical filtering methods, such as correlation-based filtering, were also used for the selection of the most relevant descriptors in DEMETRA data sets. It is also noteworthy that many of the statistical methods, that can be used to develop (Q)SAR models, can also be used to refine the data prior to constructing a model.

3. MODELS WITH CLASSIFIERS

3.1. FISs

Fuzzy logic (FL) (Zadeh, 1977; Sugeno, 1985a) mimics human reasoning in its use of approximate information and uncertainty to generate decisions about intrinsically imprecise problems. The FL concepts provide mathematical rules and functions that enable the calculation of intermediate values between 'absolutely true' and 'absolutely false' through named degrees of membership ranging from 0.0 to 1.0. The fuzzy set theory was proposed by Zadeh (1975) to provide a mathematical way for dealing with concepts in natural language. At the root of fuzzy set theory lies the idea of linguistic variable or fuzzy variable (Fig. 1) – a complex structure having a symbolic name and a set of linguistic values (fuzzy sets), over a domain of quantitative (numerical) values.

Fuzzy proposition and fuzzy rules are similar to classical ones and are obtained by combining linguistic variables using different FL connectives and fuzzy implication. A fuzzy rule-based system or fuzzy inference system (FIS)

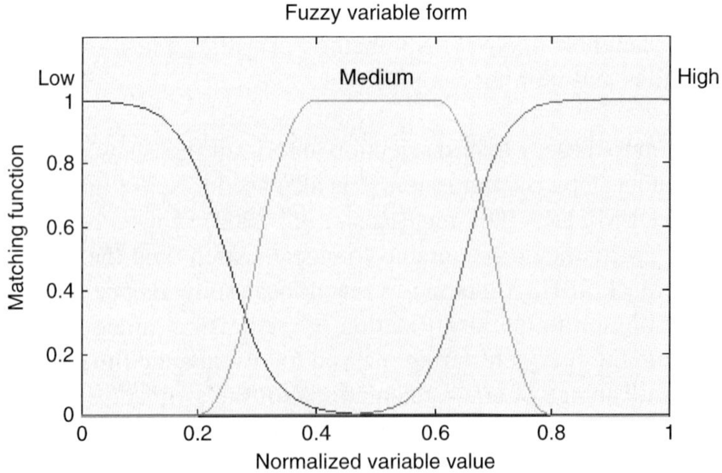

Figure 1: Linguistic variable.

is a collection of these rules that deals with approximate knowledge. There are three principal ways to obtain the rules: human experts provide the rules, rules are formed by training methods, or a combination of the two. All fuzzy rules have the following structure:

$$\text{IF(antecedent)THEN(consequent)} \tag{1}$$

The most used FISs in real-life applications are Mamdani FIS (Mamdani and Assilian, 1975) and Sugeno FIS (Sugeno, 1985b), having the structure of the fuzzy rules described by Eqs (2) and (3), respectively.

$$\text{IF}(x_1 \text{ is } A_1 \text{ AND } x_2 \text{ is } A_2 \text{ AND} \ldots \text{AND } x_d \text{ is } A_d)\text{THEN}(\mathbf{y} \text{ is } B) \tag{2}$$

$$\text{IF}(x_1 \text{ is } A_1 \text{ AND } x_2 \text{ is } A_2 \text{ AND} \ldots \text{AND } x_d \text{ is } A_d) \\ \text{THEN}(\mathbf{y} \text{ is } f(A_1, A_2, \ldots, A_d)), \tag{3}$$

where $A_1, A_2, \ldots, A_d$ and B are linguistic values (fuzzy sets), vector $\mathbf{x} = (x_1, x_2, \ldots, x_n)^{\mathrm{T}}$ is the input pattern of the fuzzy system and $\mathbf{y}$ is the output of the fuzzy system, and f is a mathematical function. If f is a constant function, then the FIS is called zero-order Sugeno FIS, and if f is a linear combination of inputs, then it is called first-order Sugeno FIS.

FISs can be used both for classification or regression. If the fuzzy system is used as a classifier, the consequence part could be '. . . THEN class is c'. In the case of numerical prediction, Wang (1992) and Kosko (1992) showed that fuzzy systems are universal approximators, with the capability to approximate any continuous function on compact sets with a degree of accuracy.

The main problem with the prediction of numerical values is the fact that the output of an FIS is in many cases a fuzzy set, and 'defuzzification' algorithms are used to translate it into a real number. There are several defuzzification methods (Fig. 2): the smallest of maximum, the largest of maximum, the bisector of area, the center of area (under the shape of the output membership function), and the mean of maximum (the average of the maximum value of the output fuzzy set).

The output of a FIS can be expressed as follows:

$$y = F(\mathbf{X}, \boldsymbol{\Theta}) = \psi\left[\bigcup_{k=1}^{N}\left(\left(\bigcap_{j=1}^{d}(\overline{x}_j \times A_j^k)\right) \circ B^k\right)\right], \tag{4}$$

where $\mathbf{X} = (x_1, x_2, \ldots, x_d)^{\mathrm{T}}$ is the input vector; $\boldsymbol{\Theta}$ is the parameters vector describing the membership functions for the linguistic variables; $\overline{x}_j$ is the fuzzified value (fuzzy transformation) of the input x_j; N is the number of fuzzy rules; A_j^k and B^k are linguistic values (fuzzy sets); $\cup$ represents the FL connective used to combine the outputs (consequents) of the rules; $\cap$ represents the FL connective used to combine the component of rule's antecedent; $\circ$ is the operation used to

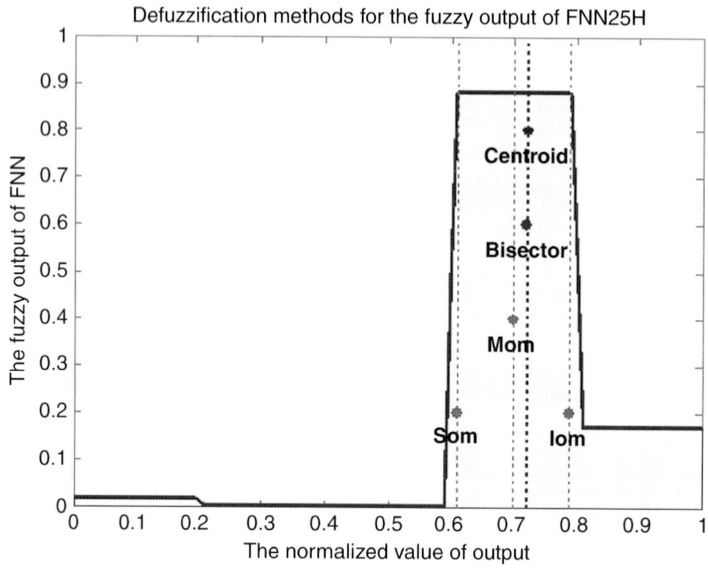

Figure 2: Defuzzification methods.

aggregate the input with the fuzzy sets corresponding to linguistic variables; $\times$ is the fuzzy implication operator; and ψ is the defuzzification function (Fig. 2).

Fuzzy set concepts are also used to generalize the traditional cluster procedures: assign for each input vector a number between 0 and 1, called the membership degree of that pattern, and a cluster can be characterized from the list of associated patterns with the highest degree. An algorithm that is used to obtain such clusters is fuzzy *c*-means for which description can be found in Bezdek (1981).

Because FL and FIS can model imprecise and qualitative knowledge, they are suitable for tackling classification and clustering problems. Despite this, very few applications based only on fuzzy sets alone can be found in predictive toxicology. Instead, they are often combined with other techniques in hybrid approaches (see Section 3.3, Chapter 5, Amaury *et al.*, this volume).

3.2. Adaptive fuzzy partition

In the context of supervised classification of biochemical data sets, the fuzzy partition algorithm (Lin and Cunningham, 1994) allows the generation of fuzzy rules from numerical data by developing two steps:

1. partitioning a working space into fuzzy subspaces and
2. defining a fuzzy rule for each fuzzy subspace.

Assuming that the working space is an N-dimensional hyperspace, defined by N molecular descriptors, each dimension i can be partitioned into L intervals I_{ij}, where j represents an interval in the partition selected. Indicating a molecular vector in the hyperspace with $P(x_1, x_2, \ldots x_n)$, a *rule* for a subspace S_k can be derived by combining N intervals I_{ij}, as defined by (Sugeno and Yasakawa, 1993):

$$\begin{aligned}&\text{if } x_1 \text{ is associated with } \mu_{1k}(x_1) \text{ and } x_2 \text{ is associated with } \mu_{2k}(x_2)\ldots\\&\quad\text{and } x_N \text{ is associated with } \mu_{Nk}(xN)\\&\quad\Rightarrow \text{the score of the activity } O \text{ for } P \text{ is } O_{k\text{P}},\end{aligned} \tag{5}$$

where x_i represents the value of the ith descriptor for the molecule P, μ_{ik} is the membership function related to the descriptor i for the subspace k, and $O_{k\text{P}}$ is the biochemical activity value related to the subspace S_k. The 'and' of the fuzzy rule is generally represented by the *Min operator* [5] that selects the minimal value amongst all the μ_{ik} components; the membership functions can be defined by triangular, trapezoidal, or gaussian shapes (Ichino, 1979; Sugeno and Yasakawa, 1993; Fritzke, 1994).

These techniques of rule generation are very simple, as all the fuzzy rules can be formulated by linguistic labels. But their performances in the data set classification depend on the choice of the partition selected. Generally, a coarse partition not only leads to a generalist system but also to a model where prediction results are too approximate. A fine partition, on the other hand, leads to a precise model of classification but also to a non-generalist system. To overcome this drawback, fuzzy classification methods have been proposed (Ishibuchi *et al.*, 1992), which simultaneously use several fuzzy partitions of different sizes in a single fuzzy rule-based classification system. Thus, the relationship (5) becomes

$$\begin{aligned}&\text{if } x_1 \text{ is associated with } \mu_{1km}(x_1) \text{ and } x_2 \text{ with associated to } \mu_{2km}(x_2)\ldots\\&\quad\text{and } x_N \text{ is associated with } \mu_{Nkm}(x_N)\\&\quad\Rightarrow \text{the score of the activity } O \text{ for } P \text{ is } O_{km\text{P}},\end{aligned} \tag{6}$$

where $m \geq 2$ and represents the amount of fuzzy subspaces according to the partition on the axes.

This approach allows achievement of a good compromise between generalist and specialist systems, improving the classification performances, as shown by the satisfactory results obtained in a previous paper studying olfactory series (Audouze *et al.*, 2000). However, the algorithm still does not allow tackling another problem related to fuzzy partition that is due to the very high number of fuzzy subspaces generated when a large set of molecular descriptors is considered. To overcome this problem, a new algorithm termed as adaptive fuzzy partition (AFP) (Ros *et al.*, 2003; Pintore *et al.*, 2003a,b), derived from a number of studies concerning fuzzy and non-fuzzy fields (Lin *et al.*, 1997; Pedrycz, 1990; Chou, 1991; Ripley, 1993), has been implemented. In this algorithm, the hyperspace partition is adapted to the training data by dividing dynamically and separately the axes associated with the N molecular descriptors.

In a first phase, the global-descriptor hyperspace is considered and cut into two subspaces where fuzzy rules are derived. These two subspaces are divided step by step into smaller subspaces until certain conditions are satisfied:

1. The number of molecular vectors, within a subspace, attains a minimum threshold number.

2. The difference between two generated subspaces is negligible in terms of chemical activities represented.

3. The number of subspaces exceeds a maximum threshold number.

The aim of this algorithm is to select the descriptor axis and the cut position, which allows the achievement of maximal difference between the two fuzzy rule scores generated by the new subspaces. The score is determined by the weighted average of the chemical activity values in an active subspace A and in its neighboring subspaces. If the number of trial cuts per descriptor is defined by the variable N_cut, the number of trial partitions equals $(\text{N\_cut}+1)N$. Only the best cut is selected to subdivide the original subspace. In Fig. 3, for example, three cuts per axis are tested from the original descriptor space with two dimensions. As cut x_1 is the best, two subspaces are generated and considered to be further divided. Then, cut y_3 is selected, but the procedure evaluates useful partitioning only in subspace S_2; finally, three subspaces are built.

The fuzzy rules are generated by the relationship (5), but the membership functions, defined by trapezoidal shapes, are based on the boundaries of the subspaces. If the width of a subspace S_k on the ith dimension, after each cut, is

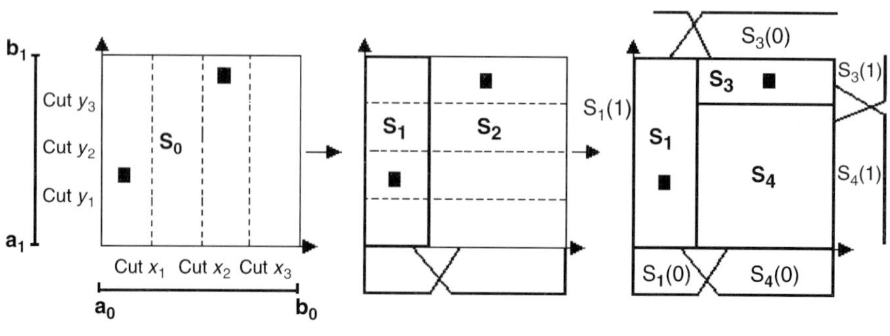

Figure 3: Example of adaptive partition on a two-dimensional space. Three cuts for each axis are evaluated, and three subspaces are generated.

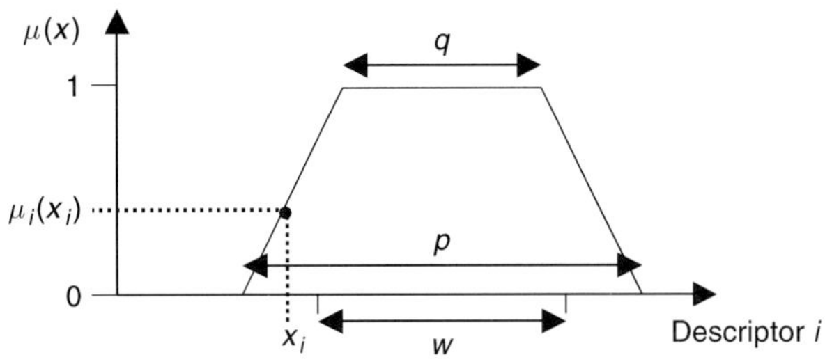

Figure 4: Representation of a trapezoidal membership function $\mu(x)$, defined on the descriptor i.

represented by w_i, the p and q parameters defining the shape of the trapezoid (Fig. 4) are calculated by

$$p = \lambda_i w_i \text{ and } q = n_i w_i, \tag{7}$$

where the two parameters λ_i and n_i vary so that $p \geq 1$ and $q \leq 1$. If $p = 1$ and $q = 1$, the membership function becomes a rectangle. All the rules created during the fuzzy procedure are considered to establish the model between descriptor hyperspace and biochemical activities.

The degree of membership of the subspace S_k can be represented by

$$O_k = \frac{\sum_{j=1}^{M} (\mathrm{Min}_i^N \mu_{ik}(x_i)_{P_j}) \cdot (A_{P_j})}{\sum_{j=1}^{M} (\mathrm{Min}_i^N \mu_{ik}(x_i)_{P_j})}. \tag{8}$$

M is the number of molecular vectors in a given subspace, N is the total number of descriptors, $\mu_{ik}(x_i)_{P_j}$ is the fuzzy membership function related to the descriptor i for the molecular vector P_j, and A_{P_j} is the experimental activity of the compound P_j. A classic procedure of centroid defuzzification (Gupta and Qi, 1991) is implemented to determine the chemical activity of a new test molecule. All the subspaces k are considered, and the general formula to compute the degree of membership of the activity O for a generic molecule P_j is

$$O(P_j) = \frac{\sum_{k=1}^{N\_subsp} (\mathrm{Min}_i^N \mu_{ik}(x_i)_{P_j}) \cdot (O_k)}{\sum_{k=1}^{N\_subsp} \mathrm{Min}_i^N \mu_{ik}(x_i)_{P_j}}, \tag{9}$$

where N_subsp represents the total number of subspaces.

Pintore *et al.* (2002) applied AFP to a data set of 235 pesticide compounds, divided into three classes according to their toxicity toward rats. AFP gave a robust and generalized performance, achieving 76% correctly predicted compounds in the test set and 86% correctly predicted compounds within the most toxic class. Moreover, the authors compared the results obtained by AFP with those obtained by other standard and widely used classification techniques, such as linear and quadratic discriminant analyses and classification and regression tree methodology. Compared to these techniques, the AFP method allowed to derive better models, improving their prediction power by between 6 and 19%.

3.3. *k*-NN methods

The *k*NN method has been the subject of many studies for nearly half a century. The basic principle behind this method is that a given instance within a data set will generally exist in close proximity with other instances sharing similar properties. Hence, additional information about an instance can be obtained by observing other instances that are close to it, that is, the NNs. The method can thus be used as a classification technique. If the instances within a data set are tagged with a classification label, then the class of a new instance can be determined by observing the classes of its NNs.

The *k*NN method has been demonstrated for its power in a number of real-world applications, for example, text categorization (Sebastiani, 2002), time series data mining (Illa *et al.*, 2004), recognition of DNA and RNA sequences (Cost and Salzberg, 1993), thyroid disease diagnosis (Kibler and Aha, 1987), clinical audiology diagnosis (Bareiss and Porter, 1987), meeting predictions (Kozierok and Maes, 1993), Internet information filtering (Payne and Edwards, 1997), and image understanding (Singh *et al.*, 1999).

For a given new instance, the *k*NN method finds its *k*NNs, each of which is tagged with a class label, traditionally using the majority-voting rule to determine its class, that is, assigning the single most frequent class label associated with the *k*NNs to the new instance. This is illustrated in Fig. 5. The two classes here are depicted by '■' and '○', with ten instances for each class. Each instance is represented by a 2D point within a continuous-valued Euclidean space. A new instance x, represented as '▲', is then presented, and the NNs are used to classify it.

In Fig. 5, for example, if $k = 5$, the class of the new instance x is determined by identifying the five NNs and is classified as '■' class.

The *k*NN method for approximating a discrete-valued function $f: \Re^n \rightarrow C$ is described as follows (Mitchell, 1997):

1. Training procedure: For each training instance $< x, f(x) >$, add the instance to the training set. $f(x)$ is of the form $f: \Re^n \rightarrow C$, where C is a finite set $\{c_1, c_2, \ldots, c_{|C|}\}$.

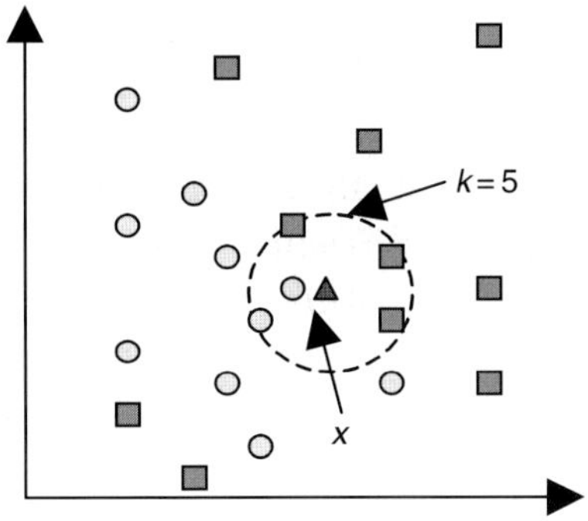

Figure 5: An example of classification of instance x when $k = 5$.

2. Classification procedure: Given a new instance x to be classified, let $x_1, x_2, \ldots, x_k$ denote the k instances in the training set that are nearest (most similar) to x.

$$f(x) \leftarrow \arg\ \max_{c \in C} \sum_{i=1}^{k} \delta(c, f(x_i)), \tag{10}$$

where $\delta(a,b) = 1$ if $a = b$ and $\delta(a,b) = 0$ otherwise.

3.3.1. Weighted k-*NN method*

In voting kNN, the k neighbors are implicitly assumed to have equal weights in decisions, regardless of their distances to an instance x to be classified. It is intuitively appealing to give different weights to the k neighbors based on their distances to x, with closer neighbors having greater weights.

In weighted k-nearest neighbor (wkNN) method, the kNNs are assigned different weights. Let Δ be a distance measure, and $x_1, x_2, \ldots, x_k$ be the kNNs of x arranged in increasing order of $\Delta(x_i, x)$. So x_1 is the first NN of x. The distance weight w_i for ith neighbor x_i is defined as follows:

$$w_i = \begin{cases} \dfrac{\Delta(x_k,x) - \Delta x_i,x)}{\Delta(x_k,x) - \Delta(x_1,x)} & if \quad \Delta(x_k,x) \neq \Delta(x_1,x) \\ 1 & if \quad \Delta(x_k,x) = \Delta(x_1,x) \end{cases}. \tag{11}$$

Instance x is assigned to the class for which the weights of the representatives among the kNNs sum to the greatest value.

3.3.2. k*NN model-based classification method*

The basic idea of kNN model-based classification method (kNNModel) (Guo *et al*., 2003) is to choose a subset of training instances with some extra information to serve as the basis for classification in order to improve the efficiency and reduce the large storage requirement of *k*NN. The extra information is obtained by inductively learning from the original training set, which can be interpreted as a type of compact information for the removed instances. It improves the effectiveness of *k*NN by keeping as much information as possible for classification during the process of instances removal. In addition, with the extra information stored in each representative, classifying a new instance simply depends on which representative covers it, thus alleviating the dependence on *k*. The detailed model construction algorithm of *k*NNModel is described as follows:

Step 1: Select a similarity measure, that is, Euclidean distance, and create a similarity matrix from a given training set *D*.

Step 2: Set to 'ungrouped' the tag of all data tuples.

Step 3: For each 'ungrouped' data tuple, find its local neighborhood.

Step 4: Among all the local neighborhoods obtained in step 2, find its global neighborhood N_i. Create a representative $<Cls(x_i), Sim(x_i), Num(x_i), Rep(x_i)>$ into *M* to represent all the data tuples covered by N_i, and then set to 'grouped' the tag of all the data tuples covered by N_i.

Step 5: Repeat steps 3 and 4 until all the data tuples in the training set have been set to 'grouped'.

Step 6: Model *M* consists of all the representatives collected from the above learning process.

In this algorithm, 'neighborhood' is a term referred to a given instance in data space. A neighborhood of a given instance is defined as the set of NNs of this instance; 'local neighborhood' is a neighborhood that covers the maximal number of instances with the same class label, and 'global neighborhood' is defined to be the largest local neighborhood among a set of local neighborhoods. Moreover, *D* represents a given training set and *M* represents the created model. The elements of representative $<Cls(x_i), Sim(x_i), Num(x_i), Rep(x_i)>$, respectively, represent the class label of x_i, the Euclidean distance of x_i to the most distant data tuple among the data tuples covered by N_i; the number of data tuples covered by N_i, and a representation of x_i itself. In step 4, if there is more than one local neighborhood having the same maximal number of neighbors, the one with minimal value of $Sim(x_i)$ is chosen as representative, that is, the one with the highest density. The classification algorithm of *k*NNModel is described as follows:

Step 1: For a new data tuple x to be classified, calculate its similarity to all representatives in the model M.

Step 2: If x is covered only by one representative $<Cls(x_j), Sim(x_j), Num(x_j), Rep(x_j)>$, that is, the Euclidean distance of x to x_j is smaller than $Sim(x_j)$, classify x as the class label of x_j.

Step 3: If x is covered by at least two representatives with a different class label, classify x as the class label of the representative with largest $Num(x_j)$, that is, the neighborhood covers the largest number of data tuples in the training set.

Step 4: If no representative in the model M covers x, classify x as the class label of a representative whose boundary is closest to x.

The Euclidean distance of x to a representative x_i's nearest boundary is equal to the difference of the Euclidean distance of x_i to x minus $Sim(x_i)$.

3.3.3. Fuzzy **k***NN model*

*k*NNModel is a novel classification method that is observed to have higher performance on average than crisp *k*NN on some public data sets (Guo *et al.*, 2003). However, because of the crisp partitioning characteristic of *k*NNModel, it becomes less effective in classification of instances located around the boundaries of clusters. The use of fuzzy partitioning is a natural way to deal with the boundary problem. In such a case, an instance, whether inside or outside a cluster, can belong to all clusters with different degrees of membership between 0 and 1. Assuming that X is a test data set and $x \in X$, for any representative $<Cls(c), Sim(c), Num(c), Rep(c)> \in M$ obtained in *k*NNModel and a set A containing all the instances covered by $<Cls(c), Sim(c), Num(c), Rep(c)>$, a classical representation of the set A can be expressed as:

$$A = \{x | d(x,c) < Sim(c)\}, \tag{12}$$

where $d(x,c)$ is a distance function used in *k*NNModel. A fuzzy representation of the set A in X is defined as a set of ordered pairs.

$$A = \{x, u_A(x) | x \in X\}. \tag{13}$$

In Eq. (13), $u_A(x)$ is called the membership function of x in A, which is defined by Eq. (12). The membership function maps each element of X to a membership grade between 0 and 1.

$$u_A(x) = \begin{cases} 1 & x \le a \\ 1-2\left(\dfrac{x-a}{b-a}\right)^2 & a \le x \le \dfrac{1}{2}(a+b) \\ 2\left(\dfrac{x-b}{b-a}\right)^2 & \dfrac{1}{2}(a+b) \le x \le b \\ 0 & x \ge b \end{cases}. \tag{14}$$

Given a and b, the Z-shaped built-in membership function in Matlab FL toolbox maps each instance of testing set to a membership grade between 0 and 1 (Fig. 6).

An interpretation of a and b in the context of fuzzy kNNModel is shown in Fig. 7, where

$$a = \frac{1}{2} Sim(x_j), \quad b = \frac{3}{2} Sim(x_j).$$

Obviously, given four test instances denoted as triangles, Δ_1, Δ_2, Δ_3, and Δ_4 in Fig. 7, the membership function $\mu_A(x)$ maps four test instances of X to membership grades between 0 and 1 as follows:

$$\mu_A(\Delta_1) = 0$$

$$\mu_A(\Delta_2) = 2\left(\frac{d(\Delta_2, x_j) - b}{b-a}\right)^2$$

$$\mu_A(\Delta_3) = 1 - 2\left(\frac{d(\Delta_3, x_j) - a}{b-a}\right)^2$$

$$\mu_A(\Delta_4) = 1$$

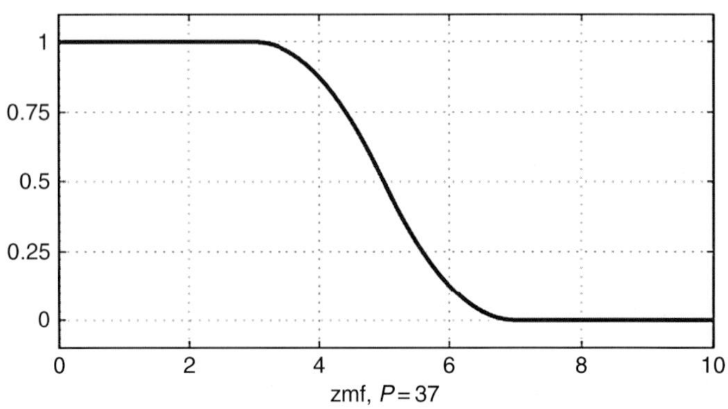

Figure 6: Z-shaped membership function.

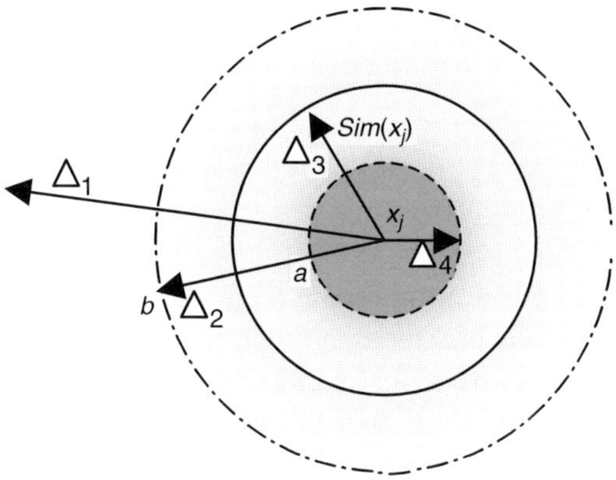

Figure 7: An example of instances distribution.

The model construction algorithm of fuzzy *k*NNModel is exactly the same as that of *k*NNModel. The detailed classification algorithm of fuzzy *k*NNModel (Guo and Neagu, 2005) is described as follows:

Input: a model M, parameter γ and a new instance x to be classified.

Output: the class of the new instance.

1. For a new instance x to be classified calculate its distance to all representatives in the model M.

2. If x is covered only by the kernel of one representative $<Cls(x_j), Sim(x_j), Num(x_j), Rep(x_j)>$, viz. the distance of x to x_j is smaller than $\gamma \times Sim(x_j)$, classify x as the class label of x_j.

3. If x is covered by at least two kernels of representatives with different class label, calculate its degrees of membership to each cluster it covers, and then sum up all the degrees of membership to clusters with the same class label, finally classify x as the class label with the largest degree of membership.

4. If no cluster in the model M covers x, calculate x's degrees of membership to each cluster in the M, and then sum up all the degrees of membership to clusters with the same class label, finally classify x as the class label with the largest degree of membership.

4. MODELS WITH REGRESSION SYSTEMS

4.1. Traditional linear regression QSAR models

Hansch and Fujita (1964) were the first to show that biological activity could be correlated linearly to free-energy-related terms (physicochemical parameters, now known as molecular descriptors). One or more of steric, hydrophobic, and electronic parameters were found to be the most consistent predictors of biological activity. Thus,

$$\mathrm{Log}\frac{1}{C} = aH + bS + cE + \text{constant},$$

where C is the molar concentration giving a constant biological response, H is the hydrophobic parameter, S is the steric parameter, and E is the electronic parameter.

This equation is a form of the classic equation for a straight line, and thus the statistical tool of choice for fitting QSAR relationships of this kind is linear regression, more specifically, MLR. Hansch and other workers rapidly realized that biological systems would not obey this linear relationship 'ad infinitum', but rather there would be optimum values for the descriptors beyond which biological activity would plateau and then decline. This 'parabolic' model of activity versus property simply introduces a squared term into the equation to transform the parabola into a straight line. Hence, the classical QSAR model became

$$\mathrm{Log}\frac{1}{C} = aH + b(H)^2 + cS + d(S)^2 + eE + f(E)^2 + \text{constant}$$

Many QSAR publications have utilized the first-order form of the equation, because it has been found that the limitations of the available data set often result in all data points falling on the linear portion of the relationship.

The basis of MLR is the replacement of an observed output (in DEMETRA, this is an LD_{50} or LC_{50}), with an estimated one, based on the relationship between LD_{50} (or LC_{50}) and one or more physicochemical properties. As this can never be done with certainty, the aim of MLR is to minimize the error due to lack of fit of some of the data points. Although higher order terms can be incorporated into the model (polynomial regression), and even interaction terms between the regressors, this often results in an equation that describes the data points but has no underlying meaning and no predictive power. In QSAR, the 'rule of parsimony' is a good watchword: the simpler the equation, the more likely it is to be 'right'. The most common approach to finding an MLR is the method of least squares. In this method, the error term is the difference between the observed and predicted activity for each data point. However, to avoid a cancelling out of positive and negative error terms, the square of the error term is used, and it is the sum of these squares that is minimized during MLR. The procedures for minimizing the errors vary with the software used, but they are iterative processes often based on steepest descent type methods to improve the errors with each iteration until the calculations converge.

Assuming that feature reduction has been carried out (see section 2.), and the 'best' descriptors are submitted for MLR, a number of strategies are available for generating MLR models. These include models with all selected descriptors, forward stepwise modeling, backward stepwise modeling, or generation of all possible models.

1. The forward selection method starts out with a model containing no features beside the intercept. New features (descriptors) are added to the model one at a time, and the *F*-statistic used to decide whether the additional descriptor improves the model.

2. Backward elimination stepwise regression starts with all the descriptors, and based on the *F*-statistic, the most useful descriptors removed one at a time. Both forward and backward stepwise regression depend on the selection of values of *F* for input and removal, respectively, for their convergence and termination.

Stepwise methods have their disadvantages, and thus with the computing power now available, it may be feasible to calculate all possible models that can then be assessed for validity and quality as described below. However, question that remains is how to know which model is the best. For this, a number of indicators of validity are available for use:

1. *s*, the standard deviation about the regression line. This is an indication of how well the model chosen predicts the observed biological activity. The smaller the value of *s*, the better is the QSAR.

2. *r*, the multiple correlation coefficient. This is a measure of how well the equation fits the data; that is, it measures how good the correlation is. More often r^2 is quoted, and this gives the fraction of the variance (usually as a percent) that is explained by the regression line.

3. *F*, value derived from the *F*-test, which indicates the probability of a real relationship or the significance level of the MLR model. The *F*-value is the ratio between explained and unexplained variance.

4. The *t*-statistic calculated for regression coefficients compares each coefficient with its standard error. To be significant, the regression coefficient should be about twice as big as its standard error, at the 5% level for five or more degrees of freedom (Livingstone, 1995).

5. Q^2, the cross-validation score to determine model robustness. Based on correlations derived from repeated determinations with one or more instances removed each time, the so-called leave one out (LOO) or leave-several-out (LSO) procedures, respectively.

6. Tests for multi-colinearity to reject models in which not all regressors are orthogonal.

7. Ratio of features to instances (descriptors to compounds). Traditionally in QSAR the number of descriptors should be less than one fifth of the number of compounds in the data set.

A more detailed discussion about validation will be presented in Chapter 6 (Benfenati *et al.*, this volume).

In the vast majority of published QSARs, the analytical method of choice is MLR, usually employing algorithms based on ordinary least squares (OLS) procedures. There are, however, some variations on this basic procedure, such as weighted least squares (WLS) and partial least squares (PLS).

WLS takes into account that not all the features (descriptors) are determined with equal accuracy, and thus some will have a greater range of errors than others. WLS takes such variation in magnitude of errors into account.

PLS is a technique that combines PCA with regression. It has particular use when there are a large number of descriptors. It is also particularly useful in data reduction steps and in predicting outliers. PLS, like PCA, produces factor scores as linear combinations of the original predictor variables rather than using the original variables.

4.2. ANNs and fuzzy neural networks

4.2.1. ANNs

Because real-life (Q)SAR problems are usually multi-dimensional and non-linear, newer mathematical algorithms are used, which can deal with the inherent uncertainty or 'noise' in biological data. Most notable of these methods are the many flavors of neural networks. These connected non-linear-processing elements operating in parallel (DARPA, 1988) have been termed 'neural networks' because of 'biological' inspiration from the function of human brain. In reality, they are another family of parameterized non-linear functions, which like polynomials, wavelets, Fourier series, radial basis functions, and splines are non-linear approximators. In the context of (Q)SAR modeling, these non-linear functions are intended to approximate the regression function of a predicted property, that is, the expectation value of the latter (viewed as a random variable), conditional to the set of variables of the model (i.e., the descriptors of the molecules in a (Q)SAR).

ANN modeling has become one of the most used branches of artificial intelligence-based machine learning in industry, education, and research. ANNs have the remarkable information-processing features; such as non-linearity, high parallelism, robustness, fault and failure tolerance, learning, ability to handle imprecise and noisy information, and the capability to generalize (Basheer and Hajmeer, 2000).

The central element of an ANN is the artificial neuron (Fig. 8) – a mathematical approximation of the biological neuron. Each biological neuron consists of a cell body, a collection of dendrites, which brings electrochemical information

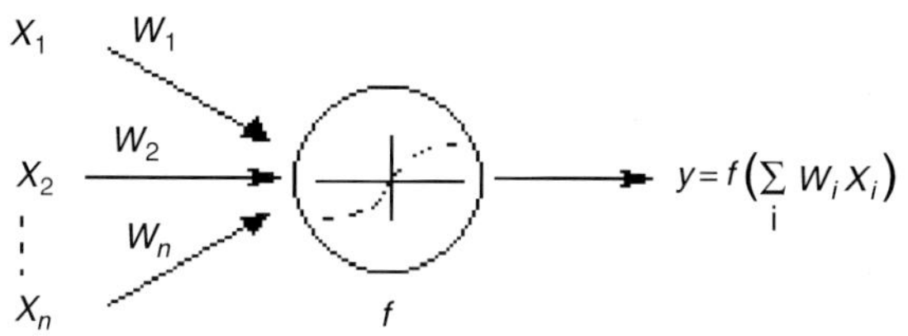

Figure 8: The artificial neuron.

into the cell and an axon, which transmits electrochemical information out of the cell. A neuron produces an output along its axon when the collective effect of its inputs reaches a certain threshold. The axon from one neuron can influence the dendrites of another neuron across junctions called synapses.

The artificial neuron receives one or more inputs and produces an output based on the calculation of a weighted sum of the inputs using a simple non-linear transfer function as a threshold or step function, which is usually a sigmoid. The artificial neurons are interconnected in a single-layer or multi-layer neural networks that are able to discern complex and hidden patterns in the information presented to it. Based on the topology, the connection between the neurons of an ANN could be feedforward and feedback. In a feedforward ANN, the connections between nodes do not form cycles. In feedback or recurrent ANN models, there are cycles in the connections. The most used ANNs are the first type, because the feedback models are usually more difficult to train (Sarle, 1994) (Fig. 8).

Recent developments have led to more complicated and flexible neurons, and ANN behavior can be tailored by replacing the threshold transfer function with more complicated functions. Many modern software packages allow ANNs to be built with a huge variety of different transfer functions; for example, the traditional threshold style where output is either on or off depending on the weighted sum of the inputs, linear function where the output is a continuous value proportional to the inputs, and sigmoidal where the output is continuous but non-linear. Many more novel and innovative transfer functions are possible depending on the specificities of data and the task to be addressed (Duch and Jankowski, 1997).

Usually, ANNs ‘learn’ from training data by means of topology or weight modifications. There are two kinds of training methods used for ANNs: supervised and unsupervised. In supervised training, the corresponded output for each input pattern is presented, and the algorithm seeks to reduce the global error modifying the connection weights between neurons. In unsupervised training, there is no explicit information about the output, and the network forms ‘natural groupings’ of the input patterns modifying the connections weights too.

The most commonly used ANN is the fully connected feedforward network with three layers: input, hidden, and output (Fig. 9). It has been shown (Funahashi, 1989) that this type of network with sufficient number of hidden units and proper

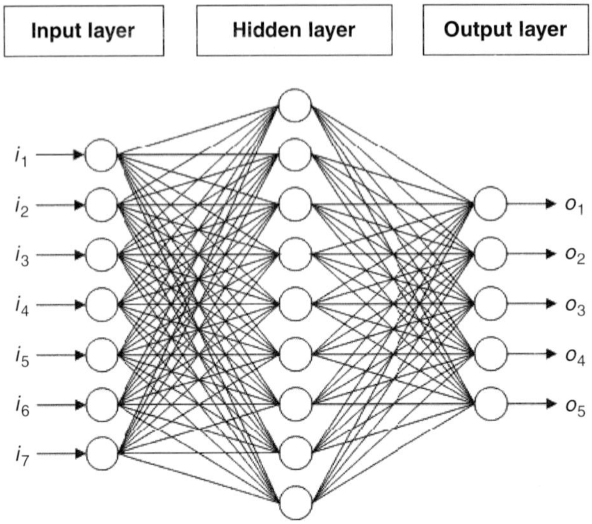

Figure 9: Fully connected three-layer feedforward network.

non-linear transfer function for hidden neurons has the ability to approximate any continuous mapping between inputs and outputs.

The strength of neural networks lies in their ability to automatically discover complicated patterns from data, where other methods may fail. ANNs can also decipher noisy data as the connections between neurons do not have to be simple threshold functions – a sigmoidal function, for example, allows dealing with 'gray' areas as opposed to 'black and white' situations. However, neural networks do have a number of disadvantages: they can be too capable of learning patterns from training data and are often subject to over-fitting, where a network learns the training examples too well and is then unable to make accurate predictions for anything other than the training data. A small error on training data and a large error on test data is a clear sign of over-fitting. Unfortunately, the optimum dimension of hidden layer (i.e., the number of hidden neurons) is not easy to establish, so, there are different methods such as GAs, heuristic approaches, or try-and-error methods to detect the appropriate size of the layer. Neural networks can also be difficult to understand, as they can often be very large and complex with hundreds, thousands, or even hundreds of thousands of connections.

To avoid over-fitting and to improve the generalization capability of ANNs, there are two prevalent techniques: early stopping and Bayesian regularization.

The output of a feedforwad three-layer network N hidden units and one output is as following:

$$\mathbf{Y} = F(\mathbf{X}, \boldsymbol{\Theta}) = \psi\left(\sum_{j=1}^{N} \omega_j \varphi\left(\sum_{p=1}^{d} \omega_{jp} x_p\right)\right), \tag{15}$$

where $\mathbf{X} = (x_1, x_2, \ldots, x_d)^T$ is the input vector; $\mathbf{\Theta} = (\omega_1, \omega_2, \ldots, \omega_N, \omega_{11}, \omega_{12}, \ldots, \omega_{Nd})$ is the weights vector with ω_j – the weight of the connection between the jth hidden unit and the output unit and ω_{jp} – the weight of the connection between the jth hidden unit and the pth input unit; and ψ and φ are the transfer functions of hidden and output layer, respectively (linear function, sigmoidal function, etc.).

One of the most popular methods for training such multi-layer networks is based on gradient descent in error and is known as the backpropagation algorithm (or generalized delta rule). Each iteration includes two steps: a feedforward step and a backpropagation step. During the feedforward step the training data set is presented to the model, and the output is computed based on the input, weights, and transfer functions. During the backpropagation step, the output error is computed, and the errors for all processing nodes and the weights are adjusted in order to reduce those errors. This iterative training process stops when the error has reached the criteria set by the network designer.

Another slightly different type of ANN is the radial basis function neural network (RBFNN) (Hagan *et al.*, 1996). ANNs based on RBF have originally been used in function interpolation (Powell, 1987) to describe an unknown function f as linear combination of basis functions depending on the norm of the difference between the training vectors and the new vector for which the approximate value is desired. The most used radial basis functions are Gaussians, and usually, Euclidian distance is used to compute the norm. An RBFNN produces a local mapping:

$$\mathbf{Y} = F(\mathbf{X}, \mathbf{\Theta}) = \psi\left(\sum_{i=1}^{N} \omega_i b_i(\mathbf{X})\right), \tag{16}$$

where $\mathbf{X} = (x_1, x_2, \ldots, x_d)^T$ is the input vector, $\mathbf{\Theta} = (\omega_1, \omega_2, \ldots, \omega_N)$ is the weights vector with ω_j – the weight of the connection between the jth hidden unit and the output unit, b_i are the radial basis functions (e.g., Gaussian) as transfer functions of the hidden layer, and ψ is the transfer function of the output unit.

An ANN that combines the unsupervised and supervised learning paradigms is learning vector quantization networks (Kohonen, 1987). These contain two modules: one competitive layer to group data in clusters and one multi-layer feedforward module to learn the associations between the data in every cluster and the real class label, known a priori and used for supervised training.

ANNs have proven useful in solving a wide variety of real-life problems, for example, pattern classification, function approximation, clustering, or prediction. They are powerful tools for non-linear modeling of complex causal-effect relationships where a functional dependence is not clear. Initially, ANNs were considered unsuitable for data mining (knowledge extraction) because of their black-box nature. ANN models can still not be easily used to elucidate the mechanistic nature of the correlation established between variables as symbolic information. However, classification rules can be generated through analysis of weights and activation values of the hidden units from a trained feedforward three-layer neural networks with a good accuracy rate, after removing the redundant connections with a pruning algorithm (Tickle *et al.*, 1998). Certain quantitative measures can be used to evaluate performance of the extracted rules (Mitra *et al.*, 2002) in terms of accuracy, fidelity, confusion, coverage, confidence, and so on. Based on these measures and the generated rules, new insights into the application domain of an ANN can be provided.

Several types of ANNs have been used in predictive toxicology to represent knowledge and to model data. They include supervised ANNs – such as those based on backpropagation (Gini *et al.*, 1999), radial basis function, generalized regression, and probabilistic algorithms (Kaiser and Niculescu, 1999, 2001; Kaiser *et al.*, 2000) – and unsupervised ANNs – such as self-organizing maps and learning vector quantization NNs (Baurin *et al.*, 2001). ANNs have also been studied for SAR applications by Adamczak and Duch (2000). Together with other methods, the authors used a smooth transformation of a multi-layer perceptron (MLP) network, which is a type of feedforward–backpropagation network, achieved during the network training by gradually increasing the slope of sigmoidal functions to obtain crisp decision regions. After training, they extracted logical rules in order to describe the data from two SAR sets: antibiotic activity of pyrimidine compounds and carcinogenicity data from the predictive toxicology evaluation project of the US National Institute of Environmental Health Science (NIEHS).

Probabilistic NNs have also formed the basis for two stand-alone toxicity computation programs of TerraBase Inc., Hamilton, ON, Canada: TerraQSAR™–FHM to compute the acute toxicity (96-h LC_{50}) of organic chemicals to the fathcad minnow (*Pimephales promelas*) and TerraQSAR™-RMIV to compute the intravenous LD_{50} of organic chemicals for rat and mouse.

Further optimization of virtually any parameter in a neural network can be achieved by the use of GA that allows the lowest error (Niculescu, 2003). For example, the use of GA can optimize the number of hidden units, the learning rates, and the input selection to improve the network performance. This has also led to the development of hybrid models that combine GA, PCA, and fuzzy methodology with neural networks (Neagu, 2002). Self-organization in networks is also a fascinating area in the neural network field. Such networks can learn to detect regularities and correlations in their input data and adapt their future response to that input. The neurons of competitive networks learn to recognize groups of similar vectors and to separate the dissimilar ones (clustering) (Kohonen, 1987; Vračko *et al.*, 2006).

In DEMETRA project, the use of ANNs certainly provided a major improvement in the reliability and performance of QSAR models as compared to those based on MLR. When the descriptors (pre-selected by GA) in MLR models were remodeled using different neural network methods, there was a distinct improvement in the model performances.

4.2.2. Fuzzy neural networks

Fuzzy neural networks are an example of a hybrid approach, which combines the learning ability of a neural network with the noise-handling capability of FL (Buckley and Hayashi, 1995; Nauck and Kruse, 1996). In their simplest form, a fuzzy neural network can be viewed as a three-layer feedforward network, with a fuzzy input layer (fuzzification), a hidden layer containing the fuzzy rules, and a final fuzzy output layer (defuzzification). Fuzzy sets are contained within the (fuzzy) connections between layers, though sometimes a five-layer network with sets contained in the second and fourth layers can be found.

The input layer represents the input membership functions for the fuzzy rules, with sufficient input causing a rule in the hidden layer to fire. The weights between the layers represent the fuzzy sets, with membership in each set determined by the relative weights – these can be altered using particular training algorithms as per a normal neural system. Transfer functions are usually continuous and pass real values through the network to the output layer to be interpreted as degrees of membership in fuzzy sets based on the firing of fuzzy rules in the hidden layer.

Fuzzy neural networks combine the strengths of both neural networks and FL, making them a very powerful hybrid tool. They allow the integration of expert knowledge into the system, and are considered inherently more understandable because of their use of human-like fuzzy inference.

In an extended study on all five DEMETRA data sets and another data set for phenols, Neagu *et al.* (2005a) compared the performance of a variety of machine-learning approaches such as support vector machine, Bayes net, decision tree (DT), instance-based learner, repeated incremental pruning to produce error reduction, and MLP. The authors have described a library of models aimed at providing a useful benchmark for researchers in this field. The development of accurate models proved difficult because of the problems posed by properties of these data sets. However, when performance of the models was compared across all data sets, some algorithms appeared more suitable for particular data sets, exhibiting greater than average accuracy. This implies that a careful selection of algorithms could enable the development of accurate models. The outcomes of this statistical study have shown that a single classifier was not discriminative enough on all DEMETRA data sets considered (Neagu *et al.*, 2005a).

A study on the performance of neural network and fuzzy neural network parts of more complex hybrid intelligent systems has also been reported by Neagu *et al.* (2005b). In another study, Guo and Neagu (2005) have proposed a new robust algorithm, fuzzy *k*NNModel, which is based on a supervised clustering method (*k*NNModel) employing fuzzy partitioning instead of crisp partitioning to group clusters. The authors have compared experimental results of fuzzy *k*NNModel on thirteen public data sets including the five DEMETRA data sets with results of fuzzy *c*-means clustering, *k*-means clustering, *k*NN, fuzzy *k*NN, and original *k*NNModel in terms of classification performance.

Fuzzy *k*NNModel combines the strengths of fuzzy *c*-means clustering and *k*NNModel. All experimental results show that fuzzy *k*NNModel obtains similar results and in some cases (mainly the toxicity data sets) outperforms other methods (Guo and Neagu, 2005). However, a main drawback for most fuzzy hybrid models, for which further research is required, regards how to choose appropriate fuzzy membership functions and specific subsequent parameters for data sets to further fine-tune the developed models.

4.3. Self-organizing statistical-learning networks

Self-organizing statistical-learning networks (SOSLNs) implement principles of self-organization of data through inheritance, mutation, and selection to generate and validate a model structure systematically (Müller and Lemke, 2000). The use of SOSLN imparts knowledge extraction to yield models with large generalization power and accuracy. SOSLN tends to see things in terms of algorithms, that is, processing the data using the algorithm to yield a result. The algorithmic core of a SOSLN lies in computational methods used to implement the search- and data-management components.

There is a major difference between interpolation tasks addressed by traditional statistical methods, artificial intelligence tools, and SOSLN. Under SOSLN, analysis of experimental data consists of two parts – data mining and knowledge extraction. The data mining part works using interpolation algorithms of artificial intelligence, without the application of self-organization of models. Self-organization is only needed in the knowledge extraction part. All score functions rank models as a function of how useful the models are, on the basis of model error. Results of data mining are valid only within the sample space of all given data. An example is pattern recognition with a given number of classes. Data mining is able to solve any interpolation task, that is, result of data mining generates a description of the given data in the form of models and patterns, which contains information of the given data set.

New information or knowledge can, however, be derived only by continuing the learning process on new data, explicitly, or by applying inductive knowledge extraction methods using score functions (selection criteria) that are based on both error and bias calculated on some objective external information. This model selection depends on dispersion of noise in the data, and the result of knowledge extraction is optimal complex models (Müller and Lemke, 2000).

In self-organizing modelling, models are generated adaptively from data by networks of active neurons in an evolutionary fashion; thus populations of competing models of growing complexity produce new generations through validation and selection, until an optimal complex model is developed that is neither too simple nor too complex. That means growing a tree-like network out of seed information (input and output variables' data) in an evolutionary fashion of pairwise combination and survival of the fittest selection from a simple single individual (neuron) to a desired final, not over-specialized behavior (model). Neither the number of neurons and layers in the network nor the actual behavior of each neuron (transfer function) is pre-defined.

Self-organizing modeling creates optimal complex models, systematically and autonomously, by employing both parameter and structure identification. An optimal complex model is a model that optimally balances model quality on a given learning data set ('closeness-of-fit') and its generalization power on new, not previously seen data, with respect to both the noise level in the data and the task of modelling (prediction, classification, modelling, for example). It thus systematically avoids 'over-fitted' models based on information from the data only, as the selection feature depends on dispersion of noise in the data.

In contrast to neural networks using GAs, for example, as an additional procedure to optimize the network architecture and to counter over-training, the SOSLN approach introduces principles of evolution – inheritance, mutation, and selection – for generating a network structure systematically enabling automatic model structure synthesis and model validation. Instead of adaptation of unknown parameters, the method uses the estimation of parameters by means of subsystems of normal equations for every neuron. Unknown parameters are estimated by means of closed form solutions (least squared error) that are used for every transfer function in every layer to estimate no more than five to seven parameters.

4.3.1. Principles of self-organization

The SOSLN approach is based on induction, which is composed of the cybernetic principle of self-organization as an adaptive formation of a network without giving subjective points. There are three principles to be fulfilled [for more details, see Müller and Lemke (2000)]:

1. There is a very simple initial organization (neuron) that enables the description of a large class of systems through its evolution;
2. There is an algorithm for the mutation of the initial- or already-evolved organizations of a population (network layer); and
3. There is a selection criterion for validation and measure of the usefulness of an organization compared with its intended task.

In SOSLN, these principles are used not only in the initial data mining process but also in several different steps of the knowledge discovery process, such as estimation of missing values, dimension reduction (integrated feature selection), and synthesis of alternative models (combining) (Lemke and Müller, 2002a).

4.3.2. Self-organizing modeling

Self-organizing modeling consists of several levels of self-organization:

1. Self-organization of transfer functions (neurons);
2. Self-organization of networks; and
3. Self-organization of systems of networks (autonomous systems of equations) for multi-output systems.

In this way, it is possible to generate from small and noisy data sets:

1. Linear/non-linear time series models,
2. Static/dynamic linear/non-linear multi-input/single-output models,
3. Systems of linear/non-linear difference equations (multi-input/multi-output models), and
4. Systems of static/dynamic multi-input/multi-output fuzzy rules.

In all four cases, the resulting models are described analytically by dynamic equations or rules. For high noise levels, non-parametric models (pattern/cluster) can be obtained by analog complexing to solve different tasks: time series prediction (AC prediction), cluster analysis (AC clustering), and classification (AC classification) (Müller and Lemke, 2000).

4.3.3. Validation and noise filtering

A key problem in data mining and knowledge discovery from data is the final evaluation of generated models. This evaluation process is an important criterion for application of models. From data mining alone, it is impossible to decide whether an estimated model can reflect the causal relationship between input and output adequately or whether it is just a stochastic model with non-causal correlations. Model evaluation needs – in addition to noise filtering to avoid over-fitting of learning data – some new external information to justify model quality in terms of both predictive and descriptive powers (DPs).

As shown by Lemke (2005), a modeller cannot decide whether a model is true or not from the information on error or closeness-of-fit. This can only be ascertained by testing the models on some other data that adds new, external information. Lemke (2005) has clearly shown that any measure of closeness-of-fit does not suffice for evaluating a model's predictive and DP, as model evaluation requires at least a two-stage validation approach.

4.3.3.1. Level 1

Noise filtering (hypothesis testing), to avoid over-fitting of learning data, is based on external information that has not been used in the 'learning' process for developing a candidate model (hypothesis). For this purpose, a tool that has been used successfully involves cross-validation based on LOO principle.

4.3.3.2. Level 2

A characteristic is required that describes the noise filtering behavior of the 'learning' process to justify model quality based on additional external information not been used in the first validation level. This characteristic can be obtained by running a Monte Carlo simulation of a corresponding data mining algorithm many times, so that it expresses a kind of new, independent 'common knowledge' that any model can be and must be adjusted with (Lemke and Müller, 2002b). The objective of a second-level validation is as follows:

1. The noise filtering implemented in level 1 is very likely to not being an ideal noise filter and thus not working properly in any case (Fig. 10).
2. To get a new model quality measure – DP – that is adjusted by the noise filtering power of the algorithm.

Fig. 10 shows a detail of the noise sensitivity characteristic for a SOSLN algorithm implemented in the KnowledgeMiner data mining software (Knowledge Miner, 2006).

The noise sensitivity characteristic expresses a virtual model quality Q_u that can be obtained when using a data set of M potential inputs of N random samples. It is a virtual model quality, because by definition, there is not any causal relationship between stochastic variables (true model quality $Q = 0$), but there are actually models of quality $Q > 0$, which, when using random samples, reflect the stochastic correlations (Lemke and Müller, 2002b). By implementing an algorithm's noise sensitivity characteristic into a data mining tool, it is possible for any given number of potential inputs M and number of samples N to calculate a threshold quality $Q_u = f(N, M)$ such that any model's quality Q must exceed Q_u to be stated valid, as it describes some relevant relationship between input and output. Otherwise, a model of quality $Q \leq Q_u$ is assumed invalid, because its quality Q can also be reached when simply using independent variables, which means that this model does not differ from a model of just stochastic correlations.

In addition to deciding whether a model appears being valid or not, the noise sensitivity characteristic is also a tool for quantifying to what extent the data is described by a causal relationship between input and output. This introduces a

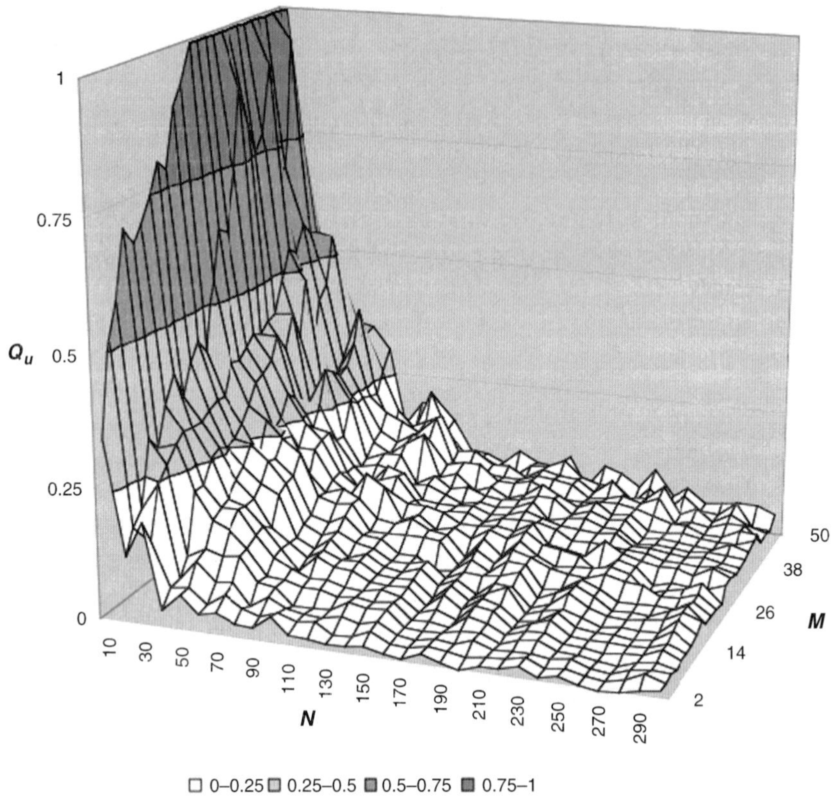

Figure 10: Noise sensitivity characteristic. M, number of potential inputs; N, number of samples; Q_u, virtual quality of a model; $Q_u = 1$, noise filtering does not work at all; and $Q_u = 0$, ideal filtering.

new, noise filtering, and model complexity-adjusted model quality measure: DP is defined as

$$DP = \begin{cases} 0 & Q \le Q_u(N,L) \\ \dfrac{Q - Q_u(N,L)}{1 - Q_u(N,L)} & Q > Q_u(N,L),\, Q_u(N,L) > 1 \end{cases}.$$

Here, Q is the measured quality of the evaluated model, and $Q_u(N,\ L)$ is the reference quality calculated from the number of samples N the model was created on and from the number of input variables L the model is actually composed of (selected relevant inputs in the model), with $L \le M$. This means that the DP measure excludes any virtual quality that may exist and that it directly allows for model complexity. For example, two models M_1 and M_2 show the same quality $Q = Q_1 = Q_2$, but M_1 uses more relevant inputs than M_2 to reach that quality Q, so, with $L_1 > L_2$, the DP of M_2 is higher than that of M_1.

The implemented two-stage model validation approach now allows, for the first time, to get an active decision support in model evaluation based on the model's DP calculated on the learning data only, for minimizing the risk of false interpreting models and using invalid models that simply reflect some non-causal correlation. A model may be invalid because of the following factors:

1. Most important input variables are not included in the data set and therefore the investigated output variable(s) cannot be sufficiently explained by an input–output model. The output variable should be considered as exogenous and should be described separately by a time series model.
2. The data set is not balanced, that is, there are more variables than observations. In addition to methods for dimension reduction (Lemke and Müller, 2002a), the quality of model results can be improved by combining.

The results obtained by models with small sample numbers are in most cases insufficient. In many fields, such as ecology or ecotoxicology, there are only a small number of observations that make the results uncertain. All methods involving automated model selection lead to a single best model, and conclusions and decisions are made as if it was the true model. However, this does not take into account a major element of uncertainty, that is, uncertainty about the model itself.

The task of combining involves seeking a prediction by means of voting or averaging (simple, weighted, and Bayesian) from a given ensemble of predictors. Combining the corresponding outputs of a number of trained networks is similar to creating a large network in which the trained networks are subnetworks operating in parallel, and the combination weights are the connection weights of the output layer. It is therefore possible to generate a combination of models (synthesis) by SOSLN algorithms itself. The big advantage of this approach is that the best (voting) or combined best models are automatically selected by self-organization linearly or non-linearly.

4.3.4. Multi-leveled self-organization workflow

Deriving knowledge from data is an interactive and iterative workflow process of various subtasks and decisions and is called knowledge discovery from databases (KDD) (Fayyad *et al.*, 1996). Usually, a single data mining process has been automated in the form of algorithms and software (neural networks, DTs, fuzzy modeling, GAs, and classical statistical methods, for instance). The remaining parts require user interaction and manual work, and they are overall most time consuming. This means that the result of knowledge discovery is very much dependent on knowledge, skills, expertise of the person carrying out the analysis, and may not be transparent or reproducible by another person. Seen from an end-user perspective, however, in many cases these are key features for generating acceptance, trust, and reliability, as in the case of QSAR modeling for prediction of chemical toxicity.

Our approach to a multi-leveled self-organization was motivated by the initial idea of KDD by making the overall workflow process more automated and more objective, and to limit the user involvement to the inclusion of well-known a priori knowledge and to manually apply some pre- and post-processing tasks that are difficult to automate. Fig. 11 shows the KDD workflow process when implementing an automated multi-leveled self-organization.

The concept of a multi-leveled self-organization starts with data pre-processing tasks that can be automated and may include the following:

1. Missing values detection and handling;
2. Further pre-selection of input variables according to some a priori (given or intended) constraints such as a variable's diversity, type (continuous or discrete), or origin;
3. Generation of additional, derived potential input variables;
4. Deterministic or stochastic subdivision of data sets; and
5. Dimension reduction in state and/or sample space.

The first two levels of self-organization have been the basic idea of SOSLNs for more than 20 years. They are built on the abovementioned main concepts. The two levels of self-organization incorporate the following essential tasks:

1. Self-organization of neuron transfer function,
2. Self-organization of the network's structure or topology by generating alternative model candidates of different input variables and of growing complexity, and
3. The first level of model evaluation and model selection.

The last step in multi-leveled self-organization is further evaluation and selection of models that pass the lower self-organization levels by calculating the models' DP as described above. This process is highly computationally intensive – the self-organization of a non-linear regression model of about 10 relevant input variables out of 1000 potential inputs and 200 samples, for example, may take up to 2 days of computing time. However, it does not require any user interaction and can run in the background. Increased transparency and reproducibility are other important features of this approach. This approach was used for the first time within the DEMETRA project.

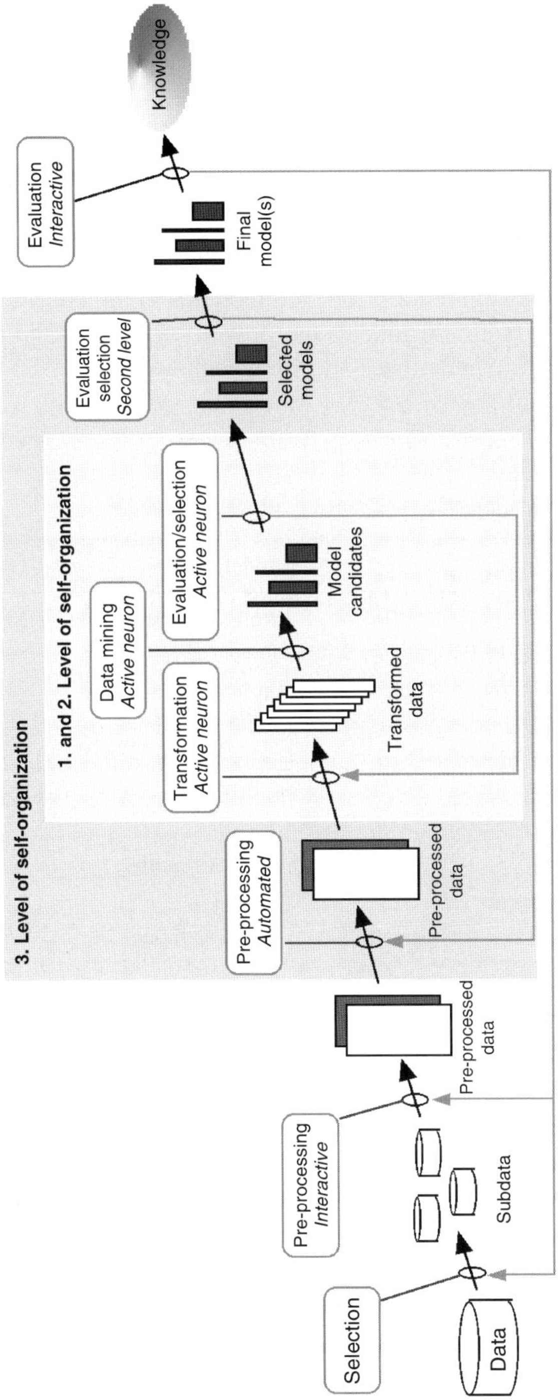

Figure 11: Multi-leveled self-organization displayed in gray box as a tool for knowledge discovery from databases (KDD) workflow processing.

To give an example on how models generated by SOSLN compare, the results of the five best individual QSAR models for the trout data set are listed in Table 1 (see also Chapter 7, Amaury *et al.*, this volume). Some QSAR models were created using 2D molecular descriptors (inputs) only, others were built on 3D or on 2D and 3D descriptors.

The *model type* column of Table 1 distinguishes between implicit and explicit regression models. Although neural networks typically distribute and hide the created model in the network, the result of multi-leveled self-organization are explicit analytical models. Fig. 12 shows, for example, the regression equation of the first model of Table 1. Neither the formal model structure nor the input variables composition was given a priori; the model is completely self-organized. This true knowledge extraction from data has proven very useful and advantageous for model interpretation, evaluation, and implementation issues. So it is possible

Table 1
Five best models for the data set D_1 – Trout – with respect to $R^2_{A,B,C}$

$R^2_{A,B,C}$	$Q^2_{A,B}$	R^2_C	m	Model type	DM method
0.67	0.69	0.59	10	Explicit linear model	Multi-leveled self-organization
0.66	0.66	0.64	15	Explicit linear model	Multi-leveled self-organization
0.65	0.66	0.63	6	Implicit nonlinear model	Neural network (GA-MLP)
0.63	0.63	0.65	8	Implicit nonlinear model	Neural network (GA-MLP)
0.63	0.71	0.64	11	Explicit nonlinear model	Multi-leveled self-organization

$N = 275$, $N_{A,B} = 229$, $N_c = 46$ and $M =$ up to 1800. $R^2_{A,B,C}$, R^2 calculated on the entire data set D; $Q^2_{A,B}$, leave one out cross-validation on the data subset $D_{A,B}$; R^2_C, R^2 calculated on the test data subset D_C, m, number of variables used in the model and M, number of potential input variables and state space dimension. Multi-leveled self-organization: high-dimensional modeling algorithn using multi-leveled self-organization with GMDH networks as active neurons and Neural network (GA-MLP): genetic algorithm for dimension reduction and multilayer percepton neural network for modeling.

$$LC_{50}\text{ (trout) [mmol/l]} = -1.6023\ (\text{C-031})^{-1} - 1.53\ \text{MATS3e} - 1.3148\ (\text{nOH})^{-1} - 27.1340\ \text{GATS3m} - 0.8957\ \text{nxch3} + 2.1469\ (\text{SEigZ})^{-1} - 0.2699\ \text{LogDpH7} + 0.7736\ (\text{D/Dr09})^{-1} - 0.0313\ \text{D/Dr03} + 5.8706\ (\text{Mp})^{-1} + 28.220$$

Figure 12: Self-organized linear regression model in chemical notation.

to implement these types of models in a MS excel sheet, automatically, for immediate use for further analysis, evaluation, or just application purposes (Knowledge Miner, 2006).

Similar results of individual models were obtained for the other four DEMETRA data sets.

In summary, experience has shown that model performance (e.g., in terms of closeness-of-fit) of SOSLN and other modeling and data mining methods, if applied appropriately, are comparable. However, the SOSLN approach has some additional distinct properties. For example, knowledge extraction in SOSLN is by self-organization. The approach only requires minimal, uncertain a priori, information about the system to model and makes it possible to model short and noisy data samples. Under SOSLN, high-dimensional modeling and model validation are integrated parts of the model building process, which eliminates the need for subdividing modeling into subtasks, such as feature selection, data mining, and model validation. The output of SOSLN is an optimal complex model, which is likely to be not over-fitted, and has a transparent explanation component.

5. CONCLUSIONS

The statistical algorithms described in this chapter are those that have been successfully used in DEMETRA project to develop predictive QSAR models for pesticide toxicity. The enormous variety of available methods that can be used for (Q)SAR applications indicates the extent of developments in the field of molecular modeling since the first use of linear algorithms in the 1960s. Indeed, the modeling strategy in DEMETRA started off with simple algorithms, such as linear regression methods, but the challenging task of modeling large and complex data sets of compounds, belonging to different chemical classes and with different modes of action, led us to explore the use of more versatile algorithms. Our experience with DEMETRA and other QSAR projects has shown that no single algorithm can be regarded as the method of choice for all (Q)SAR applications. For small, relatively simpler, and less 'noisy' data sets, linear regression-based algorithms are usually adequate. These, when combined with powerful feature selection algorithms, such as GA, can also be useful for larger data sets. The field of (Q)SAR has, however, benefited most from the advent of newer non-linear algorithms and soft-computing techniques, such as fuzzy systems, probabilistic methods, ANNs, and several combination of these methods. Further integration of these methods with powerful feature selection tools, such as GA, has proven very useful in modeling linear or non-linear relational patterns in large, imprecise, and complex data sets.

The development of hybrid systems that can combine a multitude of individual models is a further major milestone in the advancement of the new (Q)SAR paradigm. The ability to unveil the black-box nature of complex statistical relationships that constitute, for example, a neural network is also a major step forward that will bring the much needed clarity and validation to (Q)SARs developed using the new methodologies so that they become acceptable for regulatory use. All these developments, combined with availability of enormous computing power in recent years, bring closer the idea that it may be possible one day to develop more comprehensive and all-inclusive models that will be able to predict specific toxicity endpoints for any chemical, irrespective of its chemical class or mode of action.

REFERENCES

Adamczak R., Duch W. (2000) *Neural Networks for Structure-Activity Relationship Problems*, 5th Conference on Neural Networks and Soft Computing, 669–674, Zakopane, Poland.

Agrafiotis D.K., Cedeno W., Lobanov V.S. (2002) On the use of neural network ensembles in (Q)SAR and QSPR. *J. Chem. Inf. Comput. Sci.*, 42, 903–911.

Audouze K., Ros F., Pintore M., Chrétien J.R. (2000) Prediction of odours of aliphatic alcohols and carbonylated compounds using fuzzy partition and self organising maps (SOM). *Analysis*, 28, 625–632.

Bareiss E., Porter B. (1987) *Protos: An Exemplar-Based Learning Apprentice*. Proceedings of the 4th International Workshop on Machine Learning, 12–23, June, 1987, Irvine, CA, USA.

Basheer A., Hajmeer M. (2000) Artificial neural networks: fundamentals, computing, design, and application. *J. Microbiol. Methods*, 43, 3–31.

Baurin N., Marot C., Mozziconacci J.C., Morin-Allory L. (2001) *Using of Learning Vector Quantization and BCI fingerprints for the Predictive Toxicological Challenge 2000–2001*, ECML/PKDD-01 Workshop, The Predictive Toxicology Challenge (PTC), Freiburg, Germany.

Bezdek J.C. (1981) *Pattern Recognition with Fuzzy Objective Function Algorithms*. Plenum Press, New York, USA.

Buckley J.J., Hayashi Y. (1995) Neural networks for fuzzy systems. *Fuzzy Sets Syst.*, 71, 265–276.

Chou P.A. (1991) Optimal partitioning for classification and regression trees. *IEEE T. Pattern Anal. Mach. Intel.*, 13, 340–354.

Cost S., Salzberg S.A (1993) Weighted nearest neighbour algorithm for learning with symbolic features. *Machine Learn.*, 10, 57–78.

Cronin M.T.D., Livingstone D.J. (eds) (2004) *Predicting Chemical Toxicity and Fate*, CRC Press, Boca Raton, USA.

DARPA (1988) *DARPA Neural Net Study*, 60. AFCEA International Press, Fairfax, Virginia, USA.

Duch W., Jankowski N. (1997) New neural transfer functions. *Appl. Math. Comput. Sci.*, 7, 639–658.

Fayyad U.M., Platestsky-Shapiro G., Smyth. P. (1996) *Advances in Knowledge Discovery and Data Mining. From Data Mining to Knowledge Discovery: An Overview*, 1–30. AAAI Press, Menlo Park, CA, USA.

Fritzke B. (1994) Fast learning with incremental radial basis function networks. *Neural Process. Lett.*, 1, 2–5.

Funahashi K. (1989) On the approximate realization of continuous mappings by neural networks. *Neural Netw.*, 2, 183–192.

Gini G., Lorenzini M., Benfenati E., Grasso P., Bruschi M. (1999) Predictive carcinogenicity: a model for aromatic compounds, with nitrogen-containing substituents, based on molecular descriptors using an artificial neural network. *J. Chem. Inf. Comput. Sci.*, 39, 1076–1080.

Guo G., Neagu D. (2005) Fuzzy kNNModel applied to predictive toxicology data mining. *IJCIA*, 5, 321–333.

Guo G., Wang H., Bell D., Bi Y., Greer K. (2003) *kNN Model-Based Approach in Classification*. Proceedings of ODBASE 2003, 986–996, Nov. 3–7, 2003, Catania, Sicily (Italy).

Gupta M.M., Qi J. (1991) Theory of T-norms and fuzzy inference methods. *Fuzzy Sets Syst.*, 40, 431–450.

Hagan M.T., Demuth H.B., Beale M.H. (1996) *Neural Network Design*. PWS Publishing Company, Boston, MA, USA.

Hansch C., Leo A. (1979) *Substituent Constants for Correlation Analysis in Chemistry and Biology*. Wiley, New York, USA.

Hansch C., Fujita T. (1964) ρ-σ-π Analysis. A method for the correlation of biological activity and chemical structure. *J. Am. Chem. Soc.*, 86, 1616–1626.

Ichino M. (1979) A nonparametric multiclass pattern classificatier. *IEEE T. Syst. Man Cy.*, 6, 345–352.

Illa J.M.G., Alonso J.B., Marré M.S. (2004) Nearest-neighbours for time series: special issue: machine learning policies. *J. Appl. Intell.*, 20, 21–35.

Ishibuchi H., Nozaki K., Tanaka H. (1992) Distributed representation of fuzzy rules and its application to pattern classification. *Fuzzy Sets Syst.*, 52, 21–32.

Kaiser K.L.E. (2003) Neural networks for effect prediction in environmental and health issues using large datasets. *Quant. Struct.-Act. Relat.*, 22, 1–6.

Kaiser K.L.E., Niculescu S.P. (1999) Using probabilistic neural networks to model the toxicity of chemicals to the fathead minnow (*Pimephales Promelas*): a study based on 865 compounds. *Chemosphere*, 38, 3237–3245.

Kaiser K.L.E., Niculescu S.P. (2001) Modeling the acute toxicity of chemicals to Daphnia magna: a probabilistic neural network approach. *Environ. Toxicol. Chem.*, 20, 402–431.

Kaiser K.L.E., Niculescu S.P., Schultz T.W. (2000) Modeling the toxicity of chemicals to Tetrahymena pyriformis using molecular fragment descriptors and probabilistic neural networks. *Archiv. Environ. Toxicol. Chem.*, 39, 289–298.

Kibler D., Aha D. (1987) *Learning Representative Exemplars of Concepts: An Initial Case Study*. Proceedings of the 4th International Workshop on Machine Learning, 24–30, Irvine, CA, USA.

Knowledge Miner (2006) *Self-Organizing Data Mining and Prediction Tool*, available at http://www.knowledgeminer.com.

Kohonen, T. (1987) *Self-Organization and Associative Memory*. Springer-Verlag, Berlin, Germany.

Kosko B. (1992) *Fuzzy Systems as Universal Approximators*. Proceedings of the First IEEE Conference on Fuzzy Systems, 1153–1162, San Diego, CA, USA.

Kozierok R., Maes P. (1993) *A Learning Interface Agent for Scheduling Meetings*. Proceedings of the ACM-SIGCHI International Workshop on Intelligent User Interfaces, 81–88, New York, USA.

Lemke F. (2005) *Data-Driven Modeling of Acute Toxicity of Pesticide Residues as Alternative Tool Within Official Registration, Evaluation and Authorization Procedures*. Proceedings Data Mining Case Study Workshop, ICDM 05, Houston, TX, USA.

Lemke F., Müller J.-A.(2002a) *Self-Organising Data Mining*. Proceedings International Conference on Inductive Modeling, Lvov, Ukraine.

Lemke F., Müller J.-A. (2002b) *Validation in Self-Organising Data Mining*. Proceedings International Conference on Inductive Modeling, Lvov, Ukraine.

Lin Y., Cunningham G.A. III, Coggeshall S.V. (1997) Using fuzzy partitions to create fuzzy systems for input output data and set the initial weights in a fuzzy neural networks. *IEEE T. Fuzzy Syst.*, 5, 614–621.

Lin Y., Cunningham G.J. (1994) Building a fuzzy system from input-output data. *J. Intell. Fuzzy Syst.*, 2, 243–250.

Livingstone D. (1995) *Data Analysis for Chemists: Applications to QSAR and Chemical Product Design*. Oxford University Press, New York, USA.

Mamdani E.H., S. Assilian (1975) An experiment in linguistic synthesis with a fuzzy logic controller. *Intl. J. Man-Machine Stud.*, 7, 1–13.

Mitchell T. (1997) *Machine Learning*. New York, McGraw-Hill.

Mitra S., Pal S.K., Mitra P. (2002) Data mining in soft computing framework: a survey. *IEEE T. Neural Netw.*, 13, 3–14.

Müller J.-A., Lemke, F. (2000) *Self-Organising Data Mining*. Extracting Knowledge From Data. BOD, Hamburg, Germany.

Nauck D., Kruse R. (1996) Designing neuro-fuzzy systems through backpropagation, in: *Fuzzy Modelling: Paradigms and Practice*, 203–228, Pedrycz W. (Ed.). Kluwer, Boston, MA, USA.

Neagu C.D. (2002) *Toxicity Prediction Using Assemblies of Hybrid Fuzzy Neural Models*. Proceedings of the 6th International Conference on Knowledge-Based Intelligent Information & Engineering Systems – KES2002, 1093–1098, IOS Press, Crema, Italy.

Neagu D., Guo G., Trundle P., Cronin M.T. (2005a) *A Comparative Study of Machine Learning Algorithms Applied to Predictive Toxicology Data Mining*. Proceedings of the SSCT& ETS Joint Conference Chemicals, 19–21, Human and Environment, Estonian Society of Toxicology, Anne Kahru (Ed.), ISSN 9949-13-297-5, Toila, Estonia.

Neagu D., Craciun M.V., Stroia S.A., Bumbaru S. (2005b) *Hybrid Intelligent Systems for Predictive Toxicology – A Distributed Approach*. 5th International Conference on Intelligent Systems Design and Applications ISDA2005, 26–31, Sept. 8–10, 2005, IEEE Computer Society Press, Wroclaw, Poland.

Niculescu S.P. (2003) Artificial neural networks and genetic algorithms in (Q)SAR. *J. Mol. Struct.*, 622, 71–83.

Payne T., Edwards P. (1997) Interface agents that learn: an investigation of learning issues in a mail agent interface. *Appl. Artif. Intell.*, 11, 1–32.

Pedrycz W. (1990) Fuzzy sets in pattern recognition: methodology and methods. *Pattern Recogn.*, 23, 121–146.

Pintore M., Piclin N., Benfenati E., Gini G., Chrétien J.R. (2002) Data mining with adaptive fuzzy partition: application to the prediction of pesticide toxicity on rats. *Environ. Toxicol. Chem.*, 22, 983–991.

Pintore M., Piclin N., Benfenati E., Gini G., Chrétien J.R. (2003a) Predicting toxicity against the fathead minnow by adaptive fuzzy partition. *(Q)SAR Comb. Sci.*, 22, 210–219.

Pintore M., Piclin N., Benfenati E., Gini G., Chrétien J.R. (2003b) Database mining with adaptive fuzzy partition (AFP): application to the prediction of pesticide toxicity on rats. *Environ. Toxicol. Chem.*, 22, 983–991.

Powell M.J.D. (1987) Radial basis function, in: *Algorithms for Approximation*, 143–167, Mason J.C., Cox M.G. (Eds). Oxford University Press, New York, USA.

Price N.R., Watkins R.W. (2003) Quantitative structure-activity relationships (QSAR) in predicting the environmental safety of pesticides. *Pesticide Outlook*, 14, 127–129.

Ripley B.D. (1993) Statistical aspects of neural networks, in: *Networks and Chaos: Statistical and Probabilistic Aspects*, 40–123, Barndorff-Nielsen O.E., Jensen J.L., Kendall W.S. (Eds). Chapman and Hall, London, UK.

Ros F., Pintore M., Chrétien J.R. (2002) Molecular description selection combining genetic algorithms and fuzzy logic: application to database mining procedures. *Chemom. Intell. Lab. Syst.*, 63, 15–26.

Ros F., Taboureau O., Pintore M., Chrétien, J.R. (2003) Development of CNS predictive models by adaptive fuzzy partitioning. *Chemom. Intell. Lab. Syst.*, 67, 29–50.

Sarle W.S. (1994) *Neural Networks and Statistical Models*. Proceedings of 19th Annual SAS User Group International Conference, 1538–1549, Dallas, TX, USA.

Sebastiani F. (2002) Machine learning in automated text categorization. *ACM Comput. Surv.*, 34, 1–47.

Singh S., Haddon J., Markou M. (1999) *Nearest Neighbour Strategies for Image Understanding*. Proceedings of the Workshop on ACIVS'99, Nigeria Falls, Ontario, Canada.

Sugeno M. (1985a) An introductory survey of fuzzy control. *Inform. Sci.*, 36, 59–83.

Sugeno M. (1985b) *Industrial Applications of Fuzzy Control*. Elsevier Science Pub. Co., Amsterdam, The Netherlands.

Sugeno M., Yasakawa T. (1993) A fuzzy-logic-based approach to qualitative modeling. *IEEE T. Fuzzy Syst.*, 1, 7–31.

Tickle A.B., Andrews R., Golea M., Dietrich, J. (1998) The truth ill come to light: directions and challenges in extracting the knowledge embedded within trained artificial neural networks. *IEEE T. Neural Netw.*, 9, 1057–1068.

Vračko M., Bandelj V., Barbieri P., Benfenati E., Chaudhry Q., Cronin M., Devillers J., Gallegos A., Gini G., Gramatica P., Helma C., Neagu D., Netzeva T., Pavan M., Patlevicz G., Randić M., Tsakovska I., Worth A. (2006). Validation of counter propagation neural network models for predictive toxicology according to the OECD principles: a case study, *SAR & QSAR Environ. Res.*, 17, 265–284.

Wang L.-X. (1992) Fuzzy Systems are universal approximators. *IEEE T. Syst. Man. Cyb.*, SMC-7, 1163–1170.

Zadeh L.A. (1975) Fuzzy Logic and Approximate Reasoning, *Synthese*, 30, 407–428.

Zadeh L.A. (1977) Fuzzy sets and their applications to classification and clustering, in: *Classification and Clustering*, 251–299, J. Van Ryzin (Ed.). Academic Press, New York, USA.

Quantitative Structure-Activity Relationships (QSAR) for Pesticide Regulatory Purposes
Edited by Emilio Benfenati

Chapter 5

Hybrid systems

Nicolas Amaury[1], Emilio Benfenati[2], Severin Bumbaru[3], Antonio Chana[2], Marian Craciun[3], Jacques R. Chrétien[1], Giuseppina Gini[4], Gongde Guo[5], Frank Lemke[6], Viorel Minzu[3], Johann-Adolf Müller[6], Daniel Neagu[5], Marco Pintore[1], Silviu Augustin Stroia[3], Paul Trundle[5]

[1]BCX, Biochemics Consulting SAS, Orléans, France

[2]Laboratory of Environmental Chemistry and Toxicology, Istituto di Ricerche Farmacologiche "Mario Negri". Milano, Italy

[3]Department of Computer Science and Engineering, University "Dunarea de Jos", Galati, Romania

[4]Department of Electronic and Information, Politecnico di Milano, Milano, Italy

[5]Department of Computing, School of Informatics, University of Bradford, Bradford, UK

[6]KnowledgeMiner, Berlin, Germany

1. INTRODUCTION: GOALS OF THE HYBRID SYSTEMS

The term 'hybrid system' dates back to the development of expert systems (Jackson, 1999) from the initial rule-based aspect to the modern modeling and interpretation systems. Most of the accent in the beginning has been on the idea of making use of more representations of the problem, more paradigms of knowledge representation, and more algorithms to find a solution.

A seminal work by Gallant (1993) introduced a way to look together to neural networks and rule-based systems. In his approach, a net, built from data and in absence of symbolic knowledge, is used to extract rules. This idea developed in the artificial intelligence (AI) community the well-known area of integrating connectionist and symbolic systems.

The starting machine learning community developed in the same years another way to make use of data in the absence of knowledge that led to the development of inductive trees, well exemplified by C4.5 (Quinlan, 1993). Integrating different representations and solutions is a direction taken in AI in the years around 1995. The term 'expert system' in those years was almost replaced by the term 'intelligent system' or 'intelligent agent'.

Using different representations to reach a common agreement or a problem solution led to the idea of using computational different methods on different problem representations, to make use of their relative strengths. Examples are the hybrid neural and symbolic learning systems (d'Avila *et al.*, 2002). Another kind of hybrid intelligent system (HIS) is the neuro-fuzzy system (Funabashi *et al.*, 1995) that combines connectionist and symbolic features in two directions:

- Symbolic information is inserted into a neural network, then training examples are used to refine the knowledge.
- From the trained network, symbolic information, in form of rules, is extracted.

Although the neural representation offers the advantage of homogeneity, distribution, and parallelization, and of working with incomplete and noisy data, the symbolic representation brings the advantages of human interpretation and knowledge abstraction (Neagu and Gini, 2003).

A fundamental stimulus to the investigations of hybrid systems is the awareness that combined and integrated approaches will be necessary to solve real-world problems using AI tools. Recently, HISs are becoming popular due to their capabilities in handling many real-world complex problems, involving imprecision, uncertainty, and vagueness, and high dimensionality, all of them well known in domains as financial prediction (Chen and Wang, 2004). Research in this area is very active following both the various traditional directions of neural networks integration with expert systems, with fuzzy systems, or with global optimization algorithms and the more recent tracks of integration of different learning paradigms (supervised/unsupervised/reinforcement), hybrid multi-agent systems, and the hybridization of soft computing with other machine learning techniques: support vector machines (SVMs), rough sets, Bayesian networks, probabilistic reasoning, and statistical learning.

Curiously, integrating different problem representations and solutions was a new formulation of another similar evolution that started in the pattern recognition community about composing classifiers. Again in this area, most of the intuitions started with a seminal work: the bagging of classifiers (Breiman, 1996; Avnimelech and Intrator, 1999).

During the past twenty years, in a variety of application domains, researchers in machine learning, pattern recognition, and statistics have explored and developed systems combining ensembles of classifiers (Bauer and Kohavi, 1999; Dietterich, 2000; Freund *et al.*, 2004).

Combining the predictions of a set of classifiers has shown to be an effective way to create composite classifiers that are more accurate than any of the component classifiers (Jain *et al.*, 2000). This research has the potential to apply accurate composite classifiers to real-world problems by intelligently combining known learning algorithms.

Research in the past ten years showed that both classification and regression problem ensembles are often much more accurate than the individual base learners that make them up. If one considers the metaphor of a committee of experts, the idea is that for some complex problems it may be better to combine relatively simple multiple experts with somewhat differing opinions rather than designing a single complex expert.

In literature, we can find at least two main streams deriving from 'ensembles' of highly correct classifiers that disagree as much as possible, and 'mixture of experts', built on the idea to train individual networks on a sub-task, and then combine their predictions with a 'gating' function that depends on the input. Basic combinations as majority vote or average of continuous outputs are sometimes effective. Finally, it is possible to train the output classifier separately using the outputs of the input classifiers as new features.

There are many methods for combining the predictions given by component classifiers. Although in the literature terms as 'voting' (Bauer and Kohavi, 1999), 'combination' (Kittler *et al.*, 1998; Ho, 2002), 'ensemble' (Krogh and Vedelsby, 1995), and 'mixture of experts' (Jacobs *et al.*, 1991; Singer, 1997) are used to indicate sets of learning machines that work together to solve a problem, we use here the term 'hybrid systems' to include the whole aspects of integrating local experts and combining their different knowledge representations.

2. OUR HYBRID APPROACH FOR QUANTITATIVE STRUCTURE-ACTIVITY RELATIONSHIP

Quantitative structure-activity relationship (QSAR) problems do not have, in general, linear solutions, and the problem is how to model those situations, at least in the case we have enough data. Another important point is that the non-linear model should not be assumed but should emerge from data analysis. In making a good model from data, we need to carefully check our data for signal-to-noise ratio and for quality. On the other end, domain knowledge is not enough in this field (and this is an important difference from the engineering area) to predict the behavior of the system but can eventually serve to constrain the solution.

As widely known in the QSAR literature, no single method can be considered as the only way to predict toxicity (Benfenati *et al.*, 2002). Several methods can give good predictions in a comparable way, because each approach can extract and represent some parts of the knowledge. Examples of application of those concepts in chemometrics are appearing in literature (Merkwirth *et al.*, 2004).

In the present investigation, we integrate the best models individually developed for each endpoint into a hybrid system for that endpoint. This has to be flexible to accept further inputs or modules, if available. Whereas inputs to the basic models are the chemical descriptors, input to the hybrid model are the n values predicted for each molecule by the n integrated models; the output is always the toxicity for that molecule.

Starting from basic combination strategies, we extend the concept of 'ensembling' different models order to build a model with the maximum possible value for our application. We employ methods from pattern recognition to AI, including attention to the statistical meaning of the result and on the knowledge level of the proposed combination. Instead of concentrating on building the best expert, we combine some good experts that are accurate and conceptually different, so they make different errors.

We take, as the basic measure of the value of our hybrid model, the model obtained averaging the single component models. The average model is always an improvement of the basic models because it reduces the variance of the error (Bauer and Kohavi, 1999). Other kinds of hybrid models are then built and checked against the average model and retained only if they are doing better.

As an help in building and evaluating the ensemble, we used a graphical representation provided by the regression error characteristic (REC) curves and illustrated in Chapter 6 (Benfenati *et al.*, this volume).

Regarding the ensembling techniques, we follow two major streams:

- Using a symbolic system to combine the models. In this case, we try to modify the weight of the component models according to the diverse areas of the predicted toxicity.
- Using a gating network, in the line indicated in the literature about mixtures of experts, which use trained individual networks and combine their predictions with a 'gating' function that depends on the input.

In the rest of this chapter, we will illustrate the basic theory behind those combinations as well as the models obtained.

3. GATING NETWORKS

3.1. Introduction

Starting with basic combination strategies, the concept of ensemble development based on gating networks is introduced below, to include also the knowledge level of the proposed combination. Instead of focusing on building the best expert for the whole domain, we propose combinations of some good local experts that conceptually are experts on different parts of the domain.

The first choice (Benfenati *et al.*, 2002) are mixtures of experts (Jacobs *et al.*, 1991), which used trained individual networks and combined their predictions with a 'gating' function depending on the input. A strong point that makes combinations of classifiers attractive, besides the better results in classification performance, is their ability of being distributed in time and space. Subsets of the data are distributed to different processors, that apply a learning algorithm, and then results are combined to yield a single classifier. This work is motivated by two goals: to reduce the execution time and to improve accuracy.

The reduction in execution time results from the distribution of the expensive learning step to multiple processors, a research area of increasing interest because of the latest developments in parallel computing. The improvement in accuracy comes from the fact that each of the classifiers tends to make different types of errors, and the consensus opinion of multiple classifiers is often more accurate than that of a single classifier (Ho, 2002).

Harmonizing pattern recognition/AI terminology, we introduce various kinds of combinations, in particular HISs based on neuro-fuzzy architectures (Fuller, 1999). HISs combine connectionist and symbolic features in two directions:

- Symbolic information is inserted into a neural network, then training examples are used to refine the knowledge.
- From the trained network, symbolic information, in form of rules, is extracted.

As the neural representation offers the advantage of homogeneity, distribution, and parallelization, and of working with incomplete and noisy data, the symbolic representation comes with the advantages of human interpretation and knowledge abstraction. In our work, explicit and implicit rules are represented using fuzzy neural networks, hybrid neural networks (HNNs) (Buckley and Hayashi, 1995; Neagu and Gini, 2003), multi-layer perceptrons (MLPs) (Rumelhart and McClelland, 1986), or neuro-fuzzy nets (Neagu and Palade, 2000).

In recent literature, one can find different ways to use more classifiers in a recognition problem (Kittler *et al.*, 1998), at least two main streams deriving from 'ensembles' of highly correct classifiers that disagree as much as possible, and 'mixture of experts' (Jacobs *et al.*, 1991), built on the idea to train individual networks on a sub-task, and then combine their predictions with a 'gating' function that depends on the input. Basic combinations as majority vote (Bauer and Kohavi, 1999), or average of continuous outputs, are sometimes effective. Finally, it is possible to train the output classifier separately using the outputs of the input classifiers as new features, as proposed and discussed in Krogh and Vedelsby (1995), Benfenati *et al.* (2002), Chan and Stolfo (1995), and Ho (2002).

Consequently, gating networks are defined as mixtures of experts' decision outputs d_j, $j = 1 \dots L$ with a weighted voting contribution, where the weights w_j are input dependent:

$$y = \sum_{j=1}^{L} w_j d_j. \tag{1}$$

In this section, we study the development of different combining methods as gating networks. For selected real data, both supervised and unsupervised learning paradigms are applied to cluster them. Individual classifiers are trained for each class and further combined for hybrid predictions. The 'gating' function is then a classifier that learns how to allocate examples to the experts, making the assumption that a single expert is responsible for each example (Jacobs *et al.*, 1991).

3.2. Gating networks for predictive toxicology – a new approach based on descriptors clustering

In predictive data mining, the regression-based modeling problem has the goal to induce an approximation function from the feature space to the target values, based on meta-data pairs of type (<input>, <target>). The aim (Craciun *et al.*, 2004) is the approximation of a mathematical function to relate input variables to their corresponding output value:

$$f : D \subseteq \boldsymbol{R}^n \rightarrow \boldsymbol{R}, \tag{2}$$

where $n \in N$ is dimension of the input space.

There is no information whether this mapping relation is a continuous function or not. Sometimes, from theoretical and experimental considerations of the domain experts, it can be assumed that the function could be continuous on different sub-domains of the domain D; these sub-domains represent collections of similar entities (e.g., the chemical compounds in the same chemical class). The function f will be

$$f(x) = \begin{cases} f_1(x), & x \in D_1 \\ f_2(x), & x \in D_2 \\ \vdots & \\ f_K(x), & x \in D_K \end{cases} \quad (\forall) x \in D, \tag{3}$$

where $D_i \subseteq \boldsymbol{R}^n$, $D = D_1 \cup D_2 \cup \ldots \cup D_K, D_i \cap D_j = \Phi$, for $i, j = 1, \ldots, K$ and $f_i = f/D_i$, are the continuous projections of the function f on the sub-domains D_i, for $i = 1, \ldots, K$.

The original complex problem is now separated in K sub-problems: the approximation of K functions $(f_i : D_i \subseteq \boldsymbol{R}^n \rightarrow \boldsymbol{R}), i = 1, \ldots, K$, with a more comprehensible and easy to manage and interpret behavior on their domains, as the data in the same sub-domain would converge to similar characteristics.

One of the difficulties and main challenges in predictive toxicology is the increasing number of chemical compounds to be modeled and included in further models. Moreover, the chemical structure of the classified compounds can involve, for example, various radicals, on which one can assign a particular compound to a class or another. Thus, the modeling algorithm would be at most a good approximation of a dynamically changing collection of training and testing sets. Consequently, the aim of our modeling approach is that, for every such a target function f_i, $i = 1, \ldots, K$, to find a good approximation in the sense of the uniform topology on D_i with an error less than ε_i, given by a function F_i, such as

$$||f_i - F_i||_\infty = \sup_{x \in D_i} |f_i(x) - F_i(x)| < \varepsilon_i. \tag{4}$$

Therefore, based on Eqs. (3) and (4), follows

$$||f - F||_\infty = \sup_{x \in D} |f(x) - F(x)| < \varepsilon = \max_{i=1,\ldots,K} \varepsilon_i \tag{5}$$

with

$$F(x) = \begin{cases} F_1(x), & x \in D_1 \\ F_2(x), & x \in D_2 \\ \vdots & \\ F_K(x), & x \in D_K \end{cases} (\forall) x \in D. \tag{6}$$

According to Eq. (5), the function F given by Eq. (6) is an approximation for f in the sense of uniform topology on D with a global error less than ε. This function F is the global approximation modeling function we were looking for.

Unfortunately, in real-world applications, the modeling problem is more complex. There are at least two further practical difficulties to surpass: the procedures the sub-domains D_i are described by and the algorithms based on which the local approximation functions F_i are generated, $i = 1, \ldots, K$ (Fig. 1).

For the first task, the domain modeling, a possibility is (Craciun *et al.*, 2004) a self-organizing computational network-based approach to cluster similar chemical compounds of the training set in the sense of the Euclidian distance in the descriptor space. For the second task, the models development, multi-linear regression, standard feed-forward back-propagation neural networks with one hidden layer and regression trees have been used.

Three-layered feed-forward back-propagation artificial neural networks (ANNs) have the capacity to approximate with any degree of accuracy a continuous function (Funahashi, 1989), providing that a sufficient number of hidden layer neurons is used (Hornik *et al.*, 1989). This well-known ability of feed-forward connectionist structures suits the modeling constraints expressed in Eqs. (3) and (4). Regression trees have similar capabilities with ANNs for approximating discrete-valued target functions, as according to results presented in Mitchell (1997). Moreover, because any continuous function can be approximated by a discrete-valued function (e.g., step functions), it can be assumed that a regression tree could approximate any continuous function with a reasonable accuracy.

Our approach (Craciun *et al.*, 2004) comes to take into consideration also the lack of data and the imperfection (noise) of available data (Schultz and Cronin, 2003). Consequently, any machine learning algorithm is not capable to build a

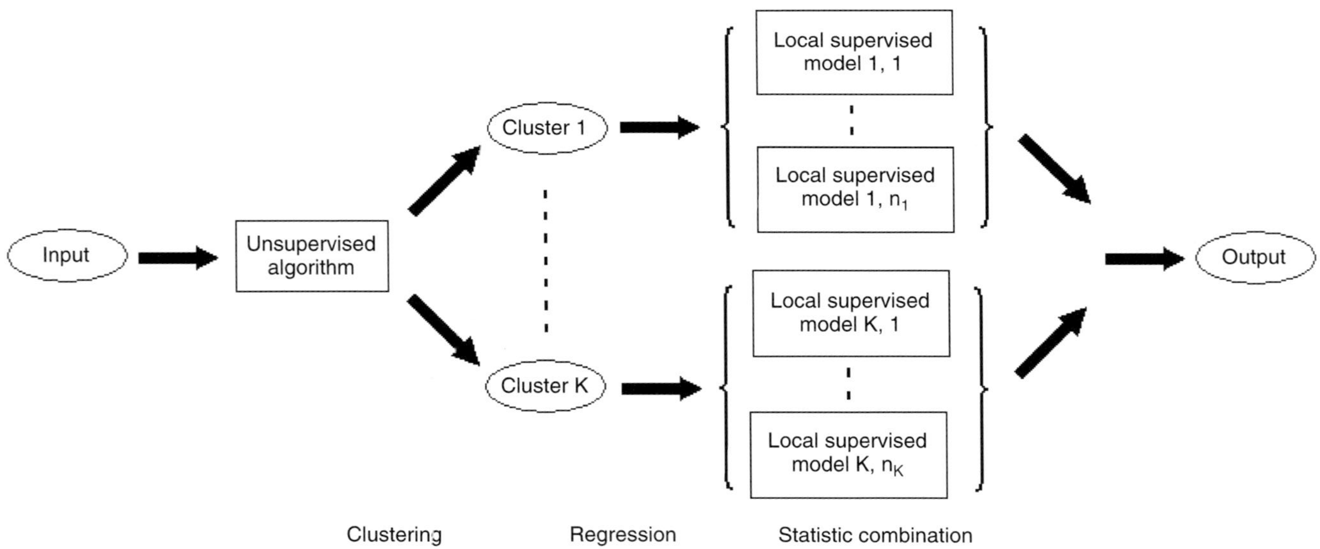

Figure 1: Model development based on supervised and unsupervised machine learning methodologies.

best approximation function. Additionally, every machine learning algorithm has its own strengths and drawbacks to increase the difficulties of building a best approximation and prediction tool and also to decrease the error and re-tuning further requirements. Results of combining two or more models built on the same training data to increase their power and to reduce their drawbacks instead on concentrating on building the best model for the whole domain or for the local sub-domains (Neagu and Palade, 2002) are reported in Sections 5.1 and 5.2 of this chapter and in Chapter 7, Amaury *et al.*, this volume.

3.3. Hybrid neural fuzzy systems

Hybrid neural/fuzzy architectures use smaller and modularized experts [neural nets and mapped fuzzy inference systems (FISs)], whose solutions are dynamically integrated through a gating network. Although multi-agent architectures developed for AI aim at better expressing knowledge, we focus more on improving performances of models.

We define (Neagu and Palade, 2000) *explicit knowledge* (EK) as a knowledge base represented by neural networks computationally identical to a fuzzy rules set and created by mapping given fuzzy rules into HNNs. The intrinsic representation of EK is based on fuzzy neurons. We define also *implicit knowledge* (IK) as a trained network, where the weights implicitly represent the knowledge gained from data.

As a standard neural network employs multiplication, addition, and sigmoid transfer function, the HNN is a neural structure with crisp signals and weights, and crisp transfer function, using T-norms to combine the inputs (usually membership degrees of a fuzzy concept), weights over the unit interval, T-conorms to aggregate the results, and any continuous function from input to output as a transfer function.

We studied various strategies to combine IK module (IKM) and EK module (EKM): fire each module (FEM), unsupervised-trained gating network (UGN), and supervised-trained gating network (SGN) (Neagu and Palade, 1999, 2000).

- Fire each module with two versions: statistical FEM (FEMS) uses a statistical function to integrate the consensus of all the implied experts, whereas fuzzy FEM (FEMF) uses a fuzzy expert, through an aggregation function, and a defuzzification process. FEM is an adapted fire each rule method (Buckley and Hayashi, 1995) for modular networks, in two versions: statistical combination of crisp outputs (FEMS) and fuzzy inference of linguistic outputs (FEMF).

- Unsupervised-trained gating network, a competitive-based aggregation of the EKM and IKM outputs. The architecture uses neural EKM and IKM, and the gating network votes for the best aggregation of fuzzy terms, computed by expert nets, to describe the linguistic output. EKMs and IKMs are already trained. The gating network is trained with the constraint to have as many output neurons as fuzzy terms describing the linguistic variable Y, the output

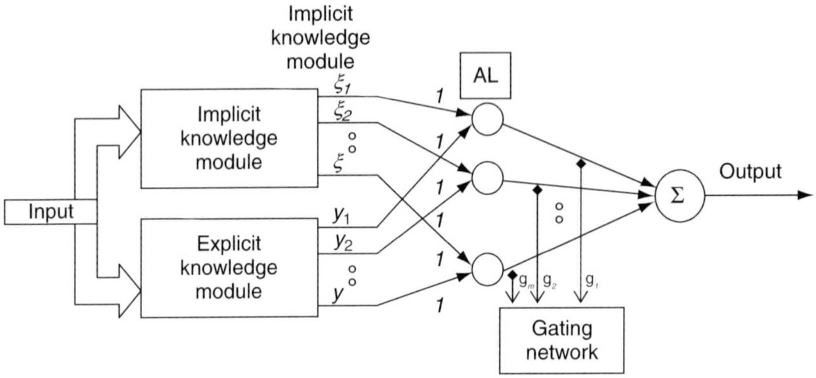

Figure 2: Integration of explicit knowledge module (EKM) and implicit knowledge module (IKM) using unsupervised-trained gating network (UGN).

of the global network. The gating network is a single layer of *m* output neurons, each one having *m* inputs. The activation function of its output neurons is *softmax* (Fig. 2).

- Supervised-trained gating network – although the second strategy proposes competitive aggregation of EKMs and IKMs, the SGN uses a supervised trained layer to process the overall output of modules (Negoita *et al.*, 2005). The structure contains expert networks represented by EKMs and IKMs solving various sub-problems of the main task and a supervised trained network mediating their outputs' combination. After training, the expert networks compute different functions, each of them mapping different regions of the input space. Each defuzzified output of expert networks is an input for the final layer. The supervised training process of the final network assures a weighted aggregation of expert networks' outputs with respect to their specialization (Fig. 3).

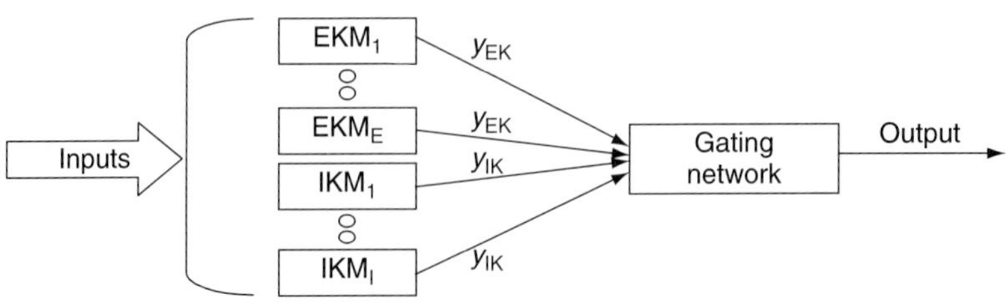

Figure 3: Integration of explicit and implicit modules using a supervised-trained gating network (SGN).

All these models are considered for knowledge representation in HISML – the mark-up language used by the prototype NIKE introduced later in Section 5 of this Chapter and defined further in Chapter 9, Benfenati *et al.*, this volume.

3.4. Gating networks as HISs – a data-driven approach

Let's consider a multi input single output (MISO) HIS with n inputs. Let also be considered $U = \prod_{i=1}^{n+1} D_i$, the universe of discourse over the application domain as the Cartesian product of sets $D_i,\ i = 1, \ldots, n+1$, having the input variables $X_i \in D_i,\ i = 1, \ldots, n$, and the output $Y \in D_{n+1}$.

An HIS as an integrated model of the problem Φ based on IKMs and EKMs is a good approximation of Φ as defined by

$$HIS = \left\{ \underset{j=1,\ldots,m}{M_j} \ / \forall \varepsilon > 0, \exists X \in \prod_{i=1}^{n} D_i, \forall Y = \Phi(X) : \left\| M_j(X) - Y \right\| < \varepsilon \right\}, \qquad (7)$$

where the knowledge modules are functional models $M_j : \prod_{i=1}^{n} D_{ij} \rightarrow D_{n+1,j}$.

The modules M_j are, in our approach, either IK or EK models $M_j \in \{M_{IKM\_CNN}, M_{IKM\_FNN}, M_{EKM\_Mamdani}, M_{EKM\_Sugeno}\}$. For any of these models, based on the connectionist homogeneous implementation of any M_j model, we propose a formal parameter-based description of HIS:

$$M_j = \langle \Theta, \Lambda, \Omega \rangle, \qquad (8)$$

where Θ is the set of *topological parameters* (i.e., number of layers, number of neurons on each layer, and connection matrices) of individual models and also of general structure (type and number of individual models and gating networks), Λ is the set of *learning parameters* (learning rate, momentum term, any early stopping attribute for IKMs, but NIL for EKMs), and Ω is the set of *description parameters* (defining for any fuzzy model number and type of fuzzy sets, and parameters of membership functions associated to linguistic variables).

Three distinctive cases to develop further integrated models are identified.

Case 1: $D_j = \prod_{i=1}^{n} D_{ij}$, for all $j = 1, \ldots, m$. The model is a modular architecture (Neagu and Palade, 2002) combining experts on the whole input domain.

Case 2: $\bigcap_{j=1}^{m} \prod_{i=1}^{n} D_{ij} = 0$ and $D_i \cap D_j = 0$, for $j, k = 1, \ldots, m$. The HIS model is a collection of m expert models on disjunctive input domains; the system is a top-down integrated decomposition model, by dividing the initial problem in separate less-complex sub-problems.

Case 3: $\bigcap_{j=1}^{m} \prod_{i=1}^{n} D_{ij} \neq 0$. The models are built on overlapping sub-domains and further algorithms to refine the problem as case 1 or 2 is required (Neagu and Palade, 2002).

4. MULTI-CLASSIFIER SYSTEMS

Decision-making occurs in a wide range of human activities. At its broadest, the term could cover any activity in which some decision or forecast is made on the basis of currently available information, and a classifier is then some formal method for repeatedly making such judgments in new situations (Michie *et al.*, 1994). Various approaches to classification have been developed and applied to real-world applications for decision-making. Examples include probabilistic decision theory, discriminant analysis, fuzzy-neural networks (Mazzatorta *et al.*, 2003), belief networks, non-parametric methods (Mitchell, 1997), tree-structured classifiers, and rough sets (Pawlak, 1991).

Unfortunately, no dominant classifier exists for all the data distributions, and the data distribution of the task at hand is usually unknown. A single classifier cannot be discriminative enough if the number of classes is huge. For applications where the classes of content are numerous, unlimited, and unpredictable, one specific classifier cannot solve the problem with a good accuracy. In predictive toxicology, data describing multiple facets of relationship between chemical structure and biological activity, from various sources, and on different chemical classes are used. Single classifiers are not necessarily able to cover the whole complexity of required models: therefore, the approach which proposes combinations of experts could discover easier ways to represent domains.

A multiple classifier system (MCS) is a powerful solution to difficult decision-making problems involving large sets and noisy input because it allows simultaneous use of arbitrary feature descriptors and classification procedures (Ho *et al.*, 1994). The ultimate goal of designing such an MCS is to achieve the best possible classification performance for the task at hand. Empirical studies have observed that different classifier designs potentially offer complementary information about the patterns to be classified, which could be harnessed to improve the performance of the selected classifier (Baykut and Ercil, 2003).

Many different approaches have been developed for classifier combination. Examples include majority voting (Nadal *et al.*, 1990), entropy-based combination (Saerens and Fouss, 2004), Dempster–Shafer theory-based combination (Zhang and Srihari, 2002; Bi *et al.*, 2004), Bayesian classifier combination (Xu *et al.*, 1992), and statistical models (Ho *et al.*, 1994).

Various modeling and prediction techniques from statistics and AI are available and applied to QSAR data: Bayesian classifiers, decision trees (DTs) and classification rules, expert systems, ANNs, FIS, or hybrid approaches like neuro-fuzzy models. There were also proposals like competitive or cooperative combinations of two or more (global or local) techniques, rather than using a single algorithm to build the solution. Starting from basic combination strategies – simple voting or averaging – researchers tried to extend the concept of ensemble development from pattern recognition to AI, including attention to the statistical meaning of the result and on the knowledge level of the proposed combination. The meta-learning concept, combining prediction of trained individual experts – or level 0 models – with a gating function – or level 1 model, is representative for the mixture of experts, adaptive techniques, or stacking strategies.

In this section, we first introduce the existing approaches in the literature for MCSs and then illustrate three key components of an MCS. Some experimental results of an implemented multiple classifier combination system applied to toxicology data mining will be reported later in Chapter 7 (Amaury *et al.*, this volume).

4.1. Approaches for multi-classifier systems

Given a set of classifiers $\boldsymbol{C} = \{\boldsymbol{C}_1, \boldsymbol{C}_2, \ldots, \boldsymbol{C}_L\}$ and a data set $\boldsymbol{D}$, each instance $\boldsymbol{x}$ in $\boldsymbol{D}$ represents as a feature vector $[x_1, x_2, \ldots, x_n]^T$, $\boldsymbol{x} \in \Re^n$. A classifier gets as its input $\boldsymbol{x}$ and assigns it to a class label from Ω, that is, $\Re^n \rightarrow \Omega$.

Four approaches are generally used to design a classifier combination system (Kuncheva, 2001) that is presented in Fig. 4.

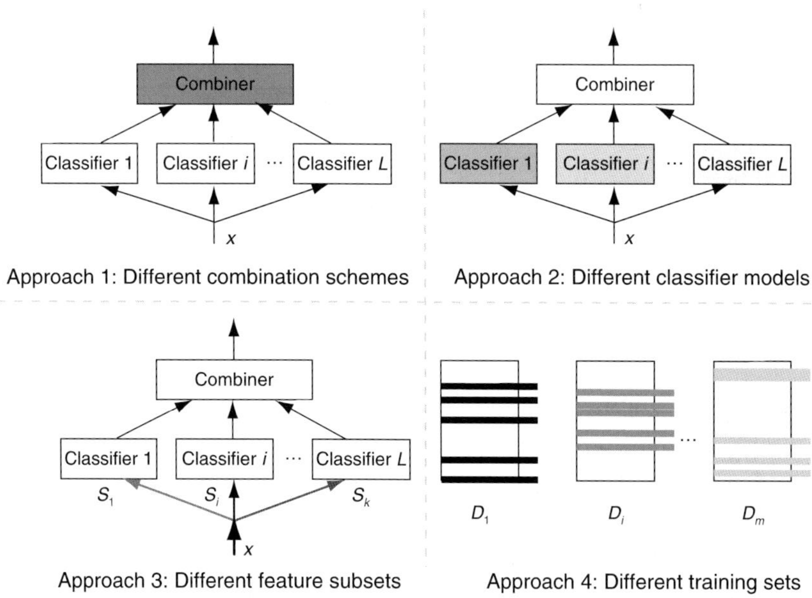

Figure 4: Four approaches to designing a classifier combination system.

Each approach is briefly described as follows (Kuncheva, 2001):

Approach 1: picking a combination scheme for $\boldsymbol{L}$ classifiers $\boldsymbol{C}_1, \boldsymbol{C}_2, \ldots, \boldsymbol{C}_L$ studied to form a combiner.

Approach 2: choosing individuals by considering the issues of similarity/diversity, homogeneous/heterogeneous, and so on.

Approach 3: building each $\boldsymbol{C}_i$ on an individual subset of features (subspace of $\Re^n$).

Approach 4: selecting training subsets $\boldsymbol{D}_1, \boldsymbol{D}_2, \ldots, \boldsymbol{D}_m$ of the data set $\boldsymbol{D}$ to lead to a team of diverse classifiers.

4.2. An architecture of MCS

An MCS generally comprises three key components: data pre-processing, data classification, and classifier combination, which are presented in Fig. 5.

Data pre-processing consists of feature selection (FS), noise removal (NR), and data normalization (DN), which will be uniformly applied to training, validation, and classification. Data classification classifies instances using each individual classifier, that is, DT, instance-based learner (IBL), repeated incremental pruning to produce error reduction (RIPPER), MLP, and SVM. Classifier combination employs different strategies to integrate various classifiers, such as majority voting-based combination (MVC), maximal probability-based combination (MPC), average probability-based combination (APC), and weighted probability-based combination (WPC) – details described below in this chapter, Section 4.4 – to combine individual classifier results to achieve potentially better performance. All three components together make the MCS a valid and practical approach (Fig. 5).

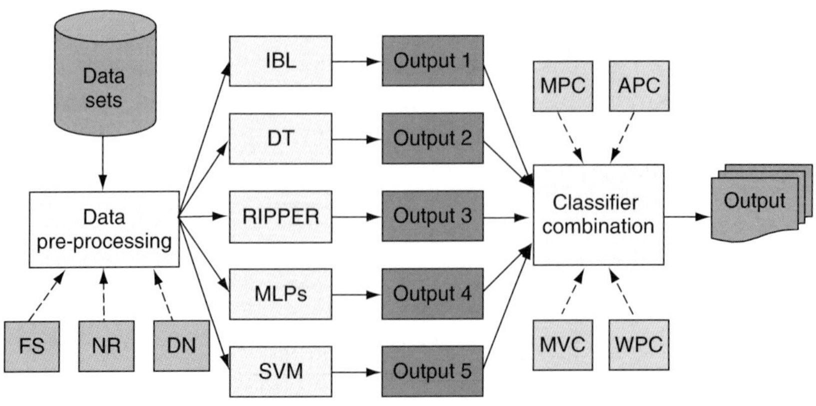

Figure 5: An architecture of multiple classifier system (MCS).

4.3. Classifiers

Many classifiers are available for decision-making such as SVMs (Vapnik, 1995), Bayesian networks (Lucas, 2001), DTs (Mitchell, 1997), ANNs (Ripley, 1996), fuzzy neural networks (Liu and Li, 2004), IBL, and rule-based learner.

Here, we give a brief introduction of five classifiers applied to predictive toxicology data mining. The experimental results will be reported in Chapter 7 (Amaury *et al.*, this volume):

- Support vector machine is based on the *structural risk minimization* principle from statistical learning theory. Given a training set in a vector space, SVM finds the best decision hyperplane that separates the instances in two classes. The quality of a decision hyperplane is determined by the distance (referred as margin) between two hyperplanes that are parallel to the decision hyperplane and touch the closest instances from each class.

- Decision tree is a widely used classification method in machine learning and data mining. The DT is grown by recursively splitting the training set based on a locally optimal criterion until all or most of the records belonging to each of the leaf nodes bear the same class label.

- Instance-based learners classify an instance by comparing it with a set of pre-classified instances and choose a dominant class of similar instances as the classification result.

- Repeated incremental pruning to produce error reduction is a propositional rule learning algorithm that performs efficiently on large noisy data sets. It induces classification (if–then) rules from a set of pre-labeled instances and looks at the instances to find a set of rules that predict the class of earlier instances. It also allows users to specify constraints on the learned if–then rules to add prior knowledge about the concepts to get more accurate hypothesis.

- Multi-layer perceptrons are feed-forward neural networks with one or two hidden layers, trained with the standard back-propagation algorithm. They can approximate virtually any input–output map and have been shown to approximate the performance of optimal statistical classifiers in difficult problems.

4.4. Combination Methods

The goal of designing an MCS is to achieve the best possible classification performance for the task at hand by using the expertise of existing individual classifiers. It has been observed that different classifier designs potentially offer complementary information about the pattern to be classified. Many classifier combination methods exist in the literature (Kuncheva, 2001). Here we only introduce four basic combination schemes for the illustration purpose.

1) Majority voting-based classifier combination.
Given a new instance x to be classified, whose true class label is $t_x \in C = \{c_1, c_2, \ldots, c_m\}$, and k predefined classifiers are denoted as $A_1, A_2, \ldots, A_k$, respectively, the classifier A_i approximates a discrete-valued function $f_{A_i}: \Re^n \rightarrow C$. The final class label of x, obtained by using majority voting-based classifier combination, is described as follows:

$$f(x) \leftarrow \arg\max_{c\in C} \sum_{i=1}^{k} \delta(c, f_{A_i}(x)), \tag{9}$$

where $\delta(a, b) = 1$, if $a = b$, and $\delta(a, b) = 0$, otherwise.
With the same aforementioned assumption, the classification result of x classified by A_j is given by a vector of probability values of x to each class, represented by $P = <P_{j1}, P_{j2}, \ldots, P_{jm}>$, where $j = 1, 2, \ldots, k$. The final class label of x can be obtained in four different ways.

2) Maximal probability-based combination:

$$f_1(x) \leftarrow \arg\max_{c_v\in C}\{\max\{P_{uv}\,|u = 1, \ldots, k\}\,|v = 1, \ldots, m\}. \tag{10}$$

3) Average probability-based combination:

$$f_2(x) \leftarrow \arg\max_{C_v\in C}\left\{\sum_{u=1}^{k}\left(\frac{P_{uv}}{k}\right)\middle|\, v = 1, 2, \ldots, m\right\}. \tag{11}$$

4) Weighted probability-based combination:

$$f_3(x) \leftarrow \arg\max_{c_v\in C,\alpha}\{\{\alpha \times \max\{P_{uv}|u = 1, 2, \ldots, k\} + (1-\alpha) \times \sum_{u=1}^{k}\left(\frac{P_{uv}}{k}\right)\{|v = 1, 2, \ldots, m\}, \quad 0 \leq \alpha 1, \tag{12}$$

where α is a control parameter used for setting the relative importance of local optimization and global optimization of combination.

In Eq. (12), if $\alpha = 0$, then $f_3(x) = f_1(x)$. Given a new instance x, the MPC only considers the fitness of each individual classifier for x and chooses the best classifier's output as the combination result. This is a local optimal combination as it only considers the best classifier for each instance to be classified.

When $\alpha = 0$, we obtain $f_3(x) = f_2(x)$. Given a new instance x, the APC averages all the classification results assigned to each class and then outputs the class label with the highest average probability as the combination result. This is a global optimal combination as it takes all the classification results into account.

When $0 < \alpha < 1$, the WPC uses α to control the balance between the local optimization and global optimization.

The experimental results of the aforementioned MCS applied to predictive toxicology data mining will be reported in Chapter 7 (Amaury *et al.*, this volume).

4.5. Distributed multi-classifier systems

A strong point that makes combinations of classifiers attractive, besides the better results in classification performance, is their ability of being distributed in time and space: inter- and intra-model parallelism and data and processing parallelism. Subsets of the data could be distributed to different processors, that apply different learning algorithms, and then results are combined to yield a single classifier.

The idea to build distributed multi-classifier system combining simple and HISs arose from the need to train ANNs (consider a heavy computing job) using ordinary, network-connected computers available in almost every laboratory. The same strategy suits very well to complex multi-classifier systems, hybrid intelligent predictive systems, and general data mining tasks, too (Neagu *et al.*, 2004). It is generally difficult for a single processor system to provide reasonable training time in the cases discussed in the Sections 3 and 4 of this chapter. Parallel processing in general and parallel data mining in special offer new alternatives as they incorporate techniques from parallel databases and parallel programming.

The proposed system (Neagu *et al.*, 2005) acts as a virtual multi-processor computer with a single entry point and concentrates on the computing power of the machines found in the network to sum up their computing power. Depending on the problem context, there are different forms of parallelism in data processing. From a view point of the model built, it can be distinguished: inter- and intra-model parallelism (Small and Edelstein, 1997). From a view point of data or process, there are data parallelism and control parallelism (Freitas, 1998). The first version of the proposed distributed engine provides the possibility of using all types of parallelism, but the user is to decide how to split his data sets and jobs into an array of individual tasks to feed the system and to receive the results. The architecture of the virtual computing machine is based on a loosely coupled multi-processor system with a 'first come first served' scheduling policy (Fig. 6).

The components of the distributed computing engine are the following (Neagu *et al.*, 2005).

- An arbitrary number of *computing units*: individual nodes of the distributed computing engine. The nodes of the system could run different operating systems on heterogeneous hardware platform.

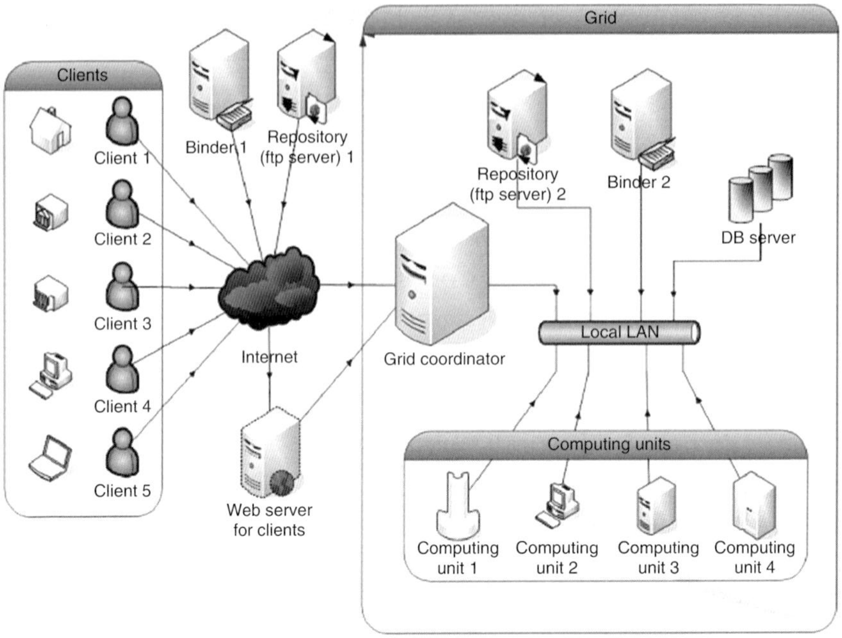

Figure 6: The architecture of the distributed computing engine.

- A *coordinator* machine – for the activity of all the computing units of the engine and to supervise the allocation of the tasks for each node. Into more complex grid architectures, it is possible to have more than one coordinator.
- The *users* of the computing engine: to provide to the system an array of synchronous or asynchronous atomic tasks. These tasks will be allocated by the coordinator to various computing units for processing. Results of each task will be collected by the coordinator and presented to the users.
- Two *binder* machines: to register the coordinators and the computing units, respectively. The coordinators have to register themselves to the correspondent binder and the client users will interrogate it to obtain the list of the coordinators. The computing units have to register themselves to the second binder and the coordinator will interrogate it to obtain the list of all available computing units.
- Two *repository* computers: the storage units where the mobile code will be stored and retrieved when necessary using various protocols (e.g., HTTP and FTP). The first repository is used to transfer the *bytecode* from users to coordinator and the second from coordinator to computing units.
- A Web server to facilitate the access of users to coordinator through a user friendly interface via a Web browser.

- A *data base server* to store useful data and activities related to users, tasks, and so on.

- An important advantage of this modular distributed approach is flexibility of developed models, able to be optimized for specific components (sub-models). Future research steps target the improvement of the overall performance of the solution, based on local developed models. Another advantage of the system is its reusability, not only in predictive toxicology, data mining, or machine learning: any synchronous or asynchronous parallel processing tasks can be easily adapted and executed.

5. NEURAL IK- AND EK-BASED SYSTEMS – INTRODUCTION OF THE PROTOTYPE NIKE

A direction in the software development during the project has paid attention to an easy and friendly integration of data and models in hybrid systems as a generalized tool to study, save and use HISs for complex model building in predictive toxicology. A general framework to allow import and export of various data sets and soft computing models has been proposed (Neagu *et al.*, 2004): two XML schemas for predictive toxicology (PToxML) and HISs (HISML) are available on the DEMETRA server and used by our integration software NIKE. Initial studies using NIKE regarding the integration of the different predictive models show a better performance of hybrid models than individual models.

Using NIKE provides the developer the opportunity to combine under the paradigm of HISs the following modules: neural networks, fuzzy neural networks, Mamdani and Sugeno FIS, and other generalized polynomial models. One of the issues addressed, given the variety of techniques and algorithms used into DEMETRA framework, is the possibility to implement QSARs using simple and complex HIS models available via our prototype. We therefore propose, for example, users to simulate QSARs developed using property software (others than those implemented in our prototype, NIKE) by training ANNs (crisp or fuzzy versions) to approximate available models. The case of importance of topological, learning, and descriptive parameters for the development of hybrid systems has been also addressed (Neagu *et al.*, 2004).

NIKE prototype interfaces are based on the following functions:

- the architecture of the entire system, based on an original approach to integrate data sets, distributed models and complex predictive hybrid systems;

- object-oriented (OO) description of core modules/classes;

- the structure of the input and output files – based on PToxML and HISML files and also available via user-friendly wizards;

- the implementation of ANNs, FIs, and neuro-fuzzy networks as simple HIS; and
- the implementation of other machine learning algorithms for complex HIS.

The software development followed an individual route, based on our original approach of modular HISs based on EKMs and IKMs (NIKE) as described in Section 3.3. This approach represents the core of this property software developed by University of Galati and University of Bradford for the DEMETRA project, based on OO analysis (OOA), OO design (OOD), and Java implementation. Each collection of knowledge algorithms are represented as text-structured PToxML and HISML files. The end-user has the opportunity to build knowledge databases, combining the project-developed algorithms as well as his/her private ones.

The management and control of the project is possible due to a list of the HISs built (and trained) using NIKE facilities and permits toxicity data and predictive modules (simple or complex HIS) loading. Also, the prototype presents a main interface between the system functionalities and the user of the system (Fig. 7).

As it was already presented in Section 4.4 of Chapter 2, Benfenati *et al.*, this volume, there is a meta-representation of the information exchanged between the NIKE modules: PToxML for toxicology data and HISML for predictive models. The information is grouped in three sub-structures:

1) data set – to contain some information about the *endpoint* and *number of molecules* about the chemical set used for training and evaluation;

2) descriptors – to resume some info about the *starting set of* descriptors and the list of *selected descriptors*;

3) models – details about *authors (and reference person)* who built the models, the name of the software used for calculus (*software* field), and few *performance* characteristics: R^2 for training and external test set, ${R_{CV}}^2$, and the *file with the output values for the model.*

As it can be easily seen, most of these structures are already comprised in the HISML structure, but some of them could be taken in consideration for future versions if it is necessary.

For a schematic description of the specifications, our software development process is based on the following.

Main objective: the development of a software system to give a quantitative prediction of the toxicity of a molecule, in particular molecules of pesticides, candidate pesticides, their derivatives, and related compounds having in mind a robust

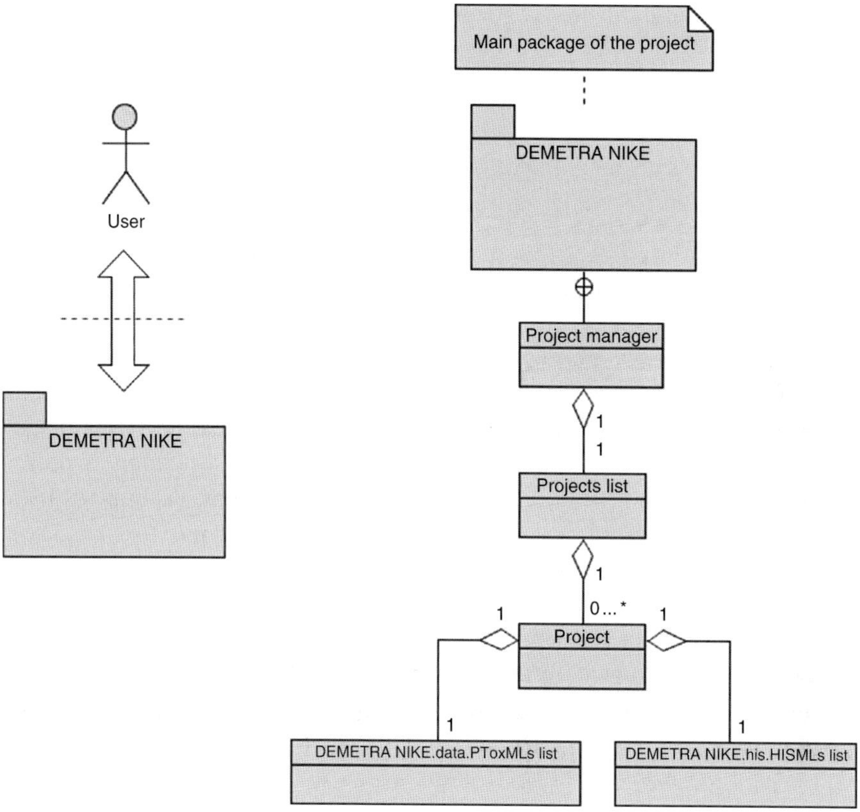

Figure 7: The management of the project – class diagram.

detailed characterization of the ecotoxicological properties of different chemical structures, their modes (mechanisms) of actions, and their effects against different organisms with a specific way of administration and a specific exposure time.

Inputs: the structure and properties of the chemical compounds (molecules) described by many molecular descriptors.

Output: the toxic dose against some life form of the chemical studied measured in milligrams per kilogram (mg/kg), milligrams per liter (mg/l), millimols per kilogram (mmol/kg), or millimols per liter (mmol/l).

Potential users: regulatory bodies, industries, non-governmental agencies, researchers from academic and non-academic areas, and general public.

Facilities:

- the prototype uses the numerical and categorical data as input;
- the prototype implements 'learning from data' paradigm – for QSAR models;
- the prototype implements 'learning from experts' paradigm – using a hybrid neuro-fuzzy integration of QSAR models;

- the software architecture is generalizable;
- the prototype implements three methods for integration: statistical, fuzzy, and connectionist;
- the prototype runs on multiple operating systems, consequence of Java implementation; and
- the prototype is easy to install and runs on a wide range of basic configurations, because of the general profile of the users.

The main interface and menus of the system are depicted in Figs. 8 and 9. It provides the user with a personalized interface to manage data from toxicology databases in a table representation of spreadsheet-like format: a normal user will copy/paste data if the original .xls version of their data is structured as presented in a specific .xls format (header, list of chemical compounds, descriptors list, and data matrix; Figs. 8–10).

The prototype NIKE offers a real opportunity to various users to integrate different models developed or available from various software packages but with respect to the same data sets (train and test data). Therefore, a major concern was the uniformity and homogeneity of data available in the input and output files of the prototype. The knowledge and data processed by NIKE regards two main sources:

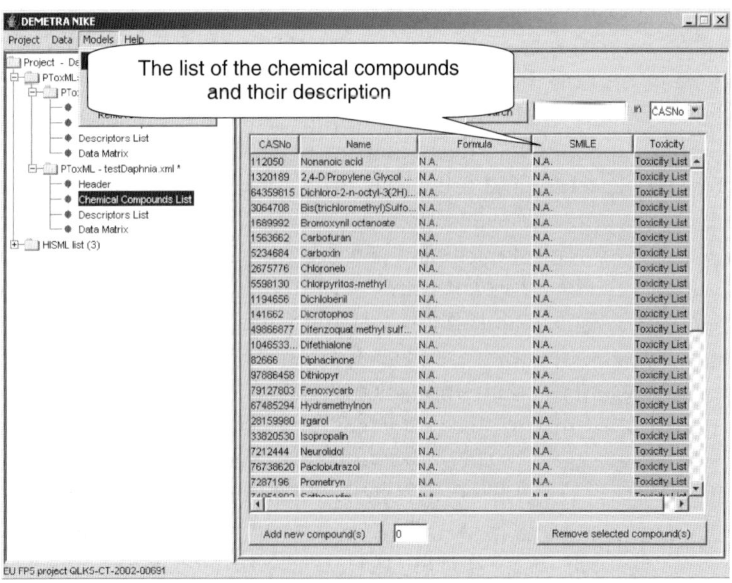

Figure 8: Interface for chemical compounds data management.

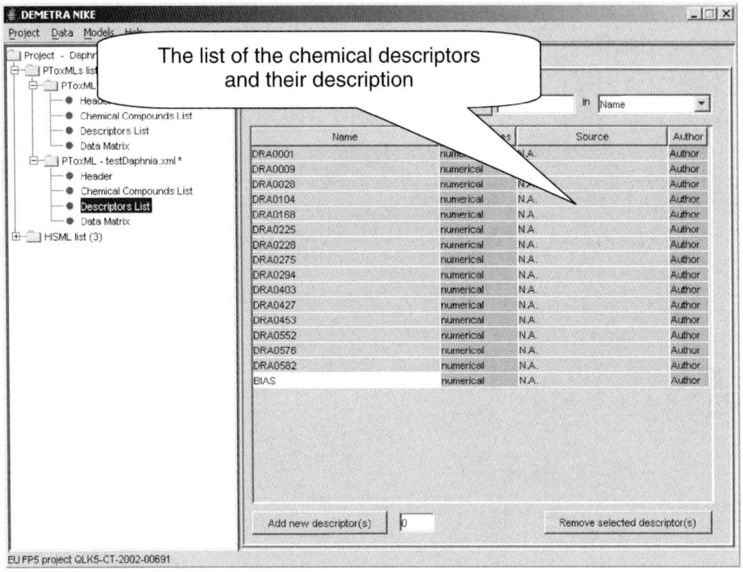

Figure 9: Interface for descriptors data management.

QSARs (as individual or integrated models) and toxicity and chemoinformatics data (chemical compounds description). For NIKE, data are available for input and output files as XML schemas: PToxML and HISML (Neagu *et al.*, 2004). The standard XML syntax for knowledge representation, data exchange, and analysis of experimental data was developed as the HISs markup language HISML. From our knowledge, this is the very first attempt to propose a standard for integrated soft computing techniques, such as HIS. The proposed HISML syntax captures the structure and parameters of modeling experiments. The information stored in a HISML document is further required to analyze and replicate the developed models.

To gather and evaluate the chemical compounds and toxicity information to build computational models capable to assess the toxic effect of chemicals using only structural descriptors is a difficult task, given distributed and heterogeneous data sources. An important step is our initiative of standardization of data and vocabulary of predictive toxicology, where XML provides a powerful way to describe objects (chemicals in our case). Thus, we proposed PToxML as an application of XML to describe chemical information related to predictive toxicology tasks. PToxML documents consist of three main sections: the header, the identity, and the data. The first section defines authoring information [version of the data set, calendar date, author/owner, initial source(s), and comments; Fig. 11], the second section relates properties to the entities (chemical compounds list and descriptors list), and the third section presents the data rows.

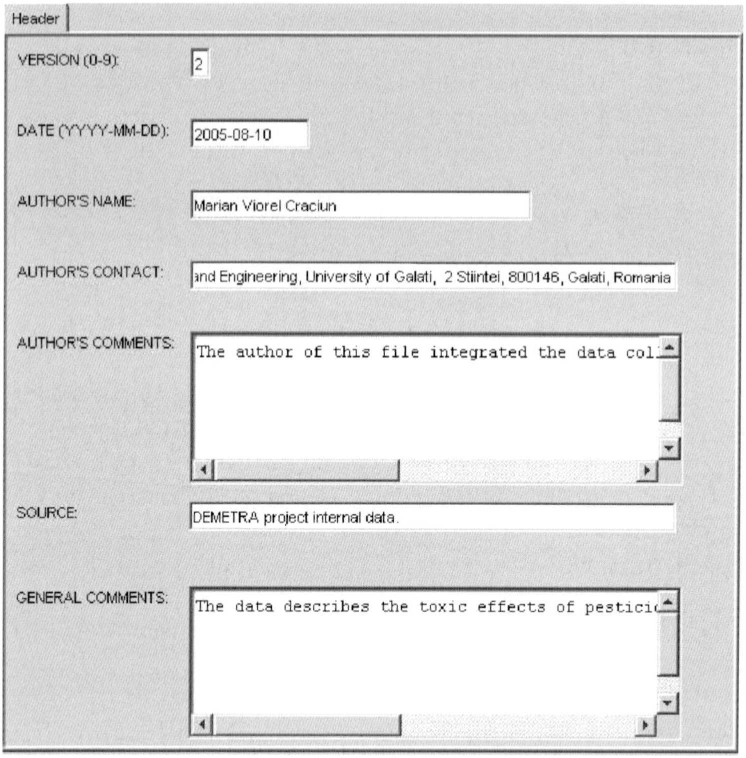

Figure 10: The main interface of the prototype NIKE and its principal functions.

```
<xsd:element name='PToxML'>
     <xsd:complexType>
        <xsd:sequence>
            <xsd:element ref='header' />
            <xsd:element ref='chemicalCompoundsList' />
            <xsd:element ref='descriptorsList' />
            <xsd:element ref='dataMatrix' />
        </xsd:sequence>
        <xsd:attribute name='version' type='xsd:string' use='required'/>
      </xsd:complexType>
 </xsd:element>

<!-- Contains general info -->
<xsd:element name='header' id='el.header'>
     <xsd:complexType>
        <xsd:sequence>
            <xsd:element ref='version' />
            <xsd:element ref='date' />
            <xsd:element ref='author' />
            <xsd:element ref='source' />
            <xsd:element ref='comments' minOccurs='0'/>
        </xsd:sequence>
    </xsd:complexType>
</xsd:element>
```

Figure 11: The main sections of PToxML documents and the header subsection.

A chemical compound is identified by a 'CASNo' and a 'name' (both required) and also chemical 'formula', a 'SMILES' code, and 'toxicity' (all optional). The toxicity structure also is one of the main features of the PToxML document. The information regarding toxicity (possibly unavailable for some chemicals) contains a list having one or more 'toxInfo' elements to describe: the species (target life form) part of the experiments, the measured toxic effects (e.g., LC_{50}) – a compulsory field, and the mechanism of action (if available), the dose (if available) and the measure unit (mg/l, mmol/l, etc.), the exposure time for that life form and the time units (seconds, minutes, hours, etc.). It is also required to supply information about the source and the calendar date when this data was published, together with some comments and their author. Important information is contained also in the list with zero or more 'prediction' fields: predicted values and info regarding the predictor (model) used to obtain this value, the source of this model, calendar date, some comments, and their author. More about the structure of the data described by PToxML schema can be found in the Web site and by request.

As said before, the basic structures of NIKE for model development are ANN – *ArtificialNeuralNetwork* and FIS – *FuzzyInferenceSystems*. The user has therefore the opportunity to develop simple models or recursive HIS structures:

- *HISs* – HIS as either *Simple HISs* – SimpleHIS or *Complex HISs* – ComplexHIS (consisting of two or more SimpleHISs or ComplexHISs).
- A *Simple HIS* is finally an *ANN* (as *IKM – CrispNeuralNetwork*) or a combination of *ANN* with *FIS* (as *IKM – Fuzzy Neural Network* or *EKM*).
- A *Complex HIS* contains also an *integration module* to combine in an intelligent way all the (simple or complex) modules inside.

The prototype NIKE provides therefore the possibility to organize toxicity data, to develop QSARs as simple neural networks, FIS, neuro-fuzzy networks, and statistical, connectionist, or fuzzy recursive combination of further simple or combined models. It proves to be a powerful tool to study toxicity data for predictive purposes.

A number of experiments regarding the performance of the NIKE tool have been developed.

5.1. Experiment 1

In this experiment, five classifiers (IBL, DT, RIPPER, MLPs, and SVM) together with four combination methods (MVC, MPC, APC, and WPC) were tested over five toxicity data sets using ten-fold cross-validation method. The experimental results are presented in Tables 1 and 2.

Table 1
Performance of five classifiers evaluated on seven data sets

Data set	IBL	*k*	DT	RIPPER	MLPs	Lr	SVM
Trout	59.93	5	55.32	56.74	58.16	0.9	62.06
Oral quail	57.76	5	62.93	60.34	51.72	0.3	65.52
Daphnia	54.17	5	50.38	50.00	53.41	0.3	54.55
Dietray quail	48.78	10	45.53	39.84	55.28	0.3	48.78
Bee	58.09	5	45.71	46.67	51.43	0.3	53.33

DT, decision tree; IBL, instance-based learner; MLP, multi-layer perceptron; RIPPER, repeated incremental pruning to produce error reduction; SVM, support vector machine.

Table 2
Performance of four combination methods evaluated on seven data sets

Data set	MVC	MPC	APC	WPC
Trout	63.12	56.38	59.22	59.93
Oral quail	62.93	56.03	60.34	60.34
Daphnia	54.17	53.78	53.78	54.92
Dietray quail	53.66	43.90	52.03	50.41
Bee	58.10	42.86	55.24	54.29

APC, average probability-based combination; MVC, majorityvoting-based combination; WPC, weighted probability-based combination; MPC, maximal probability-based combination.

5.2. Experiment 2

In this experiment, four classifiers, *k*NN, *wk*NN, *k*NNModel, and Fuzzy kNN-Model, were tested over five toxicity data sets using ten-fold cross-validation method. The experimental results are presented in Table 3.

Table 3
Performance of four classifiers evaluated on seven data sets

Data set	*k*NN	*k*	*wk*NN	*k*	*k*NNModel	ε	*N*	Fuzzy *k*NNModel	α
Trout	59.93	5	56.79	10	60.71	1	2	62.14	0.8
Oral quail	57.76	5	55.45	10	64.55	0	4	64.55	0.3
Daphnia	54.17	5	55.38	10	60.00	2	0	62.69	0.3
Dietray quail	47.97	5	50.83	10	50.83	3	0	51.67	0.8
Bee	58.10	5	58.00	10	53.00	0	0	53.00	0.7

These data show that NIKE has interesting possibilities to classify ecotoxicity properties. The advanced user can use NIKE models to develop ad hoc models.

6. RULE-BASED SYSTEMS

The rule-based approach was tested to analyze how prediction performances can be increased by combining individual models. This procedure was subdivided into the following two main steps to minimize the computational time:

- to select the best model combinations by looking for the highest average predicted values and
- to establish combination rules, for correcting the bad predictions and, above all, avoiding false negatives. They will depend on the activity range and will weight in a different way the models selected.

Selecting the best model combinations

Let us consider a set $S(m_1, \ldots, m_n)$ of n models m_i, predicting toxicity values (T); each compound c_i associates a set $V(Tm_1, \ldots, Tm_n)$ of n predicted values Tm_i. Analyzing all possible permutations of the model set S requires to evaluate $2^n - 1$ combinations. The following algorithm was used to asses all these combinations:

For i = 1 to n models

create all possible subsets of i different models (${}^nC_i$ distinct subsets)
for each subset

– compute the average predictions (Y) for all compounds

– compute the correlation coefficient R^2 and the number of molecular descriptors involved in the subset

End For

The best model subsets were selected by maximizing the R^2 coefficients. When similar results were obtained, the solutions retained were those developed by using the lowest number of molecular descriptors. This pre-processing step is easy and fast, but, generally, it does not allow to really improve the slope of the regression equation between predicted and experimental values and still keeps too many false negatives. Then, there is need to establish more sophisticated combination rules able to improve these parameters.

Combination rules

The rules were built with help of three main parameters, that is, the maximal (Max), minimal (Min), and average (Avg) prediction values computed, respectively, for each compound on the set $V(Tm_1, \ldots, Tm_n)$. A simple and linear

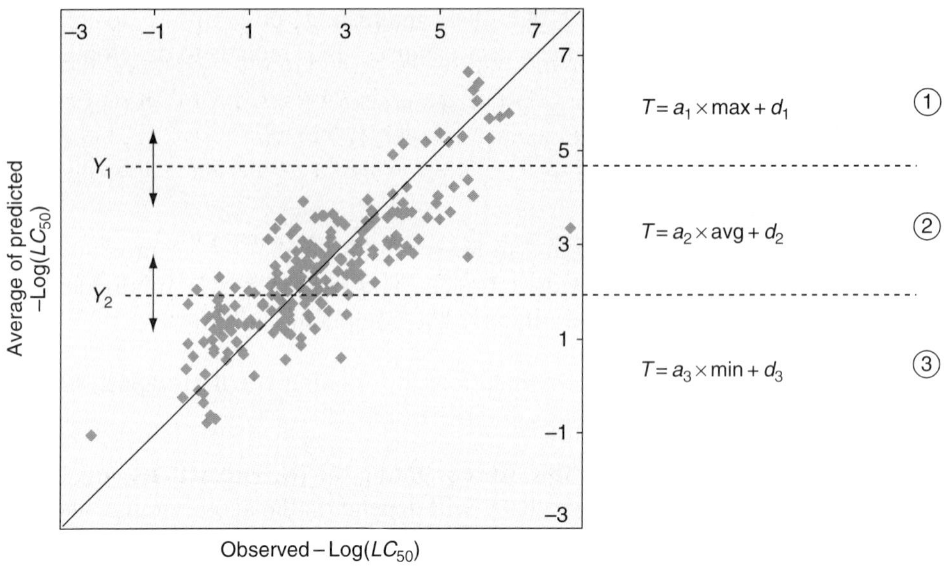

Figure 12: Example of a combination rule established on three domains, represented by three different equations. Domains are delimited by average values Y_1 and Y_2. The algorithm searches for the best rule by varying Y_i, a_i, d_i, and B parameters.

correction factor is used to improve the prediction results, according to the following equation:

$$T(c_i) = a \times B(c_i) + d, \tag{13}$$

where T is the new predicted value for the compound c_i; a and d are regression coefficients; B is one parameter selected amidst Max, Min, and Avg. It has to be underlined that Eq. 13 is not the same for all compounds. The model is partitioned into N different domains (Fig. 12), whose limits are identified by the average predictions Y_i, and each of them is represented by a different equation. For example, the general combination rule represented in Fig. 12 and established on three domains can be described as follows:

If average $> Y_1$ *Then* $T = a_1 {}^*B + d_1$

Else if $Y_2 <$ *average* $< Y_1$ *Then* $T = a_2 {}^*B + d_2$

Else if average $< Y_2$ *Then* $T = a_3 {}^*B + d_3$

The best rules are identified by an algorithm that modifies the Y_i, a_i, d_i, and B parameters till to find the highest R^2 value. A downhill simplex method (Cormen *et al.*, 2001) was implemented to reduce the computational search time. The simplex method is an optimization algorithm that, differently from the most popular methods, does not deal with first derivatives calculation. Actually, using the latter is useless for our problem due to the discontinuity associated with the function to be optimized (Fig. 12). As our search involves a maximization process, the downhill simplex method was modified to achieve an uphill search. This allows selecting parameters that optimize simultaneously the slope and interception of the regression equation.

The rule-based system used in the DEMETRA project allowed improving in a relevant way the prediction performances for all endpoints. The correlation coefficients R^2 were increased of more than 10%, and slopes and intercepts were significantly improved too.

7. SELF-ORGANIZING STATISTICAL LEARNING NETWORKS

All methods of automatic model selection lead to a single 'best' model, whereas the accuracy of model result depends on the variance of the data. A common way for variance reduction is the aggregation of similar model results following the idea: generate many versions of the same predictor/classifier and combine them in a second step. If modeling aims at prediction, it is helpful to use alternative models that estimate alternative forecasts. These forecasts can be combined using several methods to yield a composite forecast of a smaller error variance than any of the models have individually (Müller and Lemke, 2000; Elder, 2003). The desire to get a composite forecast is motivated by the pragmatic reason of improving decision-making rather than by the scientific one of seeking better explanatory models. Composite forecasts can provide more informative inputs for a decision analysis, and therefore, they make sense within decision theory, although they are often unacceptable as scientific models in their own right, because they frequently represent an agglomeration of often conflicting theories (Müller and Lemke, 2000).

Based on the five sets of individual models generated by different partners, which now serve as input information, we generated a combined model for each data set by a self-organizing statistical learning network (SOSLN) algorithm (KnowledgeMiner, 2006) (see also Section 4.3 in Chapter 4, Chaudhry *et al.*, this volume). The result is five self-selected, optimally composed linear or non-linear regression models, including their regression equation. It should be noted that the combined models are not just an equally weighted composition, or simply the mean, of, say, the five or seven best individual models of a data set but actually an a priori unknown, optimal mix of models that – combined – decrease the error variance of the combined model most.

As outlined in Section 4.3 in Chapter 4, Chaudhry *et al.*, this volume, the SOSLN approach is based on the concept of induction, which is composed of the cybernetic principle of self-organization as an adaptive creation of a network without subjective points given (Müller and Lemke, 2000). This means, no

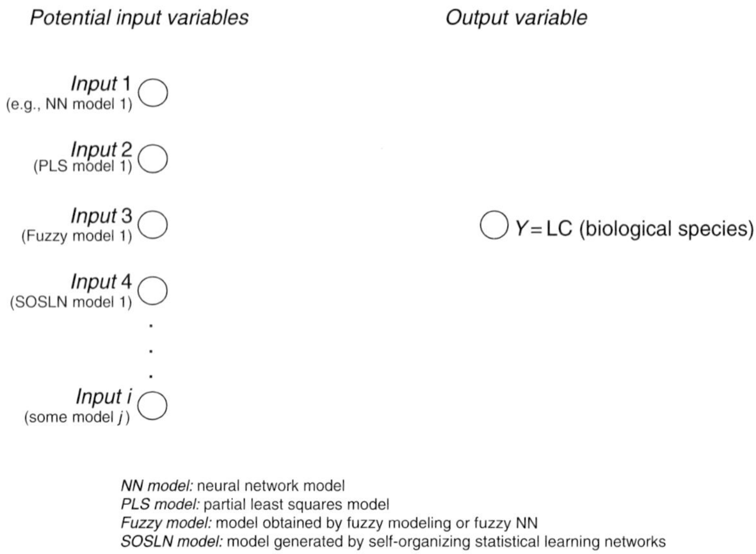

Figure 13: Network status of a self-organizing statistical learning network (SOSLN) model at the start of modeling – the network topology is undefined.

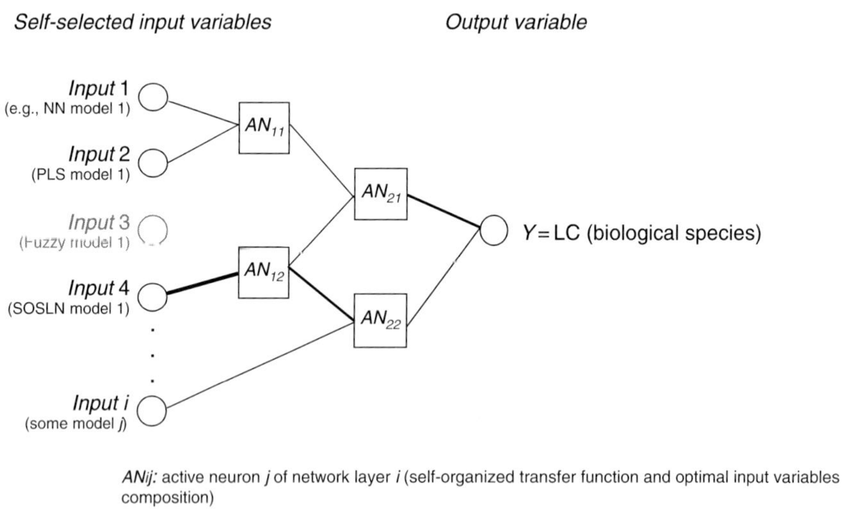

Figure 14: Network status of a possible final, optimal complex combined model, which is composed of a self-selected set of relevant individual models.

assumptions are made about model structure or network topology at start of the modeling process (Fig. 13). At the end of the self-organizing process of model evolution, model validation, and model selection, a final optimal complex and optimal composed predictive model is generated. Figure 14 shows an example of a possible combined model self-organized by SOSLN.

Table 4
Summary of nine individual models for quail (dietary exposure) used as potential inputs for building a model composite

Individual model	Model type	Input variable	Included in combined model	R^2 (training)	R^2 (test)
1	PLS	X_1	Yes	0.60	0.55
2	SOSLN	X_2	Yes	0.83	0.62
3	SOSLN	X_3	Yes	0.78	0.47
4	PLS	X_4	No	0.59	0.53
5	SOSLN	X_5	No	0.63	0.27
6	SOSLN	X_6	No	0.73	0.16
7	SOSLN	X_7	No	0.87	0.27
8	GA-MLR	X_8	No	0.68	0.00
9	PCA-NN	X_9	No	0.64	0.31

GA-MLR, combination of genetic algorithm and multi linearregression; PCA-NN, combination of principal component analysis and neural network; PLS, partial least squares method; SOSLN, self-organizing statistical learning network.

A combined model for the fourth biological endpoint (quail dietary exposure), for example, was generated by applying SOSLNs (KnowledgeMiner, 2006). In this specific case, we considered nine individual models generated earlier by different project partners using various modeling methodologies as potential input variables. Neither the network topology – number of network layers and number of neurons in a layer – nor the neuron's transfer function were defined a priori by the user/modeler. Table 4 summarizes these nine individual models.

During self-organization, the algorithm autonomously composes an optimal complex model that minimizes error variance on both learning and validation data subsets. The final combined model for dietary quail uses just three of the nine individual models and shows a coefficient of determination, R^2, of 0.79 for the training data set and of 0.69 for the test data set, and it is described by this transformed equation generated by the algorithm:

$$Y = 0.096465\ x_1 + 0.628581\ x_2 + 0.440833\ x_3 + 0.494189\ x_1 x_2 - 0.08229 x_2 x_2 - 0.25701\ x_1 x_1 + 0.00939. \quad (14)$$

The corresponding network model along with the active neuron's transfer function is shown in Fig. 15.

This example shows that SOSLNs are a powerful approach for finding an optimal composite of individual models out of a larger number of potential input candidates in a systematical, transparent, reproducible, fast, and easy way. Furthermore, the resulting model is described by an analytical equation on the fly for interpretation, implementation, and further analysis and validation purposes.

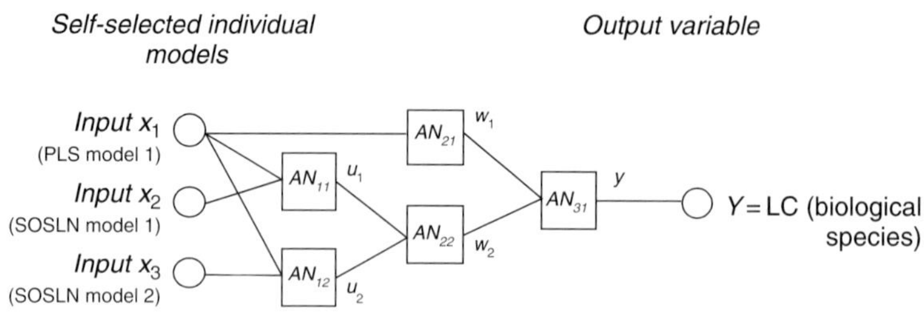

AN_{ij}: Active neuron *j* of network layer *i* (self-organized transfer function and optimal input variables composition)

AN_{31}: $y = -0.1089 \times w_1 + 0.8586 \times w_2 - 0.6107$
AN_{21}: $w_1 = 1.7304 \times x_1 + 1.0943$
AN_{22}: $w_2 = 0.5922 \times u_1 + 0.4502 \times u_2$
AN_{11}: $u_1 = 1.2360 \times x_2 + 0.3653 \times x_1 + 0.9717 \times x_1 x_2 - 0.1618 \times x_2 x_2 - 0.5054 \times x_1 x_1 + 0.7861$
AN_{12}: $u_2 = 0.2565 \times x_1 + 1.1402 \times x_3 + 0.8785$

Figure 15: Optimal complex network model of the combined model for dietary quail composed of three individual models.

If the individual input models were generated using different modeling methodologies, the final model composite can also be seen as a hybrid network model.

8. CONCLUSIONS

This chapter introduced theoretical concepts and motivations of using HISs paradigm for the development of QSAR models for complex toxicity data. Both theoretical approaches and practical implementation research efforts drove toward quality models and original algorithms and performances discussed in Chapter 7, Amaury *et al.*, this volume, for the five toxicity data sets of the project DEMETRA. The models identified, part of the objectives of the project, are furthermore results of applying the principles, techniques, and software discussed in this chapter and detailed also in the appendices.

REFERENCES

Avnimelech R., Intrator N. (1999) Boosted mixture of experts: An ensemble learning scheme. *Neural Comput.*, 11, 483–497.

Bauer E., Kohavi R. (1999) An empirical comparison of voting classification algorithms: Bagging, boosting, and variants. *Mach. Learn.*, 36, 105–139.

Baykut A., Ercil A. (2003) Towards automated classifier combination for pattern recognition, 94–105. In: *Lecture Notes in Computer Science 2709*. Windeatt T., Roli F. Eds. Springer Verlag, Berlin, Germany.

Benfenati E., Mazzatorta P., Neagu D., Gini G. (2002) Combining classifiers of pesticides toxicity through a neuro-fuzzy approach, 293–303. In: *Lecture Notes in Computer Science 2364*. Windeatt T., Roli F. Eds. Springer Verlag, Berlin, Germany.

Bi Y., Bell D., Wang H., Guo G., Greer K. (2004) Combining multiple classifiers using Dempster-Shafer's rule for text categorization, 127–138. In: *Lecture Notes in Computer Science 313/2004*. Proceedings of MDAI, Barcelona, Spain.

Breiman L. (1996) Bagging predictors. *Mach. Learn.*, 24, 123–140.

Buckley J.J., Hayashi Y. (1995) Neural nets for fuzzy systems. *Fuzzy Sets Syst.*, 71, 265–276.

Chan P., Stolfo S. (1995) Learning arbiter and combiner trees from partitioned data for scaling machine learning, 39–44. In: *Proceedings of the First International Conference on Knowledge Discovery and Data Mining (KDD-95)*. Montreal, Canada, August 20–21. AAAI Press, Menlo Park, California, USA. ISBN 0-929280-82-2.

Chen S.H., Wang P.P. (2004) *Computational Intelligence in Economics and Finance*. Springer-Verlag, Berlin, Germany.

Cormen T.H., Leiserson C.E., Rivest R.L., Stein C. (2001) *Introduction to Algorithms, Section 29.3: The Simplex Algorithm*. MIT Press, Cambridge, Mass, 790–804.

Craciun M.V., Neagu D., Craciun C.A., Smiesko M. (2004) A study of supervised and unsupervised machine learning methodologies for predictive toxicology, 61–69. In: *Intelligent Systems in Medicine*. H.N. Teodorescu Ed., Performantica, Iasi, Romania.

d'Avila Garcez A.S., Broda K., Gabbay D.M. (2002) *Neural-Symbolic Learning Systems: Foundations and Applications, Perspectives in Neural Computing*. Springer-Verlag, Berlin, Germany.

Dietterich T.G. (2000) *Ensemble Methods in Machine Learning*: http://www.cs.orst.edu.

Elder J. (2003) The generalization paradox of ensembles. *J. Comput. Graph. Stat.*, 12, 853–864.

Freitas A.A. (1998) Survey of parallel data mining, 287–300. In: *Proceedings of the 2nd International Conference on the Practical Applications of Knowledge Discovery and Data Mining*. The practical application company, London, UK.

Freund Y., Yishay M., Schapire R.E. (2004) Generalization bounds for averaged classifiers. *Ann. Stat.*, 32, 1698–1722.

Fuller R. (1999) *Introduction to Neuro-Fuzzy Systems. Advances in Soft Computing Series*. Springer-Verlag, Berlin, Germany.

Funahashi K. (1989) On the approximate realization of continuous mappings by neural networks. *Neural Netw.*, 2, 183–192.

Funabashi M., Maeda A., Morooka Y., Mori K. (1995) Fuzzy and neural hybrid expert systems: Synergetic AI. *Exp. IEEE*, 10, 32–40.

Gallant S.I. (1993) *Neural Network Learning and Expert Systems*. MIT Press, Cambridge, MA.

Ho T.K. (2002) Multiple classifier combination: Lessons and next steps. In: *Hybrid Methods in Pattern Recognition*. Kandel A., Bunke H. Eds. World Scientific. IEEE Computer Society, Los Alamitos, California, USA.

Ho T.K., Hull J.J., Srihari S.N. (1994) Decision combination in multiple classifier systems. *IEEE Trans. Pattern Anal. Mach. Intell.*, 16, 66–75.

Hornik K., Stinchcombe M., White H. (1989) Multilayer feedforward networks are universal approximators. *Neural Netw.*, 2, 359–366.

Jackson P. (1999) *Introduction to Expert Systems*, 3rd edition. Addison Wesley Longman, London, UK.

Jacobs R.A., Jordan M.I, Nowlan S.J., Hinton G.E. (1991) Adaptive mixtures of local experts. *Neural Comput.*, 3, 79–87.

Jain A.K., Duin R.P.W., Mao J. (2000) Statistical Pattern Recognition: A Review, *IEEE Transactions on Pattern Analysis and Machine Intelligence*, 22, 1, 4–37.

Kittler J., Hatef M., Duin R., Matas J. (1998) On combining classifiers. *IEEE Trans. Pattern Anal. Mach. Intell.*, 20, 226–239.

KnowledgeMiner (2006) *Self-Organizing Data Mining and Prediction Tool*: http://www.knowledgeminer.com.

Krogh A., Vedelsby J. (1995) Neural network ensembles, cross validation and active learning. In: *Advances in Neural Information Processing Systems*. Tesauro G., Touretzky D.S., Leen T.K. Eds. MIT Press, Cambridge, MA.

Kuncheva L.I. (2001) Combining classifiers: Soft computing solutions, 427–452. In: *Pattern Recognition: From Classical to Modern Approaches*. Pal S.K. Ed. World Scientific, Singapore.

Liu P., Li H. (2004) Fuzzy neural network theory and application. In: *Machine Perception and Artificial Intelligence, 59*. World Scientific, Singapore.

Lucas P.J.F. (2001) Expert knowledge and its role in learning bayesian networks in medicine: An appraisal LNAI 2101, 156–166. In: *AIME 2001*. Springer-Verlag, Berlin, Germany.

Mazzatorta P., Benfenati E., Neagu D., Gini G. (2003) Tuning neural and fuzzy-neural networks for toxicity modelling. *J. Chem. Inf. Comput. Sci.*, 43, 513–518.

Merkwirth C., Mauser H., Schulz-Gasch T., Roche O., Lengauerý T. (2004) Ensemble methods for classification in cheminformatics. *J. Chem. Inf. Comput. Sci.*, 44, 1971–1978.

Michie D., Spiegelhalter D.J., Taylor C.C. (1994) *Machine Learning, Neural and Statistical Classification*. Ellis Horwood, New York.

Mitchell T. (1997) *Machine Learning*. McGraw Hill, New York.

Müller J.A., Lemke F. (2000) *Self-Organising Data Mining*. BoD, Hamburg, Germany.

Nadal C., Legault R., Suen C.Y. (1990) Complementary algorithms for the recognition of totally unconstrained hand written numeral, 434–449. In: *Proceedings of the 10th International Conference on Pattern Recognition A*. IEEE. Atlantic City, NJ.

Neagu C.D., Palade V. (1999) Fuzzy computing in a multi purpose neural network implementation, 697–700. In: *Proceedings of International Conference on 6th Fuzzy Days in Dortmund*. B. Reusch Ed. Springer Verlag, Berlin, Germany.

Neagu C.D., Palade V. (2000) An interactive fuzzy operator used in rule extraction from neural networks. *Neural Netw. World J.*, 10, 675–684.

Neagu D., Palade V. (2002) Modular neuro-fuzzy networks used in explicit and implicit knowledge integration, 277–281. In: *Proceedings of 15th International Conference of Florida Artificial Intelligence Society – FLAIRS*. Pensacola, FL. AAAI Press, Menlo Park, California, USA.

Neagu C.D., Gini G. (2003) Neuro-fuzzy knowledge integration applied to toxicity prediction In: *Innovations in Knowledge Engineering*, Jain R., Abraham A., Faucher C., Jan van der Zwaag B. Eds. Advanced Knowledge International, Pty Ltd, Australia Magill, Adelaide.

Neagu D., Craciun M., Chaudhry Q. (2004) Knowledge specification for versatile hybrid intelligent systems, 219–224. In: *Proceedings of the 5th International Conference on Recent Advances in Soft Computing*, Lotfi A. Ed. University of Nottingham Press, Nottingham, UK.

Neagu D., Craciun M.V., Stroia S.A., Bumbaru S. (2005) Hybrid intelligent systems for predictive toxicology – a distributed approach, 26–31. In: *5th International Conference on Intelligent Systems Design and Applications*. IEEE Computer Society Press, Wroclaw, Poland.

Negoita M., Neagu D., Palade V. (2005) *Computational Intelligence – Engineering of Hybrid Systems, Series in Fuzziness*. Springer Verlag, Berlin, Germany.

Pawlak Z. (1991) *Rough Sets: Theoretical Aspects of Reasoning About Data*. Kluwer Academic Publishers, Dordrecht, The Netherlands, and Boston, MA.

Quinlan J.R. (1993) *C4.5: Programs for Machine Learning*. Morgan Kauffman, San Francisco, CA.

Ripley B.D. (1996) *Pattern Recognition and Neural Networks*. Cambridge University Press, Cambridge, UK.

Rumelhart D., McClelland J. (1986) *Parallel Distributed Processing. Explanations in the Microstructure of Cognition*. MIT Press, Cambridge, UK.

Saerens M., Fouss F. (2004) Yet another method for combining classifiers outputs: A maximum entropy approach, 82–91. In: *Proceedings of MCS'04, the 5th International Workshop on Multiple Classifier Systems*. LNCS 3077, Cagliari, Italy.

Schultz T.W., Cronin M.T.D. (2003) Essential and desirable characteristics of ecotoxicity quantitative structure-activity relationships. *J. Environ. Toxicol. Chem.*, 22, 599–607.

Singer Y. (1997) Adaptive mixtures of probabilistic transducers. *Neural Comput.*, 9, 1711–1733.

Small R.D., Edelstein H.A. (1997) *Scalable Data Mining*. Two Crows Company, MD, and Washington, DC, Potomac, Maryland.

Vapnik V. (1995) *The Nature of Statistical Learning Theory*. Springer, New York.

Xu L., Krzyzak A., Suen C. (1992) Methods of combination multiple classifiers and their applications to handwritten recognition. *IEEE Trans. Syst. Man Cybern.*, SMC-22, 418–435.

Zhang B., Srihari S.N. (2002) Class-wise multi-classifier combination based on Dempster-Shafer theory. In: *Proceedings of the VIIth International Conference on Control, Automation, Robotics and Vision, ICARV 2002*, Marina Mandarin Hotel, Singapore, December, CEDAR Publications.

Edited by Emilio Benfenati

Chapter 6

Validation of the models

Emilio Benfenati[1], Jacques R. Chrétien[2], Giuseppina Gini[3], Nadège Piclin[2], Marco Pintore[2], Alessandra Roncaglioni[1]

[1]Laboratory of Environmental Chemistry and Toxicology, Istituto di Ricerche Farmacologiche "Mario Negri", Milano, Italy

[2]BioChemics Consulting SAS, Orléans, France

[3]DEI, Politecnico di Milano, Politecnico di Milano, Milano, Italy

1. INTRODUCTION

When a qualitative or quantitative model has to be assessed, some 'good practice criteria' should be adopted taking into account three main components of a model: (1) robustness, (2) prediction ability and (3) applicability domain (AD). These components are now receiving increasing attention and consensus in different papers (Devillers, 2001; Tropsha *et al.*, 2003), but still, many studies confuse these concepts or totally neglect them. For this reason, although many international regulatory bodies recognize the potential benefits of quantitative structure-activity relationship (QSAR) techniques, e.g. in the risk assessment of industrial chemicals (CE, 1995), they are scarcely used in real applications. Some general principles have been listed, as we reported in Chapter 1 (Benfenati *et al.*, this volume), but the lack of guidelines and standardized protocols accepted and used by all research groups prevents an effective world-wide development of such strategies.

The main problem is that the proposed models in many cases may apply only within the training set that has been used to build the model. Traditionally, QSAR models have been developed to describe a phenomenon, suitable to identify a rational relationship between a given parameter and the property. For instance, the ecotoxicity phenomenon is often closely associated with log P, as described in Chapter 1 (Benfenati *et al.*, this volume). However, later on, the emphasis has been put on the use of these relationships to predict the properties of unknown compounds. This introduced the problem of checking the correctness of the predicted property. We underline that the interest has shifted from the description of the phenomenon within a series of chemicals to its prediction for new compounds. Different tools are required, based on sound statistical methods. Indeed, in many cases, especially when there are many descriptors describing the molecules and complex algorithms are used, there is a risk of over-fitting the model. This is the situation in which the model follows too closely the peculiar behaviour of the compounds in the training set, without the capacity to obtain a general model, suitable for the real cases to be evaluated.

For these reasons, the statistical tools should prove the capability of the model to be valid in a general way, i.e. to be predictive for compounds not used in development of the model. There is a debate in the scientific community on the most suitable way to assess the robustness and predictive performances of a model. The external set used for validation has been indicated as the most appropriate solution (Tropsha *et al.*, 2003). Such a set of compounds is not used in the building of the model but only to verify the predictive power of the model. However, the use of a single defined set of compounds can represent a bias, so an accurate selection of test set compounds that is representative of the chemical space of the training set is required. But, frequently, the scarce number of available compounds prevents to select a representative external test set, thus other statistical tools have been proposed. One can repeatedly split the overall set of compounds into a training and validation set, making different splits each time the procedure is repeated. The performances of the different models can then be assessed on the basis of the overall result mediated on the different subsets.

It should also be noted that the use of appropriate statistical validation tools, such as the training and test set or others, should be adopted for predictive models not only in the case of QSAR based on descriptors but also in the case of models based on rules. In other words, the rules that are defined as appropriate for predictive purposes should also be validated. Within DEMETRA, we adopted a series of statistical checks, including the use of the test set.

2. SELECTION OF THE TRAINING AND TEST SETS

It is important to rationally define a training set from which the model is built and an external test set on which to evaluate its prediction power. The object of this selection should be to generate two sets with similar molecular diversity, in order to be reciprocally representative and to cover all the main structural and physico-chemical characteristics of the global data set.

Several procedures can be adopted for the selection of the training and test sets, the latter which should contain between 15 and 40% of the compounds in the full data set. Among the most effective methods, *D*-optimal design (Marengo and Todeschini, 1992), the Kohonen artificial neural network (k-ANN) (Gasteiger and Zupan, 1993; Zupan *et al.*, 1997), the Kennard–Stone technique (Kennard and Stone, 1969) and the sphere excluder (Snarey *et al.*, 1997) can be listed.

The *D*-optimal design is an algorithm for optimal distance based on experimental design that does not require any preliminary hypothesis about a regression model. The best set of compounds is defined through a fast exchange algorithm where, in each cycle, substitution provides the maximum increase in the minimum distance between the currently selected compounds. Such an algorithm provides a final distribution of the most dissimilar compounds selected from the set of allowed candidates.

The splitting of the data set realized by k-ANN takes advantage of its clustering capabilities allowing the selection of a meaningful training set and a representative validation set. The structural information represented by the X-variables (molecular descriptors) is used as variables to build a Kohonen map. At the end of the defined epochs of the net training, similar chemicals fall within the same neuron, as they carry similar information. To select the training set of chemicals, it is assumed that the compound closest to each neuron centroid is the most representative of all the chemicals within the same neuron. Thus, the selection of the training set chemicals is performed by the minimal distance from the centroid of each cell in the top map.

The Kennard and Stone algorithm is quite similar to the maximum dissimilarity method. It is a sequential technique that maximizes the Euclidean distances between the newly selected compounds and the ones already selected. An additional compound is selected by computing its distance to each selected compound and by maximizing the distance to the closest compound. This method guarantees that the training set compounds cover the whole area of the representative points. The latter selection technique was used in DEMETRA to select training and test sets. We checked that the training and test sets had the same distribution of chemicals for the toxicity ranges of each endpoint, to verify a similar distribution of the pesticides in the property domain.

Finally, the sphere excluder is also based on dissimilarity concepts. A dissimilarity threshold is set, which can be thought of as the radius of a hypersphere in the multidimensional chemistry space. The basic algorithm proceeds by selecting a compound at each stage and then excluding from further consideration all those other compounds within the sphere centred on that compound.

3. INTERNAL VALIDATION AND ROBUSTNESS

The crucial step for evaluating if the model developed by any computational method is 'useful', consists of assessing, by several statistical parameters, its robustness and ability to accurately predict biochemical properties of new compounds (Eriksson *et al.*, 2003; Tropsha *et al.*, 2003). In DEMETRA, the stochastic methods for defining robustness involved above all calculations of the following parameters for the training set:

- The root-mean square error (RMSE), i.e. the measurement of the average difference between predicted and measured response values. It can be interpreted as the average prediction error, expressed in the same units as the original response values, and represented by the equation:

$$RMSE = \sqrt{\frac{\sum_{i=1}^{N}(A_i' - A_i)^2}{N}},$$

where A_i and A_i' are the calculated and experimental values for the molecule sample i, respectively;

- The squared correlation coefficient, R^2, between predicted and experimental values. It estimates the degree of correlation between X- and Y-variables and gives the explained variance of the model representing its goodness of fit. This coefficient, not dependent on the activity units, can be computed by the formula:

$$R^2 = \frac{\left[\sum_{i=1}^{N}(A_i - M_A)(A_i' - M_{A'})\right]^2}{\sum_{i=1}^{N}(A_i - M_A)^2 \sum_{i=1}^{N}(A_i' - M_{A'})^2},$$

where M_A and $M_{A'}$ are the average predicted and experimental values computed on all N molecules. A good correlation is characterized by scores close to 1.0;

- The cross-validated values, Q^2, by leave-one-out (LOO) and leave-several-out (LSO). The cross-validation methods consist in leaving out a given number of compounds from the training set and rebuilding the model, which is then used to predict the compounds left out. If a single compound is left out, the technique is called LOO, otherwise LSO. An LOO or LSO correlation coefficient, generally indicated with Q^2, is computed by evaluating the accuracy of these 'test' compounds prediction. The Q^2 formula is similar to the standard correlation coefficient one (R^2), but the predicted values come from models built without the predicted compound.

$$Q^2 = 1 - \frac{\sum_{i=1}^{N}(A_i' - A_i)^2}{\sum_{i=1}^{N}(A_i' - M_{A'})^2}.$$

The procedure is iterated many times, and the related model should be reasonably robust if a high average cross-validated coefficient is obtained, i.e. at least superior to 0.5.

Many discussions involved the scientific community in the reliability of Q^2 for assessing the model robustness and providing a preliminary evaluation of model predictability. It is often considered that the LOO procedure can be too optimistic in evaluating model performances so that the LSO procedure should be also employed, because it increases the perturbation inflicted to the data set. Especially with big data sets, the LOO method could give a too small perturbation to the data, so that the LOO Q^2 could approach the R^2 in fitting (Eriksson *et al.*, 1999). Moreover, other authors (Golbraikh and Tropsha, 2002) argued that cross-validated parameters are not enough to provide an idea about the predictivity for the model on external compounds;

- The scores associated with the Y-randomization test, where the dependent-variable vector, Y-vector, is randomly shuffled and a new model is established by using the same original independent-variable matrix. After repeating this test several times, the average LSO or LOO Q^2 value is expected to be low. If a high score is obtained, the original model is not acceptable, as it was due to a chance correlation or a structural redundancy in the training set.

It has to be underlined that all these tests have to be simultaneously positive for defining a model as robust. In reality, many researchers are satisfied with their models if a statistical value is good, e.g. $R^2 > 0.9$ or $Q^2 > 0.5$, whereas the latter are necessary, they are not sufficient on their own to ensure that a model is robust. For example, having very different R^2 and Q^2 values can hide model weaknesses that require further investigations; actually, only if this gap is small, it is possible to affirm that the model is not significantly perturbed by eliminating a proportion of the training information and that it represents all different SARs associated with the data set well.

4. EXTERNAL VALIDATION

Another critical issue present in many published models is the linguistic and, above all, conceptual misunderstanding that exists between robustness and prediction ability. Obtaining a robust model means that all good practices in building it, i.e. the tests listed in the previous paragraph, were fulfilled but does not give real information about its prediction power as briefly discussed above. This is evaluated by predicting the compounds included in a sufficiently large test set. In particular, the prediction power can be better characterized by assessing (1) the determination coefficient between experimental and predicted values; (2) the Y-intercept of the regression lines, i.e. q and q', the first one associated with predicted versus observed values and the second one with the reverse case; (3) coefficients of determination R_0^2 and R'^2_0. The latter coefficients characterize linear regression with Y-intercept set to 0, the first one associated with predicted versus observed values, the second one related to observed versus predicted values, and (4) slopes k and k' of the regression lines, relating predicted versus observed, and observed versus predicted values, respectively.

The following conditions, according to the work proposed by Golbraikh *et al.* (2003), were adopted in DEMETRA to conclude that a QSAR model has an acceptable prediction power if

1) $Q^2 > 0.5$,

2) $R^2 > 0.6$,

3) $(R^2 - R^2_0)/R^2 < 0.1$ and $0.85 < k < 1.15$

   or

   $(R^2 - R'^2_0)/R^2 < 0.1$ and $0.85 < k' < 1.15$ and

4) $|R^2_0 - R'^2_0| < 0.3$

The reason to use R^2_0 and require k values that are close to 1 is that when actual versus predicted activities are compared, an exact fit is required, not just a correlation. It is preferable that the third condition is satisfied by using both R^2_0 and R'^2_0 coefficients, but this requirement is extremely stringent as confirmed by all the DEMETRA work, and the less stringent condition 4 was adopted. It must be stressed that all the relevant validation parameters, R'^2_0, R^2_0, k, and q, were checked for both the training and the test sets.

Additionally, as for the training set, RMSE scores were also verified for the test set, together with other important parameters, such as the Fisher test and PRESS/SSY values. Also in this case, of course, all these criteria have to be simultaneously evaluated for assessing the real prediction ability of a model.

Fisher test determines if the correlation is significant for at least $x\%$ compounds. More particularly, if the Fisher coefficient

$$Fc = \frac{\sum_{i=1}^{N}(A_i - M_A)^2}{\sum_{i=1}^{N}(A'_i - M_{A'})^2}$$

is superior to a given threshold value (quartile $x\%$), the correlation is verified for $x\%$ of the reference compounds, whereas an extrapolation is made for the other ones. Hence, the higher the Fisher coefficient, the larger the samples for which the correlation is verified.

PRESS/SSY represents the fraction of residual (or unexplained) variance over the total variance, and it is computed by

$$\frac{PRESS}{SSY} = \frac{\sum_{i=1}^{N}(A'_i - A_i)^2}{\sum_{i=1}^{N} A_i^2}.$$

In a reasonable model, PRESS/SSY should be smaller than 0.4, and values close to 0.1 or lower indicate excellent models.

5. VALIDATION PARAMETERS FOR CLASSIFIERS: MATRIX OF CONFUSION

Most of the previous validation methods, such as cross-validation, Y-randomization, and test set evaluation, adapted to regression modelling algorithms, can be also extended to classifiers after slight modifications based on the use of the 'confusion matrix'. The latter contains information about actual and predicted classifications and allows several statistical parameters to be derived to assess the performances of the classification systems. An example of such a matrix is represented in Table 1 for a simple two-class ('negative' and 'positive') case, where A and D represent the number of correct predictions that an instance is 'negative' and 'positive', respectively; in return, B and C indicate the number of incorrect predictions that an instance is 'positive' and 'negative', respectively.

The main classification parameter is the accuracy (AC), i.e. the proportion of the total number of predictions that were correct. It is determined using the equation

$$AC = \frac{A+D}{A+B+C+D}.$$

The cross-validation and Y-randomization methods work on this statistical parameter accordingly to the same rules listed above. Simple derivations of the AC formula allow deriving other very useful parameters, such as

Table 1
Example of a confusion matrix for a two-class classifier discriminating 'positive' and 'negative'

		Predicted	
		Negative	Positive
Experimental	Negative	A	B
	Positive	C	D

- True positive and negative rate (*TP* and *TN*), i.e. the ratio of positive and negative cases that were correctly identified, respectively, calculated using the equations

  $$TP = \frac{D}{C+D}; \quad TN = \frac{A}{A+B}.$$

  The latter parameters are also known as 'sensitivity' and 'specificity', respectively.

- False positive and negative rate (*FP* and *FN*), i.e. the ratio of negative and positive cases that were incorrectly classified as positive and negative, respectively, computed by the following equations

  $$FP = \frac{B}{A+B}; \quad FN = \frac{C}{C+D}.$$

- Precision (*P*), i.e. the ratio of the predicted positive cases that were correct, evaluated by

  $$P = \frac{D}{B+D}.$$

The accuracy determined by the above equation may not be an adequate performance measure when the number of negative cases is much greater than the number of positive cases (Kubat *et al.*, 1998); the accuracy remains very high even though, e.g. the classifier predicts the negative cases well but totally misses all the positive ones. In this case, other parameters account better for this behaviour, such as geometric mean (*g*-mean) and *F* measure (*F*):

$$g-mean = \sqrt{TP \times TN}; \quad F = \frac{(\beta^2+1) \times P \times TP}{\beta^2 \times P + TP},$$

where β has a value from 0 to infinity and is used to control the weight assigned to *TP* and *P*.

In fact, any classifier evaluated using these equations will have a performance value of 0, if all positive cases are classified incorrectly.

6. GRAPHICAL EVALUATION OF THE MODELS: THE RECEIVER OPERATING CHARACTERISTIC AND REGRESSION ERROR CHARACTERISTIC CURVES

Receiver operating characteristic (ROC) graphs represent an alternative way to confusion matrices, to examine the classifier performances, by plotting *TP* versus

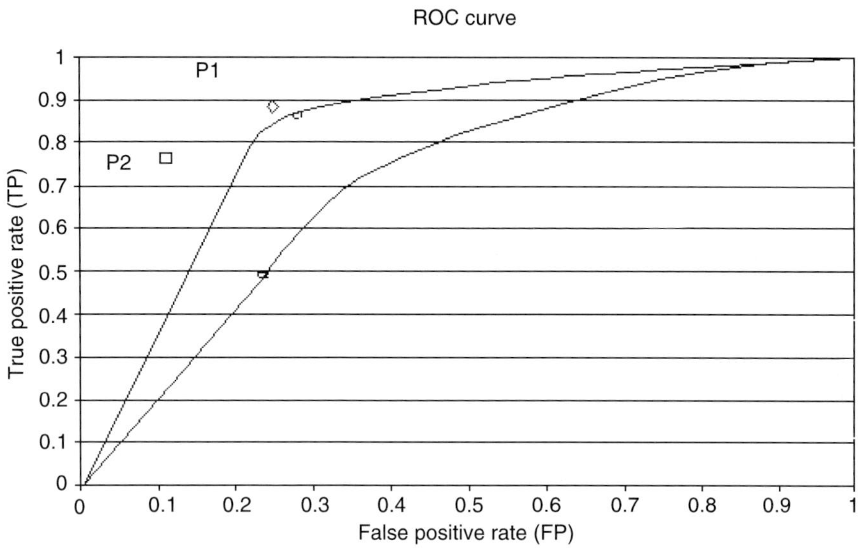

Figure 1: Example of an receiver operating characteristic (ROC) curve.

FP as represented in Fig. 1. ROC curves have proven to be a valuable way to evaluate the quality of a discriminant function (classifier). For two-class discrimination problems, the ROC curve is constructed by varying the threshold or probability used to discriminate between classes for that function. A classifier performs well if the ROC curve climbs rapidly towards the upper left-hand corner. The expected performance of a classifier can be characterized by the area under the ROC curve (AUC): the AUC for a perfect classifier is 1. Random guessing would yield a diagonal line with the AUC equal to 0.5, so valid classifier should have an AUC greater than 0.5. The point (0,1) is the perfect classifier, as all positive and negative cases are predicted correctly. The points (0,0) and (1,1) represent a classifier that predicts all cases to be negative and positive, respectively, whereas the point (1,0) is associated with a classifier that predicts ever wrongly. In many cases, a classifier's parameters can be adjusted to increase *TP* at the cost of an increased *FP* or decrease *FP* at the cost of a decrease in *TP*. Then, each parameter setting provides a (*FP*, *TP*) pair, and a series of such pairs can be used to plot an ROC curve (Fig. 1).

The main advantages in using the ROC graphs are

- each point is not dependent on the class distribution or error costs;
- they incorporate all information contained in the confusion matrix, because *FN* is the complement of *TP* and *TN* is the complement of *FP*; and

- provide a visual tool for examining the trade-off between the ability of a classifier to correctly identify positive cases and the number of negative cases that are incorrectly classified.

It must be stressed that the AUC can be used as an alternative measure of accuracy, above all for comparing classifiers with imprecise cost and class distribution information.

Another way to derive accuracy from ROC graphs is equating it (AC_d) with the Euclidian distance from the perfect classifier, point (0,1), on the graph. A weight factor W is also included to define relative misclassification costs, if such information is available.

$$AC_d = 1 - \sqrt{W(1-TP)^2 + (1-W)FP^2},$$

where the weight W is ranging from 0 to 1 and that is used to assign relative importance to false positives and false negatives. AC_d varies from 0, the perfect classifier, to $\sqrt{2}$ for a classifier that classifies all cases incorrectly. It differs from g-mean and F as it is equal to 0 only if all cases are classified correctly. So, a classifier evaluated using AC_d obtains some credits for correct classification of negative cases, regardless of its accuracy in correctly identifying positive cases.

ROC curves for classifiers have been exemplified in the recent predictive toxicology challenge (Helma and Kramer, 2003; Toivonen *et al.*, 2003).

Another paper (Bi and Bennett, 2003) devised a methodology for regression problems with similar benefits to those of ROC curves. In regression, existing measures of residuals such as mean squared error, mean absolute deviation, R^2, and Q^2 provide only a single snapshot of the performance of the regression model, whereas regression error characteristic (REC) curves have comparable benefits of ROC.

REC plots the error tolerance on the x-axis versus the percentage of points predicted within the tolerance on the y-axis (accuracy). The resulting curve estimates the cumulative distribution function of the error. The error can be defined as the difference between the predicted value $f(x)$ and actual value y of response for any point (x, y), or the squared residual $(y - f(x))^2$. Accuracy is defined as the percentage of points that fit within the tolerance. If we have zero tolerance, only those points that the function fits exactly would be considered accurate. If we choose a tolerance that exceeds the maximum error observed for the model on all the data, then all points would be considered accurate. The concept of error tolerance is appealing because it gives a measure useful for the human expert. So as the tolerance increases, the accuracy also increases and eventually goes to 1.

The range of the tolerance adjusts the appearance of REC curves. As an example, we have applied REC to basic and combined models. Starting from the basic models, we computed the mean model by averaging, and then plotted this model on the REC box. Then, we produced more complex hybrid models and again plotted them on the same REC area. The x-axis of the box starts with 0 and ends as the largest value of the errors obtained by the average model on the sample data. In ROC curves, the AUC provides an estimate of the expected accuracy. Correspondingly the area over the REC curve (AOC) is a measure of the expected error for a regression model (Bi and Bennett, 2003).

For example, considering the quail (oral exposure) endpoint, we obtained a mean model with values: R^2_{train} (mean) $= 0.69$; Q^2_{train} (mean) $= 0.67$; R^2_{test} (mean) $= 0.45$.

If we build a stacking ensemble based on a simple ANN, we improve the mean model. We can visualize the models in the REC curve, where we see that the ANN model dominates the mean model. There is still a way to improve this ensemble model, for instance using Bayesian regularization. If we want to compare the two ANN models, we can draw together their REC curves on the training and on the test set, as we see in Figs. 2 and 3. Here, we see that the second model is better that the first: it reaches unitary accuracy before and dominates the other model.

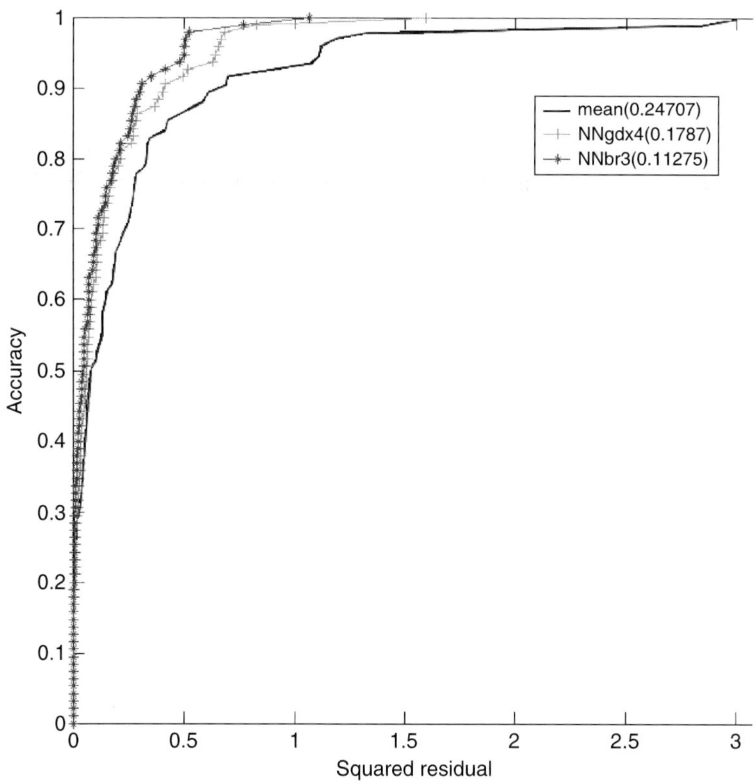

Figure 2: The regression error characteristic (REC) curves for three quail (oral exposure) hybrid models for the mean and two ANN for the training set. In the box, we see the value of the area over the REC curve (AOC).

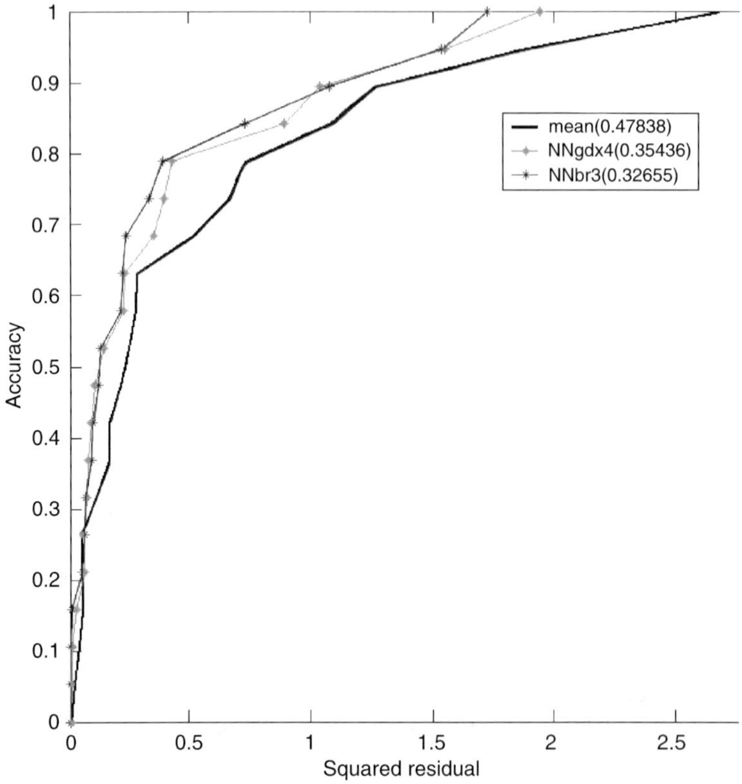

Figure 3: The regression error characteristic (REC) curves for three quail (oral exposure) hybrid models for the mean and two ANN for the test set. In the box we see the value of the area over the REC curve (AOC).

7. HOW TO DEAL WITH FALSE NEGATIVES/FALSE POSITIVES

The QSAR models have traditionally been used to identify good regression with a given phenomenon, such as acute fish toxicity, in the case of models for toxicity. The quality of a given QSAR model has been measured since the beginning on the basis of how good the calculated values were compared with the experimental ones. As we have seen in the previous sections, to state that a given model is predictive requires additional statistical checks, to be sure that it is not over-fitted, i.e. that the values are fitted in a good way simply thanks to the powerful software which can nowadays mimic the pattern of any set of values, especially using non-linear systems and a large set of parameters.

However, careful attention should be given to the intended use of a given model. As we commented in a previous paper (Benfenati, 2004), there are different interests in developing/using a QSAR model. The complexity of the scientific issues involved in the QSAR model development is not sufficient to take into account all different perspectives in the use of a model. Typically, a modeller is interested in developing a new chemical descriptor or a new algorithm. A new model, improved on the basis of the performances, has in many cases been a valid academic interest. The issue, as we said, is complex on a scientific point of view because the modeller should take care of many factors, related to different fields: toxicology, biochemistry, theoretical chemistry, statistics, computer science, etc. But besides these scientific aspects, the real use of a given model should face other aspects, related to the possible application of the values, the interest of a given endpoint, the reliability of the results, the possibility to use it in a given context, etc.

In particular, in the case of DEMETRA, and more in general in the case of assessment to be used for environmental and toxicological aspects, regulators are very careful in avoiding the risk of contamination and exposure to a chemical which is harmful: if the regulator approves a given use for a chemical, its use has to be safe, in the defined context. This means that the QSAR model should avoid as much as possible to predict a pesticide as being safe, when in fact it is harmful. This prediction is called false negative. The opposite case, to predict a pesticide as being toxic when in fact it is not (false positive), represents a minor mistake for the regulator.

Although in the case of classification problems, especially binary classifiers, it is common to assess false positive and false negative, and to measure them separately (see Section 5), in the case of regression algorithms, the appreciation of false positive and false negative is not done. Thus, this requires a conceptual modification of the way to measure the performances of the models, which is not the typical case of QSAR models. When we discuss the results of our models (Chapter 7, Amaury *et al.*, this volume), we will list and distinguish the false positive and the false negative predictions for the reason here discussed. This is another example of the novelty of our approach in this project.

Aside from being a way to express and discuss the results, the relevance of false negative may induce a different strategy in the modelling philosophy compared with a 'neutral', simply mathematical point of view. We already discussed in Chapter 5 (Amaury *et al.*, this volume) the development of a new approach to combine individual models forcing the results in a way to reduce false negative predictions. A similar attention has to be put in all steps of the modelling process, and in the evaluation of the results, to meet the requests of the regulators, who have to avoid in particular the risk to allow the use of pesticides that are unexpectedly toxic.

8. THE APPLICABILITY DOMAIN

The concept of AD has shown an increasing consideration in the QSAR field due to the need of better defining areas where it is possible to practically use the models with an increased confidence about the prediction so obtained (Netzeva *et al.*, 2005).

Many methods have been recently evaluated for possible use in the AD definition, and a complete overview is out of the scope of this chapter. However, the concept that has been most widely used in addressing this topic is that of similarity (Sheridan *et al.*, 2004). The goal is to assess how similar two compounds are or, better, how similar a given compound is to a population of compounds. In the case of the possible application of a model, the discussion is complicated, because the real problem would be to assess if a new compound to be evaluated by the model is similar and fits well into the population of chemicals used to build up the model.

Several metrics can be used to assess similarity (Makara, 2001; Besalú *et al.*, 2002), and some have been proposed to address the AD definition. Among the available approaches, there are a series of chemometric tools based on the comparison of the descriptors used to develop the models, for the new molecules to be tested with the descriptors distribution for the molecules in the training set. A review on these approaches has been recently published (Jaworska *et al.*, 2005). Furthermore, these methods have been implemented in a freely available software: AMBIT Disclosure (AMBIT, 2006).

Another approach foresees the comparison of the structural features of the compounds in an aprioristic way, without necessarily using the descriptors selected in the models. In this case, structures are encoded in fingerprints or by taking into account relevant fragments and using them to assess the similarity with the training set. A similar concept has been also used to address the problem of outliers for the DEMETRA hybrid models. Of course, depending on the definition for the AD adopted in each situation, the outcome may vary.

An important issue is related to the probabilistic meaning of the AD definition: if the compound belongs to the training set population, it is possible to refer to the statistical measurement to describe its similarity with the original data set. This does not necessarily mean that if the compound is out of the AD the predicted value will be wrong and vice versa. The result will be a kind of measurement of the probability for a given compound to belong to this population. If the compound does not belong to this population, we do not know if the results will be correct. If the compound is inside the AD, it means that its structural features are similar to those of the training set, but again it can be placed in some areas of weakness of the model and this possibility has to be taken into account for evaluating the prediction reliability.

REFERENCES

AMBIT (2006) http://ambit.acad.bg/.

Benfenati E. (2004) Modelling aquatic toxicity with advanced computational techniques: procedures to standardize data and compare models. In: *Lecture Notes in Computer Science Volume 3303/2004 'Knowledge Exploration in Life Science Informatics'*, López J.A., Benfenati E., Dubitzky W., Eds. Springer Verlag, Berlin, Germany, 235–248.

Besalú E., Gironés X., Amat L., Carbó-Dorca R. (2002) Molecular quantum similarity and the fundamentals of QSAR. *Acc. Chem. Res.*, 35, 289–295.

Bi J., Bennett K.P. (2003) Regression error characteristic curves. In: *Proceedings of the Twentieth International Conference on Machine Learning (ICML-2003)*, AAAI Press, Menlo Park, CA.

CE (1995) 95/365/CE: commission decision of 25 July 1995 establishing the ecological criteria for the award of the Community eco-label to laundry detergents. *Official J. L.*, 217, 0014–0030.

Devillers J. (2001) QSAR modeling of large heterogeneous sets of molecules. *SAR QSAR Environ. Res.*, 12, 515–528.

Eriksson L., Jaworska J., Worth A., Cronin M., McDowell R.M., Gramatica P. (2003) Methods for reliability, uncertainty assessment, and applicability evaluations of regression based and classification QSARs. *Environ. Health Perspect.*, 111, 1361–1375.

Eriksson L., Johansson E., Kettaneh-Wold N., Wold S. (1999) *Introduction to Multi and Megavariate Data Analysis Using Projection Methods (PCA & PLS)*. Umetrics AB, Umeå, Sweden.

Gasteiger J., Zupan J. (1993) Neural networks in chemistry. *Angew. Chem. Int. Ed. Engl.*, 32, 503–527.

Golbraikh A., Shen M., Xiao Z., Xiao Y.D., Lee K.H., Tropsha A. (2003) Rational selection of training and test sets for the development of validated QSAR models. *J. Comput. Aided Mol. Des.*, 17, 241–253.

Golbraikh A., Tropsha A. (2002) Beware of q2! *J. Mol. Graph. Model.*, 20, 269–276.

Helma C., Kramer S. (2003) A survey of the predictive toxicology challenge 2000-2001. *Bioinformatics*, 19, 1179–1182.

Jaworska J., Nikolova-Jeliazkova N., Aldenberg T. (2005) QSAR applicability domain estimation by projection of the training set in descriptor space: a review. *Altern. Lab. Anim.*, 33, 445–459.

Kennard R.W., Stone L.A. (1969) Computer aided design of experiments. *Technometrics*, 11, 137–148.

Kubat M., Holte R.C., Matwin S. (1998) Machine learning for the detection of oil spills in satellite radar images. *Mach. Learn.*, 30, 195–215.

Makara G.M. (2001) Measuring molecular similarity and diversity: total pharmacophore diversity. *J. Med. Chem.*, 44, 3563–3571.

Marengo E., Todeschini R. (1992) A new algorithm for optimal distance – based experimental design. *Chemom. Int. Lab. Syst.*, 16, 37–44.

Netzeva T.I., Worth A.P., Aldenberg T., Benigni R., Cronin M.T.D., Gramatica P., Jaworska J.S., Kahn S., Klopman G., Marchant C.A., Myatt G., Nikolova-Jeliazkova N., Patlewicz G.Y., Perkins R., Roberts D.W., Schultz T.W., Stanton D.T., van de Sandt J.J.M, Tong W., Veith G., Yang, C. (2005) Current status of methods for defining the applicability domain of (quantitative) structure-activity relationships. *Altern. Lab. Anim.*, 33, 155–173.

Sheridan R.P., Feuston B.P., Maiorov V.N., Kearsley S.K. (2004) Similarity to molecules in the training set is a good discriminator for prediction accuracy in QSAR. *J. Chem. Inf. Comput. Sci.*, 44, 1912–1928.

Snarey M., Terrett N.K., Willett P., Wilton D.J. (1997) Comparison of algorithms for dissimilarity-based compound selection. *J. Mol. Graph. Model.*, 15, 372–385.

Toivonen H., Srinivasan A., King R.D., Kramer S., Helma C. (2003) Statistical evaluation of the predictive toxicology challenge 2000-2001. *Bioinformatics*, 19, 1183–1193.

Tropsha A., Grammatica P., Gombar V.K. (2003) The importance of being earnest: validation is the absolute essential for successful application and interpretation of QSPR models. *QSAR Comb. Sci.*, 22, 69–77.

Zupan J., Novic M., Ruisánchez I. (1997) Kohonen and counter propagation artificial neural networks in analytical chemistry. *Chemom. Int. Lab. Syst.*, 38, 1–23.

Quantitative Structure-Activity Relationships (QSAR) for Pesticide Regulatory Purposes
Edited by Emilio Benfenati

Chapter 7

Results of DEMETRA models

Nicolas Amaury[1], Emilio Benfenati[2], Elena Boriani[2], Mosè Casalegno[2], Antonio Chana[2], Qasim Chaudhry[3], Jacques R. Chrétien[1], Jane Cotterill[3], Frank Lemke[4], Nadège Piclin[1], Marco Pintore[1], Chiara Porcelli[2], Nicholas Price[5], Alessandra Roncaglioni[2], Andrey Toropov[2]

[1]BioChemics Consulting SAS, Orléans, France

[2]Laboratory of Environmental Chemistry and Toxicology, Istituto di Ricerche Farmacologiche "Mario Negri", Milano, Italy

[3]Central Science Laboratory, Sand Hutton, York, United Kingdom

[4]KnowledgeMiner, Berlin, Germany

[5]Central Science Laboratory Fellow, Technology for Growth, York, UK

1. OVERVIEW OF RESULTS WITH THE REGRESSION APPROACH

At the basis of the DEMETRA models there are many steps, aimed to optimize the final model. The overall process, as described in the Chapters 1 to 6, involves a careful selection of the data, a check of the chemical structures, the calculation of thousands of descriptors and fragments and on that basis a development of hundreds of models. Current computer techniques allow the exploration of a huge space of possibilities in a short time, facilitating our task. What we did was to explore a full battery of models, as described in Chapter 5 (Amaury *et al.*, this volume). As expected, many of the models were not valid, and the performances were poor. However, a certain number of models gave interesting results. Table 1 summarizes some of the best models.

We obtained good results with the use of different models and different chemical descriptors. Models using three-dimensional (3D)-based descriptors were not better than models based simply on two-dimensional (2D) descriptors. This fact is not completely unexpected. Indeed, also models for fish toxicity using a large data set of chemicals reached similar results (Netzeva *et al.*, 2005). One likely reason is that the variability of the experimental data, which is quite large compared to the variability of 2D descriptors, does not allow full advantage to be taken of the theoretically larger information present in 3D descriptors. Indeed, in case of quite complex situations, as this one studied for pesticides, we can surely expect variability for the numerous factors involved. The 3D descriptors are suited

Table 1
Detailed report of results for the five end points

Models*	Type of descriptors*	Des[a]	Method[a]	R^{2a} training	Q^2(LSO)[a]	RMSE[a]	MFE[a]	R^2Test[a]	K^a	R^{2a}_0	$(R^2-R^2_0)/R^{2a}$	$R^2-R^{2a}_0$	RMSE[a]	MFE[a]
Model1	**DRA/ACD**	**3(2D)**	**AFP**	**0.74**	**0.68**			**0.76**						
Model2	COD/DRA/MDL/ACD	14 (2D)	GMDH-NN	0.70	0.7 (LOO)	0.87	0.67	0.62	0.71	0.46	0.26	0.16	1.04	0.76
Model3	DRA/DFG/CAC	6 (2D)	MLP-NN	0.66	0.66	0.90	0.68	0.63	0.68	0.46	0.27	0.17	0.99	0.78
Model4	MDL/Pallas	11 (2D)	RBF	0.73	0.57			0.73	0.69	0.41	0.44	0.32		
Model5	**DRA/ACD**	**11 (2D)**	**PLS**	**0.67**	**0.60**	**0.91**	**0.70**	**0.59**	**0.65**	**0.37**	**0.37**	**0.22**	**1.07**	**0.77**
Model6	COD/DRA/ACD	11 (2D)	PLS	0.56	0.61	0.92	0.70	0.63	0.62	0.37	0.41	0.26	0.99	0.73
Model7	**DRA/ACD**	**10 (2D)**	**PLS**	**0.66**	**0.62**	**0.93**	**0.69**	**0.61**	**0.64**	**0.35**	**0.43**	**0.26**	**1.06**	**0.75**
Model8	COD/DRA/MDL/ACD	11 (2/3D)	GMDH-NN	0.71	0.71 (LOO)	0.83	0.64	0.64	0.67	0.46	0.28	0.18	0.97	0.77
Model9	DRA	7 (2/3D)	MLR	0.63	0.53 (LOO)	0.95	0.74	0.42	0.43	−0.3	1.74	0.73	1.23	0.95
Model10	REC/CAC/ACD	7 (2/3D)	PLS	0.61	0.56	0.97	0.72	0.59	0.74	0.48	0.19	0.11	1.09	0.83
Trout endpoint														
Model11	DRA/REC/CAC	8 (2D/3D)	GA-MLR	0.66	0.63	0.94	0.71	0.65	0.68	0.43	0.34	0.22	1.02	0.76
Model12	DRA/COD	7 (2/3D)	GA-MLR	0.57	0.54	1.06	0.82	0.45	0.56	0.10	0.78	0.35	1.24	1.02
Model14	COD	4 (3D)	ANFIS	0.66		0.90	0.67	0.49	0.61	0.18	0.63	0.31	1.23	0.96
COMB Model1	Three models	2D		0.71		0.84	0.63	0.71	0.71	0.57	0.20	0.14	0.87	0.65
COMB Model2	Seven models			0.81		0.69	0.54	0.70	0.74	0.56	0.20	0.14	0.91	0.74

COMB Model3	Four models	20 (2D)		0.74		0.96	0.73	0.72	1.02	0.70	0.03	0.02	1.14	0.84
COMB Model3	Three models	20 (2D)		0.74		0.96	0.72	0.70	1.02	0.69	0.01	0.01	1.17	0.86
Daphnia endpoint														
Model1	**DRA/ACD**	**6 (2D)**	**AFP**	**0.78**	**0.69**			**0.61**						
Model2	**DRA**	**11 (2D)**	**MLP-NN**	**0.74**	**0.74 (LOO)**	**0.92**	**0.70**	**0.65**	**0.58**	**0.40**	**0.38**	**0.25**	**1.06**	**0.83**
Model3	**DRA**	**15 (2D)**	**MLP-NN**	**0.77**	**0.76 (LOO)**	**0.88**	**0.67**	**0.59**	**0.59**	**0.35**	**0.41**	**0.24**	**1.14**	**0.89**
Model4	**DRA**	**15 (2D)**	**PLS**	**0.66**	**0.59**	**1.06**	**0.83**	**0.72**	**0.54**	**0.39**	**0.46**	**0.33**	**1.01**	**0.82**
Model5	**DRA**	**11 (2D)**	**PLS**	**0.65**	**0.59**	**1.08**	**0.86**	**0.65**	**0.56**	**0.44**	**0.32**	**0.21**	**1.08**	**0.84**
Model6	DRA	6 (2D)	GMDH-NN	0.62	0.62 (LOO)	1.09	0.89	0.53	0.52	0.12	0.77	0.41	1.24	1.02
Model7	DRA/COD	12 (2D)	GMDH-NN	0.7	0.7 (LOO)	0.98	0.77	0.48	0.40	−0.20	1.42	0.68	1.30	1.02
Model8	DRA	8 (2/3D)	MLR	0.55	0.5 (LOO)	1.19	0.96	0.38	0.39	−0.32	1.84	0.70	1.40	1.06
Model9	DRA/COD	13 (2/3D)	GMDH-NN	0.71	0.71 (LOO)	0.95	0.75	0.44	0.42	0.09	0.80	0.35	1.33	1.06
Model10	DRA/COD	12 (2/3D)	GMDH-NN	0.67	0.67 (LOO)	1.02	0.80	0.42	0.47	0.05	0.88	0.37	1.36	1.05
Model11	DRA/MDL/ COD/REC	10 (2/3D)	GMDH-NN	0.79	0.79 (LOO)	0.80	0.65	0.52	0.53	0.27	0.48	0.25	1.22	0.92
Model12	CAC/COD/ DRA/REC	21 (3D)	ANN	0.67		1.02	0.82	0.55	0.58	0.3	0.45	0.25	1.12	0.95
COMB Model1	Five models			0.81		0.77	0.61	0.61	0.6	0.38	0.38	0.23	1.12	0.85
COMB Model2	Seven models			0.86		0.67	0.52	0.76	0.71	0.62	0.18	0.14	0.89	0.69
COMB Model3	Three models	16 (2D)		0.76		1.20	0.96	0.73	0.89	0.68	0.07	0.05	1.09	0.96
COMB Model4	Three models	16 (2D)		0.76		0.90	0.71	0.72	0.65	0.51	0.29	0.21	0.97	0.84

(*Continued*)

Table 1
(*Continued*)

Models*	Type of descriptors*	Des[a]	Method[a]	R^{2a} training	Q^2(LSO)[a]	RMSE[a]	MFE[a]	R^2 Test[a]	K^a	R^{2a}_0	$(R^2-R^2_0)/R^{2a}$	$R^2-R^{2a}_0$	RMSE[a]	MFE[a]
COMB Model5	Two models	15 (2D)		0.73		1.20	0.95	0.70	0.86	0.64	0.09	0.06	1.11	0.97
Quail (oral exposure) endpoint														
Model1	**DRA**	**7 (2D)**	**AFP**	**0.88**	**0.72**			**0.63**						
Model2	MDL	6 (2D)	MLP-NN	0.67	0.65	0.52	0.40	0.27	0.35	0.24	0.11	0.03	0.83	0.67
Model3	**DRA**	**10 (2D)**	**PLS**	**0.6**	**0.51**	**0.56**	**0.42**	**0.53**	**0.31**	**0.29**	**0.45**	**0.24**	**0.74**	**0.60**
Model4	**DRA**	**12 (2D)**	**PLS**	**0.69**	**0.54**	**0.50**	**0.40**	**0.59**	**0.45**	**0.57**	**0.03**	**0.02**	**0.64**	**0.56**
Model5				0.78		0.42	0.33	0.61	0.48	0.5	0.18	0.11	0.65	0.52
Model6	DRA/ MDLKeys	10 (2/3D)	MLP-NN	0.9	0.9 (LOO)	0.28	0.22	0.49	0.53	0.49	0.00	0.00	0.69	0.60
Model7	COD/REC/ DRA	8 (2/3D)	GMDH-NN	0.72	0.72 (LOO)	0.48	0.39	0.24	0.35	0.2	0.17	0.04	0.87	0.74
COMB Model1	Four models	2D		0.85		0.34	0.27	0.62	0.54	0.59	0.05	0.03	0.61	0.48
COMB Model2	Two models	2D		0.82		0.45	0.36	0.71	0.64	0.45	0.37	0.26	0.64	0.48
COMB Model3	Four models	2D		0.77		0.45	0.37	0.45	0.41	0.43	0.04	0.02	0.72	0.58
COMB Model4	Four models			0.91		0.28	0.22	0.54	0.56	0.53	0.02	0.01	0.66	0.57

Quail (dietary exposure) endpoint														
Model 1	**DRA**	**5 (2D)**	**AFP**	**0.75**	**0.58**			**0.8**						
Model2	DRA	10 (2D)	MLP-NN	0.64	0.62 (LOO)	0.44	0.35	0.31	0.41	0.03	0.90	0.28	0.65	0.48
Model3	**DRA**	**10 (2D)**	**PLS**	**0.59**	**0.51**	**0.47**	**0.36**	**0.53**	**0.58**	**0.31**	**0.42**	**0.22**	**0.53**	**0.36**
Model4	COD/DRA	10 (2D)	GMDH-NN	0.77	0.77 (LOO)	0.35	0.28	0.47	0.57	0.29	0.38	0.18	0.56	0.36
Model5	CAC/REC/ DRA	6 (3D)	MLP-NN	0.78	0.77 (LOO)	0.35	0.27	0.49	0.56	0.31	0.37	0.18	0.55	0.39
Model6	COD/DRA/ REC	23 (2/3D)	GMDH-NN	0.8	0.8 (LOO)	0.34	0.26	0.28	0.39	0.06	0.79	0.22	0.66	0.47
COMB Model1	Three models			0.79		0.34	0.26	0.69	0.73	0.62	0.10	0.07	0.43	0.28
COMB Model2	Five models			0.88		0.26	0.21	0.71	0.79	0.69	0.03	0.02	0.41	0.30
COMB Model3	Seven models			0.89		0.25	0.20	0.73	0.77	0.71	0.03	0.02	0.40	0.31
Honey bee endpoint														
Model1	**DRA**	**5 (2D)**	**AFP**	**0.74**	**0.57**			**0.71**						
Model2	**DRA/COD**	**7 (2D)**	**PCA-NN**	**0.71**	**0.7 (LSO)**	**0.76**	**0.60**	**0.80**	**1.05**	**0.8**	**0.00**	**0.00**	**0.72**	**0.58**
Model3	**DRA**	**10 (2D)**	**PLS**	**0.63**	**0.51**	**0.86**	**0.68**	**0.78**	**0.94**	**0.71**	**0.09**	**0.07**	**0.87**	**0.62**
Model4	DRA	12 (2D)	PLS	0.66	0.55	0.83	0.63	0.69	0.78	0.56	0.19	0.13	0.85	0.65
Model5	DRA/COD	13 (2D)	GMDH-NN	0.7	0.7 (LOO)	0.78	0.63	0.61	0.87	0.56	0.08	0.05	1.01	0.78
Model6	**DRA/COD**	**15 (2D)**	**GMDH- NN**	**0.69**	**0.69 (LOO)**	**0.80**	**0.65**	**0.62**	**0.88**	**0.6**	**0.03**	**0.02**	**0.95**	**0.75**
Model7	DRA	10 (3D)	PLS	0.61	0.52	0.88	0.71	0.52	0.80	0.51	0.02	0.01	1.06	0.77
Model8	DRA/COD	7 (3D)	MLP-NN	0.81	0.81 (LOO)	0.61	0.50	0.57	0.84	0.55	0.04	0.02	1.01	0.81
Model9	COD/REC/ DRA/CAC	11 (3D)	GMDH-NN	0.73	0.73 (LOO)	0.72	0.56	0.61	0.62	0.45	0.26	0.16	0.84	0.63

(*Continued*)

Table 1
(*Continued*)

Models*	Type of descriptors*	Des[a]	Method[a]	R^{2a} training	Q^2(LSO)[a]	RMSE[a]	MFE[a]	R^2 Test[a]	K^a	R^{2a}_0	$(R^2-R^2_0)/R^{2a}$	$R^2-R^{2a}_0$	RMSE[a]	MFE[a]
COMB Model1	Two models			0.8		0.62	0.51	0.78	0.93	0.78	0.00	0.00	0.67	0.48
COMB Model2	Five models			0.83		0.57	0.46	0.78	0.92	0.78	0.00	0.00	0.66	0.48
COMB Model3	Two models	19 (2D)		0.76		0.72	0.58	0.82	1.25	0.82	0.00	0.00	0.93	0.72
COMB Model4	Two models	19 (2D)		0.74		0.78	0.62	0.84	0.97	0.78	0.07	0.06	0.77	0.63
COMB Model5	Two models	17 (2D)		0.76		0.78	0.58	0.84	1.21	0.83	0.01	0.01	0.72	0.58

AFP, adaptive fuzzy partition; ANFIS, Adaptive neuro fuzzy inference system; ANN, artificial neural network; 2D, two-dimensional; 3D, three-dimensional; GMDH, Group method of data handling type (neural network); LOO, Leave one out; LSO, Leave several out; MLP, Multi-layer perceptron (neural network); MLR, multi linear regression; PLS, Partial least squares, RBF, Radial basis function.

[a] Columns report:

Models: If the model was an individual (Model n) or a combined one (COMB Model n).

Type of descriptors: DRA=DRAGON, COD=CODESSA, Combined models are more detailed in Section 4.

Des: The number of descriptors and, in parenthesis, whether is based on 2D or 3D structure.

Method: The algorithm used for individual models.

R^2 training: R^2 on the training set.

Q^2: Q^2 using LSO (unless specified LOO).

RMSE: Root mean square error on the training set.

MFE: Mean fitting error on the training set.

R^2test: R^2 on the test set.

K: The slope of the linear fitting between experimental versus calculated toxicities.

R^2_0: the R^2 of the test set forcing the equation to pass through zero.

$(R^2-R^2_0)/R^2$: Measure of the identity between the two measures. The ideal value is 0.

$(R^2-R^2_0)$: Distance between both measures, a value under 0.3 is acceptable.

RMSE: Root mean square error on the test set.

MFE: Mean fitting error on the test set.

The best models are indicated in bold characters.

to identify fine features in case of small perturbations of the system, such as in the case of more local models, but when the number of factors increases, the system complexity hides the minor phenomena possibly described by 3D descriptors.

This fact is of valuable help in the simplification of the final model, because there is no need of the time-consuming 3D calculations, which also bring a subjective component to the model, because the optimization of the 3D-based structure in most of the cases is done manually by expert modellers. The advantage gained is not only the achieved simplicity but also a greater reproducibility, which is a fundamental aspect for models to be used for regulatory purposes. This is one of the differences between quantitative structure-activity relationships (QSAR) models for regulatory purposes and those for research, where the researcher may explore personal pathways.

Another interesting point is that models based on fragments as produced by typical commercial software in our hands gave lower quality models, compared to those using chemical descriptors (we have, however, to remember that most of the software we used to calculate chemical descriptors, such as CODESSA and DRAGON, calculate a number of simple fragments and constitutional descriptors). The possible reason for this is that, on the basis of the present data set, global descriptors are more suitable to capture important general behaviour of the pesticides. Indeed, on the basis of the relatively reduced set of compounds (a few hundreds, in the most favourable situation) the system may not have enough cases to learn the great number of processes caused by the many fragments. It is possible that starting from a much larger set it may yield better results. However, we obtained some preliminary good results using an in-house set of fragments (see Section 4.2.5, this Chapter), within a spin-off activity studying fragments. The development of specific descriptors was not a planned activity within DEMETRA, because of the limited time frame of the project.

We have to mention here that some interesting results have also been obtained using Comparative Molecular Field Analysis (CoMFA), but we have not used these models within the final integrated models for the reason we already mentioned: CoMFA models are strongly dependent on the alignment of the molecules, which is done using optimization processes based on the manual expertise of the modeller.

Considering the high number of descriptors we used as a starting point, it was fundamental to operate a good reduction of the descriptors. We remember that many descriptors are similar or related. One selection method can choose a given descriptor, while another selection method can select a related descriptor. We also note that the algorithms we used may achieve similar results, using different methods, such as partial least squares (PLS) and artificial neural network (ANN). The heterogeneity of the methodologies increases the robustness of the results, once comparable results are obtained. Indeed, one model can support the other, especially when the starting point and methodology are different.

2. OVERVIEW OF THE PREDICTION RESULTS OBTAINED BY CLASSIFICATION METHODS

2.1. Data sets and toxicity intervals

The five data sets of pesticide compounds, related to toxicity against trout, daphnia, quail (oral and dietary exposure) and bee, were used to develop the classification models (see also Chapter 2, Benfenati *et al.*, this volume). Three toxicity classes were generated for each data set according to the intervals established by the EC regulation (ECC, 1992), after regrouping the toxicity classes as too weakly represented (Table 2). All data sets were split into training and test series. The training set for each species was selected in a rational way according to the Kennard and Stone method (Kennard and Stone, 1969) and represents about 80% of the whole data set.

2.2. Descriptors selection and classification results

Several sets of molecular descriptors were tested and compared to build the structure-activity models on the five endpoints, divided into two main categories: (1) parameters derived from 2D structures and (2) parameters computed on optimized 3D structures (see also Chapter 3, Benfenati *et al.*, this volume).

The first step of the data mining strategy consisted, for each endpoint and set of descriptors, on selecting the most relevant parameters by hybrid selection algorithm (HSA) (see also Chapter 4, Chaudhry *et al.*, this volume).

Then, all adaptive fuzzy partition (AFP) models were established on the training set compounds distributed in the hyperspace derived from the descriptors selected. The best models retained to predict later the test compounds were those joining the highest cross-validation leave several out (LSO) scores and lowest number of AFP rules.

2.2.1. Aquatic species

The most relevant parameters selected for the two aquatic species, whatever the descriptor set used, included in all cases one or more lipophilicity parameters (Table 3). Actually, lipophilicity is a very important property involved in the mechanism of molecular accumulation into the fish body and, more generally, plays a fundamental role in all aquatic organisms (Hermens, 1990). Furthermore, other descriptors strictly related with an aquatic environment, such as H-bonds and polarity parameters, were often selected.

A general analysis of all models, when the best results are compared, underlines that the robustness and the prediction abilities obtained by using descriptors computed on 2D structures are similar to the 3D ones. In return, the models established on the fingerprints show in any case the worst performances.

The best validation results related to daphnia were slightly lower than those associated with the trout model, above all for the test set prediction

Table 2
Compound distribution for the aquatic (trout and daphnia) and terrestrial (quail and bee) species, respectively

	Toxicity	Total compounds	Training set	Test set
Trout	LC50 (mg/l)			
Class 1	<1	128	108	20
Class 2	[1–10]	89	74	15
Class 3	≥ 10	64	53	11
All classes		**281**	**235**	**46**
Daphnia	LC50 (mg/l)			
Class 1	<1	121	104	17
Class 2	[1–10]	65	52	13
Class 3	≥ 10	77	64	13
All classes		**263**	**220**	**43**
Quail (oral)	LD50 (mg/kg body weight)			
Class 1	<50	32	27	5
Class 2	[50–500]	24	19	5
Class 3	≥ 500	60	51	9
All classes		**116**	**97**	**19**
Quail (dietary)	LC50 (mg/kg diet)			
Class 1	<1000	47	39	8
Class 2	[1000–5000]	54	45	9
Class 3	≥ 5000	22	19	3
All classes		**123**	**103**	**20**
Bee	LD50 (μg/bee)			
Class 1	<1	36	30	6
Class 2	[1–50]	40	34	6
Class 3	≥ 50	29	24	5
All classes		**105**	**88**	**17**

Class by class and total (in bold) number of compound is given, together the training and test set repartition.

(65–70 vs. 70–75%), whereas the cross-validation scores were quite similar, i.e. 68–70%. All the models were also submitted to a Y-randomization test, by shuffling 50% of the training set compounds; this process was repeated five times. The final average LSO score always fell to about 30–35% and confirmed that no chance correlation or compound redundancy was present in the structure-toxicity relationships developed.

Table 3

Statistical values (ratio of the compounds correctly predicted to the total number of compounds) defining the robustness and the prediction ability of the best aquatic AFP models established on several descriptor sets computed on 2D and 3D structures

Software	Descriptors selected by HSA	Training (%)	LSO (%)	Test (%)
	Trout models			
MDL (2D)	Five descriptors: SsOH_acnt; SssS_acnt; SHBint2_acnt; Qsv; Log *D*-pH9	74	67	71
DRAGON (2D)	**Three descriptors: Mp; nS; Log *D*-pH7**	**74**	**68**	**76**
CODESSA (3D)	Seven descriptors: FPSA-2; HACA-2/SQRT(TMSA); LUMO; HBCA; Tmol-2-centRes E/Atnum; Log *D*-pH5 ; Log *D*-pH7.4	76	61	65
DRAGON (3D)	Five descriptors: HATS4p; R1p+; RDF045p; L3s; Log *D*-pH9	71	60	70
MDL Keys	**Eight fingerprints**	64	57	57
	Daphnia models			
MDL (2D)	Seven descriptors: SssO; SssS; SsssCH_acnt; Xvp3; SHBint4_acnt; Log *D*-pH6.5; ΔLog *D*(6.5–7.4)	79	61	63
DRAGON (2D)	Six descriptors: Sv; Mp; RBN; NCOOR; Log *D*-pH7.4; Log *D*-pH9	70	69	66
CODESSA (3D)	Three descriptors: COD158; COD262; Log *D*-pH 9	76	64	67
DRAGON (3D)	**Seven descriptors: qnmax; Mor04m; Mor17p; HATS1m; RDF080v; Vu; Log *D*-pH9**	**80**	**70**	**70**
MDL Keys	Eight fingerprints	72	60	61

The best model for each endpoint is indicated in bold characters. The most relevant parameters are also indicated.

MDL Keys

AFP, adaptive fuzzy partition; 2D, two-dimensional; 3D, three-dimensional; HSA, hybrid selection algorithm; LSO, leave several out.

By focusing on the trout AFP models, Table 3 summarizes the best results that were obtained with the DRAGON set calculated on 2D structures. The robustness of this model was chiefly confirmed by the LSO score that reached 68%. More importantly, this value was not too dissimilar to the prediction scores associated with the training set; this means the model was not significantly perturbed by eliminating 10% of training set information and well-represented all the different structure-toxicity relationships associated with the trout data set. The prediction ability of this model is reported in Table 4, which shows the percentage of test compounds rightly predicted, globally and class by class. The average score of 76% indicates that the AFP model is undoubtedly predictive. Furthermore, this value is similar to the score obtained by testing the training set compounds, showing that the model developed is general. It has also to be underlined that Class 1, represented by high-toxicity compounds, was the best predicted class, with a non-error of 85%. This result is very important for real applications of these models, because predicting as being toxic a compound which is not (false positive) is preferable to predicting the inverse. Finally, the 2D DRAGON model was generated with a highly reduced number of descriptors. The simplicity of these parameters, which characterize lipophilicity, polarizability and the number of sulphur atoms, can help the modeller to derive an easier interpretation of the model.

Table 4
Class by class and global (in bold) prediction abilities, represented by the ratio of the compounds correctly predicted to the total number of compounds, for the best aquatic models reported in Table 3

Class	LC50 (mg/l)	ACD/LogD descriptors		Pallas descriptors	
		Training%	Test%	Training%	Test%
		2D Dragon trout model			
1	<1	75	85	80	80
2	[1; 10]	73	67	73	67
3	≥ 10	74	73	74	73
All classes		**74**	**76**	**77**	**74**
		3D Dragon daphnia model			
1	<1	84	77	86	77
2	[1; 10]	71	62	65	69
3	≥ 10	81	69	83	69
All classes		**80**	**70**	**80**	**72**

The validation scores obtained with lipophilicity parameters computed by ACD/logD or PALLAS are compared.
2D, two-dimensional; 3D, three-dimensional.

In the daphnia case, the best AFP model was still derived by using DRAGON parameters but computed on 3D structures. The model is robust (LSO = 68%), and the prediction scores evaluated on the test set, globally and class by class, remain satisfactory (Table 4). Furthermore, the most toxic class is always the best-predicted one, with a validation score of 77%. However, the model derived by CODESSA descriptors could be a good choice too. Actually, the different validation and prediction scores are only slightly weaker (2–3%), but this model was established with only three descriptors, instead of seven, and its applicability and interpretability are then clearly superior.

The last analyses performed on the trout and daphnia models examined the dependence of their performances on the software used to compute the lipophilicity parameters. The ACD/log *D* descriptors were replaced by the PALLAS ones, and the two best models were rebuilt. Table 5 shows that the global and class-by-class validation scores are very similar, with negligible differences inferior to 2–3%. These results demonstrate it is possible to use lipophilicity parameters computed by different software, while keeping similar robustness and prediction power.

2.2.2. Terrestrial species

The descriptor sets and the validation results associated with the best AFP models established on quail and bee are included in Table 5. A main difference with the aquatic models can be emphasized immediately; lipophilicity descriptors were never present amongst the most relevant parameters selected, except one case. In return, H-bonds and polarity parameters were retained frequently, like the aquatic case.

The comparison between the models using descriptors computed on 2D and 3D structures, respectively, confirms that the 3D contribution is not relevant in improving the model quality. On the contrary, the 3D results are clearly lower, above all in the test prediction, with an average gap of about 10%. About fingerprints, the validation results are similar to those obtained by the 3D set, thus largely inferior to the 2D one.

The analysis of the single species shows that the best quail model (oral exposure) was developed with MDL descriptors. Both cross-validation and test set prediction are very satisfactory, with scores of 73 and 74%, respectively; moreover, the class-by-class scores included in Table 6 underline that, once more, the most toxic compounds are the best-predicted ones. Very similar results and trends were obtained by the best-dietary quail model, generated by 2D DRAGON descriptors; the only difference is the LSO score, slightly lower than the oral one.

Finally, the best bee model was established on 2D DRAGON descriptors. Its robustness, defined by LSO = 57%, is inferior to the other terrestrial and aquatic species, whereas the prediction ability on the test set is quite similar, i.e. 71%. The relative weakness of the bee model with regard to the other endpoints is confirmed by the class-by-class validation (Table 6), where the best predicted class is the least toxic one.

Table 5
Statistical values defining the robustness and the prediction ability of the terrestrial adaptive fuzzy partition (AFP) models established on different work descriptor sets

Software	Descriptors selected by HSA	Training (%)	LSO (%)	Test (%)
	Quail models (oral exposure)			
MDL (2D)	**Six descriptors: SssNH; SsCH3_acnt; SdO_acnt; SdsssP_acnt; xvp7; W**	**85**	**73**	**74**
DRAGON (2D)	Seven descriptors: nP; C-008; X5v; CIC5; BELv3; BELp4; ATS8p	87	72	63
CODESSA (3D)	Eight descriptors: Pol/d^2;At Charg Net Min; DPSA-2; RPCG; HBSA; HBCA; HDCA-2;C at Max Val	82	67	53
DRAGON (3D)	Seven descriptors: qpmax; Mor18e; Mor27e; H5e; RDF100p; G1m; Te	93	68	58
MDL Keys	Four fingerprints	77	65	58
	Quail models (dietary exposure)			
MDL (2D)	Five descriptors: SssssC_acnt; SdsssP_acnt; nxpc4; xvpc4; nelem	66	62	59
DRAGON (2D)	**Five descriptors: nAB; nO; nP; nCt; nPhX**	**76**	**66**	**73**
CODESSA (3D)	Six descriptors: nH; nPRel; nBonds; K&H-I02; IC01; $\log D(7.4)$	75	62	70
DRAGON (3D)	Six descriptors: qpmax; W3D; R5u; R2m+; RDF050u; RDF115v	81	64	65
MDL keys	Six fingerprints	74	66	65
	Bee models			
MDL (2D)	Six descriptors: SssCH2; SdsCH; SssO; SsCl; idcbar; SHBint2	77	57	69
DRAGON (2D)	**Five descriptors: Mv; nP; nCOOR; nRORPh; nHDon**	**74**	**58**	**71**
DRAGON (3D)	Six descriptors: ADDD; G(O..S); Mor04v; SHP2; ISH; RDF015v	75	51	53
MDL keys	Four fingerprints	83	57	53

The best models for each endpoint is indicated in bold characters. The most relevant parameters are also indicated.
2D, two-dimensional; 3D, three-dimensional.

Table 6

Class by class and global (in bold) prediction abilities, represented by the ratio of the compounds correctly predicted to the total number of compounds, for the best terrestrial models reported in Table 5

Class	Toxicity	Training set	Test set Q85
	2D MDL quail model (oral exposure)		
1	<50	85%	80%
2	[50; 500]	58%	80%
3	≥500	94%	67%
All classes		**85%**	**74%**
	2D Dragon quail model (dietary exposure)		
1	<1000	74%	88%
2	[1000; 5000]	80%	67%
3	≥ 5000	63%	50%
All classes		**76%**	**73%**
	2D Dragon bee model		
1	<1	60%	67%
2	[1; 50]	85%	67%
3	≥50	75%	80%
All classes		**74%**	**71%**

2D, two-dimensional.

2.3. Conclusions about classification results

Globally, the best models developed gave good classification results for each endpoint with cross-validation and test set scores that reached values of about 70–75%. Moreover, the Y-randomization test showed that the models are significant with a high degree of certitude, and no chance correlation or training compound redundancy is hidden in them. It has to be also underlined that the most toxic class, which can be considered as the most important for real application of the models, was generally the best-predicted one, with validation ratios of 75–85%. The only exception to this general trend was represented by the bee models, which showed lower robustness and worse class-by-class predictions. Owing to the high variability affecting the experimental procedures in the area of ecotoxicity and the complexity of the phenomena related, these results are very encouraging.

Finally, the comparison between the best AFP models established by descriptors computed on 2D and 3D structures, respectively, gave similar validation and prediction scores. This underlines that probably, in ecotoxicity, where a large number of complex and interactive mechanisms define the biological phenomena, the role of the 3D descriptors in generating SAR models is less important with regard to other fields. By focusing on the DRAGON parameters, it is interesting to observe that relevant parameters selected in the 2D models were often not retained in the 3D case, so weakening the related models. This behaviour is associated with the increased complexity of the 3D set, where the parameters double; the HSA procedure must then explore a larger hyperspace, and the difficulties to converge to the good solutions increase in proportional way.

Another set of descriptors tested, i.e. fingerprints, gave generally results inferior to the other 2D and 3D sets; this trend should be inherent to the very limited contribution of pure structural fragments to the global molecular information or bioactivity of any molecule.

The performances of the different descriptor software were also compared. Generally, the models established by DRAGON gave the best models, except in the case of oral quail endpoint. However, the average differences in the prediction ability were not relevant and show the possibility to use different descriptor packages for obtaining satisfactory ecotoxicity models. Similar conclusions were formulated by replacing, in the aquatic models, the lipophilicity parameters computed by ACD/logD with the PALLAS ones; the differences in the global and class-by-class prediction scores were never more than 2–3%.

3. OVERVIEW OF RESULTS OF LOCAL MODELS

3.1. Chemical classes

The goal of the DEMETRA project was to build a single, combined QSAR model for a given biological endpoint on a set of chemical compounds, which span various very different chemical classes according to a number of chemical properties. In contrast to these so-called global models, we also followed for the trout endpoint a supplemental approach by building QSAR models on clearly defined chemical classes (sub-models).

There have been several experiments (Gini *et al.*, 1998, 2004), which employ data-driven algorithms of clustering such as self-organizing maps, neural networks or principal component analysis to find some similarities between the chemical compounds and to finally define chemical classes in this way. However, seen from a general perspective, sub-dividing (classifying) a given data set into subsets (chemical classes) introduces another model (classification model) into the overall process of building QSAR models from data. This classification model also introduces error and uncertainty into the overall model results, which, of course, has to be considered when evaluating the results of the final QSAR models. This effect is widely underestimated when using data-driven classification models. The quality and performance of both classification and QSAR models is important for the overall predictive power of a solution, finally. We followed a theory-driven approach for building chemical classes, because by using well-defined theoretical domain knowledge as classification rules we get reliable and highly predictive classification models, which are also valid on new chemical compounds. In this way, the misclassification rate of the classification model is reduced to a minimum.

The division of the entire trout data set into chemical classes was done by theoretical chemical domain knowledge, and the resulting chemical classes describe the most general identity of the compounds of a given class. Every compound was classified according to any particular fragment present in this compound, which means that a compound could belong to more than just one chemical class. For example, compounds containing one pyrethroid ring plus one heterocycle were classified into both classes a priori.

The chemical classes approach we started for the trout data set divides all compounds into 11 main chemical classes specific for pesticides: organo phosphorous, pyrethroids, triazines, heterocycles, aromatics, halogenated, organometallic, carbamates, ureas, small compounds and mixed compounds.

The following section outlines three different QSAR modelling approaches and their results:

1. Using high-dimensional descriptors sets and self-organizing statistical learning networks;
2. Using CODESSA descriptors and best multi-linear regression; and
3. Using a combined model for the heterocyclic class based on multiple toxicity values.

3.1.1. Using high-dimensional descriptors sets and self-organizing statistical learning networks

Here, the initial data set was composed of all 2D descriptors obtained by ACD, CODESSA, DRAGON, MDL and PALLAS that defines a descriptor space dimension of 1061. After removing highly redundant descriptors (similarity $>98\%$) from this data set, the data set D_1 of 647 descriptors and 281 compounds was obtained and used for building QSAR models for the given chemical classes. Due to the, sometimes, very small number of compounds the classes consist of, we did not sub-divide the data set into learning and testing subsets, i.e. all compounds of a class was used to build a model. This is a well-known issue for QSAR models built on classes, which means they have to be handled carefully. As model performance metrics, we used descriptive power as described in Section 4.3.3.2, Chapter 4 (Chaudhry *et al.*, this volume), the coefficient of determination (R^2) and the mean absolute percentage error (MAPE) as follows:

$$R^2 = 1 - \delta^2, \delta^2 = \frac{\sum_{i \in N} (y_i - \hat{y}_i)^2}{\sum_{i \in N} (y_i - \bar{y})^2} \leq 1, \quad \text{MAPE} = \frac{\sum_{i \in N} |y_i - \hat{y}_i|}{\sum_{i \in N} |y_i|} \times 100\%, \tag{1}$$

where y_i, $\hat{y}_i$ and $\bar{y}$ are the true, estimated and mean values of the output variable, respectively, and δ^2 is the approximation error variance criterion (Müller *et al.*, 2000).

The models obtained from the data set D_1 using self-organizing statistical learning networks (see Section 4.3 in Chapter 4, Chaudhry *et al.*, this volume) are given in Table 7. To have an indication, at least, on how valid these models might be, Table 7 also summarizes the model results when applying a model on all non-class member compounds.

Table 7
Models on specific classes of pesticides

Class	Class members				Non-class members	
	Model	R^2	DP	MAPE	R^2	MAPE
Organo phosphorous ($n = 66$)	Model 1: linear. Descriptors: 9. Noise immunity: good	0.87	61%	12%	0.0	88%
	Model 2: linear. Descriptors: 13. Noise immunity: sufficient	0.93	71%	9%	0.0	76%
Mixed ($n = 16$)	Model 1: linear. Descriptor: 1. Noise immunity: good	0.43	4%	46%	0.0	72%
	Model 2: linear. Descriptors: 5. Noise immunity: very poor	0.85	42%	24%	0.0	106%
Pyrethroids ($n = 24$)	Model 1: linear. Descriptors: 2. Noise immunity: good	0.97	81%	4%	0.0	117%
One or more HetArRing ($n = 91$)	Model 1: linear. Descriptors: 14. Noise immunity: good	0.81	53%	22%	0.13	44%
	Model 2: non-linear. Descriptors: 4. Noise immunity: good	0.75	46%	24%	0.0	>150%
	Model 3: non-linear. Descriptors: 7. Noise immunity: very poor	0.81	46%	22%	0.0	70%
One or more HetCyclic ($n = 99$)	Model 1: linear. Descriptors: 14. Noise immunity: good	0.83	56%	20%	0.0	54%
	Model 2: linear. Descriptors: 17. Noise immunity: sufficient	0.90	66%	17%	0.0	49%

(*Continued*)

Table 7
(*Continued*)

Class	Class members				Non-class members	
	Model	R^2	DP	MAPE	R^2	MAPE
Aromatics ($n = 223$)	Model 1: linear: Descriptors: 15 : Noise immunity: very good	0.78	52%	20%	0.0	56%
	Model 2: non-linear. Descriptors: 12. Noise immunity: sufficient	0.82	53%	18%	0.0	78%
Small Compounds ($n = 19$)	Model 1: linear. Descriptors: 2. Noise immunity: sufficient	0.89	60%	25%	0.0	67%
Halogenated ($n = 219$)	Model 1: linear. Descriptors: 17. Noise immunity: good	0.79	53%	19%	0.3	44%
	Model 2: non-linear. Descriptors: 12. Noise immunity: sufficient	0.79	48%	20%	0.0	>150%
Organo metallic ($n = 7$)	Model 1: linear. Descriptors: 2. Noise immunity: very poor	0.99	93%	1%	0.0	>150%
Carbamates ($n = 37$)	Model 1: linear. Descriptors: 2. Noise immunity: good	0.72	43%	15%	0.0	56%
	Model 2: linear. Descriptors: 6. Noise immunity: sufficient	0.89	61%	10%	0.0	56%
Ureas Imidourea	Model 1: linear. Descriptors: 2. Noise immunity: sufficient	0.60	27%	35%	0.0	67%
Cyclourea ($n = 23$)	Model 2: linear. Descriptors: 7. Noise immunity: poor	0.95	70%	10%	0.0	73%

Prediction of trout toxicity. Results using self-organizing statistical learning networks. R^2 according to Eq.1, this Chapter. DP = Descriptive power (see Chapter 4, Chaudhry *et al.*, this volume). MAPE according to Eq.1, this Chapter.

3.1.2. Using CODESSA descriptors and best multi-linear regression (reported class with better results)

We also used best multi-linear regression to model the sub-classes. Table 8 tabulates the best results obtained. In this case, we used a training set and a test set.

Table 8
Models on specific classes of pesticides

Class	Model	R^2	Q^2	s^2	R^2 test
Aromatic ($n = 138$) (aromatic compounds in the training set)	Model 3: CODESSA Descriptors + Log *P* (KowWin), Best multi-linear regression	0.65	0.61	0.80	0.59
Organophosphorous ($n = 27$) (organophosphorous compounds in the training set)	Model 3: CODESSA Descriptors + Log *P* (KowWin), Best multi-linear regression	0.81	0.76	0.38	0.75
Carbamates ($n = 15$) (carbamates compounds in the training set)	Model 3: CODESSA Descriptors + Log *P* (KowWin), Best multi-linear regression	0.75	0.64	0.17	0.71

Predictions of trout toxicity. Results using multi-linear regression.
s^2: standard error of the regression.

Tables 7 and 8 summarize that good results on some restricted chemical classes are obtained.

3.1.3. Using a combined model for the heterocyclic class based on multiple toxicity values

For a number of chemical compounds of the official DEMETRA project data set on trout, there are different experimental toxicity values available, which were all obtained by bioassays that strictly followed the testing guidelines defined by the OECD. Therefore, these sets of different experimental toxicity values have to be taken as equally valid toxicity values. A common practice, which we also followed in this project for setting up the five official project data sets (see also Chapter 2, Benfenati *et al.*, this volume), is using only the single experimental value from the set of available toxicity values of a given compound that expresses the most toxic effect on a biological species to build up data sets for modelling. The information – and, finally, the a priori given uncertainty on the system to be modelled – of the other existing toxicity values of a specific compound is not considered in this way.

Here, we followed another approach in creating two data sets for modelling. Based on the official data set for trout, D_1, the first data set of the heterocyclic class, $D_{1,HC1}$, is composed in the traditional way by using only a single – the most toxic – experimental value for any chemical compound. This results in a learning data set of $N_{AB,HC1} = 91$ compounds and a testing data set of $N_{C,HC1} = 15$ compounds. This means that 106 of the total 281 compounds of the trout data set D_1 fit into the heterocyclic chemical class, whereas 175 compounds are classified as being non-heterocyclic.

The second data set of the heterocyclic chemical class, $D_{1,HC2}$, is formed using all the given valid experimental toxicity values for heterocyclic compounds. The resulting training data set now extends to $N_{AB,HC2} = 156$ samples, and the test data set contains $N_{C,HC2} = 25$ samples.

For the two data sets $D_{1,HC1}$ and $D_{1,HC2}$, we calculated 2D DRAGON and CODESSA descriptors to form the input data space of about 350 input variables.

We built in totally 15 linear and non-linear regression models by multi-levelled self-organization, which performs feature selection, modelling and validation in a single, autonomously running process [Section 4.3 in Chapter 4 (Chaudhry *et al.*, this volume)], on the two heterocyclic-class training data sets (see Table 7, this Chapter), and we then applied these models on the given test data sets, accordingly. From these 15 individual regression models, which specifically describe the behaviour of heterocyclic compounds, an optimal complex, combined model of self-selected relevant individual models was generated by a self-organizing statistical learning network [see Section 4.3 in Chapter 4 (Chaudhry *et al.*, this volume) and see Section 6 in Chapter 5 (Amaury *et al.*, this volume)]. All 15 individual models as well as the combined model are described analytically by a linear or non-linear regression equation.

The results obtained by this approach are shown in Fig. 1, indicating that good results were achieved. Indeed, the R^2 on the training set was 0.85 and on

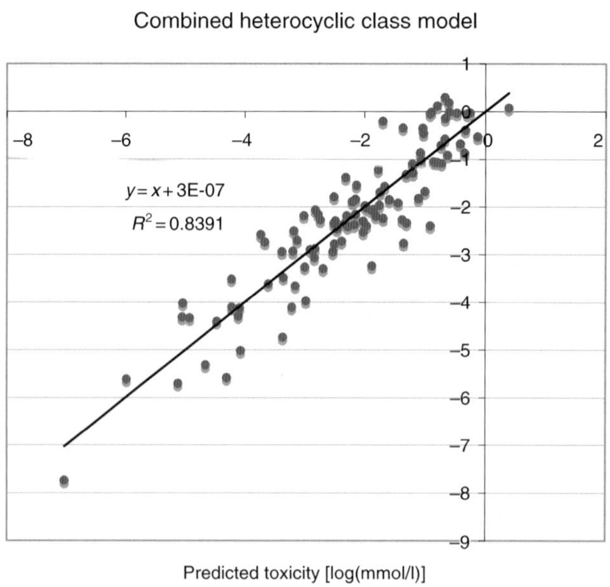

Figure 1: The predicted toxicity values – given in log(mmol/l) – for trout, using pesticides with at least a heterocyclic ring ($n = 106$ compounds) and all toxicity values, even multiple values for the same pesticide.

the test was 0.71. Overall, the R^2 on both training and test sets was 0.84. The prediction for the chemicals without heterocyclic ring was $R^2 = 0.37$ indicating that the model is specific.

4. OVERVIEW OF RESULTS OBTAINED WITH THE HYBRID MODELS

In Chapter 5 (Amaury *et al.*, this volume) we described how the integration of different models can improve the results, compared to the results obtained from a single one. The basic idea is that each model may bring a different content of the complex system, which is modelled. Thus, if the basis of the different models is different, the possibility to obtain a larger knowledge is increased. For this reason, it is not always true that the models to be combined have to be only the best ones. A model with lower performances may improve the overall performances of a hybrid model because it brings different information; vice versa, combining a set of good models, which are very similar, may result in a poor improvement of the overall results. The differences in the models can be on the way to describe the chemical information, or on the algorithm used to develop the model. We considered these aspects in our general approach.

Several strategies were used as described in Chapter 5 (Amaury *et al.*, this volume). In all cases, the outputs of the individual models are used as inputs of the hybrid model. The approaches that gave better results are those based on SOSLN and on rule-based approach giving weights for different parts of the model curve. SOSLN, as better described in Section 4.3, Chapter 4 (Chaudhry *et al.*, this volume), train the network to optimize the results, whereas in the case of the rule-based hybrid model, the approach is simpler.

Five hybrid models were built, one for each endpoint. Model selection was based on the performances of the individual models, but we also tried to use simpler models in order to facilitate the use of the final models. Thus, models based on 2D descriptors were preferred, because we obtained similar performances on the basis of the individual models using 2D and 3D descriptors, and thus there was no reason to use 3D descriptors, which take longer to obtain, require more experience and introduce variability. Whenever possible, we also tried to limit the different software necessary to calculate the chemical descriptors, preferring solutions based on a more limited number of software programs involved.

We underline that the models which were selected had, in many cases, performances similar to the performances from other hybrid models, as described later. This proves that the achieved results were robust, and starting from the available material results cannot be significantly improved.

We remember that the DEMETRA task was to develop an algorithm using chemical descriptors available from sources outside the consortium, public or commercial. The ways to calculate the chemical descriptors used within

DEMETRA are detailed in Chapter 3 (Benfenati *et al.*, this volume). In some cases, we found better results combining models that adopted innovative chemical descriptors, developed within the DEMETRA project as a spin-off activity. In some cases, we developed the software to freely calculate these descriptors. However, we had no resources to develop public software to calculate many descriptors that proved very useful in improving the results in case of the trout model (see Section 4.2.5, this Chapter). Thus, these models, even if superior, were not selected, for the fact that the user could not use them.

4.1. Hybrid model for rainbow trout

The trout hybrid model that gave best performances has been obtained with SOSLN. Inputs for this model were a PLS model and two ANNs.

The performances of this model are resumed in Table 9 and Fig. 2. In order to offer a clear vision of the predicted versus experimental values, the plot is built in weight unit, −log(mg/l), while for a deeper evaluation of the statistical performances the values in the Table 9 refer to the units used for the model building and optimization, −log(mmol/l).

In Table 9 the slope, intercept and the squared correlation coefficient are expressed both for the fitting ‘calculated versus experimental values’ (K, q, R^2) and the fitting ‘experimental versus calculated values’ (K', q', R^2). Moreover, in both cases all these parameters are calculated also forcing the intercept to zero. These values are important for the evaluation of the model, in fact the optimal performances should be close to the ideal ones in both cases, i.e. $K \approx K' \approx 1$, $q \approx q' \approx 0$; $R^2 \approx R^2{}_0 \approx 1$. From a theoretical point of view, the two ways to present the results address the same problem of optimization: the calculated versus experimental fitting shows how much the predicted values are similar to the observed, while the experimental versus calculated fitting shows how the experimental values are explained by the prediction, i.e. by a function of the descriptors. As it is deducible, for example, from the difference between the squared correlation coefficients ($R^2 - R^2{}_0 = 0.07 \approx 0$), the hybrid model shows good performances.

Table 9
Statistical parameters of the hybrid model for trout toxicity

	NC	ND	R^2	K	q	R_0^2	K_0 ($q_0 = 0$)	K'	q'	$R^{2\prime}{}_0$	$K_0{}'$	RMSE	Outliers
Training	229	20	0.71	0.70	0.61	0.64	0.88	1.01	0.06	0.71	1.04	0.84	6+; 4–
Test	46	20	0.71	0.71	0.89	0.57	0.97	1.00	−0.19	0.71	0.94	0.87	1+; 3–

NC, number of compounds; ND, number of descriptors; K, slope; q, intercept.
outliers +, predicted value (mg/l)/expected value (mg/l) > 50;
outliers –, expected value (mg/l)/predicted value (mg/l) > 50.

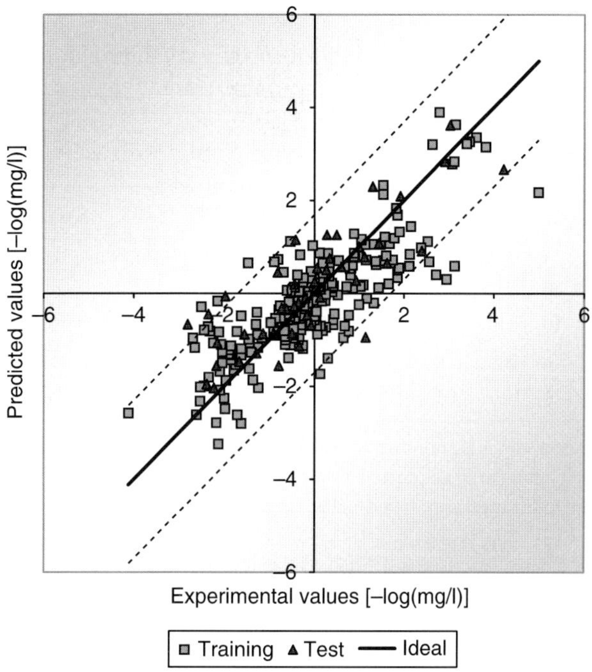

Figure 2: Predicted versus experimental values for trout toxicity using the DEMETRA hybrid model. Dotted lines indicate a factor of 50 in mg/l.

We also checked results of the individual models. They are reported in Table 10. One of the three individual models was built on 229 compounds instead of 235, because a few chemical compounds were not included due to unavailable values for some descriptors. This is the reason why these compounds were excluded also from the set of the final model. The performances of the individual models were checked by means of cross-validation and of course prediction on the test set. It is clear from Tables 9 and 10 that the performances of the hybrid model are superior to those obtained using any individual model.

The descriptors we used are listed in Table 11. As expected, log *P* and log *D* at pH7, octanol–water partition coefficient and the effective partition coefficient for dissociative systems, respectively, are present in the hybrid model.

As it is explained in Chapter 4 (Chaudhry *et al.*, this volume) and Chapter 5 (Amaury *et al.*, this volume), we calculated the descriptors with many software packages, and we used different methods to select the best variables able to describe the system. This hybrid model is a quite complex model in terms of the amount of software required to describe the structures, but on the other hand, once the 2D structure is available the calculation of the 2D descriptors is very fast.

Hydrophobic mechanism plays a key role as described by log *P* and log *D*. Of course, reactivity is also important for these compounds, as expected.

Table 10
Statistical parameters of the three individual models combined in the hybrid model for trout toxicity

		NC	ND	Q^2	R^2	K	q	R^2_0	K_0 ($q_0 = 0$)	K'	q'	$R^{2\prime}_0$	K_0'	Outliers
PLS	Training	235	11	0.61	0.66	0.66	0.82	0.54	0.90	1.00	0	0.66	1.00	9+; 8–
	Test	46	11	–	0.63	0.62	1.55	0.37	0.96	1.02	−0.29	0.63	0.93	2+; 4–
NN	Training	235	10	0.69	0.69	0.69	0.74	0.60	0.91	1.00	0	0.69	1.00	3+; 7–
	Test	46	10	–	0.59	0.68	1.06	0.41	0.99	0.87	0.07	0.59	0.89	2+; 4–
NN	Training	229	6	0.66	0.66	0.66	0.82	0.54	0.90	1.01	−0.03	0.66	1.00	6+; 8–
	Test	46	6	–	0.63	0.68	1.04	0.45	0.98	0.94	−0.09	0.63	0.91	1+; 4–

NC, number of compounds; ND, number of descriptors; K, slope; q, intercept.
outliers +, predicted value (mg/l)/expected value (mg/l) > 50;
outliers –, expected value (mg/l)/predicted value (mg/l) > 50.

Table 11
Descriptors used in the hybrid model for trout toxicity

Software	Descriptor symbol	Correlation[a]	Definition
MDL	nxch3	–	Number of 3-membered rings
ACD	LogDpH7	–	Log *D* at pH7
CACHE	Log *P*	–	Log *P*
DRAGON	Mp	–	Mean atomic polarizability (scaled on carbon atom) – constitutional descriptors
DRAGON	nOH	+	Number of total hydroxyl groups – functional groups
DRAGON	nOHPh	–	Number of phenols – functional groups
DRAGON	nHDon	+	Number of donor atoms for H-bonds (with N and O) – functional groups
DRAGON	nHAcc	–	Number of acceptor atoms for H-bonds (N O F) – functional groups
DRAGON	C-031	–	X–CR–X – atom-centred fragments[b]
DRAGON	PSA	–	Fragment-based polar surface area – properties
DRAGON	SEigZ	–	Eigen value sum from Z weighted distance matrix (Barysz matrix) – topological descriptor

Table 11
(*Continued*)

Software	Descriptor symbol	Correlation[a]	Definition
DRAGON	D/Dr03	–	Distance/detour ring index of order 3 – topological descriptor
DRAGON	D/Dr09	–	Distance/detour ring index of order 9 – topological descriptor
DRAGON	SRW05	–	Self-returning walk count of order 05 – molecular walk counts
IN HOUSE	DFG0047	–	Number of [≥C-OH] – fragment descriptor
DRAGON	MATS3e	–	Moran autocorrelation – lag 3/weighted by atomic Sanderson electronegativities – 2D autocorrelations
DRAGON	MATS8e	+	Moran autocorrelation – lag 8/weighted by atomic Sanderson electronegativities – 2D autocorrelations
DRAGON	GATS2m	+	Geary autocorrelation – lag 2/weighted by atomic masses – 2D autocorrelations
DRAGON	GATS3m	+	Geary autocorrelation – lag 3/weighted by atomic masses – 2D autocorrelations
DRAGON	nCOOR	–	Number of esters (aliphatic) – functional groups

[a] Correlation between the toxicity measured as concentration versus the descriptor value. Positive means a direct relationship: if the descriptor value increases, the predicted concentration increases (lower toxicity) and vice versa.

[b] R represents any group linked through carbon; X represents any electronegative atom (O, N, S, P, Se, halogens)

The polarizability, the number of donor/acceptor atoms for H-bonds and the polar surface area indicate this feature. As log *P* is a general parameter for the likely specific interaction with the cellular membrane, these parameters are related to specific mechanisms in which the reactivity can be measured through these descriptors.

The descriptor correlations remark these features. There is an inverse relationship between the increasing presence of electronegative atoms, particularly oxygen. We observe as well the same behaviour with the hydrophobicity represented as log *P* and log *D*, reinforced by the importance of rings distribution.

On the other hand, we found a positive correlation among the descriptors that count the number of bonds and rings, as a measure of the molecular size and branching.

Such features show how solubility decreases the toxicity, while branching, not influenced by rings presence, increases it. Thus, compounds able to be barely soluble of medium/high size appear like the more toxic for this endpoint.

4.2. Outliers and the applicability domain

The hybrid model for trout showed some outliers. We defined outliers as those compounds with a ratio between expected and predicted values, both in mg/l, higher than 50. More precisely if the ratio 'predicted value/expected value' is higher than 50 the compound is a false negative, while if the 'expected value/predicted value' is higher than 50 the compound is a false positive.

All outliers are listed in Table 12. Outliers are important signals for understanding the applicability domain of the model. We addressed this issue, which

Table 12
Outliers in the hybrid model for trout toxicity

Name	CAS	Experimental value (mg/l)	Predicted value (mg/l)	Training/ Test set	False negative/false positive
Antimycin A	1397-94-0	0.00001	0.00670	Training	False negative
Fenpyroximate	134098-61-6	0.0011	0.496	Training	False negative
Pyridaben	96489-71-3	0.00072	0.258	Training	False negative
Rotenone	83-79-4	0.0019	0.395	Training	False negative
1,2-Benzenedicarboxaldehyde	643-79-8	0.072	8.780	Test	False negative
Dichloro-2-*n*-octyl-3(2H)-isothiazolone	64359-81-5	0.0027	0.209	Training	False negative
Bioban	2224-44-4	0.75	53.88	Training	False negative
Difenzoquat methyl sulfate	43222-48-6	491.9	9.121	Training	False positive

Table 12
(*Continued*)

Name	CAS	Experimental value (mg/l)	Predicted value (mg/l)	Training/ Test set	False negative/false positive
Ethylene dichloride	107-06-2	225	2.806	Test	False positive
Hexadecadienol, acetate	53042-79-8	94.9	1.154	Test	False positive
Diflubenzuron	35367-38-5	140	1.471	Training	False positive
Spinosad	131929-60-7	30	0.222	Training	False positive
Trisulfuron methyl	126535-15-7	648	4.650	Test	False positive
Z-11-Hexadecanal	53939-28-9	320	1.938	Training	False positive

CAS, Chemical Abstracts Service Number.

is an important one for any model, as documented by the relevance given to it within the OECD principles, using four approaches:

1. checking, once again, the experimental toxicity values of the outliers;
2. using chemiometric tools;
3. using rules derived by chemical knowledge (human expert); and
4. using rules derived by toxicological knowledge (human expert).

4.2.1. Comparison of the data: data variability

The procedure which involved the use of the first three databases (EPA-OPP, SEEM and BBA) in creating the DEMETRA data sets was explained in detail in Chapter 2 (Benfenati *et al.*, this volume).

After the generation of the models and the hybrid system, a further evaluation of the toxicity for some of the compounds was conducted, in particular for some of the outliers.

This evaluation was focused on the rainbow trout acute toxicity LC_{50} 96-h data, as it is one of the endpoints with the largest amount of data. All databases accessible through the Internet were used to search for additional data (ECOTOX, PAN database, EXTOXNET and AGRITOX). RIVM data were also considered.

Some examples are reported in Figs. 3–8. In all the figures, the dotted line represents the ecotoxicity value used in the DEMETRA data set. When

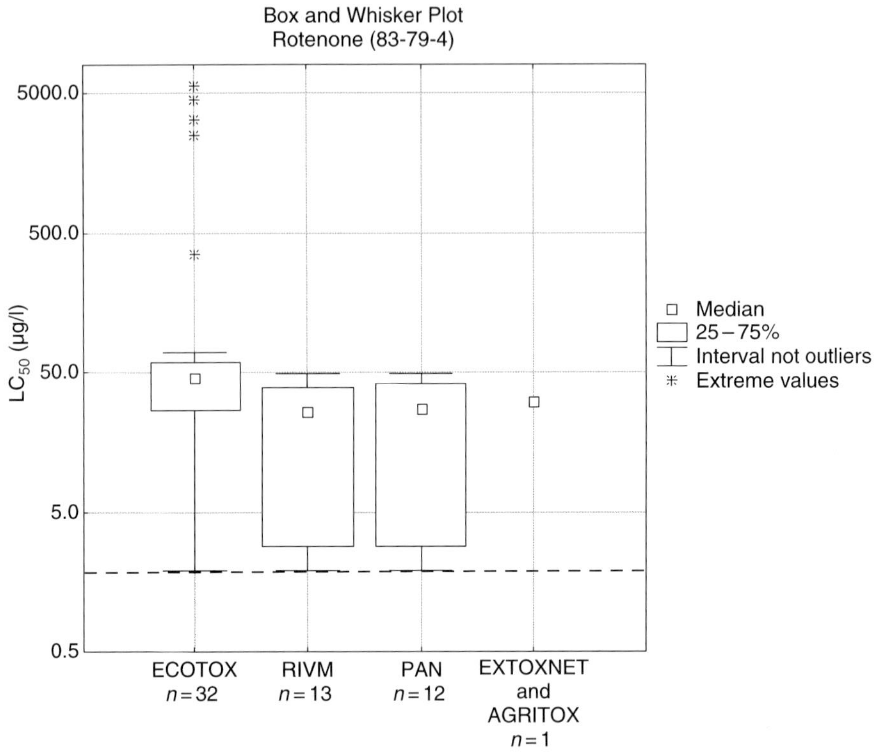

Figure 3: Distribution of activity data for Rotenone in five databases with available data. The number of studies available in each database is also reported.

a large number of data points were available in some of the databases instead of a histogram representation (used for Figs. 4 and 5) the Box and Whisker plot was chosen as it is more able to describe the overall distribution of the data. With this plot the median of the data distribution for each database is displayed along with a box encoding for the data contained within the interquartile range (identified by the first and third quartiles: the median of the lower or upper half of the data, respectively). The lower and upper lines define the interval representing non-outlier range (including data within 1.5 times the interquartile range from the end of the box). All points exceeding this range are outliers, and if they exceed more than three times the interquartile range from the end of the box, they are called extreme values.

For Rotenone (Fig. 3), there is a very wide spread of toxicity values with a general agreement on the lower. Some data points in the ECOTOX database are extreme values even if the source for these data (Marking *et al.*, 1976) quotes other values hundreds of times lower.

In the case of Antimycin A, all data are within a 10-fold variability except for an extreme value present in all databases. The source quoted for this data point in

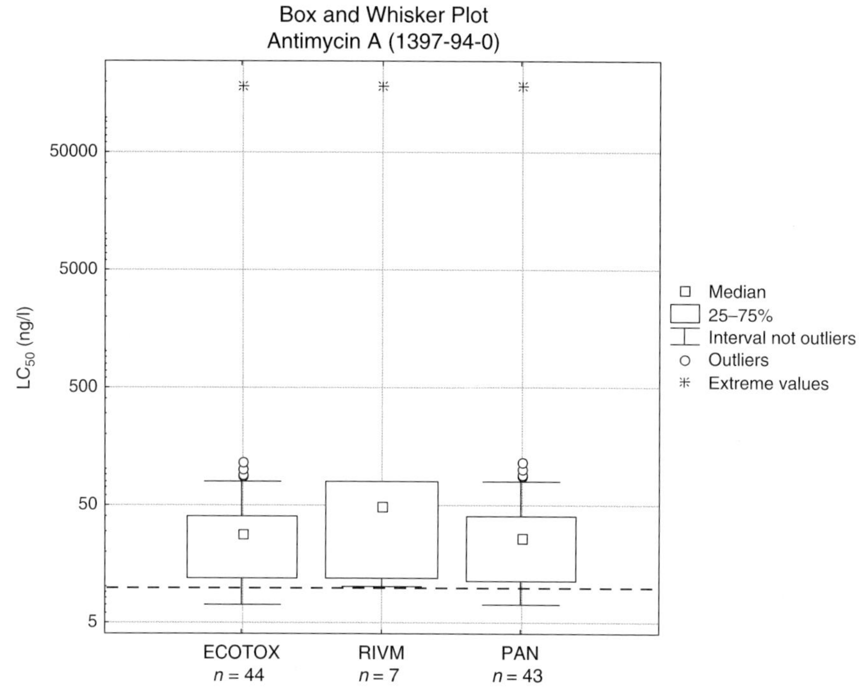

Figure 4: Distribution of activity data for Antimycin A in three databases with available data. The number of studies available in each database is also reported.

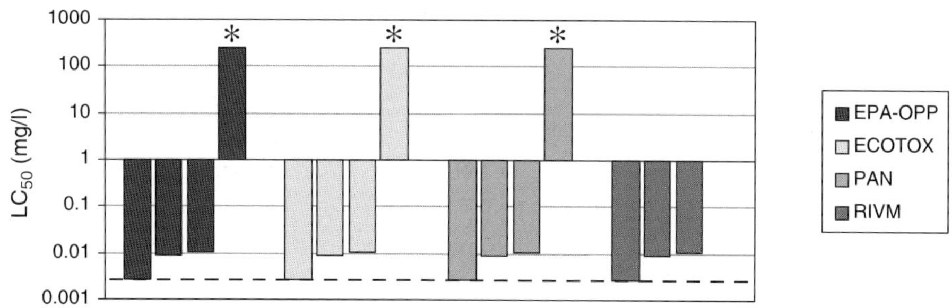

Figure 5: Ecotoxicity data for Kathon 930 (64359-81-5) in four databases, including EPA-OPP. For the data point indicated with asterisk the EPA-OPP database refers this value to a degradation product.

all databases is the EPA-OPP database, but in the original file of EPA-OPP it is also specified that this experiment was conducted on material presenting only a 10% of active substance (AS). This piece of information is not reported in the other databases because usually the purity grade is not reported but for in the PAN database. In this latter source, anyway, the chemical purity for this experiment is not stated.

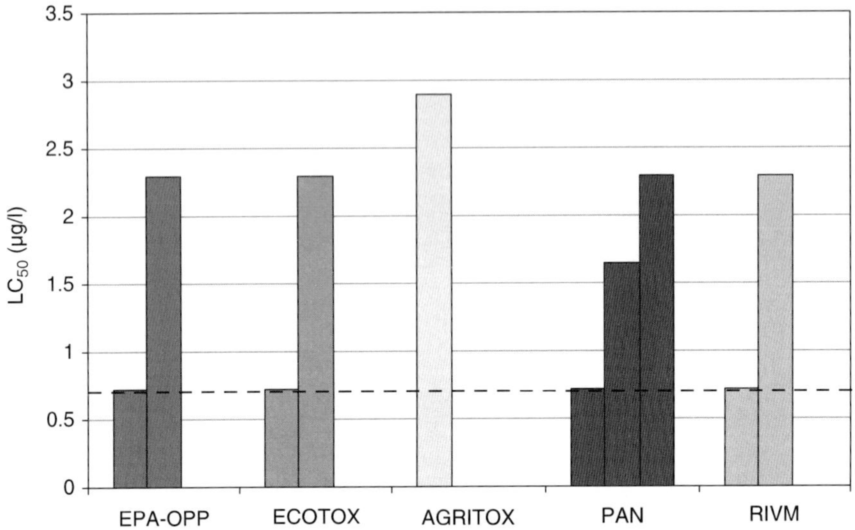

Figure 6: Ecotoxicity data for Pyridaben (96489-71-3) in five databases, including EPA-OPP.

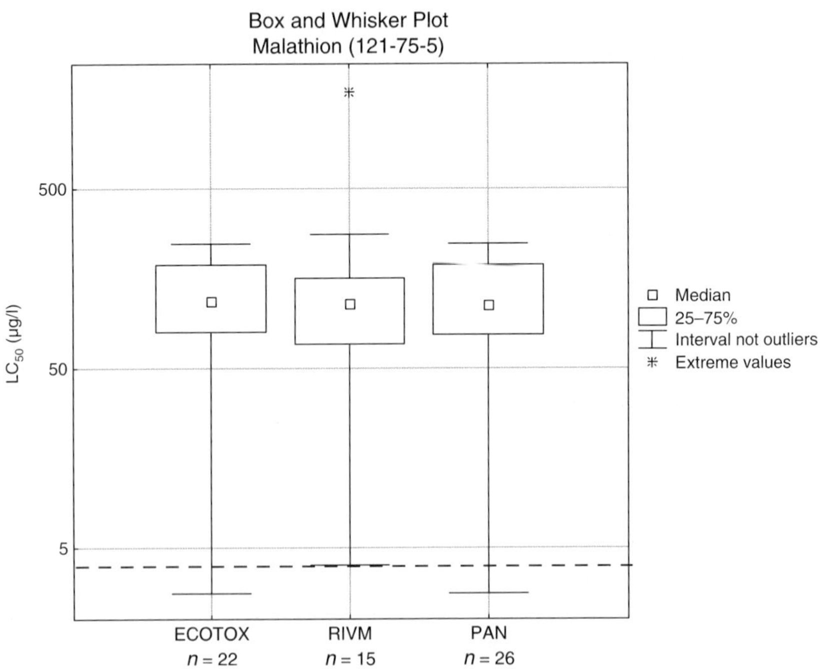

Figure 7: Distribution of activity data for Malathion in three databases with available data. The number of studies available in each database is also reported.

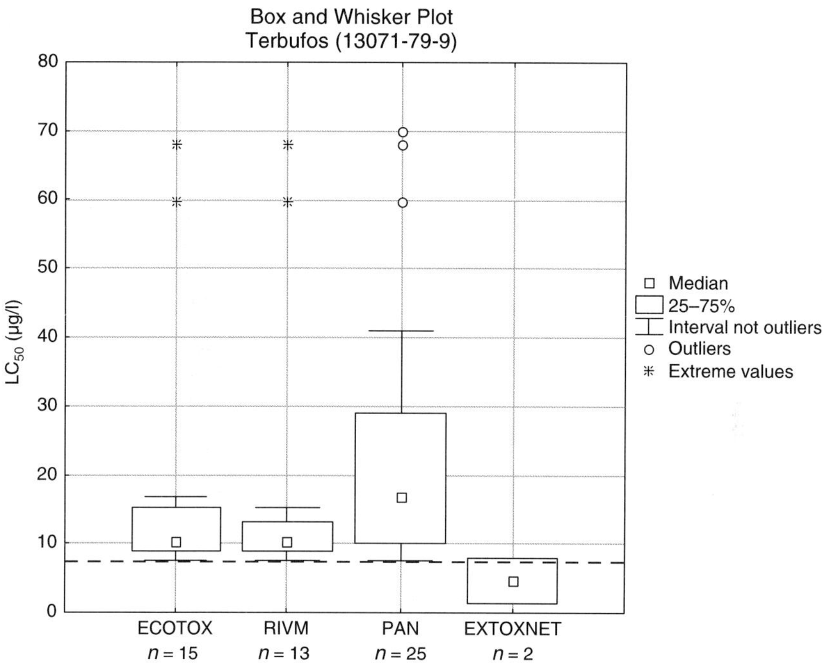

Figure 8: Distribution of activity data for Terbufos in four databases with available data. The number of studies available in each database is also reported.

For Kathon 930 (Fig. 5) and Pyridaben (Fig. 6), there were only a few data points.

From the histogram representation for Kathon, it is clearly visible that all databases report the same values and they quote as reference source the EPA-OPP one. Interestingly, three out of four data have a reduced range of variability, whilst one of the data points is five orders of magnitude bigger. The EPA-OPP database refers this data point as degradation product, but this piece of information is not available in the other databases that report this value.

For Pyridaben, the situation is similar: data reported in all databases are derived from EPA-OPP, confirming an overlapping of different sources of data, except for AGRITOX data. A very low variability (toxicity range from 0.72 to 2.9 μg/l) is observed for the four available data.

For Malathion, a larger variability in the spread of toxicity values is observed. One of the reasons is that some of the data are referred to a low purity grade for this AS, but this observation is not sufficient to justify all data points. The extreme value present in the RIVM database quotes again the source as EPA-OPP, but it is referred to a different species of trout (*Oncorhynchus clarki*).

For Terbufos, a low variability is observed in the data coming from ECOTOX and RIVM database, while a large spread of the values can be found in the PAN database. Some of the data points are outliers or extreme values of the distribution but with a restricted variability range (within a factor of 10), and they are referred to a low purity grade for the AS.

Interestingly, it seems that EXTOXNET quotes the lowest toxicity value of 1.3 μg/l from a literature source (Mayer *et al.*, 1986). The same source is contained in ECOTOX and PAN databases where the minimum for this group of experiments is settled to 7.6 μg/l, and another data point is at 13 μg/l. Quite probably a typo error occurred in one of the two cases.

From this comparison of available sources of ecotoxicity data for pesticides, some conclusions can be drawn: generally the value selected for the DEMETRA data set coincides with the lowest or is near to the lowest. This selection is based on a precautionary approach which takes care that it would be better to overestimate pesticide toxicity than the converse. Moreover in many cases, we saw that some of the studies reporting a lower toxicity for the compound are associated with a low purity grade for the chemical itself introducing a bias in the toxicological data.

The scrutiny of different databases increases the confidence in the selected data; thus, this procedure was also adopted in generating DEMETRA data sets, but an intrinsic limitation to this approach is due to the overlapping of the data sources quoted in the databases. Thus, often new information is not introduced, but replicates are considered.

Finally, from this analysis, it appeared that the EPA-OPP database contains a good level of details in reporting the experiments (e.g. AS percentage and indication that some of the data may refer to degradation products). This increased our confidence in using it as primarily sources of data for DEMETRA data sets. Only PAN database generally also states the AS percentage but usually with less details.

We also checked the toxicity values of the outliers using other databases and sources of data as the Pesticide Manual (Tomlin, 1997). Then we used AMBIT (AMBIT, 2006), available in the Web to define the applicability domain. The results of these approaches did not give any consistent explanation of the prediction of the outliers. Finally, human experts analysed the outliers one by one.

4.2.2. Visual inspection

On the basis of the analysis of the training set compounds, we identified some common characteristics of the outliers by visual inspection of the chemical structures.

Considering the mechanistic basis we found that the model does not reliably predict toxicity of compounds with specific modes of action, for example, involving electron transfer as the Antimycin A, Rotenone and Fenpyroximate (Rule 1).

Furthermore, we analysed the chemical structures and found five fragments which are related to an error of a factor of 50:

- Ethylene with one heteroatom in alpha (Rule 2).
- CHO (Rule 3).

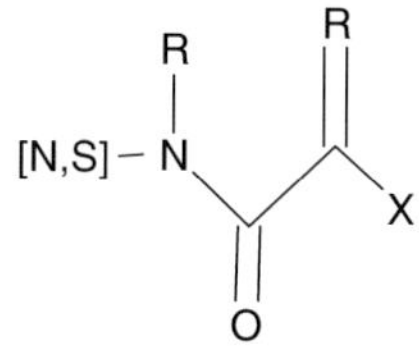

Figure 9: General sub-structure that can be found in outliers found by the rule 4.

- Fragment 1 (Fig. 9) (Rule 4).
- CH–NO_2 (Rule 5).
- Oxime linked to a benzyl at the O (Rule 6).

While the first characteristic of the outliers, the interference with electron transport, is based on the mechanism of action, and therefore not predictable a priori from the structure, the other characteristics are based on the chemistry of the compounds and become rules to identify the weaker region of the applicability domain of the model.

Once detected the rules for outliers, the applicability domain of the model has been reduced, and all the molecules presenting those five fragments were considered out of the applicability domain (see Table 13).

It is important to underline that these rules are intended to *exclude* the molecules with a weaker prediction. In this sense, the model maintains the general

Table 13
Compounds out of the applicability domain in the hybrid model for trout toxicity

Name	CAS	Detected rule	Training/ Test set	Comments on the compounds out of domain
Antimycin A	1397-94-0	–CHO	Training	False negative
Fenpyroximate	134098-61-6	Ar–CH–O–N=	Training	False negative
Pyridaben	96489-71-3	Fragment 1	Training	False negative
Rotenone	83-79-4	=CH_2	Training	False negative
1,2-Benzenedi-carboxaldehyde	643-79-8	–CHO	Test	False negative
Dichloro-2-*n*-octyl-3(2H)-isothiazolone	64359-81-5	Fragment 1	Training	False negative
Bioban P-1487	2224-44-4	–CH–NO_2	Training	False negative

(*Continued*)

Table 13
(*Continued*)

Name	CAS	Detected rule	Training/ Test set	Comments on the compounds out of domain
Octhilinone	26530-20-1	Fragment 1	Training	Predictive value >10 experimental value
Trifloxystrobin	141517-21-7	Ar–CH–O–N=	Training	Predictive value >10 experimental value
Ethalfluralin	55283-68-6	$=CH_2$	Training	Predictive value >10 experimental value
Nerolidol	7212-44-4	$=CH_2$	Training	Good prediction
(E)-(3,3-Dimethyl-cyclohexylidene) acetaldehyde	26532-25-2	–CHO	Test	Good prediction
Imazalil	35554-44-0	$=CH_2$	Test	Good prediction
Norflurazon	27314-13-2	Fragment 1	Training	Good prediction
(Z)-11-Hexadecenal	53939-28-9	–CHO	Training	False positive

CAS, Chemical Abstracts Service Number.

applicability. Of course, more specific and more accurate local models can be built in order to better predict these weak regions (e.g. one model only for aldehydes), but this is out of the scope of DEMETRA.

The performances of the new model considering all the rules are summarized in Fig. 10 and Table 14. In this model, there are not false negatives predicted with a factor higher than 50 with respect to the experimental value.

In a deeper evaluation, we also considered the minor errors, and we analysed also the structure of all compounds with a predicted value in mg/l higher than a factor of 10 with respect to the experimental value. Table 15 lists these outliers.

There are several fragments, which are probably related to an enhanced or anyway unusual reactivity in the molecule not identified by the modelling tools, or because their occurrence is low, or because the fragment is more complicated than those used within the QSAR models.

Most of these fragments characterize a reduced performance of the model. As it is shown above, only a few of them are likely related with a possible error higher than a factor of 50 in the case of false negative. Then other fragments are identified and considered, jointly with the previous five, as responsible for a possible error higher than a factor of 10. The model is working even if these

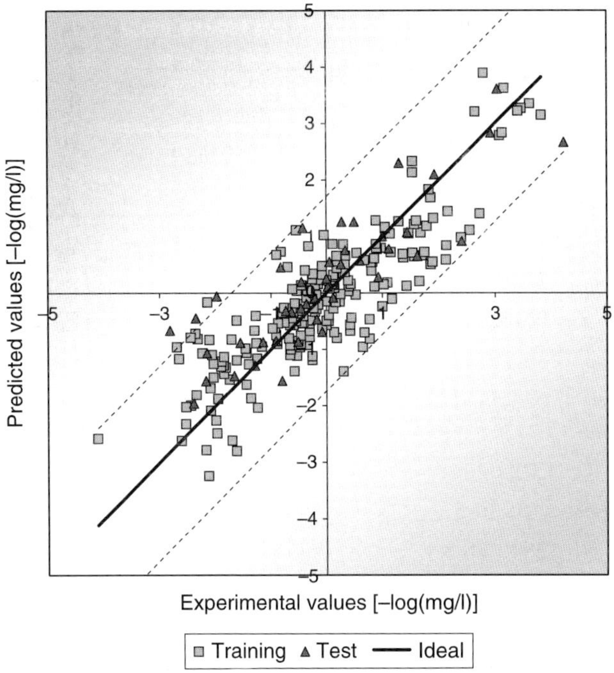

Figure 10: Predicted versus experimental values for trout toxicity hybrid model within the applicability domain; false negative higher than a factor of 50 explained and excluded (see text). Dotted lines indicate the factor of 50 in mg/l.

Table 14

Statistical parameters of the hybrid model for trout toxicity within the applicability domain (factor of 50)

	NC	ND	R^2	K	q	R^2_0	K_0 ($q_0 = 0$)	K'	q'	$R^{2\prime}_0$	K_0'	*RMSE*	*Outliers*
Training	213	20	0.76	0.77	0.51	0.72	0.92	1.00	0.04	0.76	1.01	0.72	2–
Test	43	20	0.75	0.70	0.99	0.58	0.99	1.07	–0.46	0.74	0.93	0.83	3–

NC, number of compounds; ND, number of descriptors; K, slope; q, intercept.
outliers +, predicted value (mg/l)/expected value (mg/l) > 50;
outliers –, expected value (mg/l)/predicted value (mg/l) > 50.

fragments are present, but the error can vary depending on which fragment is present. Thus, the user can choose the level of confidence.

In this section we discuss these fragments, first those mainly related to false negatives, then those related to false positive.

Table 15
List of compounds considered as outliers in the model for trout toxicity using a factor of 10 for false negatives and a factor of 50 for false positives

Name	CAS	Experimental value (mg/l)	Predicted value (mg/l)	Training/ Test set	False negative/false positive
Antimycin A	1397-94-0	0.00001	0.0067	Training	False negative
Fenpyroximate	134098-61-6	0.0011	0.50	Training	False negative
Pyridaben	96489-71-3	0.00072	0.26	Training	False negative
Rotenone	83-79-4	0.0019	0.39	Training	False negative
Dichloro-2-*n*-octyl-3(2H)-isothiazolone	64359-81-5	0.0027	0.21	Training	False negative
Bioban	2224-44-4	0.75	53.88	Training	False negative
Dichlorprop(2,4-DP)	120-36-5	0.50	24.57	Training	False negative
Sodium dichloro-*s*-triazinetrione	2893-78-9	0.25	8.99	Training	False negative
Propachlor	1918-16-7	0.17	6.48	Training	False negative
Octhilinone	26530-20-1	0.047	1.61	Training	False negative
Terbufos	13071-79-9	0.0076	0.26	Training	False negative
Trifloxystrobin	141517-21-7	0.014	0.46	Training	False negative
Trifluralin	1582-09-8	0.022	0.62	Training	False negative
Fluvalinate	69409-94-5	0.0029	0.077	Training	False negative
DDT	50-29-3	0.0041	0.10	Training	False negative
Pival	83-26-1	0.21	5.026	Training	False negative
Methidathion	950-37-8	0.014	0.31	Training	False negative
Phorate	298-02-2	0.013	0.28	Training	False negative
Ethalfluralin	55283-68-6	0.037	0.75	Training	False negative
Chlorfenapyr	122453-73-0	0.0074	0.14	Training	False negative
3-Iodo-2-propynyl butylcarbamate	55406-53-6	0.067	1.28	Training	False negative
Dichlorvos	62-73-7	0.10	1.74	Training	False negative
Benomyl	17804-35-2	0.17	2.92	Training	False negative
Sodium fluroacetate	62-74-8	54	629	Training	False negative
Glyphosate	1071-83-6	134	1751	Training	False negative
Carbofuran	1563-66-2	0.38	4.69	Training	False negative
Fonofos	944-22-9	0.020	0.22	Training	False negative
Lindane	58-89-9	0.018	0.19	Training	False negative
Difenzoquat methyl sulphate	43222-48-6	711	9.12	Training	False positive

Table 15
(*Continued*)

Name	CAS	Experimental value (mg/l)	Predicted value (mg/l)	Training/ Test set	False negative/false positive
Diflubenzuron	35367-38-5	140	1.47	Training	False positive
Spinosad	131929-60-7	30	0.22	Training	False positive
Z-11-Hexadecanal	53939-28-9	320	1.94	Training	False positive
1,2-Benzenedi-carboxaldehyde	643-79-8	0.072	8.78	Test	False negative
Tefluthrin	79538-32-2	0.00006	0.0021	Test	False negative
Malathion	121-75-5	0.004	0.12	Test	False negative
Ethylene dichloride	107-06-2	225	2.81	Test	False positive
Hexadecadienol, acetate	53042-79-8	94.0	1.15	Test	False positive
Trisulfuron methyl	126535-15-7	648	4.65	Test	False positive

CAS, Chemical Abstracts Service Number.

These fragments have been identified upon visual inspection. Parallel data mining tools have identified part of them, or smaller moieties, which however are also present in a large part of chemicals, which are not outliers. To have a better overview of the chemical typical groups involved, we list below again the fragments already introduced related to the highest error (factor 50) (see also Table 16).

Table 16
List of compounds selected for removal within the trout hybrid model after the application of the high confidence rules

Name	CAS	Detected rule	Training/ Test set	Comments
Antimycin A	1397-94-0	–CHO	Training	Outlier
Fenpyroximate	134098-61-6	Ar–CH–O–N=	Training	Outlier
Pyridaben	96489-71-3	Fragment 1	Training	Outlier
Rotenone	83-79-4	Terminal ethylene with N or O in C-alpha	Training	Outlier
Dichloro-2-*n*-octyl-3(2H)-isothiazolone	64359-81-5	Fragment 1	Training	Outlier

(*Continued*)

Table 16
(*Continued*)

Name	CAS	Detected rule	Training/ Test set	Comments
Bioban	2224-44-4	CH–NO_2	Training	Outlier
Dichlorprop(2,4-DP)	120-36-5	COOH with O, N or F in C-alpha	Training	Outlier
Sodium dichloro-*s*-triazinetrione	2893-78-9	X–CO–NH–C	Training	Outlier
Propachlor	1918-16-7	N–CO–CH_2–Cl	Training	Outlier
Octhilinone	26530-20-1	Fragment 1	Training	Outlier
Terbufos	13071-79-9	P(=S)S–CH–X	Training	Outlier
Trifloxystrobin	141517-21-7	Ar–CH–O–N=; methyl ester	Training	Outlier
Trifluralin	1582-09-8	Para-trifluoro-aminobenzene	Training	Outlier
Fluvalinate	69409-94-5	Para-trifluoro-aminobenzene	Training	Outlier
DDT	50-29-3	n°Cl $\geq$ 5	Training	Outlier
Pival	83-26-1	$(C{=}O)_3$–CH	Training	Outlier
Methidathion	950-37-8	P(=S)S–CH–X; methyl ester	Training	Outlier
Phorate	298-02-2	P(=S)S–CH–X	Training	Outlier
Ethalfluralin	55283-68-6	Terminal ethylene with N or O in C-alpha; Para-trifluoro-aminobenzene	Training	Outlier
Chlorfenapyr	122453-73-0	Ar–CH–O (Ar fully substituted)	Training	Outlier
3-Iodo-2-propynyl butylcarbamate	55406-53-6	Triple bond with O in C-alpha; X–CO–NH–C	Training	Outlier
Dichlorvos	62-73-7	–CH=CCl	Training	Outlier
Benomyl	17804-35-2	X–CO–NH–C; methyl ester	Training	Outlier
Sodium fluroacetate	62-74-8	COOH with O, N or F in C-alpha; n°C < 3	Training	Outlier
Glyphosate	1071-83-6	COOH with O, N or F in C-alpha; C–P	Training	Outlier

Table 16
(*Continued*)

Name	CAS	Detected rule	Training/ Test set	Comments
Carbofuran	1563-66-2	X–CO–NH–C	Training	Outlier
Fonofos	944-22-9	C–P	Training	Outlier
Lindane	58-89-9	n°Cl ≥ 5	Training	Outlier
Acetochlor	34256-82-1	N–CO–CH_2–Cl	Training	Acceptable prediction
Aldicarb	116-06-3	X–CO–NH–C	Training	Acceptable prediction
Methomyl	16752-77-5	X–CO–NH–C	Training	Acceptable prediction
Neurolidol	7212-44-4	Terminal ethylene with N or O in C-alpha	Training	Acceptable prediction
Benfluralin	1861-40-1	Para-trifluoro-aminobenzene	Training	Acceptable prediction
Methyl isothiocyanate	556-61-6	n°C < 3	Training	Acceptable prediction
Formetanate hydrochloride	23422-53-9	X–CO–NH–C	Training	Acceptable prediction
Dowicil	4080-31-3	–CH=CCl; charge	Training	Acceptable prediction
Heptachlor	76-44-8	n°Cl ≥ 5	Training	Acceptable prediction
Fluthiacet methyl	117337-19-6	Methyl ester	Training	Acceptable prediction
Bifenthrin	82657-04-3	–CH=CCl	Training	Acceptable prediction
Chloropicrin	76-06-2	n°C < 3	Training	Acceptable prediction
Kresoxim methyl	143390-89-0	Methyl ester	Training	Acceptable prediction
Carbaryl	63-25-2	X–CO–NH–C	Training	Acceptable prediction
Clodinafop-propargyl	105511-96-4	Triple bond with O in C-alpha	Training	Acceptable prediction
Alachlor	15972-60-8	N–CO–CH_2–Cl	Training	Acceptable prediction

(*Continued*)

Table 16
(*Continued*)

Name	CAS	Detected rule	Training/ Test set	Comments
Pentachlorophenol	87-86-5	n°Cl $\geq$ 5	Training	Acceptable prediction
Oxamyl	23135-22-0	X–CO–NH–C	Training	Acceptable prediction
Endrin	72-20-8	n°Cl $\geq$ 5	Training	Acceptable prediction
Lambda-Cyhalothrin	91465-08-6	–CH=CCl	Training	Acceptable prediction
Diflufenzopyr-sodium	109293-98-2	X–CO–NH–C	Training	Acceptable prediction
Diuron	330-54-1	X–CO–NH–C	Training	Acceptable prediction
Beta cypermethrin	66841-24-5	–CH=CCl	Training	Acceptable prediction
Phosmet	732-11-6	P(=S)S-CH-X	Training	Acceptable prediction
Dicofol	115-32-2	n°Cl $\geq$ 5	Training	Acceptable prediction
Trimethacarb	2686-99-9	X–CO–NH–C	Training	Acceptable prediction
Fenoxycarb	72490-01-8	X–CO–NH–C	Training	Acceptable prediction
Dimethenamid	87674-68-8	N–CO–CH_2–Cl	Training	Acceptable prediction
Metolachlor	51218-45-2	N–CO–CH_2–Cl	Training	Acceptable prediction
Norflurazon	27314-13-2	Fragment 1	Training	Acceptable prediction
Linuron	330-55-2	X–CO–NH–C	Training	Acceptable prediction
Cymoxanil	57966-95-7	X–CO–NH–C	Training	Acceptable prediction
Methyl anthralinate	134-20-3	Methyl ester	Training	Acceptable prediction
Siduron	1982-49-6	X–CO–NH–C	Training	Acceptable prediction

Table 16
(*Continued*)

Name	CAS	Detected rule	Training/ Test set	Comments
Azoxystrobin	131860-33-8	Methyl ester	Training	Acceptable prediction
DDAC	7173-51-5	Charge	Training	Acceptable prediction
Chlorpropham	101-21-3	X–CO–NH–C	Training	Acceptable prediction
Dichloropropene	542-75-6	–CH=CCl	Training	Acceptable prediction
Aldoxycarb	1646-88-4	X–CO–NH–C	Training	Acceptable prediction
Chlorimuron ethyl	90982-32-4	X–CO–NH–C	Training	Acceptable prediction
Phenmedipham	13684-63-4	X–CO–NH–C; methyl ester	Training	Acceptable prediction
Iprodione	36734-19-7	X–CO–NH–C	Training	Acceptable prediction
MCPP Acid	7085-19-0	COOH with O, N or F in C-alpha	Training	Acceptable prediction
Cyfluthrin	68359-37-5	–CH=CCl	Training	Acceptable prediction
2,4-D Acid	94-75-7	COOH with O, N or F in C-alpha	Training	Acceptable prediction
Ethion	563-12-2	P(=S)S–CH–X	Training	Acceptable prediction
Thiazopyr	117718-60-2	Methyl ester	Training	Acceptable prediction
Chlorflurenol methyl	2536-31-4	Methyl ester	Training	Acceptable prediction
Kepone	143-50-0	n°Cl $\geq$ 5	Training	Acceptable prediction
Bis(trichloromethyl) Sulfone	3064-70-8	n°Cl $\geq$ 5; n°C $<$ 3	Training	Acceptable prediction
Methyl chloroform	71-55-6	n°C $<$ 3	Training	Acceptable prediction
Triclopyr acid	55335-06-3	COOH with O, N or F in C-alpha	Training	Acceptable prediction

(*Continued*)

Table 16
(*Continued*)

Name	CAS	Detected rule	Training/ Test set	Comments
Metalaxyl	57837-19-1	methyl ester	Training	Acceptable prediction
Primisulfuron-methyl	86209-51-0	X–CO–NH–C; methyl ester	Training	Acceptable prediction
Imidacloprid	105827-78-9	N–NO_2	Training	Acceptable prediction
Imazethabenz	81405-85-8	Methyl ester	Training	Acceptable prediction
Difenzoquat methyl sulphate	43222-48-6	Charge	Training	Outlier
Diflubenzuron	35367-38-5	X–CO–NH–C	Training	Outlier
Spinosad	131929-60-7	MW > 700	Training	Outlier
Z-11-Hexadecanal	53939-28-9	–CHO; long chain	Training	Outlier
1,2-Benzenedi-carboxaldehyde	643-79-8	–CHO	Test	Outlier
Tefluthrin	79538-32-2	–CH=CCl; Ar–CH-O (Ar fully substituted)	Test	Outlier
Malathion	121-75-5	P(=S)S–CH–X	Test	Outlier
Flumetralin	62924-70-3	Para-trifluoro-aminobenzene	Test	Acceptable prediction
Acetaldehyde	26532-25-2	–CHO	Test	Acceptable prediction
Propargite	2312-35-8	Triple bond with O in C-alpha	Test	Acceptable prediction
Bendiocarb	22781-23-3	X–CO–NH–C	Test	Acceptable prediction
Fluazinam	79622-59-6	Para-trifluoro-aminobenzene	Test	Acceptable prediction
Imazalil	35554-44-0	Terminal ethylene with N or O in C-alpha	Test	Acceptable prediction
Dieldrin	60-57-1	n°Cl $\geq$ 5	Test	Acceptable prediction
Chlorophacinone	3691-35-8	$(C{=}O)_3$–CH	Test	Acceptable prediction
Potassium salt of oleic acid	143-18-0	Long chain	Test	Acceptable prediction

Table 16
(*Continued*)

Name	CAS	Detected rule	Training/ Test set	Comments
Diethyl ethyl	38727-55-8	N–CO–CH_2–Cl	Test	Acceptable prediction
Terbacil	5902-51-2	X–CO–NH–C	Test	Acceptable prediction
Mesotrione (AMBA)	104206-82-8	$(C{=}O)_3$–CH	Test	Acceptable prediction
Diphacinone	82-66-6	$(C{=}O)_3$–CH	Test	Acceptable prediction
Bromacil	314-40-9	X–CO–NH–C	Test	Acceptable prediction
PCNB	82-68-8	n°Cl ≥ 5	Test	Acceptable prediction
Tridiphane	58138-08-2	n°Cl ≥ 5	Test	Acceptable prediction
Dienochlor	2227-17-0	n°Cl ≥ 5	Test	Acceptable prediction
Tebuthiuron	34014-18-1	X–CO–NH–C	Test	Acceptable prediction
DCPA (Dacthal)	1861-32-1	Methyl ester	Test	Acceptable prediction
Ethylene dichloride	107-06-2	n°C < 3	Test	Outlier
Hexadecadienol acetate	53042-79-8	Long chain	Test	Outlier
Trisulfuron methyl	126535-15-7	X–CO–NH–C; methyl ester	Test	Outlier

There are some common features in these fragments. Several of them are related to a carbon-hybridized sp^3, bearing a hydrogen, linked to an electronegative atom on one side and to a carbon aromatic, sp^2 or sp. The hydrogen is likely reactive. In other cases, the hydrogen is linked to a carbon-hybridized sp or sp^2.

- Thus, one of these fragments is an acidic group, with an electronegative atom (oxygen, nitrogen and halogen atoms) in alpha. Examples of outliers with this fragment are Dichlorprop (ID 84) and Sodium fluoroacetate (ID 230). The structures of all pesticides are shown in Appendix B, this volume.

- A related fragment, which is likely involved in an excess toxicity, is the group N–CO–CH_2–Cl; also in this case, there is a hydrogen activated. An example of outliers with this fragment is Propachlor (ID 216).

- Somehow related to this reactivity is the fragment showing a terminal ethylene, with again a heteroatom such as oxygen or nitrogen in the carbon in position alpha to it. In this case, there are two outliers: Rotenone (ID 224) and Ethalfluralin (ID 110).

- Toxicity can be higher also in case of a fragment with a triple bond linked to a carbon atom with an oxygen in alpha position (C≡C–C–O). A pesticide with this fragment is 3-Iodo-2-propynyl butylcarbamate (ID 277).

- Another fragment with double bond is the chloro-ethylene: –CH=CCl–. This fragment is present in two outliers: Dichlorvos (ID 85) and Tefluthrin (ID 339).

- A somehow related fragment is this: –CHO. Also in this case, there is a hydrogen linked to the carbon-hybridized sp^2. This fragment likely plays a role in two false negatives: Antimycin A (ID 28) and 1,2-Benzenedicarboxaldehyde (ID 274).

- Another fragment likely involved in an excess toxicity is P(=S)S–CH–X, in which X is an heteroatom, such as S or N, or an esteric group. Examples of the outliers with this group are Terbufos (ID 240) and Phorate (ID 204).

- Another fragment with phosphorus may be involved in the excess toxicity (even if the extent of this excess toxicity is limited): phosphorus directly linked to carbon. There are two examples of outliers with this group: Glyphosate (ID 143) and Fonofos (ID 138).

- Also some benzylic groups are likely related to excess reactivity, confirming the critical issue of the hydrogen reactivity. We identified two cases: a benzylic hydrogen linked to oxygen, which is part of an oxime group, Ar–CH–O–N =. There are two examples of this fragment in the outliers: Fenpyroximate (ID 310) and Trifloxystrobin (ID 342).

- In the second case, the fragment is Ar–CH–O in which the aromatic ring is fully substituted (no hydrogen), and most of the substituents are electronegative groups. There are two examples of this fragment: Chlorfenapyr (ID 293) and Tefluthrin (ID 339).

- Another fragment involved in excess toxicity is the fragment 1 (illustrated in Fig. 9). Some examples of false negatives with this fragment are given in Table 13.

- A further fragment with a likely role in excess toxicity with activated hydrogen is the group CH–NO_2. In this case the carbon is aliphatic, and there is only one case in the set of chemicals that shows excess toxicity: Bioban (ID 287).

- A further fragment likely related to excess toxicity is *para*-trifluoro-aminobenzenes. There are three cases of pesticides with excess toxicity with this group: Ethalfluralin (ID 110), Trifluralin (ID 260) and Fluvalinate (ID136).

- A fragment which is likely related to wrong-calculated toxicity is X–CO–NH–C, in which X is oxygen or nitrogen. There are four false negatives with this group: Carbofuran (ID 46), Sodium dichloro-*s*-triazinetrione (ID 346), Benomyl (ID 31) and 3-Iodo-2-propynyl butylcarbamate (ID 277).

- An excess toxicity is likely related to a high number of chlorine in the molecule (≥ 5), as in the case of DDT (ID 76) and Hexachlorocyclohexane (ID 142).

- Finally, compound Pival (ID 208) shows an excess toxicity which is likely related to the presence of three carbonyl groups linked to a CH residue: this hydrogen is indeed in a peculiar position and is very labile.

In total, these fragments are used to avoid the 31 false negatives predicted by the model with a factor higher than 10. Some chemicals which were correctly predicted are also eliminated, but in general these fragments are quite specific. See Table 16 for the complete list of compounds considered.

As we said, most of these fragments characterize a reduced performance of the model. Only a few of them are likely related with a possible error higher than a factor of 50 in the case of false negative.

Thus, the user has two choices: adopt stricter rules, in which case the false negatives should not exceed the factor of 10, or use a limited number of rules, if the factor of 50 is acceptable.

In the trout hybrid model, there are also a number of false positives. The above-mentioned fragments are also capable to avoid seven false positives. However, in order to increase the performances of the model and avoid stronger false positives, we list these conditions:

- Molecular weight – There are a few chemicals with a relatively large molecular weight. This factor is likely related to an increased complexity of the molecule. We suggest not using the model if molecular weight exceeds 700. We found one false positive in this case: spinosad (ID 231).

- Similarly, by far, most of the chemicals present a number of carbon atoms higher than two. Thus, in case of simple molecule (number of carbon two or less), the model should not be used. We found a positive outlier with this condition: ethylene dichloride (ID 115).

- Another unusual situation is when there is a long chain (16 or more). The model should not be used. We found two false positives in this case: hexadecadienol acetate (ID 145) and *Z*-11-hexadecenal (ID 2). The likely reason is that the log *P* programs to calculate the value may be less reliable in this case.
- The presence of a charge in the molecule is also quite unusual. The model may be less reliable in this case, and indeed, we found a false positive in this case: difenzoquat methyl (ID 92)
- A fragment which is likely related to wrong-calculated toxicity is $N-NO_2$. There is one false positive in the data set with this fragment: imidacloprid (ID 185).
- Methyl esters are often badly predicted. We found three positive outliers with this fragment: metalaxyl (ID 162), imazethabenz (ID 147) and DCPA (ID 60).

In conclusion, with these conditions the model (see Fig. 11) has an error which should not exceed a factor of 10 for false negatives and a factor of 50 for false positives, in our experience.

This is an important feature of the model which represents a useful starting point for the regulatory use of QSAR models. The statistical parameters of this model are good, $R^2_{\text{Train}} = 0.87$ and $R^2_{\text{Test}} = 0.89$ (see Table 17). Table 17 summarizes the statistical parameters of the final model.

4.2.3. Other hybrid models for rainbow trout

The final hybrid model, as well as the individual models, represents one of the many possible solutions, which quite probably give similar results. Indeed, we also tested many other hybrid models. Results of the best two, without the exclusion of the outliers, are given in Table 18 where the alternative models 1 and 2 were built by means of the rule-based approach.

Among the three combined models, we choose the one with the lower root mean square error and the one with less outliers.

4.2.4. Check of solubility

In order to increase the quality of data, the solubility issue has been analysed as well. In fact, some molecules in the data set were reported with a discrepancy between the toxicity datum and the solubility datum. In the cases where the solubility is low, the common procedure is that a surfactant/solvent is added during the experiment. To avoid the possible confounding factor related to this issue, a new model has been built without those compounds with a toxicity value higher

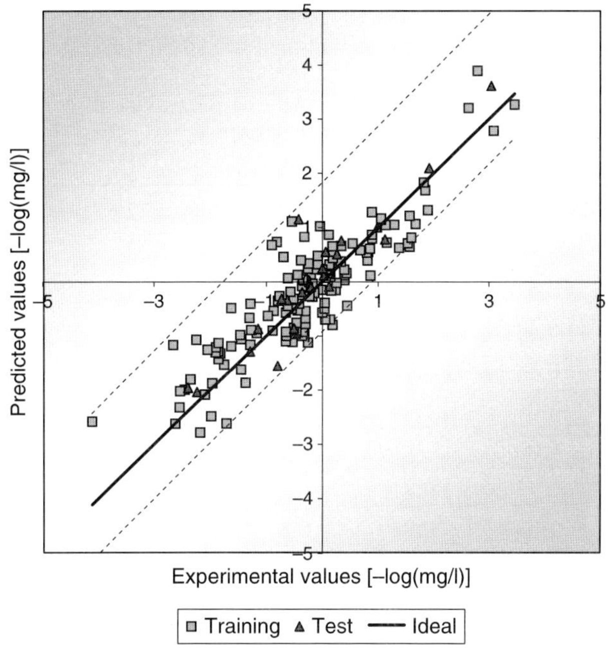

Figure 11: Predicted versus experimental values for trout high confidence hybrid model. The dotted lines show a confidence interval of a factor of 10 in case of false negatives and a factor of 50 in case of false positives.

Table 17

Statistical parameters of the hybrid model for trout toxicity taking out outliers at factor 10 for false negatives and at factor 50 for false positives

	NC	ND	R^2	K	q	R^2_0	K_0 ($q_0 = 0$)	K'	q'	$R^{2\prime}_0$	K_0'	RMSE	Outliers
Training	141	20	0.82	0.84	0.42	0.78	0.99	0.97	−0.02	0.82	0.96	0.57	–
Test	21	20	0.89	1.01	0.17	0.89	1.07	0.88	0.1	0.89	0.92	0.48	–

NC, number of compounds; ND, number of descriptors; K, slope; q, intercept.

than the solubility. The results of the model, based on a neural network, are comparable with the final hybrid model, but the performances are better compared to the single individual models (Table 19).

Because for a new compound, which eventually is not yet synthesizsed, the experimental solubility is unknown, we preferred to leave the model as it is, without excluding compounds which have been tested after the addition of the solvent to increase the pesticide solubility. Indeed, the improvement of the new model is very limited. However, in case of trout solubility, the user should keep

Table 18
Statistical parameters of the alternative combined models for trout toxicity

	Trout hybrid model		Alternative model 1		Alternative model 2	
	Training	Test	Training	Test	Training	Test
R^2	0.71	0.71	0.74	0.72	0.74	0.7
K	0.7	0.71	0.99	1.02	1.01	1.02
q	0.61	0.89	0.2	0.49	0.12	0.45
$R^2{}_0$	0.64	0.57	0.73	0.7	0.74	0.69
K_0	0.88	0.97	1.05	1.16	1.04	1.16
K'	1.01	1.00	0.74	0.7	0.74	0.69
q'	0.06	–0.19	0.49	0.33	0.54	0.4
$R^{2\prime}{}_0$	0.71	0.71	0.7	0.7	0.7	0.68
$K_0{}'$	1.04	0.94	0.87	0.78	0.87	0.78
Outliers	6+; 4–	1+; 3–	7+; 14–	1+; 6–	9+; 15–	1+; 5–
Mean fold error	0.63	0.65	0.73	0.84	0.72	0.86
RMSE	0.84	0.87	0.96	1.14	0.96	1.17

Table 19
Statistical parameters for the model for trout toxicity built only with soluble compounds

	NC	ND	R^2	K	q	$R^2{}_0$	K_0 ($q_0 = 0$)	K'	q'	$R^{2\prime}{}_0$	$K_0{}'$	RMSE	Outliers
Training	210	7	0.71	0.70	0.73	0.62	0.91	1.01	–0.02	0.71	1.00	0.87	7+; 6–
Test	38	7	0.76	0.75	0.85	0.65	0.99	1.02	–0.25	0.75	0.94	0.80	1+

NC, number of compounds; ND, number of descriptors; K, slope; q, intercept.
outliers +, predicted value (mg/l)/expected value (mg/l) > 50;
outliers –, expected value (mg/l)/predicted value (mg/l) > 50.

in mind that the predicted toxicity value can be lower than the solubility. It means that the toxic effect can be achieved only in the presence of substances increasing the pesticide solubility. This warning applies to all predictive methods for aquatic substances, not only to the DEMETRA models. What is new in any case is that we verified that our model is not affected by the presence of insoluble compounds. This improves its reliability.

4.2.5. Further models with in-house produced descriptors

Among the large number of models obtained to predict trout toxicity, it is interesting to note that the best results have been obtained combining the selected hybrid

Table 20
Statistical parameters of the hybrid model for trout toxicity using Fragment descriptors

	NC	ND	R^2	K	q	R^2_0	K_0 ($q_0 = 0$)	K'	q'	$R^{2\prime}_0$	K_0'	RMSE	Outliers
Training	228	Fragment	0.85	0.81	0.40	0.83	0.93	1.06	–0.08	0.85	1.03	0.60	3+; 4–
Test	45	Fragment	0.77	0.78	0.66	0.71	0.97	1.00	–0.10	0.77	0.97	0.74	1–

NC, number of compounds; ND, number of descriptors; K, slope; q, intercept.
outliers +, predicted value (mg/l)/expected value (mg/l) > 50;
outliers –, expected value (mg/l)/predicted value (mg/l) > 50.

model with a new model, developed with particular chemical fragments (Table 20). A special approach has been developed to cut chemical into fragments and then develop a QSAR model based on these fragments. This approach gave results superior to those obtained with other commercial software using fragments. The likely explanation of the better results achieved when combining models based on more classical chemical descriptors and fragments is that the model using fragments brings some new information not present in the classical descriptors, and thus a more comprehensive and deeper description of the chemical compound is produced, affording better results. As we said, as the software on fragment is still an internal version, under development, we could not make it freely available. We are working to produce an open version of it.

4.3. Hybrid model for water flea (*Daphnia magna*)

The more suitable hybrid model for this endpoint has been built using the rule-based approach, where the inputs are two ANNs and a single PLS models. The hybrid model was built following the usual procedure for QSAR approaches expressing the toxicity data as −log(mmol/l) as presented in Table 21, while Fig. 12 shows values in −log(mg/l) units.

Table 21
Statistical parameters of the hybrid model for daphnia toxicity

	NC	ND	R^2	Q_2	K	q	R^2_0	K_0 ($q_0 = 0$)	K'	q'	$R^{2\prime}_0$	K_0'	RMSE	Outliers
Training	220	16	0.76	0.74	0.78	0.69	0.70	0.96	0.97	−0.05	0.76	0.95	0.96	11+; 11–
Test	43	16	0.72	–	0.65	1.07	0.51	0.93	1.11	−0.50	0.70	0.96	1.17	2+; 5–

NC, number of compounds; ND, number of descriptors; K, slope; q, intercept.
outliers +, predicted value (mg/l)/expected value (mg/l) > 50;
outliers –, expected value (mg/l)/predicted value (mg/l) > 50.

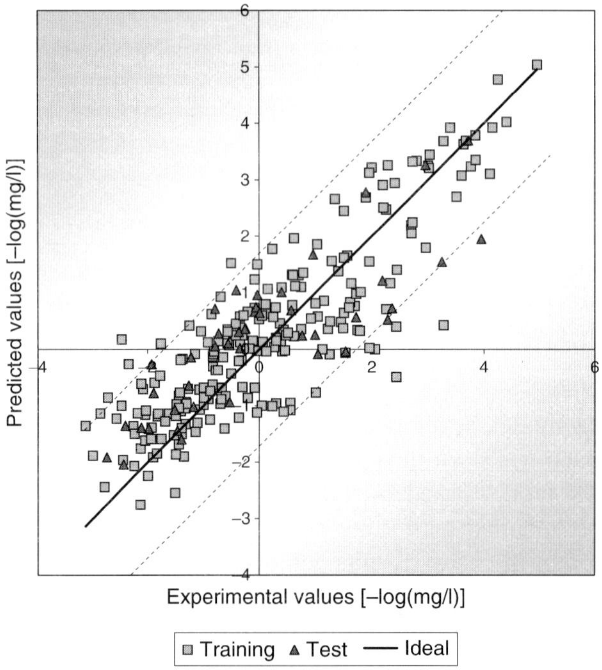

Figure 12: Predicted versus experimental values for daphnia toxicity using the DEMETRA hybrid model. Dotted lines indicate a factor of 50 in mg/l.

The chemicals in the training set appear to approach properly to the ideal model, as there are not remarkable differences between the values of K, K', K_0 and K_0' and the corresponding Y interceptions. The same behaviour has been observed for values of the external validation, a fact that shows the adequate predictive power of the model.

The evaluation of the general performance of the individual models, taken as starting point for building the hybrid model, gives indeed a more general insight of the good behaviour of the selected model with regard to the individual ones improving in general all the model parameters. The results for the individual models are reported in Table 22.

The selected hybrid model is clearly an improvement with reference to the individual models, both in performance and number of outliers, while maintaining, or even improving, the predictive ability of the individual systems. Descriptors selected for this endpoint are shown in Table 23.

Important descriptors are log P, H-bond involved atoms and mean polarizability. Additionally, many of the selected fingerprints descriptors are based in the number of oxygen atoms; a fact that can be regarded as a measure of the general electronegativity of the molecule opposed to lipophilicity.

Table 22
Statistical parameters of the individual models used in the hybrid model for daphnia toxicity

		NC	ND	Q^2	R^2	K	q	R^2_0	K_0 ($q_0 = 0$)	K'	q'	$R^{2'}_0$	K'_0	Outliers
NN	Training	220	10	0.74	0.74	0.74	0.67	0.68	0.91	1.01	−0.02	0.74	1.00	8+; 14–
	Test	43	–	–	0.65	0.58	1.10	0.40	0.87	1.12	−0.37	0.64	1.00	6+; 2–
NN	Training	220	15	0.76	0.77	0.75	0.63	0.71	0.92	1.01	−0.03	0.77	1.00	12+; 9–
	Test	43	–	–	0.59	0.59	1.16	0.35	0.90	1.01	−0.17	0.59	0.96	6+; 4–
PLS	Training	220	10	0.59	0.65	0.65	0.91	0.52	0.88	1.00	$-2E-07$	0.65	1.00	22+; 19–
	Test	43	–	–	0.65	0.56	0.97	0.44	0.82	1.14	−0.24	0.64	1.06	7+; 1–

NC, number of compounds; ND, number of descriptors; K, slope; q intercept.
outliers +, predicted value (mg/l)/expected value (mg/l) > 50;
outliers −, expected value (mg/l)/predicted value (mg/l) > 50.
NN, neural network; PLS, partial least squares.

Table 23
List of descriptors employed within the hybrid model for daphnia toxicity

Software	Descriptor Symbol	Correlation	Definition
DRAGON	MW	+	Molecular weight
DRAGON	Mp	+	Mean atomic polarizability (scaled on carbon atom)
DRAGON	nP	+	Number of phosphorus atoms
DRAGON	nNR2Ph	+	Number of tertiary amines (aromatic)
DRAGON	nHAcc	–	Number of acceptor atoms for H-Bonds (NOF)
DRAGON	O-057	–	Phenol/enol/carboxyl OH
DRAGON	O-060	+	Al–O–Ar/Ar–O–Ar/ R..O..O/R-O-C=X
DRAGON	S-107	+	R2S/RS-SR
DRAGON	MLOGP	+	Moriguchi octanol-water partition coefficient (log P)
DRAGON	WA	+	Mean Wiener Index
DRAGON	IDE	+	Mean information content on the distance equality

(Continued)

Table 23
(*Continued*)

Software	Descriptor Symbol	Correlation	Definition
DRAGON	IC2	–	Information content index (neighborhood symmetry of 2-order)
DRAGON	Eig1p	+	Leading eigen value from polarizability weighted distance matrix
DRAGON	T(F..Cl)	+	Sum of topological distances between F···Cl
DRAGON	SRW05	+	Self-returning walk count of order 0.5
DRAGON	BEHm1	+	Highest eigen value of Burden matrix weighted by atomic masses

It is remarkable that, in general, this model shares some characteristics with the trout descriptors. The main difference between daphnia and trout endpoints is whether the chemicals have rings in the molecular structure or not.

In any case we can observe the direct relationship between the presence of electronegative atoms and toxicity, while log *P* has a negative effect likely due to the solubility effect. Thus, the main toxic effect is related to the solubility of the compounds while lipophilicity decreases it, a fact strengthened by the relative importance of the molecular size. Branching, like the presence of rings for trout, seems to be another factor related to toxicity.

The results for this hybrid system is only one of the possible solutions, and we have developed indeed other hybrid models. Table 24 summarizes the ones with better performances.

These alternative models 1 and 2 were built by means of the rule-based approach. They present either an equivalent collection of descriptors and similar outliers, a fact that supports the idea that the best possible model has been selected. Table 25 summarizes the major outliers for this endpoint.

In particular, we list compounds with a predicted toxicity value 50 times above or below the experimental value. Following the same scheme that we carried out for the trout endpoint, we have identified some characteristics common to the outliers present in the training set in order to establish which molecules should be considered as non-applicable to the chosen model at this level of confidence.

On the basis of the training set compounds, we have identified some common sub-structures of the outliers.

Table 24
Comparison of the performance for the selected model and other hybrid models for daphnia toxicity

	Daphnia hybrid model		Alternative model 1		Alternative model 2	
	Training	Test	Training	Test	Training	Test
r^2	0.76	0.72	0.76	0.73	0.73	0.7
K	0.78	0.65	1.07	0.89	1	0.86
q	0.69	1.07	0.33	0.76	0.46	0.77
$r^2{}_0$	0.7	0.51	0.76	0.68	0.71	0.64
K_0	0.97	0.93	1.15	1.1	1.12	1.06
K'	0.97	1.11	0.71	0.82	0.73	0.82
Q'	0.05	0.5	0.38	0.03	0.36	0.11
$R_0^{2'}$	0.76	0.7	0.75	0.73	0.71	0.7
$K_0{}'$	0.95	0.96	0.79	0.83	0.81	0.84
MFE	0.71	0.84	0.96	0.96	0.95	0.97
RMSE	0.9	0.97	1.2	1.09	1.2	1.11

Table 25
Outliers of the hybrid model for daphnia toxicity using a factor of 50

Name	CAS	Experimental value (mg/l)	Predicted value (mg/l)	Training/ test set	False Negative/ False Positive
Rotenone	83-79-4	0.0037	3.14	Training	False negative
Pyridaben	96489-71-3	0.00053	0.375	Training	False negative
Methomyl	16752-77-5	0.0088	1.004	Training	False negative
Diflubenzuron	35367-38-5	0.0037	0.398	Training	False negative
Hexaflumuron	86479-06-3	0.000111	0.011	Test	False negative
Propoxur	114-26-1	0.011	0.944	Training	False negative
Sodium dichloro-*s*-triazinetrione	2893-78-9	0.0990	5.879	Training	False negative
Dichloro-2-*n*-octyl-3(2H)-isothiazolone	64359-81-5	0.00522	0.298	Test	False negative
Ethofenprox	80844-07-1	0.00057	0.028	Test	False negative

(*Continued*)

Table 25
(*Continued*)

Name	CAS	Experimental value (mg/l)	Predicted value (mg/l)	Training/ test set	False Negative/ False Positive
Pendimethalin	40487-42-1	0.28	12.08	Training	False negative
Difethialone	104653-34-1	0.0044	0.189	Test	False negative
Bendiocarb	22781-23-3	0.0292	1.155	Training	False negative
Carbofuran	1563-66-2	0.029	1.080	Test	False negative
Carbaryl	63-25-2	0.0056	0.196	Training	False negative
Sulfluramid	4151-50-2	0.37	12.85	Training	False negative
Aldoxycarb	1646-88-4	0.28	9.033	Training	False negative
Dichlobenil	1194-65-6	6.2	0.192	Test	False positive
Tricosene	27519-02-4	1.08	0.032	Training	False Positive
Bensulide	741-58-2	0.58	0.017	Training	False Positive
Etridiazole	2593-15-9	4.9	0.118	Training	False positive
Carboxin	5234-68-4	84.4	1.849	Test	False positive
Dipropyl isocinchomeronate	136-45-8	18	0.380	Training	False positive
Biobor	2665-13-6	730	13.8	Training	False positive
Fosamine ammonium	25954-13-6	1372	22.8	Training	False positive
Diphenamid	957-51-7	58	0.793	Training	False positive
Oxadixyl	77732-09-3	530	7.184	Training	False positive
Fenoxaprop-ethyl	66441-23-4	3.18	0.030	Training	False positive
Imazethabenz	81405-85-8	220	1.829	Training	False positive
Fomesafen	72178-02-0	294	0.668	Training	False positive

CAS, Chemical Abstracts Service Number.

1. Carbamates.
2. Carbamides.
3. Sulphanyl-acrylamide derivatives.
4. Polycyclic compounds with at least one heterocyclic ring and two or more aromatic rings.

The general sub-structures found for categories 1, 2 and 3 are depicted in Fig. 13.

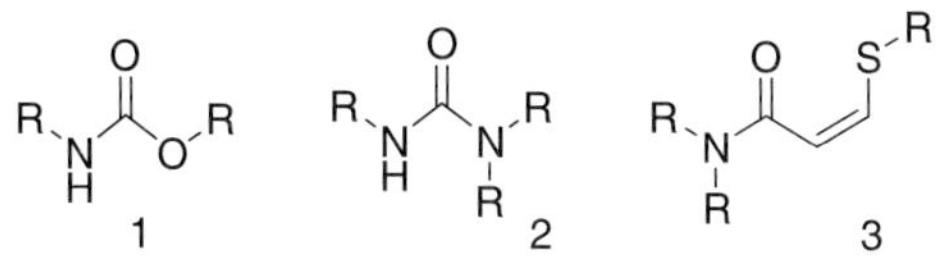

Figure 13: General sub-structures applicable for rules 1, 2 and 3.

The list of compounds removed from the model and the rules fulfilled by them are given in Table 26.

Derived for the application of these rules we notice that carbamates, Table 27, represent a large part of outliers.

Table 26
Compounds removed from the final model for daphnia toxicity using a factor of 50

Name	CAS	Detected rule	Training/test set	Comments
Carboxin	5234-68-4	3	Test	Outlier
Tebuthiuron	34014-18-1	2	Training	Acceptable prediction
Bromacil	314-40-9	2	Training	Acceptable prediction
Chlorsulfuron	64902-72-3	2	Training	Acceptable prediction
Rimsulfuron	122931-48-0	2	Training	Acceptable prediction
Chlorpropham	101-21-3	1	Training	Acceptable prediction
Thiophanate-methyl	23564-05-8	1	Training	Acceptable prediction
Terbacil	5902-51-2	2	Training	Acceptable prediction
Asulam	3337-71-1	1	Training	Acceptable prediction
Brodifacoum	56073-10-0	4	Test	Acceptable prediction
Fenoxycarb	72490-01-8	1	Test	Acceptable prediction
Pymetrozine	123312-89-0	2	Training	Acceptable prediction
Thidiazuron	51707-55-2	2	Training	Acceptable prediction

(*Continued*)

Table 26
(*Continued*)

Name	CAS	Detected rule	Training/test set	Comments
Fluometuron	2164-17-2	2	Training	Acceptable prediction
Bifenazate	149877-41-8	1	Training	Acceptable prediction
Cymoxanil	57966-95-7	2	Training	Acceptable prediction
Aldicarb	116-06-3	1	Training	Acceptable prediction
Diflufenzopyr-sodium	109293-98-2	2	Training	Acceptable prediction
Octhilinone	26530-20-1	3	Training	Acceptable prediction
Carbendazim	10605-21-7	1	Training	Acceptable prediction
Formetanate hydrochloride	23422-53-9	1	Training	Acceptable prediction
Methiocarb	2032-65-7	1	Test	Acceptable prediction
Trimethacarb	2686-99-9	1	Training	Acceptable prediction
Aldoxycarb	1646-88-4	1	Training	Acceptable prediction
Carbaryl	63-25-2	1	Training	Acceptable prediction
Carbofuran	1563-66-2	1	Test	Acceptable prediction
Bendiocarb	22781-23-3	1	Training	Acceptable prediction
Difethialone	104653-34-1	4	Test	Acceptable prediction
Sodium dichloro-*s*-triazinetrione	2893-78-9	2	Training	Acceptable prediction
Propoxur	114-26-1	1	Training	Outlier
Hexaflumuron	86479-06-3	2	Test	Outlier
Diflubenzuron	35367-38-5	2	Training	Outlier
Methomyl	16752-77-5	1	Training	Outlier
Pyridaben	96489-71-3	3	Training	Outlier
Rotenone	83-79-4	4	Training	Outlier

CAS, Chemical Abstracts Service Number.

Table 27
Statistical parameters for the hybrid model for daphnia toxicity after outliers removal

	NC	ND	R^2	K	q	R^2_0	K_0 ($q_0 = 0$)	K'	q'	$R^{2\prime}_0$	K_0'	RMSE	Outliers
Training	193	16	0.80	0.82	0.26	0.77	0.84	0.97	–0.22	0.79	0.94	0.84	7–
Test	36	16	0.79	0.75	0.35	0.72	0.71	1.05	–0.43	0.71	1.02	0.96	–

NC, number of compounds; ND, number of descriptors; K, slope; q, intercept.
outliers +, predicted value (mg/l)/expected value (mg/l) > 50;
outliers –, expected value (mg/l)/predicted value (mg/l) > 50.

Such a procedure does not mean a significant loss in the number of compounds for both training and test sets; this procedure induces an actual improvement of the final model performance.

The model performance after removal of outliers (Table 26) is depicted in Fig. 14 whose comparison with Fig. 12 clearly shows how the selected rules make a clear improvement on the region occupied by the more toxic false negatives.

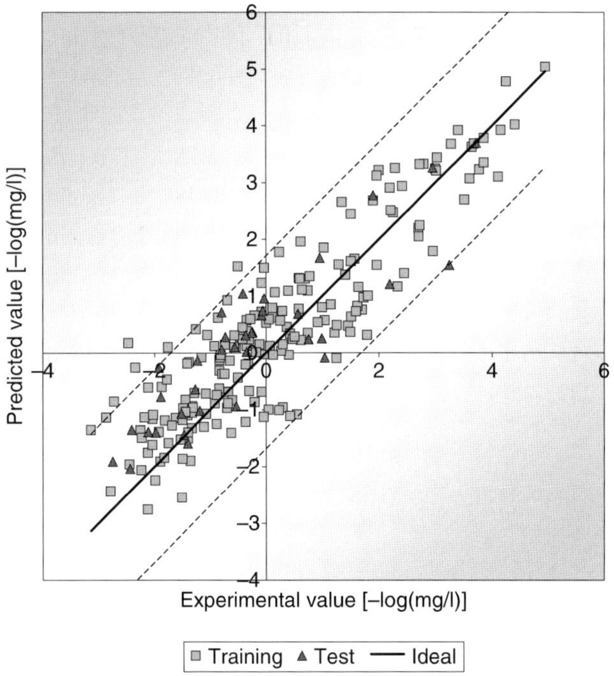

Figure 14: Predicted versus experimental values of the hybrid model for daphnia after compounds removal using a factor of 50 for outliers. Dotted lines indicate the factor of 50 in mg/l.

Seven false positives still remain (only within the training set), while only one compound of the test set is close to being predicted as false negative.

In a similar way to the trout endpoint, we consider a confidence interval around the models of 50 as safe enough for general purposes, but offering the chance of selecting a more restrictive level might be an excellent functionality for models applied to risk assessment. Therefore, we have enlarged the chemical rules to reduce the model uncertainty. Considering this safety factor the list of outliers increases (see Table 28).

Such rules have been developed by visual inspection of common chemical features of the outliers considering a factor of 15 as acceptable. The typical variability for daphnia experiments is 8 (see Chapter 2, Benfenati *et al.*, this volume); thus, for daphnia, we think that a factor of 15 is acceptable.

Table 28
List of compounds of the model for daphnia toxicity considered as outliers using a factor of 15

Name	CAS	Experimental value (mg/l)	Predicted value (mg/l)	Training/ test set	False negative/false positive
Rotenone	83-79-4	0.0037	3.15	Training	False negative
Pyridaben	96489-71-3	0.00053	0.38	Training	False negative
Methomyl	16752-77-5	0.0088	1.00	Training	False negative
Diflubenzuron	35367-38-5	0.0037	0.40	Training	False negative
Hexaflumuron	86479-06-3	0.000111	0.011	Test	False negative
Propoxur	114-26-1	0.011	0.94	Training	False negative
Sodium dichloro-*s*-triazinetrione	2893-78-9	0.0990	5.88	Training	False negative
Dichloro-2-*n*-octyl-3(2H)-isothiazolone	64359-81-5	0.00522	0.30	Test	False negative
Ethofenprox	80844-07-1	0.00057	0.028	Test	False negative
Pendimethalin	40487-42-1	0.28	12.08	Training	False negative
Difethialone	104653-34-1	0.0044	0.19	Test	False negative
Bendiocarb	22781-23-3	0.0292	1.16	Training	False negative
Carbofuran	1563-66-2	0.029	1.08	Test	False negative

Table 28
(*Continued*)

Name	CAS	Experimental value (mg/l)	Predicted value (mg/l)	Training/ test set	False negative/false positive
Carbaryl	63-25-2	0.0056	0.20	Training	False negative
Sulfluramid	4151-50-2	0.37	12.85	Training	False negative
Aldoxycarb	1646-88-4	0.28	9.03	Training	False negative
DDAC	7173-51-5	0.01624	0.48	Training	False negative
Trimethacarb	2686-99-9	0.018	0.46	Training	False negative
3-Chloro-p-toluidine hydrochloride	7745-89-3	0.05567	1.23	Training	False negative
DCDMH	118-52-5	0.5	9.17	Training	False negative
2-Benzyl-4-chlorophenol	120-32-1	0.59	10.22	Training	False negative
Azinphos-methyl	86-50-0	0.0011	0.016	Training	False negative
DDT	50-29-3	0.0047	0.07	Training	False negative
Methiocarb	2032-65-7	0.019	0.27	Test	False negative
1,2-Benzene dicarboxal dehyde	643-79-8	0.09	1.21	Test	False negative
Formetanate Hydrochloride	23422-53-9	0.07453	1.00	Training	False negative
Cyprodinil	121552-61-2	0.032	0.43	Training	False negative
Simazine	122-34-9	1.1	13.33	Training	False negative
Fludioxonil	131341-86-1	0.9	10.00	Training	False negative
Fenpyroximate	134098-61-6	0.0036	0.039	Training	False negative
Endothall	145-73-3	32.5	350	Training	False negative
Phostebupirim	96182-53-5	0.000078	0.00078	Training	False negative
Irgarol	28159-98-0	5.3	0.52	Test	False positive
Captan	133-06-2	8.4	0.76	Training	False positive
Hexazinone	51235-04-2	151.6	13.62	Training	False positive
Flumetsulam	98967-40-9	254	22.61	Test	False positive
Napropamide	15299-99-7	14.3	1.20	Training	False positive
Dithiopyr	97886-45-8	17	1.38	Test	False positive

(*Continued*)

Table 28
(*Continued*)

Name	CAS Q92	Experimental value (mg/l)	Predicted value (mg/l)	Training/ test set	False negative/false positive
Sethoxydim	74051-80-2	78.1	6.00	Test	False positive
Rimsulfuron	122931-48-0	1000	76.55	Training	False positive
Fenitrothion	122-14-5	0.011	0.00075	Training	False positive
Cycloate	1134-23-2	24	1.47	Training	False positive
Fentin hydroxide	76-87-9	0.01	0.00060	Training	False positive
Cyproconazole	94361-06-5	26	1.55	Training	False positive
N,N-Diethyl-meta-toluamide (DEET)	134-62-3	75	3.99	Training	False positive
Sulfentrazone	122836-35-5	60.4	2.98	Training	False positive
Dienochlor	2227-17-0	1.2	0.06	Training	False positive
Tetramethrin	7696-12-0	0.045	0.00022	Training	False positive
Oxydemeton-methyl	301-12-2	0.24	0.0109	Training	False positive
Chlorsulfuron	64902-72-3	370	16.87	Training	False positive
Methyl chloroform	71-55-6	11.2	0.48	Training	False positive
Bromacil	314-40-9	121	5.10	Training	False positive
Sodium 2-mercapto benzothiolate	2492-26-4	2.56	0.0899	Test	False positive
Tebuthiuron	34014-18-1	297	10.08	Training	False positive
2,4-D Butoxyethanol Ester	1929-73-3	7.2	0.24	Training	False positive
N-methyl neodecanamide	105726-67-8	130	4.12	Training	False positive
Dichlobenil	1194-65-6	6.2	0.19	Test	False positive
Tricosene	27519-02-4	1.08	0.0317	Training	False positive
Bensulide	741-58-2	0.58	0.02	Training	False positive
Etridiazole	2593-15-9	4.9	0.12	Training	False positive
Carboxin	5234-68-4	84.4	1.85	Test	False positive

Table 28
(*Continued*)

Name	CAS	Experimental value (mg/l)	Predicted value (mg/l)	Training/ test set	False negative/false positive
Dipropyl isocinchome-ronate	136-45-8	18	0.38	Training	False positive
Biobor	2665-13-6	730	13.76	Training	False positive
Fosamine ammonium	25954-13-6	1371	22.79	Training	False positive
Diphenamid	957-51-7	58	0.79	Training	False positive
Oxadixyl	77732-09-3	530	7.18	Training	False positive
Fenoxaprop-ethyl	66441-23-4	3.18	0.0301	Training	False positive
Imazethabenz	81405-85-8	220	1.83	Training	False positive
Fomesafen	72178-02-0	294	0.67	Training	False positive

CAS, Chemical Abstracts Service Number.

In that way, the rules for adapting the model to this new safety level are increased eliminating pesticides with the following fragments:

- Carbamates and carbamides.
- Acrylamide or thioacrylic ester derivatives with four substituents different from hydrogen in the positions corresponding to the double bond. Such functionality is not only related with false negatives but also with false positives like in the case of Tetramethin (ID 414) and Bromacil (ID 35).
- An unsubstituted cyclopropyl group, like Irgarol (ID 151), which is mainly related with false positives.
- *p*-Methyl aniline, either in the form of primary, secondary or tertiary amines. These are the cases of 3-Chloro-*p*-toluidine (ID 14) or Pendimethalin (ID 200).
- Sulphonamides either alkyl sulphonamides such as Fomesafen (ID 381) or aromatic ones such as Bensulide (ID 32), which usually gives false positives.
- A non-substituted cyclohexyl amine group like Cycloate (ID 67) and Hexazinone (ID 385).
- Benzyl ethers.

So we have established the new rules as

1. acrylamide or thioacrylic ester derivatives,
2. non-substituted cyclopropyl groups,
3. *p*-methyl anilines,
4. sulphonamides,
5. carbamates or carbamides,
6. terminal *N*-cyclohexylamine groups,
7. benzyl ethers and
8. polycyclic compounds with at least one heterocyclic ring and two or more aromatic rings.

The compounds finally selected for removal are listed in Table 29.

Table 29
List of compounds selected for removal in the daphnia toxicity model after the application of applicability domain rules using a factor 15

Name	CAS	Detected rule	Training/ Test set	Comments
Rotenone	83-79-4	6,8	Training	Outlier
Pyridaben	96489-71-3	1	Training	Outlier
Methomyl	16752-77-5	5	Training	Outlier
Diflubenzuron	35367-38-5	5	Training	Outlier
Hexaflumuron	86479-06-3	5	Test	Outlier
Propoxur	114-26-1	5	Training	Outlier
Sodium dichloro-s-triazinetrione	2893-78-9	5	Training	Outlier
Dichloro-2-*n*-octyl-3(2H)-isothiazolone	64359-81-5	1	Test	Outlier
Ethofenprox	80844-07-1	7	Test	Outlier
Pendimethalin	40487-42-1	3	Training	Outlier
Difethialone	104653-34-1	1,8	Test	Outlier
Bendiocarb	22781-23-3	5	Training	Outlier
Carbofuran	1563-66-2	5	Test	Outlier
Carbaryl	63-25-2	5	Training	Outlier
Sulfluramid	4151-50-2	4	Training	Outlier

Table 29
(*Continued*)

Name	CAS	Detected rule	Training/ Test set	Comments
Aldoxycarb	1646-88-4	5	Training	Outlier
Trimethacarb	2686-99-9	5	Training	Outlier
3-Chloro-p-toluidine hydrochloride	7745-89-3	3	Training	Outlier
DCDMH	118-52-5	5	Training	Outlier
Methiocarb	2032-65-7	5	Test	Outlier
Formetanate hydrochloride	23422-53-9	5	Training	Outlier
Cyprodinil	121552-61-2	2	Training	Outlier
Fenpyroximate	134098-61-6	7	Training	Outlier
Amitraz	33089-61-1	3	Training	Acceptable prediction
Trifloxystrobin	141517-21-7	7	Training	Acceptable prediction
Pyraclostrobin	175013-18-0	7	Training	Acceptable prediction
Carbendazim	10605-21-7	5	Training	Acceptable prediction
4,5-Dichloro-1,2-dithio-3-one	1192-52-5	1	Training	Acceptable prediction
Diflufenzopyr-sodium	109293-98-2	5	Training	Acceptable prediction
Aldicarb	116-06-3	5	Training	Acceptable prediction
Cymoxanil	57966-95-7	5	Training	Acceptable prediction
Kresoxim methyl	143390-89-0	7	Training	Acceptable prediction
Fluometuron	2164-17-2	5	Training	Acceptable prediction
Oryzalin	19044-88-3	4	Training	Acceptable prediction
MTI	82633-79-2	1	Training	Acceptable prediction

(*Continued*)

Table 29
(*Continued*)

Name	CAS	Detected rule	Training/ Test set	Comments
Thidiazuron	51707-55-2	5	Training	Acceptable prediction
Piperonyl butoxide	51-03-6	7	Training	Acceptable prediction
Fenoxycarb	72490-01-8	5	Test	Acceptable prediction
Brodifacoum	56073-10-0	7,8	Test	Acceptable prediction
Asulam	3337-71-1	4	Training	Acceptable prediction
Terbacil	5902-51-2	1,5	Training	Acceptable prediction
Phosphamidon	13171-21-6	1	Training	Acceptable prediction
Thiophanate-methyl	23564-05-8	5	Training	Acceptable prediction
Chlorpropham	101-21-3	5	Training	Acceptable prediction
Irgarol	28159-98-0	2	Test	Outlier
Hexazinone	51235-04-2	6	Training	Outlier
Flumetsulam	98967-40-9	4	Test	Outlier
Sethoxydim	74051-80-2	1	Test	Outlier
Rimsulfuron	122931-48-0	4,5	Training	Outlier
Cycloate	1134-23-2	6	Training	Outlier
Cyproconazole	94361-06-5	2	Training	Outlier
Sulfentrazone	122836-35-5	4	Training	Outlier
Tetramethrin	7696-12-0	1	Training	Outlier
Chlorsulfuron	64902-72-3	4,5	Training	Outlier
Bromacil	314-40-9	1,5	Training	Outlier
Tebuthiuron	34014-18-1	5	Training	Outlier
Bensulide	741-58-2	4	Training	Outlier
Carboxin	5234-68-4	1	Test	Outlier
Fomesafen	72178-02-0	4	Training	Outlier

CAS, Chemical Abstracts Service Number.

The final model can be depicted as in Fig. 15, where some outliers still appear, but the performance of the model has been clearly increased, as indicated by the statistical parameters reported in Table 30.

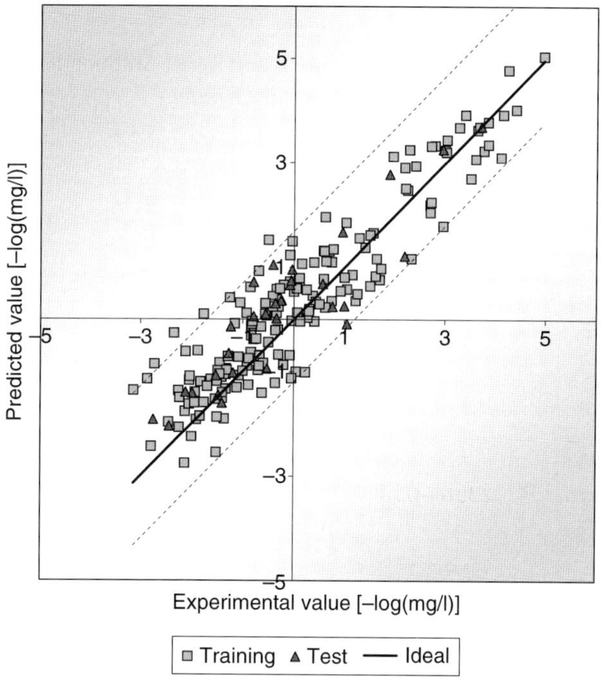

Figure 15: Predicted versus experimental values for daphnia high confidence hybrid model. The dotted lines show a confidentiality of a factor of 15 in case of false negatives and a factor of 50 in case of false positives.

Table 30

Statistical parameters of the hybrid model for daphnia toxicity taking out outliers using a factor of 15

	NC	ND	R^2	K	q	R^2_0	K_0 ($q_0 = 0$)	K'	q'	$R^{2\prime}_0$	K_0'	RMSE	Outliers
Training	176	16	0.85	0.85	0.26	0.83	0.86	1.00	−0.2	0.83	0.97	0.73	18+; 9−
Test	36	16	0.82	0.84	0.36	0.81	0.77	1.02	−0.3	0.79	0.98	0.81	3+; 2−

NC, number of compounds; ND, number of descriptors; K, slope; q, intercept.
outliers +, predicted value (mg/l)/expected value (mg/l) > 50;
outliers −, expected value (mg/l)/predicted value (mg/l) > 50.

4.4. Hybrid model for quail: oral exposure

Quail, orally exposed, has been a particularly difficult endpoint. Actually, only the use of hybrid approach has been able to build a good model. Figure 16 shows the predicted versus experimental values of the model.

The selected approach has been the rule-based approach. The statistical characteristics of this model are listed in Table 31. Table 32 presents the results obtained for the selected individual models.

One of the models, Model 5, has been built with a novel methodology, based on the presence of invariants. These invariants (Toropov *et al.*, 2006) can be

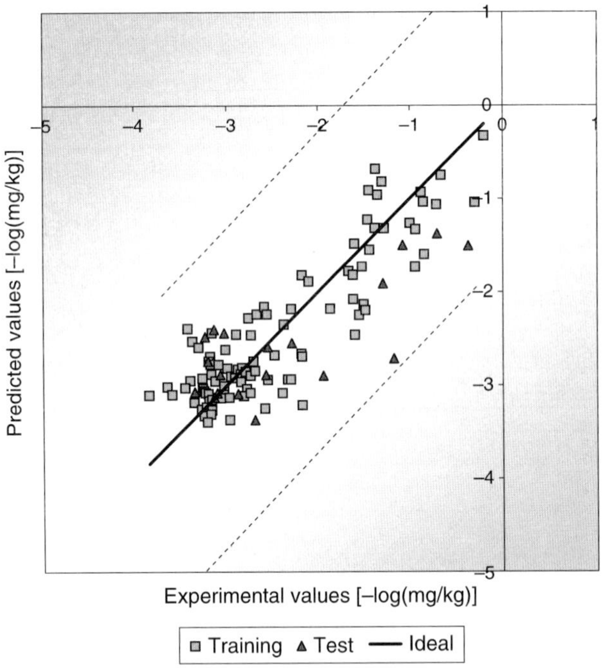

Figure 16: Predicted versus experimental values of the hybrid model for quail, oral toxicity. Dotted lines indicate a factor of 50 in mg/kg.

Table 31
Statistical parameters of the hybrid model for the toxicity towards quail, oral exposure

	NC	ND	R^2	Q^2	K	q	R^2_0	K_0 ($q_0 = 0$)	K'	q'	$R^{2\prime}_0$	K_0'	RMSE	Outliers
Training	96	13	0.82	0.70	0.99	−0.17	0.78	0.99	0.83	0.14	0.79	0.80	0.45	–
Test	19	13	0.71	–	0.63	−0.37	0.45	0.64	1.12	0.42	0.57	0.89	0.64	–

NC, number of compounds; ND, number of descriptors; K, slope; q, intercept.

Table 32
Statistical parameters of the individual models selected for the hybrid model for the toxicity towards quail, oral exposure

		NC	ND	R^2	Q^2	K	q	$R^2{}_0$	K_0 ($q_0=0$)	K'	q'	$R^{2\prime}{}_0$	K'_0	Outliers
PLS	Training	96	12	0.69	0.51	0.69	–0.01	0.69	0.69	1.00	0.00	0.69	1.00	3+; 1–
	Test	19	12	0.59	–	0.44	–0.09	0.57	0.44	1.34	0.11	0.58	1.31	2+
G Inv	Training	96	1	0.78	–	0.78	–0.00	0.78	0.78	1.00	6E-05	0.78	1.00	1+; 1–
	Test	19	1	0.61	–	0.48	–0.20	0.49	0.48	1.27	0.25	0.55	1.14	2–

NC, number of compounds; ND, number of descriptors; K, slope; q, intercept.
outliers +, predicted value (mg/l)/expected value (mg/l) >50;
outliers –, expected value (mg/l)/predicted value (mg/l) >50.

regarded as molecular fragments whose presence on the training set is weighted in order to relate their presence on the structure with the biological effect.

Table 33 lists the descriptors used for the model towards quail toxicity, oral exposure.

For this endpoint, we found that the size of the molecule is the most remarkable property related with toxicity. All topological descriptors present in the model for this endpoint are related to the number of bonds and their nature. In this case, neither branching nor rings appear as determinant factors although they have a general influence. Nevertheless, the absence of an explicit measure of branching appears to indicate that the more important molecular feature in this model is the number of bonds but not where they are placed.

Some molecular fragments are present as well in this model as molecular descriptors. Particularly the presence of oxygen appears to be constantly related with toxicity for every endpoint studied.

Another interesting feature of this model is the presence of molecular invariants. These descriptors are based on the fitting of different weights to some molecular features depending on all the possible invariants present within the training set.

For these descriptors, we found that the presence of increasing numbers of O, S, Cl and especially Br increases the toxicity, while N, rings and carbonyl groups decrease its value. Furthermore, some combinations of them have great effect, like the combination of O, N, S, branching and double bonds.

In any case, this feature substitutes specific molecular fragments in the general set of descriptors which yields the best model, absent in the other individual model.

Only few compounds are predicted with a difference to the experimental value of one order of magnitude (see Table 34).

Anyway for this model, we have not found real false negative outliers as none of them have a ratio of experimental/predicted values higher than 50. Therefore, no rules for the applicability domain are needed for this endpoint, and

Table 33
List of descriptors used for the hybrid model towards quail toxicity, oral exposure

Software	Descriptor Symbol	Correlation	Definition
DRAGON	C-005	+	CH_3X
DRAGON	C-032	–	X–CX–X
DRAGON	O-060	+	Al–O–Ar/Ar–O–Ar/R..O..R/R–O–C=X
DRAGON	TIE	–	E-state topological parameter
DRAGON	X1Av	+	Average valence connectivity index chi-1
DRAGON	IDDE	+	Mean information content on the distance degree equality
DRAGON	TIC0	+	Total information content index (neighbourhood symmetry of 0-order)
DRAGON	T(S..S)	+	Sum of topological distances between S···S
DRAGON	BEHm3	+	Third highest eigen value of Burden matrix weighted by atomic masses
DRAGON	JGI6	+	Mean topological charge index of order6
DRAGON	MATS2e	+	Moran autocorrelation – lag 2/weighted by atomic Sanderson electronegativities
DRAGON	GATS1e	+	Geary autocorrelation – lag 1/weighted by atomic Sanderson electronegativities
Toropov Applet SMILES	Torop	+	Weighted presence of local and global invariants.

Table 34
Worst predicted compounds of the model for quail toxicity, oral exposure

Name	CAS	Experimental Value	Predicted value	Training/Test Set
4-Aminopyridine	504-24-5	15	734	Test
Dinoseb acid	88-85-7	40	1313	Training
Cyproconazole	94361-06-5	150	2315	Training
Propachlor	1918-16-7	88	1156	Test

CAS, Chemical Abstracts Service Number.

it is not possible to establish general rules for the applicability domain with this training set.

Considering a safety interval of 10 for the ratio experimental/predicted values, four outliers appear (see Table 34). Unfortunately, the different chemical features present in each of them prevents us finding rules that can avoid such misleading prediction.

4.5. Hybrid model for quail: dietary exposure

The hybrid model for quail, dietary exposure, has been obtained with the SOSLN approach. Inputs for this model were a model based on PLS and four models based on GMDH.

The performances of this model are given in Table 35 and Fig. 17. Like in the other cases, the plot is built in weight unit, –log(mg/kg), while, for

Table 35
Statistical parameters of the hybrid model for the toxicity toward quail, dietary exposure

	NC	ND	R^2	Slope K	Intercept q	R^2_0	K_0 ($q_0 = 0$)	K'	q'	$R^{2\prime}_0$	K_0'	RMSE	Outliers
Training	98	43	0.88	0.86	–0.09	0.87	0.92	1.02	0.013	0.88	1.01	0.26	–
Test	20	43	0.71	0.79	–0.12	0.69	0.86	0.90	–0.06	0.71	0.94	0.41	–

NC, number of compounds; ND, number of descriptors; K, slope; q, intercept.

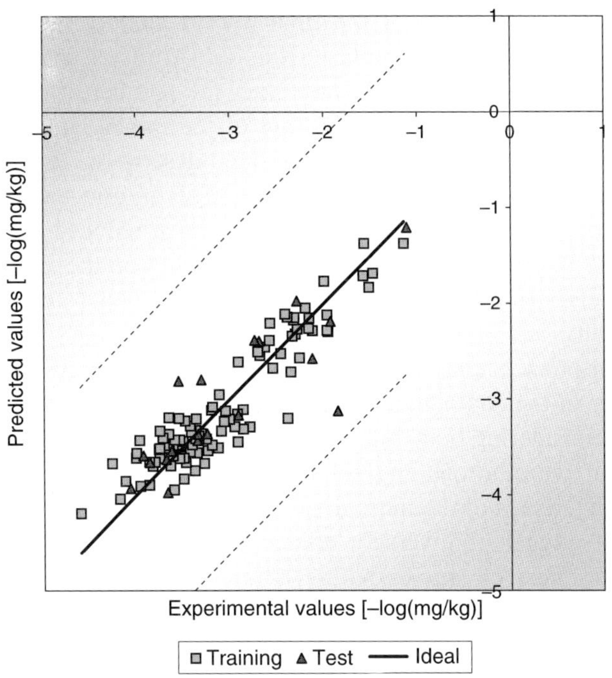

Figure 17: Predicted versus experimental values of the hybrid model for quail, dietary exposure. Dotted lines indicate a factor of 50 in mg/kg.

a deeper evaluation of the statistical performances, the values in the Table 35 refer to the units used for the model building and optimization, −log(mmol/kg). All the statistical parameters for the five individual models are summarized in Table 36.

These results on the individual models do not show good performances on the test set and they provide an example that in some situations Q^2 (always more than 0.5) is not sufficient for estimating the prediction performances on external data. On the other hand, part of the information can be present also in a model with low prediction capability, and the SOSLN approach is able to extract it efficiently, as it can be seen from the good results of the hybrid model on the test set.

The complete list of descriptors used by the model is given in Table 37. This is the most complicated model in terms of the number of descriptors, but only two software packages are required for the calculation of the 44 descriptors: DRAGON and CODESSA.

The number of descriptors used in the selected hybrid model for this endpoint is rather high and somehow reiterative. For instance, we can find that the autocorrelation of atomic properties, taken as edges, has an extensive presence within this model. This abnormal feature for a QSAR model is due to the hybridation approach, where the best regions of the individual models are selected to represent the different regions of the final hybrid model. Therefore, given that several models were selected to build it, it is not surprising that the final number of descriptors has become so big. In any case, the number of descriptors present in this model does not mean an over-fitting of the model, and it does not mean either that the model has many highly correlated descriptors to describe the same structural feature in a meaningless expression. This hybrid model calculates the value of toxicity for every individual model using a reduced set of descriptors, not 44, and the final response is modulated depending on the values given by the individual models, mainly influenced by the model of better performance in such a range of toxicity.

The descriptors' nature shows that there is a clear significance of the topological diameter of the compounds. This means a high influence of the number of bonds without considering branching or rings in order to reach the higher toxicity. The presence of phosphorous and sulphur and their relative positions are other important characteristics of the more toxic compounds. As for the other endpoints, the presence of oxygen is inversely related to the toxicity. Precisely, the large number of descriptors of similar nature among the different individual models shows the importance of these properties to describe the toxicity.

Only one compound is predicted with a difference to the experimental value of one order of magnitude (see Table 38). Anyway, none of the compounds have a ratio of experimental/predicted values higher than 50, and only one of them has a ratio of 10 times higher. Therefore the creation of safety rules is pointless for this particular endpoint.

Table 36
Statistical parameters of the individual models used for the construction of the hybrid model for toxicity towards quail, dietary exposure

		NC	ND	Q^2	R^2	K	q	R^2_0	K_0 ($q_0 = 0$)	K'	q'	$R^{2\prime}_0$	K_0'	Outliers
PLS	Training	102	11	0.52	0.6	0.6	−0.24	0.49	0.76	1.00	0.00	0.60	1.00	2+; 3−
	Test	20	11	–	0.55	0.57	−0.35	0.32	0.80	0.96	0.08	0.55	0.89	2+
NN	Training	98	10	0.63	0.63	0.63	−0.23	0.54	0.78	1.00	0	0.63	1.00	1+
	Test	20	10	–	0.25	0.41	−0.38	0.01	0.66	0.66	−0.19	0.22	0.77	2+; 1−
NN	Training	98	12	0.73	0.72	0.71	−0.18	0.67	0.83	1.00	0	0.72	1.00	–
	Test	20	12	–	0.17	0.33	−0.44	0.17	0.61	0.51	−0.25	0.11	0.72	2+; 1−
NN	Training	98	21	0.87	0.88	0.87	−0.09	0.87	0.93	1.01	0.02	0.88	0.99	–
	Test	20	21	–	0.27	0.42	−0.33	0.08	0.62	0.64	−0.21	0.23	0.82	2+; 1−
NN	Training	98	10	0.83	0.85	0.84	−0.09	0.84	0.90	1.00	−0.01	0.85	1.01	–
	Test	20	10	–	0.62	0.87	−0.08	0.62	0.93	0.71	−0.16	0.59	0.80	1+

NC, number of compounds; ND, number of descriptors; K, slope; q, intercept.
outliers +, predicted value (mg/l)/expected value (mg/l) >50;
outliers −, expected value (mg/l)/predicted value (mg/l) >50.

Table 37
List of descriptors used for the hybrid model towards quail toxicity, dietary exposure

Software	Descriptor symbol	Correlation	Definition
DRAGON	nN	−	Number of nitrogen atoms
DRAGON	nP	+	Number of phosphorous atoms
DRAGON	nF	+	Number of fluorine atoms
DRAGON	nRORPh	−	Number of ethers (aromatic)
DRAGON	nRSR	+	Number of sulphurs
DRAGON	nPhX	−	Number of X–C on aromatic ring
DRAGON	H-052	+	H attached to C0(sp3) with 1X attached to next C
DRAGON	MSD	−	Mean square distance index (Balaban)
DRAGON	TI2	−	Second Mohar index TI2
DRAGON	MAXDN	+	Maximal electrotopological negative variation
DRAGON	X1A	−	Average connectivity index chi-1
DRAGON	X3A	−	Average connectivity index chi-3
DRAGON	X4A	−	Average connectivity index chi-4
DRAGON	X4v	+	Valence connectivity index chi-4
DRAGON	PW4	+	Path/walk 4–Randic shape index
DRAGON	VEA1	+	Eigen vector coefficient sum from adjacency matrix
DRAGON	VEA2	−	Average eigenvector coefficient sum from adjacency matrix
DRAGON	VRA2	−	Average Randic-type eigenvector-based index from adjacency matrix
DRAGON	T(O..S)	+	Sum of topological distances between O···S
DRAGON	T(O..P)	+	Sum of topological distances between O···P
DRAGON	T(O..Cl)	−	Sum of topological distances between O···Cl
DRAGON	SRW05	+	Self-returning walk count of order 05
DRAGON	SRW07	+	Self-returning walk count of order 07
DRAGON	SRW09	+	Self-returning walk count of order 09
DRAGON	BEHm1	+	First highest eigen value of Burden matrix weighted by atomic masses
DRAGON	BEHm8	+	Eighth highest eigen value of Burden matrix weighted by atomic masses
DRAGON	GGI10	−	Topological charge index of order 10
DRAGON	JGI4	+	Mean topological charge index of order 4

Table 37
(*Continued*)

Software	Descriptor symbol	Correlation	Definition
DRAGON	ATS7e	+	Broto–Moreau autocorrelation of a topological structure – lag7/weighted by atomic Sanderson electronegativities
DRAGON	ATS1p	+	Broto–Moreau autocorrelation of a topological structure – lag1/Weighted by atomic polarizabilities
DRAGON	MATS2e	+	Moran autocorrelation – lag 2/weighted by atomic Sanderson electronegativities
DRAGON	MATS5e	–	Moran autocorrelation – lag 5/weighted by atomic Sanderson electronegativities
DRAGON	MATS3p	+	Moran autocorrelation – lag 3/weighted by atomic polarizabilities
DRAGON	GATS2e	+	Geary autocorrelation – lag 2/weighted by atomic Sanderson electronegativities
CODESSA	relative nC	–	Relative number of C atoms
CODESSA	nF	+	Number of F atoms
CODESSA	relative nF	+	Relative number of F atoms
CODESSA	relative nP	+	Relative number of P atoms
CODESSA	nRings	+	Number of rings
CODESSA	GIdx (all bonds)	+	Gravitation index (all bonds)
CODESSA	Ridx0	+	Randic index (order 0)
CODESSA	K&HIdx1	+	Kier&Hall index (order 1)
CODESSA	K&HIdx3	+	Kier&Hall index (order 3)
CODESSA	KshIdx2	+	Kier shape index (order 2)

Table 38
Outliers over a factor of 10 in the model towards quail, dietary exposure

Name	CAS	Experimental value (mg/kg)	Predicted value (mg/kg)	Training/test set
Aldicarb	116-06-3	71	1326	Test

CAS, Chemical Abstracts Service Number.

4.6. Hybrid model for acute contact toxicity of honey bee

The bee endpoint is the third endpoint where the rule-based approach was applied. The general features of this model are depicted in Table 39 and Fig. 18.

The hybrid model presents, in general, good features and, the most important point, does not present any outliers as false negatives at the safety level of 50

Table 39
Statistical parameters for the hybrid model for toxicity towards honey bee

	NC	ND	R^2	K	q	R^2_0	K_0 ($q_0 = 0$)	K'	q'	$R^{2\prime}_0$	K_0'	RMSE	Outliers
Training	88	19	0.74	0.70	0.85	0.54	0.99	1.06	−0.43	0.69	1.12	0.83	13+
Test	17	19	0.84	0.97	0.60	0.78	1.18	0.87	−0.24	0.84	0.80	0.91	3+

NC, number of compounds; ND, number of descriptors; K, slope; q, intercept.
outliers +, predicted value (mg/l)/expected value (mg/l) > 50;
outliers −, expected value (mg/l)/predicted value (mg/l) > 50.

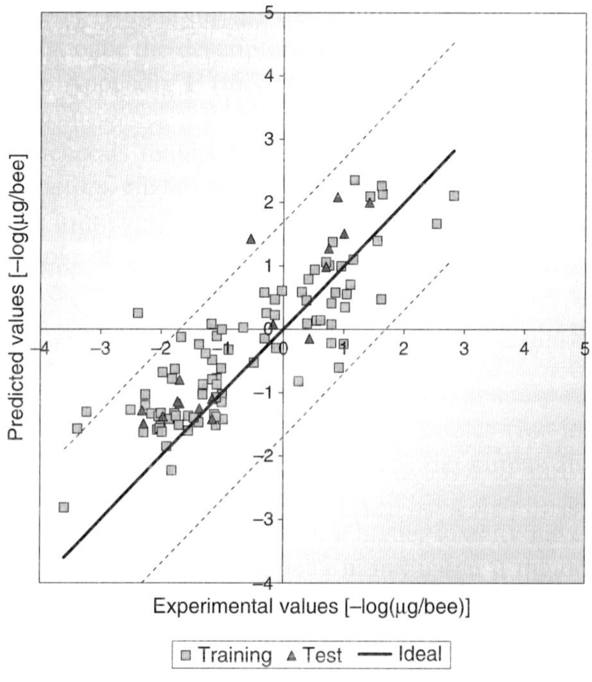

Figure 18: Predicted versus experimental values for the bee hybrid model. Dotted lines indicate a factor of 50 in µg/bee.

times. This is the model of all endpoints which is farther, to some extent, from the ideal model both in slope and *Y* interception, particularly for the last one; but the good behaviour in predicting the external test makes it a valuable model. The statistical features of the individual models used to build the hybrid model are summarized in Table 40.

These values are particularly remarkable, because they are a clear example of how the hybrid approach can improve the results with regard to the individual models, especially in avoiding the presence of false negatives (Table 41).

The bee model seems to be highly influenced by the electronegativity of the molecules. Indeed lipophility – measured as number of bonds, branching and number of double bonds – in this case decreases the toxicity, as well as the presence of phosphorous and sulphur. The presence of oxygen, number of bonds and molecular size are directly related with the toxicity.

One conclusion that can be extracted from the descriptors selection is that non-globular molecules of high size that present regions with high negative electrostatic potential are among the more toxic ones.

This endpoint does not present any real false negative outliers measuring the ratio of experimental/calculated values of 50 times. Nevertheless, some clear outliers appear when the factor of 10 times is considered, so we have developed some rules of chemical applicability like that of the trout and daphnia endpoints. The compounds listed as possible outliers at a factor 10 of acceptability are listed in Table 42.

The common chemical features found for considering a compound out of the applicability domain are as follows:

- Aromatic rings completely substituted. This is the case of Chlorfenapyr (ID 293) and PCNB (ID 329).

Table 40

Statistical parameters of the individual models used for the construction of the hybrid model for toxicity towards honey bee

		NC	ND	Q^2	R^2	K	q	R^2_0	K_0 ($q_0 = 0$)	K'	q'	$R^{2\prime}_0$	K'_0	Outliers
NN	Training	88	7	0.7	0.71	0.69	0.58	0.62	0.88	1.03	−0.06	0.71	1.01	7+; 7−
	Test	17	7	–	0.80	1.05	−0.00	0.80	1.05	0.76	0.37	0.77	0.88	1+; 1−
PLS	Training	88	12	0.55	0.66	0.66	0.63	0.56	0.87	1.00	−3E-08	0.66	1.00	6+; 10−
	Test	17	12	–	0.69	0.78	0.78	0.56	1.06	0.88	−0.12	0.69	0.84	3+

NC, number of compounds; ND, number of descriptors; K, slope; q, intercept.
outliers +, predicted value [mg/l]/expected value [mg/l] > 50;
outliers −, expected value [mg/l]/predicted value [mg/l] > 50.
NN, neural network; PLS, partial least square

Table 41
List of descriptors used within the model for toxicity towards honey bee

Software	Descriptor Symbol	Corr	Definition
DRAGON	MW	+	Molecular weight
DRAGON	nBM	–	Number of multiple bonds
DRAGON	nO	+	number of oxygen atoms
DRAGON	nCOOR	+	number of esters (aliphatic)
DRAGON	C-032	–	X–CX–X
DRAGON	H-049	+	H attached to C3(sp3)/C2(sp2)/C3 (sp2)/C3(sp)
DRAGON	O-058	+	O =
DRAGON	O-060	+	Al–O–Ar/Ar–O–Ar/R···O···R/R-O-C=X
DRAGON	P-117	+	X3-P = X (phosphate)
DRAGON	Jhetv	–	Balaban-type index from van der Waals weighted distance matrix
DRAGON	X0Av	+	Average valence connectivity index chi-0
DRAGON	X5Av	–	Average valence connectivity index chi-5
DRAGON	CSI	+	Eccentric connectivity index
DRAGON	IDDE	+	Mean information content on the distance degree equality
DRAGON	HVcpx	+	Graph vertex complexity index
DRAGON	LP1	+	Lovasz-Pelikan index (leading eigenvalue)
DRAGON	VEA1	–	Eigen vector coefficient sum from adjacency matrix
DRAGON	T(S..P)	–	Sum of topological distances between S···P
DRAGON	SRW05	+	Self-returning walk count of order 05
DRAGON	AMW	+	Relative molecular weight

Table 42
Outliers with predictions over a factor of 10 of the model for bee toxicity

Name	CAS	Experimental value (mg/l)	Predicted value (mg/l)	Training/ Test set	False negative/ false positive
Chlorfenapyr (Pirate)	122453-73-0	0.12	4.054	Training	False negative
PCNB	82-68-8	0.1	1.703	Training	False negative
Pyridaben	96489-71-3	0.024	0.332	Training	False negative
Lindane	58-89-9	0.56	6.601	Training	False negative

Table 42
(*Continued*)

Name	CAS	Experimental value (mg/l)	Predicted value (mg/l)	Training/ Test set	False negative/ false positive
Methomyl	16752-77-5	0.16	1.651	Training	False negative
Daminozide	1596-84-5	205	19.13	Test	False positive
Oxythioquinox	2439-01-2	66.47	6.041	Training	False positive
Carboxin	5234-68-4	180	15.77	Training	False positive
Bensulide	741-58-2	24	1.728	Training	False positive
Trichlorfon	52-68-6	59.8	4.250	Training	False positive
D-Phenothrin	26002-80-2	0.067	0.0044	Training	False positive
Tralomethrin	66841-25-6	0.129	0.0082	Test	False positive
Chlorothalonil	1897-45-6	181.29	10.68	Training	False positive
Glufosinate-ammonium	77182-82-2	315.8	18.58	Training	False positive
Propargite	2312-35-8	15	0.8263	Training	False positive
Dimethenamid	87674-68-8	94	4.778	Training	False positive
Cimecticarb	95266-40-3	47	1.326	Training	False positive
Diphenamid	957-51-7	2432	36.93	Training	False positive
Acetochlor	34256-82-1	1715	20.03	Training	False positive
Allethrin	584-79-2	3.40	0.0374	Test	False positive
Phenmedipham	13684-63-4	241.7	0.5568	Training	False positive

CAS, Chemical Abstracts Service Number.

- Acrylamide derivatives with four substitutions on the double bond with two highly electronegative substituents on *Z* configuration.
- Compounds which present a 2-chloro-acetamides such as Acetochlor (ID 20) or Dimethamid (ID 95).

Therefore, the compounds selected for removal from this model are listed in Table 43. The final model is depicted in Fig. 19.

It is still possible to find some outliers, all of them as false positives, but the chemical functionalities found among them are quite diverse making it difficult to create accurate rules.

In any case, the performance of the model is increased after application of these rules (Table 44), so the final model can be considered as good enough for the proposed purposes.

Table 43

List of compounds selected for removal after the application of applicability domain rules within the model for bee toxicity

Name	CAS	Detected rule	Training/ Test set	Comments
Chlorfenapyr (Pirate)	122453-73-0	1	Training	Outlier
PCNB	82-68-8	1	Training	Outlier
Pyridaben	96489-71-3	2.3	Training	Outlier
Dithiopyr	97886-45-8	1	Training	Acceptable prediction
Picloram (Acid)	1918-02-1	1	Test	Acceptable prediction
Phosphamidon	13171-21-6	3	Test	Acceptable prediction
Carboxin	5234-68-4	2	Training	Outlier
Chlorothalonil	1897-45-6	1	Training	Outlier
Dimethenamid	87674-68-8	3	Training	Outlier
Acetochlor	34256-82-1	3	Training	Outlier

CAS, Chemical Abstracts Service Number.

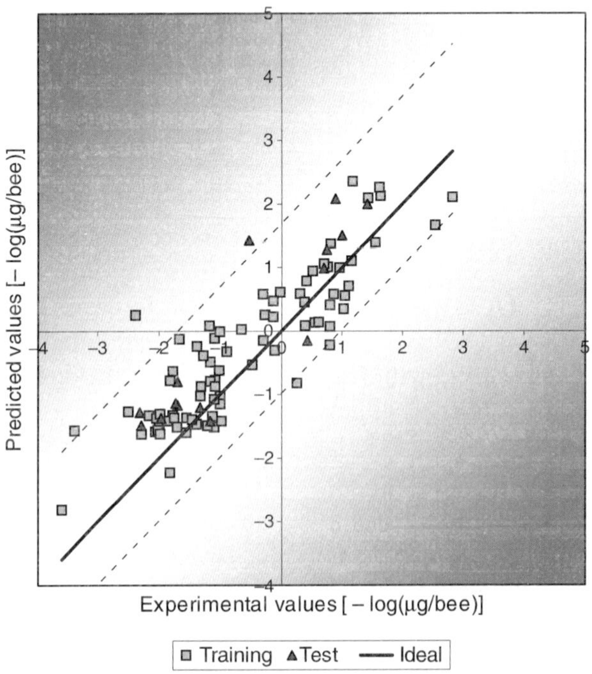

Figure 19: Predicted versus experimental values for bee high confidence hybrid model. The dotted lines show a confidentiality of a factor of 10 in case of false negatives and a factor of 50 in case of false positives.

Table 44
Statistical parameter of the hybrid model for bee toxicity after outliers removal at a factor of 10

	NC	ND	R^2	K	q	R^2_0	K_0 ($q_0 = 0$)	K'	q'	$R^{2\prime}_0$	K_0'	RMSE	Outliers
Training	80	16	0.77	0.75	0.16	0.75	0.71	1.03	−0.29	0.72	1.09	0.72	2+; 9–
Test	15	16	0.83	0.95	0.55	0.7	0.78	0.87	−0.59	0.62	0.88	0.84	3–

NC, number of compounds; ND, number of descriptors; K, slope; q, intercept.
outliers +, predicted value (mg/l)/expected value (mg/l) >50;
outliers –, expected value (mg/l)/predicted value (mg/l) >50.

5. CONCLUSIONS

DEMETRA project has developed a number of predictive models for pesticides in general. The model performances have proved to be superior to those reported for prediction of ecotoxicity properties as in a recent review (Hansen, 2004). The poor performances achieved with common predictive tools based on log *P* (actually they are simple correlation attempts in some cases) should not surprise. Indeed, pesticides are complex molecules presenting many chemical functional groups along with different reactivities, and typically more than one of them are present at the same time in the pesticide structure. Taking advantage of such features, pesticides provoke toxic effects through series of toxic mechanisms, most of them still unknown. Thus, the task to predict ecotoxicity of pesticides in general is a rather complex one. Nevertheless, we achieved good results. The predictive power of the five models is about 75–80% both on the training and test sets, which proves the model robustness.

We have dedicated particular attention to the false negatives, in order to produce a predictive model, which avoids as much as possible false negatives. Specific rules have been implemented for constructing hybrid models. Additionally, we have checked outliers, and we have identified chemical rules to characterize them. In this way, we have defined two different levels of model restrictions where the model performances have not been optimal.

In the case of trout and daphnia two kinds of restrictions were introduced, characterising the model uncertainty. In one case, the restriction allowed to avoid false negatives higher than a factor of 50; in another case, introducing additional restrictions, the model uncertainty is even lower, no more than a factor of 10. The cost of the higher number of rules is a reduced applicability domain although the DEMETRA models are the same. The application of the different rules is related to a certain level of reliability. These rules have to be adopted by the user visually inspecting the molecule of interest to check for the presence of the fragment(s). We plan in future versions to add this feature in an automatic way. It is up to the user to decide which level of confidence is preferred. The user may decide whether to use a lower number of restrictions, increasing the applicability domain of the model, or whether to screen for all the rules, and at this point, the prediction has a lower possible error in a more reduced domain.

For terrestrial endpoints, the number of outliers is much smaller, and thus, the rules are also less numerous. Except for bee endpoint, there are not enough outliers for allowing the design of rules. In any case, the performance of these models is close to the safest standards.

We remember that these rules are based on safety criteria and that the presence of a given fragment does not necessarily mean that the prediction is wrong, but the uncertainty is larger. The uncertainty of the model prediction should be compared with the typical variability of the experimental models. This variability accounts for a large part of the model uncertainty. The variability of the descriptor calculation is a minor component within the overall model. This fact is quite a general rule, and it applies in particular in our specific case. Indeed, we have chosen to use descriptors based on 2D structures, which are less affected by variability, as the manual optimization of the 3D conformation is avoided.

We underline the probabilistic nature of the prediction, which is characterized by the percentage of correct predictions and by the maximum expected error, depending on the presence of certain residues. Individual models contribute to the final hybrid model for each endpoint. Besides the predicted value obtained by the final model, DEMETRA models also report clearly the minimum and maximum values of these individual models. This can also be used to assess the uncertainty of the predictions, which is at the basis of the hybrid models. In this way, the uncertainty of the prediction is compound-specific. The user may decide to use the lower value of the individual models, if it is below the value predicted by the hybrid model. This fact enables the regulator to opt for a more conservative use of the DEMETRA model. We must remember that the hybrid model is not the average between different models but a new model using as inputs the predicted values obtained by the individual models.

The basis of the model is the number of the pesticides used to build up the model itself. Such a number represents a large part of the commercialized pesticides. Unfortunately, for terrestrial endpoints, the available data were scant, because traditionally these endpoints have been less studied. For this reason, aquatic models may be somehow more robust. On the other hand, terrestrial endpoints had a more limited range of toxicity values, and this reduces the possible model error. The models' statistical performances are quite similar. Aquatic models, however, have a larger number of outliers than the others, likely given by the larger range of toxicity values and the wider chemical variability.

ACKNOWLEDGEMENTS

Authors acknowledge ICPS, EPA-OPP, BBA and RIVM for kindly providing their database and their use within DEMETRA project.

REFERENCES

AMBIT (2006) http://ambit.acad.bg/.

Directive 92/32/ECC (1992) The seventh amendment to Directive 67/548/ECC, OJL 154 of 5.VI.92, p 1.

Gini G., Craciun M.V., König C., Benfenati E. (2004) Combining unsupervised and supervised artificial neural networks to predict aquatic toxicity *J. Chem. Inf. Comput. Sci.* 44, 1897–1902.

Gini G., Testaguzza V., Benfenati E., Todeschini R. (1998) HyTEx (Hybrid Toxicology Expert system): architecture and implementation of a multi-domain hybrid expert system for toxicology. *Chemometr. Intell. Lab. Syst.*, 43, 135–145.

Hermens J.L.P. (1990) Practical applications of quantitative structure-activity relationships (QSAR). In: *Environmental Chemistry and Toxicology*. W. Karcher, J. Devillers (eds.), Kluwer Academic, Dordrecht, The Netherlands, 263–280.

Kennard R.W., Stone L.A. (1969) Computer aided design of experiments. *Technometrics*, 11, 137–148.

Marking L.L., Bills T.D. (1976). Toxicity of Rotenone to fish in standardized laboratory tests. Invest. Fish Control No.72, Fish Wildl. Serv., Bur. Sport Fish, 11. Wildl., U.S.D.I., Washington, D.C.

Mayer F.L., Ellersieck M.R. (1986) Manual of Acute Toxicity: Interpretation and Database for 410 Chemicals and 66 species of Freshwater Animals. USFWS Publication No. 160. EPA MRID 40098401.

Müller J.A., Lemke F. (2000) *Criteria of Model Selection Self-Organising Data Mining*. Taylor & Francis, Hamburg.

Netzeva T.I., Aptula A.O., Benfenati E., Cronin M.T.D., Gini G., Lessigiarska I., Maran U., Vraèko M., Schüürmann G. (2005) Description of the electronic structure of organic chemicals using semiempirical and ab initio methods for development of toxicological QSARs. *J. Chem. Inf. Model*, 45, 106–114.

Tomlin C. (1997) The Pesticide Manual. (11th edn.). British Crop Protection Council, Farnham, UK.

Toropov A.A., Benfenati E. (2006) QSAR models for Daphnia toxicity of pesticides based on combinations of topological parameters of molecular structures. *Bioorg. Med. Chem.*, 14, 2779–2788.

Witten I.H., Frank G. (2000). *Data Mining: Practical Machine Learning Tools with Java Implementations*. Morgan Kaufmann, San Francisco, CA.

Edited by Emilio Benfenati

Chapter 8

The quality criteria of the DEMETRA models for regulatory purposes

Specificity, general lessons and future perspectives

Emilio Benfenati

Laboratory of Environmental Chemistry and Toxicology, Istituto di Ricerche Farmacologiche "Mario Negri", Milano, Italy

1. THE OECD GUIDELINES FOR QSAR MODELS

1.1. Introduction

Recently, the OECD discussed the issue of validation of quantitative structure-activity relationships (QSARs) for regulatory purposes, as mentioned in Chapter 1 (Benfenati *et al.*, this volume). The attention given by the OECD to QSAR models for regulatory purposes demonstrates the need for special consideration for this kind of model. In particular, five principles have been identified (OECD, 2005):

A (Q)SAR model for regulatory purposes should be associated with the following information:

1) A defined endpoint,

2) An unambiguous algorithm,

3) A defined domain of applicability,

4) Appropriate measures of goodness-of-fit, robustness, and predictivity, and

5) A mechanistic interpretation, if possible.

In our opinion so far, QSAR models have been developed using common QSAR criteria and then eventually evaluated for regulatory purposes. But generally, models have not been developed from the very beginning aimed at a specific regulatory context, as we did within DEMETRA. DEMETRA started before the publication of the OECD criteria, and this required us to develop independent criteria. Here, we discuss these specific criteria, which have been considered within the EC-funded project DEMETRA and listed in Table 1 in this chapter. Many of these criteria are similar or related to those listed by OECD. In addition to the quality criteria referred to the QSAR model itself, also addressed by the OECD, we dedicated great attention to the context in which the specific model should be used, which involves the inputs and the outputs of the models, and how they can be used. In the discussion below (Sections 1.5, 1.8, 1.14, 1.17, and 3.2), we will also refer to these criteria. In this way, we hope, we will contribute to the discussion of the QSAR for regulatory purposes, providing the example of the DEMETRA models. Some of the criteria discussed below do not necessarily apply to all QSAR models but only to models for regulatory purposes (see Table 1, this chapter).

1.2. The identification of the regulation

Within DEMETRA, we referred to the Directive 91/414 (see Chapter 1, Benfenati *et al.*, this volume). We believe that in the case of QSAR model dedicated to regulatory applications, it is very useful to identify the specific regulation which is addressed.

Thus, this should eventually be a further criterion to be listed in case of QSAR models for regulatory purposes.

1.3. The criteria for the endpoint selection

Once the regulation is defined, the following step is the definition of the exact endpoint(s). Indeed, more than one endpoint can be contained in the regulation. Within the Directive 91/414, several endpoints are addressed. Here, there are two possibilities: one, opportunistic, to choose the endpoint which is more suitable for the modelers (for instance, because they already have experience with it) and the other to consider the most useful models on the basis of their application.

Because DEMETRA was aimed to develop QSAR models to be used (not only formally, but in practice) within the Directive 91/414, a crucial point of the project involved a deep evaluation of criteria for its use. To develop QSAR models which will be used by industry and regulators, it is necessary to start from the most important endpoints for the specific regulation. For this, we developed criteria for evaluation of more useful endpoints, we distributed a questionnaire to regulatory and industrial end-users, and finally we evaluated the most commonly used endpoints. These activities and the approach we adopted is described in Chapter 1 (Benfenati *et al.*, this volume). Of course, this is not commonly done because it is time-consuming.

The criteria which have been identified by the potential users are

- the frequency each endpoint is requested,
- the number of animal used in the study,
- the severity of the test (vertebrates),

- the proportion of toxicity-exposure ratios (TERs) that do not trigger further testing,
- the data availability, and
- an assessment of feasibility (e.g., a single clear endpoint).

These points have been discussed in more in detail in Chapter 1 (Benfenati *et al.*, this volume). We notice that typically only the data availability is considered by the modeler.

1.4. The model utility

It is clear that the criteria as listed in the previous paragraph are specific for the intended use of the toxicity data, and, indeed, they are specific for the regulation the model is intended for. This is an important point, because it could happen that the model is developed for an endpoint which is of limited use or that the output of the model is not appropriate for its successive use. Typical is the case of the TER values in the listed criteria or the frequency of the use of the endpoint.

Thus, another point which is not explicitly mentioned in the OECD criteria is the model utility. We wanted to develop models that could be used as much as possible. In order to develop QSAR models that are effectively used, we seriously considered the end-user requirements. This approach is opposite to that adopted for many QSAR developments, where scientists work on a technique, and then apply it to a data set, which is not the main subject of their interest. Thus, the scientific effort starts from a perspective which is not the application, and only at a second stage, the developed model is proposed as suitable.

However, in the case of QSAR for regulatory purposes, the philosophy of the QSAR modeling activities should be reversed: the main target is the development of a QSAR model to satisfy the requirements of regulators and end-users, and the interests of the developers are of secondary importance.

But to correctly address the issue of usability for a given model, we should also think about the extent a given model can be used. We could have developed a model only for triazines, for instance, still addressing all the previous criteria: high-quality data, definition of the guidelines, definition of the regulation, etc. However, in this case the utility of the model would be very limited, because the pesticide industry is most probably not interested in this very specific case. It is clear that the utility of the obtained model goes beyond the simple selection of the endpoint. Thus, model utility is an additional evaluation criterion for QSAR.

1.5. The endpoint selection: identification of the guidelines

Once the criteria for the model have been defined, we used them to select the endpoints. We addressed the first five endpoints.

The identification of the endpoints provides the way to identify the guidelines relative to the experimental protocols defining at the best the endpoint. DEMETRA clearly fulfills these criteria. The guidelines are listed in Chapter 1 (Benfenati *et al.*, this volume).

OECD principle number 1 requires a clearly defined endpoint for a given QSAR model. Thus, DEMETRA fulfills OECD principle number 1 and introduces deeper details linking the endpoint to the legislation, the selection of useful endpoints, and the necessary guidelines.

1.6. The accordance of the toxicity data to the guidelines

Once the endpoint(s) and the relative guidelines are identified, the QSAR model should use only data produced according to the guideline specific for that endpoint, unless it is not available. We are not aware of QSAR models that used only data produced according to official guidelines and legislation. However, if modelers want to obtain the acceptability of their QSAR models in the regulatory framework, they have to accept what is defined in the regulation. *Dura lex, sed lex*, as the Romans said. Also here, we state that this does not refer to all QSAR models. Other models, for different purposes, can be equally valid on purely scientific criteria.

1.7. The check of quality data

Furthermore, we increased the overall quality of the used data comparing values present in three high-quality databases, to increase the reliability of the data. Again, this is unique in the scenario of QSAR models: indeed, to compare data requires more time and effort and reduces the number of chemicals. In contrast, what is occasionally done is to merge data from different sources (Vighi *et al.*, 1991; Devillers *et al.*, 2002). Further quality control was done later on for the validation of the models (see Chapter 7, Amaury *et al.*, this volume).

We notice here that the comparison we did is not the common procedure for the evaluation of pesticides for regulatory purposes. Values from a single source are enough. In this sense, the quality of our data is superior to those typically requested for Dossier evaluation.

1.8. The definition of the model components. OECD principle number 2: an unambiguous algorithm

Within DEMETRA, we took maximum care in making all model components defined. Furthermore, thanks to the fact that the project involved several partners, we checked critical points internally between several partners.

We think that this closely corresponds to the OECD principle number 2 (see Section 1.1). We believe that the OECD in this case wanted to stress the importance of an overall procedure which can be detailed in a way to make it understandable and reproducible. The concern is probably to have a model that is understood in all its components, without gray areas. The conceptual characterization we adopted for the model is the following and basically corresponds to the chapter structure of this book.

- The toxicity data.
- The chemical names and structures.
- The algorithms.

1.9. The selection of the toxicity values of the data set

Once the endpoints have been defined together with the availability of high-quality data (see Section 1.7, this chapter), the following step is to choose the correct values. Indeed, what we found, and it is not uncommon, was that in many cases there was more than one toxicity value for the same compound. As discussed, (eco)toxicity experiments are affected by a large variability, partly related to the natural variation between individuals, and partly by the differences in the experimental conditions used. The latter is responsible for the experiment uncertainty. Within DEMETRA, the procedure of quality check, comparison, and selection of the data in the same database or from different sources was defined according to a written protocol (see Chapter 2, Benfenati *et al.*, this volume). Moreover, we used all pieces of information, including the variability of the toxicity data which, if greater than four orders of magnitude, caused us to eliminate the compound (see Chapter 2, Benfenati *et al.*, this volume).

1.10. The characterization of the uncertainty of the experimental data

It is well known that the error of the final value obtained with a given algorithm is fundamentally dependent on the uncertainty of its components. But if the uncertainty of the original data is unknown or is disregarded, how can we evaluate the uncertainty of the final value? We addressed this fundamental point as mentioned, and this is again unique in the scenario of QSAR models. Most typically a single toxicity value is used, without considering the possibility of multiple values. When this topic was addressed, a single value was kept, which can be the minimum (Benfenati *et al.*, 1999) or the median (Russom *et al.*, 1997), but without using the information on the uncertainty. Of course, this uncertainty is fundamental to be known if we want to assess in an unambiguous way the final values of the model. Thus, we believe that this information should be given for all QSAR for regulatory purposes.

Also in this case, we notice that our approach is more restricted than guidelines for pesticide evaluation within Directive 91/414, which do not specify a level for acceptability of the toxicity value. The human expert who evaluates the studies has the responsibility of accepting or rejecting the studies.

1.11. The chemical structures

The absolute chemical identity of compounds in the data set is a fundamental issue. The toxicological assessment should be done with a precise knowledge of the chemical used. This is more typical in the case of experiments done according to official guidelines. However, it is not always true that experiments done according to official guidelines are done only with pure compounds. For instance, experiments can be done on the formulated pesticide, which includes additives. Thus we have discounted several experiments from those listed in the high-quality databases, for insufficient purity of the test material (Chapter 2, Benfenati *et al.*, this volume). More subtle is the case of mixtures of stereoisomers, which also have to be fully evaluated and described for the experiment. On the basis of the exact composition of the chemical(s) used for the toxicological experiment, the chemical identification has to be reported, and the work on the chemical description for modeling purposes can start.

We checked the chemical identity of each compound using more than one data sources, as described in Chapter 3 (Benfenati *et al.*, this volume). This was done to check if the data on the chemical identification were consistent. The name, CAS number, and molecular weight were checked. Very few mistakes in the original sources were found and reported to the databases' managers.

Great attention was also given to the chemical structure. Structures were drawn by two different project groups and then compared. If the structure file format was later modified into a second format, the structure was checked again. This check is important, as on converting chemical structures some changes are possible, which may result in incorrect structures.

For each chemical we checked the possible existence of isomers and tautomers. If different isomers were possible, we checked which ones were used for the toxicity experiment, and the structure was referred to that used. For tautomers we calculated the most stable one, and this was used for successive modeling. To calculate the most stable one, we used the conformation from the crystallographic database (see Chapter 3, Benfenati *et al.*, this volume). If the same compound was not present, we considered the most similar one. Ab initio calculation of the conformation was done on the different tautomers, and energies were compared. This procedure allows us to achieve the most likely preferred tautomer, in a rigorous way. This was done to use the most reliable data for the training set, and this is not the procedure most commonly used by QSAR modelers, because it takes much longer. The main reason to keep into consideration tautomers is that chemical descriptors change for the different tautomers. This happens also for some simple bidimensional descriptors, such as number of double bonds and number of OH. Thus, this is a source of irreproducibility of the QSAR model. As the main issue is the reproducibility of the model, other possible ways can be defined to achieve reproducible tautomers. We do not think that ab initio calculation is necessarily a requirement for QSAR models.

Three-dimensional (3D) structures were obtained as described in this section starting from the crystallographic data.

Thus, the following quality criteria for chemical structures were adopted.

- Definition of the chemical identity for the chemical used in the animal experiment.
- Check of the chemical identity in different sources.
- Check for stereoisomers.
- Check for tautomers.

1.12. The chemical descriptors

Under our conditions, the modeling results obtained with two-dimensional (2D) or 3D descriptors were similar. The combined use of 2D and 3D descriptors did not improve the results. Thus, we used only descriptors based on 2D descriptors, because they were more reproducible. 3D descriptors often depend on the optimization of the conformation, which is manually done by the operator (Benfenati *et al.*, 2000). 2D descriptors have a further great advantage: they are calculated much more rapidly.

The reproducibility of the chemical descriptors is not considered for QSAR models, but we believe it is an important point in case of models for regulatory purposes. The exact version of the software and the specific parameters used have to be given. The availability and ownership of the software have to be defined. This point is very important if the model has to be widely used.

1.13. The algorithms

As in the case of the software for chemical descriptors, the detailed specification of the programs used in developing the QSAR models should be made explicit. The availability and ownership of the software have to be defined. Again, this point is very important if the model has to be widely used. To improve the reliability and reproducibility of the algorithms, xml format was preferred. In this way, the exact mathematical notation is defined. Xml was also used for the chemical and toxicological data.

1.14. The performances of the model

Historically, many QSAR models have been done simply by fitting the data with simple equations. However, for many years, it has been recognized that to assess the predictivity of a model more stringent approaches have to be conducted to avoid chance correlations. This point is clearly mentioned in the OECD guidelines, principle 4 (OECD, 2005). In particular for non-linear systems, such as neural networks, the risk that the model will simply reproduce the behavior of the compounds in the training set, without extracting any general feature, is very high. This phenomenon is called over-fitting (see Chapter 6, Benfenati *et al.*, this volume). However, we underline here that in the case of QSAR models for toxicological and ecotoxicological properties, the regulators make a clear difference between false positive and false negative, and thus we carefully addressed this aspect. Again, this is something unique in the case of QSAR.

In Chapter 6 (Benfenati *et al.*, this volume), we also discussed the different possibilities, for internal and external validation. Within DEMETRA, we used a large battery of tools, in order to check the model performances.

1.15. The reproducibility of the models

Another important related issue is the model reproducibility. In the case of QSAR models for regulatory purposes, we believe that the obtained values have to be the same, independently on the user. This is a fundamental premise in the use of our models, as well as of the use of any method, in vivo or in vitro. If the results from the same method/model vary with the user, this means that the developer should improve the method/model description. Of course in case of in vivo and in vitro methods, a given variability is common. Whereas in silico models are affected to a lesser extent by variability, in any case models for regulatory purposes should specify their reproducibility.

For this reason, we fixed the parameters of our models, and we did not allow the user to modify them. These parameters have been optimized by us as described in Chapter 5, Amaury *et al.*, this volume, and the model is tested in these conditions. Several QSAR models are affected by variability, especially when the expert modeler has to optimize some parameters. For this reason, we avoided the use of 3D descriptors (also because we had the same results using 2D descriptors), and we discounted the use of models such as CoMFA, which gave in our hands promising results (data not reported), but required previous careful alignment of the molecules.

To further fix the parameters, we adopted the xml format for the algorithms, defining in a unique way the mathematical terms.

1.16. The false-negative issue

We underline that a specific request of regulators is to avoid as much as possible false negatives. In general, the issue of false negative is related to the wish to have a conservative assessment, which is the common basis of the EU regulation

(Rudén and Hansson, 2003). This issue should be considered in all steps of the modeling procedure. The procedure we adopted is as follows:

- Toxicity data were selected, keeping the most toxic ones (after careful check of the quality, as discussed in Chapter 1, Benfenati *et al.*, this volume).
- The models were developed introducing an innovative algorithm to integrate different models in order to reduce false negative in the hybrid model (Chapter 5, Amaury *et al.*, this volume).
- We assessed the presence of false negatives, evaluating very carefully these outliers (Chapter 7, Amaury *et al.*, this volume).
- We introduced specific rules to reduce false negatives and positives (Chapter 7, Amaury *et al.*, this volume).

We will discuss other aspects of the false negative issue Section 2, this chapter.

1.17. The applicability domain

The boundaries of a given model are linked to the conditions defined on the basis of the criteria identified during the model development and those identified considering the results of the model. Indeed, at the beginning of the modeling activities, some limits may be identified, because, for instance, the descriptors for a compound containing a heavy atom cannot be calculated. In this case, it is very simple and automatic to define the boundaries for these a priori conditions. We defined in Chapter 3 (Benfenati *et al.*, this volume) the rules we adopted for the inclusion and exclusion of chemicals.

Once the results of the model are known, the following steps were done within DEMETRA (Chapter 7, Amaury *et al.*, this volume). (1) Check of the correctness of the experimental data of the outliers. (2) Identification of possible reasons for the outliers, done by human experts. (3) Use of chemometric tools to identify the applicability domain on the basis of chemical descriptors. (4) Use of chemical rules to identify outliers, done by human experts. These steps have to be done after the model has been defined. For this reason, we preferred to discuss first principle 4 of the OECD guidelines (which characterizes the performances of the chosen model) and now the principle 3, which characterizes the domain of the applicability, because if the model has not been developed, the descriptors are not identified, and thus the appropriate chemometric techniques cannot be used.

Whenever possible, we defined plausible rules to identify the boundaries of the model. Thus, in Chapter 7 (Amaury *et al.*, this volume) outliers have been identified on a chemical basis (carbamates are not well predicted in the daphnia model), while others are related to specific mechanism of action: several compounds that are outliers in the trout model are active in the electron transfer in the mitochondria. These considerations have been used to identify criteria for the definition of the model boundaries, clearly stating when the model may fail in its predictions. As in the other parts of the modeling procedure, we preferred to be conservative, and thus to classify as possible outliers some chemicals, reducing the domain of applicability of the model. However, we defined different rules to identify outliers with different errors, and distinguish false positives and negatives. Thus the user is informed and can choose the uncertainty factor. This is something new for QSAR. Furthermore, when the applicability domain is defined, a single threshold is defined, thus a compound is inside or outside the domain. This is clearly a crude approximation. We defined rules to measure how much a compound can be affected by an error (Chapter 7, Amaury *et al.*, this volume), introducing another novelty in QSAR.

1.18. The quality control

An important point in the development of QSAR for regulatory purposes is the quality of the overall procedure. We recommend great care in controlling all the steps in the modeling process, even if this takes time and resources. All data should be kept in files, with back up copies. The protocols used for all steps have to be written, indicating full details used, in order to allow obtaining the same results in a reproducible, traceable way. This step serves to improve the quality control of the procedure.

Other quality criteria were used, and so, the toxicity values that were digitized into our data sets were checked by a second person, as a quality check against typing mistakes. The same was done for the chemical structures.

Further points specific to the single aspects have been already described (see Sections 1.7 and 1.11).

1.19. The use of the model

The final point to be described in the model is how to use it. We have described in detail in the case of the DEMETRA models in the Chapter 9 (Benfenati *et al.*, this volume).

The description should address the ownership of the model. We preferred to put the DEMETRA models freely available on the internet, in consideration of the public funds received from the EC and to promote their use.

In Table 1, we list the criteria adopted within DEMETRA.

2. THE SPECIFICITY OF THE QSAR MODELS FOR REGULATORY PURPOSES

There has been a wide discussion on the perspectives of QSAR models in the present situation (Benfenati, 2004; Veith, 2004; Worth *et al.*, 2004). As we

Table 1
The criteria adopted within DEMETRA

Criterion	Required or recommended[a]	Chapter[b]
The identification of the regulation	Required	1
The criteria for the identification of suitable endpoints	Required	1
The model usability	Recommended	1
The definition of the endpoint and guideline	Required	1
The accordance of the toxicity data to the guidelines	Required	2
The check of the toxicity data	Recommended	2
The selection of the toxicity values of the data set	Required	2
The characterization of the variability of the experimental data	Required	2
The chemical structures	Required	3
The chemical descriptors	Required	3
The algorithms	Required	4, 5
The performances of the model	Required	6, 7
The model reproducibility	Required	6, 7
The false negatives	Required	6, 7
The domain of applicability	Required	7
The quality control	Recommended	8
The use of the model	Recommended	9

[a] Some criteria are only recommended, while others should be requested in case of QSAR for regulatory purposes. These criteria do not necessarily apply to all QSAR. In particular, those relative to the regulation and to the guidelines refer to the specific regulatory use.
[b] We report here the chapters in this volume in which we describe the approach we used.

discussed in Chapter 1 (Benfenati *et al.*, this volume) for the case of pesticides, no good specific models have been published for several reasons, including the difficulty of the prediction of pesticide ecotoxicity. Indeed, simple models based on log *P*, which has been used to develop other ecotoxicological models, gave poor performances (Hansen, 2004).

Also on a more general point not just for pesticides, existing QSAR models can be generally considered as not fully suitable for regulatory purposes for one or more reasons. Basically, a given model, not only in silico, but also in vivo or in vitro, is accepted when it is based on a technique which is mature and a certain degree of quality and standardization has been proved. So far, tens of thousands of

QSAR models have been published and listed. This shows the huge amount of work done, which in recent years has also seen an increasing interest and use of advanced computational techniques such as artificial neural networks and machine learning (Chapter 1, Benfenati *et al.*, this volume). However, to cope with the requirements posed by the legislation, further steps have to be completed, as discussed in this section.

QSARs are complex studies, because they are related to different scientific disciplines. Experimental sciences, such as biology, toxicology, environmental sciences, and medicine, provide the data and the context where QSAR should refer. Another fundamental discipline for QSAR is chemistry, due to the basic QSAR hypothesis that the observed phenomenon is intrinsically related to the chemical structure. Finally, the third fundamental discipline refers to mathematical/statistical/information technology. Most of the QSAR models have been done with the aim to explore a single aspect of the QSAR model, such as a new chemical descriptor or algorithm. Typically, QSAR studies can be classified as academic ones, and the aim has been to show improvement of the methodology with a new introduced technique.

So far, there has been quite a mixture of purposes in QSAR studies (Benfenati, 2004). Most of the studies dealing with applications have indicated them as a possibility, without fully validating this. To fully address all criteria for a system to be practically used, it may be necessary to commit significant resources and time, and this is one of the reasons that only a few examples of practical, used applications of QSAR are available. Furthermore, for the practical use of QSAR for regulatory purposes, other issues should be addressed, and some further criteria apply. This involves all aspects of the modeling process, from the fundamental ones, such as the choice of the data, to more technical ones, such as the format of the final results. For instance, for the format, typically QSAR models use toxicity data expressed as moles, because it is assumed that a given activity is due to the interaction between a molecule and the active site, while regulators typically use weight (for instance mg).

We already mentioned in Chapters 2 and 3, Benfenati *et al.*, this volume the great care which has to be given to the data quality. But, there are also other relevant issues in the case of models for regulatory purposes. We discussed the point of false positive and false negative in Chapter 6 (Benfenati *et al.*, this volume). Regulators want to avoid false negative. An error in the opposite direction (false positive) has not the same relevance. Thus, the model evaluation is skewed. If we consider a graphical representation of the results, as in Figure 2 in Chapter 7, Amaury *et al.*, this volume, representing the predicted values on the y, against the experimental values on the x, expressed as negative log of the toxicity, the values below the ideal curve dividing the quadrant should be avoided. Actually, the false negatives are more risky when we are considering chemicals with a high toxicity value, thus chemicals with higher values in the x, in case of Figure 2 in Chapter 7, Amaury *et al.*, this volume. Indeed, if the chemical has a predicted toxicity of

100 g/l, for instance, but the actual toxicity is 1000 g/l, the error is less relevant than when the toxicity is predicted 0.1 instead than 0.01 g/l, for the intrinsic low toxicity of the compound.

An easy way to simply reduce the false negative would be to introduce a weight, shifting all prediction of a given value, in order to increase the calculated toxicity value. But this approach would fail, because in this case most of the pesticides would be predicted as toxic, and thus a model, which almost always predicts a high toxicity is useless. The approach we used introduced different rules depending on the toxicity value, and this improved the accuracy of the prediction, without introducing non-reasonable penalties.

3. THE PROBABILISTIC MEANING OF THE MODEL, THE PREDICTION OF THE EFFECT, AND THE PREDICTION OF THE MECHANISM

3.1. The probabilistic nature of the models

What human experts do is to learn from examples and identify characteristics of chemicals, in order to derive new pesticides. The approach we used relies much more on automatic processes to encode the information present in the data. As we discussed in Chapter 5 (Amaury *et al.*, this volume) and Chapter 7 (Amaury *et al.*, this volume) what we found is not one unique solution to the problem of toxicity prediction. Nowadays, it is not believed that a single program, or a single descriptor, can be the final solution, rather we believe that a range of different models can achieve similar results as demonstrated in the case of DEMETRA. In Chapter 7 (Amaury *et al.*, this volume), we identified a selected hybrid model for each endpoint, and we described how similar hybrid models produce equivalent solutions. The selected model has been chosen on the basis of some specific criteria, because the performances were slightly better than other ones (but not necessarily statistically significant) or because the descriptors were easier to be calculated by the user. Thus, the ease of use was one of the criteria, when performances were equivalent (Chapter 7, Amaury *et al.*, this volume).

We believe that, on the basis of the data we used, our results are very close to the maximum of the performance that can be achieved with the data set. What we achieved can be obtained through different models. On one hand this supports the robustness of our results, because different combinations of different models produce the same results. If different results were obtained with different combined models, the reliability of the results of these models would be questionable. The basic idea of the hybrid model, which merges different points of view and combines the capabilities of different algorithms, is that different models can be more or less powerful in one or another aspect, and that they can be improved by combining positive performances of individual models. But the same combination can be obtained in different ways. The target of the combined model is to cover as much as possible the space of the chemicals it has to model, reducing the mistakes.

Using so many chemical descriptors, it is reasonable that many of them will be similar and that a given descriptor may be represented by a second one or by a combination of two others.

Considering the algorithms, several mathematical possibilities are present, which also may provide somehow related results.

Due to the complexity and variability of the studied phenomenon, it would be very difficult to achieve a more detailed description of the situation capable of predicting all the different pesticides. We already discussed the poor results obtained using log *P* and simple chemical descriptors, but by increasing the complexity of the chemical descriptors used, we significantly improved the prediction. We will discuss more in detail this point in section 3.2, this chapter.

In QSAR, as in many other scientific fields, the availability of complex tools proves to be capable of increasing knowledge of the studied phenomenon, using non-deterministic ways. The concept of probability, so common in many fundamental theoretical disciplines (physico-chemistry, for instance, the Pauli principle), in recent fields (fuzzy logic) and applications to decision processes, is becoming valuable in the field of chemoinformatics. Regulators are well aware of the meaning of probability, because they have to assess chemicals on the basis of data, evaluating the different perspectives and evidences. Our models may provide help, producing in a reproducible way an evaluation, which is based on the codified, scrutinized data already produced for many other pesticides.

3.2. The mechanistic basis of the models

The Directive 91/414 does not require any mechanistic basis when a given toxicity value is defined using the experimental methods indicated in the guidelines. For this reason, we have not focused our attention on this aspect.

QSAR models can be more devoted to predict an effect, such as the toxicity, or can be more devoted to evaluate a mechanism, to study a biochemical process. It is also possible that a model covers both these aspects. However, these two directions, to predict the effect and to study the mechanism, can be independent. For instance, we can imagine that we already know the toxicity of a complete chemical category of compounds (all chlorinated benzenes) but we want to understand the mechanism, then in this case an in silico model can provide help, but there is no need to predict the effect, in contrast, we are more interested in the prediction of the mechanism of chlorinated benzenes.

We also notice here that in any case the QSAR model can only suggest the mechanism, but the real proof has to come from real experiments.

In the models we developed, some chemical descriptors are informative, while others are less clearly related to a possible mechanism. Thus, as we discussed in Chapter 7 (Amaury *et al.*, this volume), some descriptors present in our models have been indicated linked to a given toxic mode of action, such as log *P* in the case of trout toxicity.

However, due to the occurrence of many toxicity mechanisms in the highly heterogeneous sets of pesticides we used, it is unlikely that the QSAR models we developed are highly informative, from a mechanistic point of view. A more suitable approach to explore the toxic mechanism is related to the rules we described for some specific subsets of pesticides. In Chapter 7 (Amaury *et al.*, this volume) we indicated some rules to be used as warning and to identify potential outliers. These rules have been used to characterize a restricted group of chemicals, sharing a specific moiety or chemical components. It is interesting to note that in this case the rules are quite specific, generally common to a limited number of chemicals. The relationship between a rule, chemical in this case, and the effect is more clearly identified when some chemicals are more toxic than others, and the reason for this is quite simple. For instance, a chemical rule characterizing pyrethroids can easily define a subset of pesticides with a high toxicity toward trout. This example shows that the occurrence of clear explicit chemical rules to be put in relationship with a given effect is more suitable within a limited domain of pesticides.

3.3. The final model and the ways to obtain it

Another important concept, which follows from what we discussed in the previous chapters, is that the final hybrid model is relatively simple, even if the overall procedure we used to identify it is complex. Even if more than 1000 chemical descriptors have been calculated to develop the models, those used in the final model are about 20, depending on the model. Similarly, even if thousands of models have been developed to identify the best solution, for the final hybrid model we used only very few of them. A lot of work has been done to identify the best solution, but the user has a much simpler task.

This point has implications on the quality process and validation assurance. What we optimized is the final model. It is common in QSAR that several attempts are made to find the most suitable solution for a given problem. We believe that the evaluation criteria have to be applied to the final model, which has to be checked, proved, and tested for its reproducibility. This also refers to the OECD criterion number 4.

4. THE BENEFITS OF THE DEMETRA MODELS

We believe that the DEMETRA models will have benefits. Regulators and industry will benefit from the availability of publicly accessible tools to predict ecotoxicity.

The larger use of these models we envisage is in the case of metabolites and transformation products. Another case is when there is availability of toxicity values from experiments done on other animals, different from those modeled within

DEMETRA, so that DEMETRA can provide additional data. Finally, DEMETRA models can be used by the industry of plant protection products to gain better estimation of the ecotoxicological risk of future pesticides, during the development phase, avoiding structures which are more risky.

To conduct animal experiments on chemicals is expensive, time-consuming, and poses ethical problems. Thus, for several reasons, laboratory experiments are limited to cases when it is absolutely necessary. DEMETRA models may provide an alternative way to get insights into the toxic properties of pesticides and related compounds.

DEMETRA models are associated with a specific uncertainty, which has been characterized (Chapter 7, Amaury *et al.*, this volume). Using this uncertainty as a safety factor allows a more precautionary use of the calculated values.

Our models should not be seen as an alternative approach to experimental models. Instead, we believe that the major use of the DEMETRA models will be as supportive tool to increase the basis for the overall assessment. The two strategies are the following:

1) to evaluate a pesticide using a set of results from the experimental tests and

2) to evaluate a pesticide using a set of results from the experimental tests and additionally the predictions from the DEMETRA models. DEMETRA models can provide data from animals not used in the laboratory tests and on metabolites.

This strategy goes along the modern evolution of the use of information technologies. Nowadays, the competition between human and computer has evolved into the study on how human and computers can better 'collaborate'. More in particular in the case of QSAR models, the discussion is on how to use in silico tools within a broader decision support system.

Of course in silico models operate within a different plane than the experimental one. But the experimental studies are encoded into the in silico models. In silico models do not provide the same robustness of in vivo studies. But it would be a mistake to think that for this the information value of in silico model is zero. What we basically demonstrated with our results is that DEMETRA models are informative regarding the ecotoxicological properties of pesticides.

Furthermore, the DEMETRA project has clearly indicated the general criteria we followed. We hope that these criteria will be discussed and used for similar initiatives.

5. FUTURE PERSPECTIVES

Finally, we discuss the perspectives of the DEMETRA models and how they could be further improved. DEMETRA had a specific aim as defined by the contract of the European Commission, which made possible the work done within DEMETRA. The main target of DEMETRA was to assess the possibility to predict the ecotoxicity of pesticides. It was not foreseen to develop tools to calculate chemical descriptors, because for this there are already programs available, and because it would take a lot of effort to replicate what available. However, once a limited number of chemical descriptors have been identified, useful for the specific models, it is more convenient and feasible to develop specific algorithms for them.

The availability of these specific chemical descriptors together with the DEMETRA models, within a unified structure, would surely increase the access to the DEMETRA models. Currently, the users have to calculate a list of chemical descriptors and then they can use the DEMETRA model. It would be much better for the users to simply draw the chemical structure of the pesticide they want to study and immediately achieve the result. This is feasible and would require some limited effort. For instance, another EC-funded project, OpenMolGRID, showed that it is possible to develop QSAR models which automatically produce results, simply starting from the 2D chemical structure.

Thus, we think that a very useful extension of DEMETRA would be to develop freely available tools to automatically calculate the selected chemical descriptors and eventually to combine the tool for the chemical descriptor calculation directly with the DEMETRA models.

A further advantage of this extension is that in this way the DEMETRA models would be more stable. Indeed, the developed models are optimized with the descriptors we indicated in Chapter 3 (Benfenati *et al.*, this volume). If the company which developed the software to calculate the chemical descriptors wants to change something in its program, the chemical descriptors may change.

Finally, the availability of specific software to calculate chemical fragments may improve the performances of the models, as additional activity. We will discuss this point more in detail below.

We mentioned in Section 3.2, this chapter that the DEMETRA models are the optimal solutions based on the material we used. To improve the overall performances of these models, we have to increase the knowledge basis of the model. This can be done in one of the following ways.

- Introducing more high-quality experimental data.
- Introducing more detailed chemical information as chemical rules related to fragments/features.
- Focusing the attention to specific groups of chemicals.

The first point is a fundamental one. The best way to improve the current models would be to add toxicity values of high quality for new compounds, and the pesticide industry may contribute to this. This attitude is solicited also within other frameworks, such as the REACH legislation. If we want to increase the knowledge on the pesticides poorly represented in the current data sets, we need more examples. An open-minded attitude of the chemical industry would be beneficial for the industry itself.

Another way to increase the basis of our models is to modify the basis of the chemical information. We listed in Chapter 3 (Benfenati *et al.*, this volume) the ways to describe the chemical structures used within DEMETRA. We used a very large battery of tools, including 2D- and 3D-based chemical descriptors, fragments produced by commercial programs, and more human-based methods, such as CoMFA methods and in house programs to optimize specific fragments, to be put in relationship with the toxic activity. On the basis of the commonly available software to calculate the descriptors, we believe that no further useful chemical information can be extracted for our models. Subjective models, such as CoMFA, were excluded because of their poor reproducibility, and thus lack of suitability for regulatory purposes. There are other possibilities of ad hoc procedures to further identify automatically useful fragments/chemical features.

An alternative way which already proved to improve the results is using chemical features identified by human experts. As described in Chapter 7 (Amaury *et al.*, this volume), we identified specific chemical features present in compounds which all have high toxicity values; thus, these chemical features can be related to a given toxicity value, as a threshold. So far, we produced written rules to be used offline as check. However, dedicating some limited efforts, these rules can be encoded into a program within a new version of the DEMETRA models.

A final way to improve the models is to add some more focused models addressing specific chemical classes, such as carbamates and chlorinated compounds. We preferred to have a general model (as this was the target of the project), but we also developed some local models (Chapter 7, Amaury *et al.*, this volume). We mention that industry was not interested in simpler models for specific organophosphorus compounds, for instance, because in most of the cases these groups of compounds refer to quite old pesticides, for which there is quite a lot of information. However, an integrated strategy combining a general model and, whenever achievable, some more focused models could offer overall better performances.

As suggested above, there are ways to improve with further activities the DEMETRA models. An important way will be to gather comments and inputs from the users who want to contribute in the strategy. We believe that the better overall strategy will be to merge contributions from different sources and strategies. As discussed in Section 1 of Chapter 7, Amaury *et al.*, this volume, there is no single, simple solution to the complex problem of the prediction of ecotoxicity for pesticides. The optimal way is to start a process where information of different chemical, biochemical, and toxicological nature is integrated within a powerful computerized system, producing a reasonable prediction of the properties of interest. DEMETRA indicated the direction and the feasibility. We hope that the pathway will be found useful by others who wish to contribute to our experience.

REFERENCES

Benfenati E., Pelagatti S., Grasso P., Gini G. (1999) COMET: the approach of a project in evaluating toxicity. In: *Predictive Toxicology of Chemicals: Experiences and Impact of AI Tools.* AAAI 1999 Spring Symposium Series; Gini, G. C.; Katritzky, A. R., Eds.; AAAI Press, Menlo Park, CA; pp 40–43.

Benfenati E., Piclin N., Roncaglioni A., Varì M.R. (2000) Factors influencing predictive models for toxicology. *SAR and QSAR in Environmental Research*, 12, 593–603.

Benfenati (2004) Modelling aquatic toxicity with advanced computational techniques: procedures to standardize data and compare models. In: *Knowledge Exploration in Life Science Informatics, Lecture Notes in Computer Science 3303*; López, J. A.; Benfenati, E.; Dubitzky, W., Eds.; Springer-Verlag GmbH; pp235–238.

Devillers J., Pham-Delegue M.H., Decourtye A., Budzinski H., Cluzeau S., Maurin G. (2002) Structure-Toxicity modeling of pesticides to honey bees. *SAR and QSAR in Environmental Research*, 13, 641–648.

Hansen O.C. (2004) Quantitative Structure-Activity Relationships (QSAR) and Pesticides. Danish Environmental Protection Agency, Pesticides Research No. 94.

OECD (2005). OECD Principles for the Validation, for Regulatory Purposes, of (Quantitative) Structure-Activity Relationship Models. Paris, France. http://www.oecd.org/document/23/0,2340,en_2649_34365_33957015_1_1_1_1,00.html.

Rudén C., Hansson S.O. (2003) How accurate are the European Union's classifications of chemical substances. *Toxicology Letters*, 144, 159–172.

Russom C.L., Bradbury S.P., Broderius S.J., Hammermeister D.E., Drummond A. (1997) Predicting modes of toxic action from chemical structure: acute toxicity in the fathead minnow (*Pimephales promelas). Environmental Toxicology and Chemistry/SETAC*, 16, 948–967.

Veith G.D. (2004) On the nature, evolution and future of quantitative structure-activity relationships (QSAR) in toxicology. *SAR and QSAR in Environmental Research*, 15, 323–330.

Vighi, M., Masoero, M., Calamari G., Calamari D. (1991) QSARs for toxicity of organophosphorous pesticides to Daphnia and honeybees. *The Science of the Total Environment*, 109–110, 605–622.

Worth A.P., Van Leeuwen C.J., Hartung T. (2004) The perspectives for using (Q)SARs in a changing political environment – high expectations and a key role for the European Commission's Joint Research Centre. *SAR and QSAR in Environmental Research*, 15, 331–343.

Quantitative Structure-Activity Relationships (QSAR) for Pesticide Regulatory Purposes
Edited by Emilio Benfenati

Chapter 9

The use of the DEMETRA models

Emilio Benfenati[1], Marian Craciun[2], Daniel Neagu[3]

[1]Laboratory of Environmental Chemistry and Toxicology, Istituto di Ricerche Farmacologiche "Mario Negri", Milano, Italy

[2]Department of Computer Science and Engineering, University "Dunarea de Jos", Galati, Romania

[3]Department of Computing, School of Informatics, University of Bradford, Bradford, UK

1. INTRODUCTION

The DEMETRA models originate from the debate on the possible use of quantitative structure-activity relationships (QSAR) models as an alternative to the use of animals. This debate is of major interest in these days also in consideration of the recent EU REACH legislation for the industrial chemicals. This legislation does not refer to pesticides, because the current European legislation for pesticides already details the procedure to submit the requested material for the evaluation of the chemical to be used as pesticides. In the new REACH legislation QSAR is mentioned as a valuable alternative.

We already deeply discussed the endpoints requested by the Directive on pesticides in Chapter 1 (Benfenati *et al.*, this volume) and the currently used QSAR models. Here we only remember that keeping in mind the regulatory frameworks and the tendency to explore alternative approaches, DEMETRA proposed five models for the specific target of the pesticide evaluation, in particular for the endpoints listed in Chapter 2 (Benfenati *et al.*, this volume): trout, daphnia, quail (oral and dietary exposure), and bee. In Chapters 1 and 2 (Benfenati *et al.*, this volume), the full details of the procedure to select the endpoints are given, and general comments on this and the other criteria we used are discussed in Chapter 8 (Benfenati, this volume).

Each DEMETRA model, for each of the five endpoints, is an intelligent integrated model, also called hybrid model, including other simpler models.

2. THE USERS OF THE DEMETRA MODELS

The major user category we have identified within the DEMETRA project is the regulatory body. Officers involved in the evaluation of the pesticides for regulatory purposes may get benefits for the availability of the DEMETRA models. DEMETRA models can be used for parent compounds and metabolites. This increases the scenario of the evaluation. Indeed, to make animal experiments for all the metabolites and transformation products, which typically occur, requires resources, and this limits the data availability. Furthermore, regulators may find useful one of the DEMETRA model even for parent compound, if they have data on a related endpoint but not for what modeled within DEMETRA (for instance duck experimental value is available but not for quail).

DEMETRA has been built up since its planning in close contacts with end-users, which is a major difference compared with the current QSAR models. This has been discussed in detail in Chapter 8 (Benfenati, this volume). Also, the final output of the models is given in weight, for instance mg/l. Non-governmental agencies are also major potential users of the models, for issues similar to those of regulators.

Another important category is industry in the field of plan protection products. In this case, the possible use can be anticipated even in the very early phase of the planning and development of a new pesticide. DEMETRA models work on the basis of the pure chemical structure, and thus, it is possible to use them without the synthesis or preparation of the candidate pesticide, which can be costly and time consuming. In this way, the DEMETRA models can be used as a pre-screening, to identify potential weakness of the planned compounds.

For industry, we developed some dedicated tools. Although an applet does not require sending data over Internet as running locally (user-end application), we developed a version of the software that is intended just for local use and does not require connection to the web. In this way, the stand-alone Java program can be used without any further connection to Internet once downloaded from our project web site. Furthermore, the software can calculate ecotoxicity for many compounds, for a faster use. In this way, we hope we will facilitate a wider use of the DEMETRA models. The offline models are exactly equivalent to those present on the Internet.

Another major category of users is academy and research institutes. DEMETRA has been able to develop innovative tools on a scientific point of view, both for the chemical information, introducing some new descriptors and on a computational point of view, in particular, in the field of the hybrid models and systems for QSAR. DEMETRA developed a multipurpose predictive toolbox (see Chapter 5, Amaury *et al.*, this volume), useful for many applications, not only for pesticides. Furthermore, new specific models have been developed, as already mentioned.

3. OWNERSHIP OF THE SOFTWARE

The DEMETRA project has been funded by the EC within the Fifth Framework Programme. The QSAR models that have been developed within DEMETRA are freely available for non-commercial use. The use is regulated by the GNU license, indicated in Appendix F (this volume).

The software developed within DEMETRA is of two types: (1) dedicated hybrid models and (2) a general purpose hybrid system.

Five hybrid models are dedicated for regulatory purposes, for the prediction of five ecotoxicological endpoints, as specified in Chapter 2 (Benfenati *et al.*, this volume): trout, daphnia, quail (oral and dietary exposure), and bee. In addition, a flexible hybrid system has been developed, as a general tool to build up new models, taking advantage of the advanced algorithms integrated in such a tool (see Chapter 5, Amaury *et al.*, this volume).

Thus, the five specific models are fixed, intended for regulatory purposes, and they have to provide a single value for a single pesticide, regardless of the user, because the same predicted value has to be obtained by all users. The five hybrid models have been developed according to the criteria described within Chapter 8 (Benfenati, this volume), to fulfill quality assurance and control steps necessary for a regulatory purpose. These criteria refer to the toxicity data used, to the chemical structures and descriptors, to the validation process, etc. The validity of the obtained models has been checked, assessed, and described in Chapter 7 (Amaury *et al.*, this volume).

Vice versa, the hybrid system that has been developed is a flexible tool, useful for the development of new models, in a simple but powerful way. It takes advantage of innovative algorithms, but the obtained models are not necessarily suitable to their use within regulatory purposes. For this specific purpose, all the criteria defined in Chapter 8 (Benfenati, this volume) have to be addressed.

The five hybrid models have been developed using the toxicity data kindly provided by the US EPA-OPP. At the end of the project, these data became publicly available. The toxicity data we used are listed in Appendix B (this volume).

The descriptors used for the hybrid models are listed in Chapter 7 (Amaury *et al.*, this volume). In addition to these chemical descriptors, selected as suitable for modeling purposes, many more descriptors have been calculated. The list of descriptors is available at the DEMETRA web site: http://www.demetra-tox.net; there is no space here to report all of them.

The software that is used to calculate these descriptors is listed in Chapter 4 (Chaudhry *et al.*, this volume). We have not made a detailed comparison of the different programs available, checking whether they produced similar or different descriptors. This was not our target. Indicating the software used to calculate the used descriptors, we simply identify the programs we used, not necessarily because they are the best programs. The software to calculate chemical descriptors is proprietary software. The user who wants to calculate the chemical descriptors selected for the hybrid models has to obtain the necessary software. We tried to use as much as possible software, which is freely available and a limited number of programs, to facilitate the use of the hybrid models. However, at the moment for chemical descriptors, the user has to use software external to DEMETRA. Within DEMETRA, it was not foreseen to develop specific software to calculate chemical descriptors but simply to show that it is possible to predict ecotoxicological properties using advanced QSAR models. As future activities, we aim to develop free software to calculate the necessary chemical descriptors (see Chapter 8, Benfenati, this volume).

To develop the QSAR models and to select the most suitable chemical descriptors, we used proprietary software, as described in Chapter 4 (Chaudhry *et al.*, this volume) and Chapter 5 (Amaury *et al.*, this volume). Some programs are freely available, whereas others are commercial ones. In any case, the user does not need any of these programs to run the five hybrid models.

4. USING DEMETRA MODELS

The practical guide for the DEMETRA models, as also reported in the web site, is given in Appendix F (this volume). Briefly, the user needs the DEMETRA models, the Java applet (or the stand-alone version, both available at the DEMETRA site), and the programs to calculate the descriptors. The DEMETRA models are available at the web site: http//:www.demetra-tox.net/.

The only inputs of the five hybrid models are the chemical descriptors. They are listed in Chapter 7 (Amaury *et al.*, this volume), for the different models. The DEMETRA models, once the descriptors are available, is simplified for easy use, as described in the Appendix F (this volume), given the purpose-built designed interfaces.

Some experience is required to use the commercial programs to calculate the chemical descriptors. However, we used only relatively simple models for our purposes; thus, the user should not face problems also in this case.

As an example of use of the descriptor calculation, we give some explanations on the ways to make the calculation in the Appendix C (this volume). However, the modeler should make reference to the manuals of the programs for descriptor calculation.

The outputs of the five hybrid models are the ecotoxicity values. The value is given in weight (for instance mg/l) and in millimoles (for instance mmol/l). The model works using millimole and then gives the output also in milligram, using the molecular weight given by the user. In these cases, the value is given as co-logarithm, and thus, compounds with a higher toxicity have a higher value, whereas in the case of toxicity values given in milligram if the dose is lower, the compound is more toxic.

This value refers to the final output of the hybrid model. For transparency and for possible use by the assessor, we also give the range of the toxicity values of the individual models. We remember that each hybrid model is an integration of other individual models. The hybrid model is not a simple average of the individual models, and thus, the value of hybrid model is generally different by the average.

The model keeps into account the range used for training. If the predicted toxicity value is external to this range, the model gives a warning to the user and gives the prediction as more or less than.

In all cases, the minimum uncertainty of the predicted value is a factor of four, which is the threshold we fixed for the acceptability of the experimental toxicity data.

For the aquatic toxicity models (trout and daphnia), the user should keep in mind that the predicted value in milligram can be higher of the actual solubility of the compound, meaning that this value cannot be reached without solvents or surfactants.

5. CHEMICAL RESTRICTIONS OF THE DEMETRA MODELS

In Chapter 3 (Benfenati *et al.*, this volume), we identified the chemicals for which the model can be used. These rules were based on the availability of suitable programs for descriptor calculations. Then, on the basis of the model results, we defined a posteriori rules to identify areas where the models gave lower performances. The modeler should be aware of these restrictions. We summarize them. The models are not suitable for

- Inorganic compounds, for example, copper sulfate.
- Mixtures of more than one active substance.
- Mixtures of chemicals with different molecular weights.
- Complexes such as those with iodine.

Furthermore, as typical for QSAR modeling, some complex compounds have to be simplified. This applies to the following cases:

- Salts have to be modeled without the inorganic ion. For example, in the case of a HCl salt, it has to be used in its basic form, and in the case of a K salt, the K atom has to be replaced with a H atom.
- Hydrates have to be modeled in the anhydrous form.

Furthermore, as explained in Chapter 7 (Amaury *et al.*, this volume), we identified chemical rules for chemicals that were predicted not in an optimal way. The user should be aware that if one of the following fragment/rules is verified, the model may be less valid (see Chapter 7, Amaury *et al.*, this volume).

6. THE FORMAT FOR MODEL PRESENTATION FOR DEMETRA: HISML

Solving problems using artificial intelligence (AI) techniques requires various specialized languages, such as Lisp and Prolog for Logic Programming, or programming languages, such as Matlab®, JavaTM, C/C++, to implement machine learning algorithms. Moreover, another issue in AI applications started to claim more attention nowadays: data and knowledge representation. A unified way to propose a framework for general standard specification of data contents and structure is eXtenisble Markup Language (XML). This opens the possibilities to develop domain dictionaries, robust data processing, and validation by metadata definition. There are some attempts so far to encapsulate knowledge, rules or data which could be further used to create AI systems.

Tabet *et al.* (2000) show that the input and output, and even the rules themselves, from an AI application can be given as XML files: this allows reducing considerable time and effort in building conversion procedures. They work to develop Universal Rule Markup Language (URML) (URML, 2006) with the goal to promote the development of standards for rule markup using XML.

An original communication language and protocol for knowledge exchange for intelligent information agents, Knowledge Query and Manipulation Language (KQML) offers (KQML, 2006) a useful abstract level for definition of distributed AI systems. KQML can be used as a language for an application program to interact with an intelligent system, as well as for two or more intelligent systems to share knowledge for cooperative problem solving (Finin *et al.*, 1997). Formal Language for Business Communication (FLBC), a competitor to KQML, is an XML-based formal language proposed by Scott A. Moore, which one can use for automated electronic communication (FLBC, 2006). The DARPA Agent Markup Language (DAML), developed as an extension to XML and Resource Description Framework (RDF), provides a basic infrastructure that allows a machine to make the same sorts of simple inferences that human beings do (DAML, 2006). Case Based Markup Language (CBML) is an XML application for data represented as cases to facilitate knowledge and data markup readily reusable by intelligent agents (Hayes *et al.*, 1998). Another effort in this direction is Artificial Intelligence Markup Language (AIML) (AIML, 2006), an XML-based language used in ALICE, a chat-bot. This markup language offers a simple yet specialized open-source representation alternative for conversational agents. For specialized Machine Learning techniques, the Predictive Model Markup Language (PMML) is a language proposed to describe Statistical and Machine Learning – Data Mining models (PMML, 2006). PMML describes the inputs to data mining models, the transformations used prior to the preparation of data for data mining, and the parameters that define the models themselves.

As a contribution in the DEMETRA project, a markup language for integrated structures of Hybrid Intelligent Systems based on multi-classifiers and gating strategy (HISML) was proposed (Neagu *et al*., 2004). This is an XML application introduced to fill the gap between existing Predictive Toxicology simple models and complex models based on explicit and implicit knowledge represented as modular hybrid intelligent structures. The proposed HISML syntax captures the structure and parameters of modeling experiments (Fig. 1). The information stored in a HISML document is further required to analyze and replicate the models.

The seed of the common format to manipulate as well as to store the data is captured by the concept of the HISML element. Such an element has a required attribute *version* and contains two other sections: a header with authoring information and section HIS to define recurrent intelligent component systems (Table 1).

One of the main features the HISML syntax takes into consideration is the recurrent organization of imbricate modules. The need to propose a standard format for data storage in HIS development using the HISML syntax is justified by various methods involved and the aim of automation for many of design steps.

A HIS element comes with a *name* attribute and might be: a simpleHIS (basic atoms IKM – Implicit Knowledge Module and EKM – Explicit Knowledge Module) or a complexHIS (any other HIS, including simple ones), adding an information about its performance (Table 2).

A simpleHIS element (Fig. 1) comes with a *name* attribute and might be an IK or EK module. A simpleHIS is finally an artificial neural network (as IKM-CNN) or a fuzzy neural network (combination of ANN with FIS) (as IKM-FNN or EKM) (Neagu *et al*., 2005).

The basic types are ANN and FIS. An ANN comes with *train* (has been the net trained?) and topological data attributes (inputs number, hidden layers list, neurons data: activation functions, weights matrix to current layer, training algorithm). A FIS includes essential information to identify fuzzy inference systems: the type of the system, also the number of inputs and outputs, the number of rules, the methods for and/or implication, aggregation, and defuzzification, the rules represented as matrix and the list of fuzzy variables for input and output (Table 3).

A complexHIS (Fig. 1) consists of two or more (simple or complex) HIS elements, and a gating module (GM) or a choosing algorithm choice module (CM) (either a statistical or fuzzy inference approach) as integration module to combine using an intelligent way all the modules inside. A GM can be a (supervised or unsupervised trained) CNN or a statistical combining algorithm (FEMS, majority voting, max, min, average, etc.). A CM is a (supervised or unsupervised trained) ANN of any complexHIS types.

Different types of Simple Intelligent predictive modules could also be further considered, and possible ways to integrate the modules inside a Complex Intelligent System will be investigated. Moreover, implicit or explicit knowledge representation based on other Machine Learning techniques from Statistics and AI can be introduced – regression models, decision trees, and decision rules – and

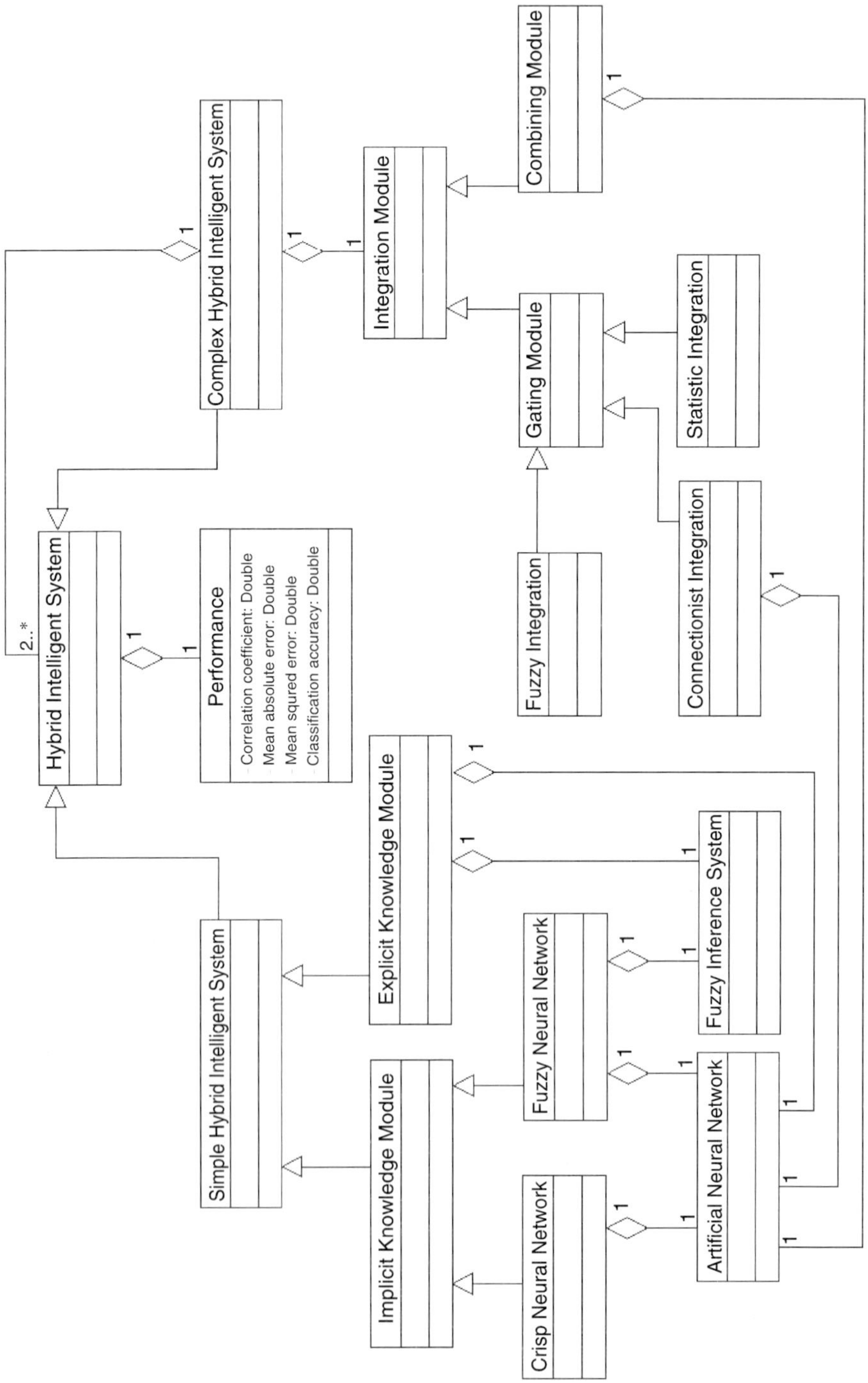

Figure 1: HISML class diagram.

Table 1
The main structure of HISML

```
<!– Root element –>
<xsd:element name=‘HISML’>
  <xsd:complexType>
    <xsd:sequence>
      <xsd:element ref=‘header’/>
      <xsd:element ref=‘HIS’/>
    < /xsd:sequence>
    <xsd:attribute name=‘version’ type=‘xsd:string’ use=‘required’/>
  </xsd:complexType>
</xsd:element>

<!– Header Description –>
<xsd:element name=‘header’ id=‘el.header’>
  <xsd:complexType>
    <xsd:sequence>
      <xsd:element ref=‘author’/>
      <xsd:element ref=‘date’/>
      <xsd:element ref=‘comments’ minOccurs=‘0’/>
    </xsd:sequence>
</xsd:complexType>
</xsd:element>
```

Table 2
The main structure of HIS

```
<!– Hybrid Intelligent System module description–>
<xsd:element name=‘HIS’ id=‘el.HIS’>
  <xsd:complexType>
    <xsd:sequence>
      <xsd:element ref=‘performance’ minOccurs=‘0’/>
      <xsd:choice>
        <xsd:element ref=‘simpleHIS’/>
        <xsd:element ref=‘complexHIS’/>
      </xsd:choice>
    </xsd:sequence>
    <xsd:attribute name=‘name’ type=‘xsd:string’ use=‘required’/>
  </xsd:complexType>
</xsd:element>
```

(*Continued*)

Table 2
(*Continued*)

```
<!– HIS performance –>
<xsd:element name='performance' id='el.performance'>
    <xsd:complexType>
      <xsd:sequence>
      <xsd:element ref='correlationCoefficient' minOccurs='0'/>
      <xsd:element ref='meanAbsoluteError' minOccurs='0'/>
      <xsd:element ref='meanSquredError' minOccurs='0'/>
      <xsd:element ref='classificationAccuracy' minOccurs='0'/>
    </xsd:sequence>
  </xsd:complexType>
</xsd:element>
```

Table 3
The main structure of fuzzy inference system (FIS)

```
<!– Fuzzy Inference System Type –>
  <xsd:complexType name='FIS' id='ct.FIS'>
    <xsd:sequence>
      <xsd:element ref='dataDescription'/>
      <xsd:element ref='typeOfFIS'/>
      <xsd:element ref='numberOfInputs'/>
      <xsd:element ref='numberOfOutputs'/>
      <xsd:element ref='numberOfRules'/>
      <xsd:element ref='andMethod'/>
      <xsd:element ref='orMethod'/>
      <xsd:element ref='impMethod'/>
      <xsd:element ref='aggMethod'/>
      <xsd:element ref='defuzzMethod'/>
      <xsd:element ref='inputsList'/>
      <xsd:element ref='outputsList'/>
      <xsd:element ref='rules'/>
    </xsd:sequence>
  </xsd:complexType>
```

more aggregation strategies can be evaluated. The schema proposed is used for hybrid intelligent models management by the prototype NIKE (see Chapter 5, Amaury *et al.*, this volume).

REFERENCES

AIML (2006) http://www.oasis-open.org/cover/aiml-ALICE.html.
DAML (2006) http://www.daml.org/.

Finin T., Labrou Y., Mayfield, J (1997) KQML as an agent communication language, in Jeff Bradshaw (ed.), Software Agents, MIT Press, Cambridge, 291–316.

FLBC (2006) http://www.oasis-open.org/cover/flbc.html.

Hayes C., Cunningham P. (1998) Distributed CBR using XML, Proc. Intell. Sys. Electr. Comm. Int. Workshop, Bremen.

KQML (2006) http://www.cs.umbc.edu/kqml/.

Neagu D., Craciun M., Chaudhry Q. (2004) Knowledge Specification for Versatile Hybrid Intelligent Systems, Proc. 5th Int. Conf. Recent Advances in Soft Computing RASC 2004, 16–18 Dec. 2004, Nottingham, UK, ISBN 1-84233-110-8, pp. 219–224.

Neagu D., Craciun M.V., Stroia S.A., Bumbaru S. (2005) Hybrid Intelligent Systems for Predictive Toxicology – A Distributed Approach, Proc. 5th Int. Conf. Intelligent Systems Design and Applications ISDA2005, Wroclaw, Poland (Halina Kwasnicka, Marcin Paprzycki eds.), IEEE Computer Society Press, ISSN 0-7695-2286-6, pp. 26–31.

PMML (2006) http://www.dmg.org/index.html.

Tabet S., Bhogaraju P., Ash, D. (2000) Using XML as a Language Interface for AI Applications, Proc. Int. Conf. PRICAI 2000, LNCS 2112, Springer, 103–110.

URML (2006) http://home.comcast.net/~stabet/urml.html.

Appendices

Quantitative Structure-Activity Relationships (QSAR) for Pesticide Regulatory Purposes
Edited by Emilio Benfenati

Appendix A

Summary of responses to DEMETRA survey

Mark Clook[1], **Steven Fryday**[2], **Andy Hart**[2]

[1]Pesticides Safety Directorate, Department for Environment, Food and Rural Affairs, Mallard House, York, UK

[2]Central Science Laboratory, Department for Food and Rural Affairs, Sand Hutton, York, UK

A summary of the aims of DEMETRA and a list of questions designed to elicit the views of end-users were sent to the 'indicative list of experts' as defined in Annex 1. Very few replies were received and the responses to each question by those that did reply is given below.

(1) What type of end-point would you like to be addressed within a QSAR model (e.g. acute, chronic, which species)?

'The end-points required for all categories of biocidal product (the so-called core data set) are given in chapter 2 at the following link: http://ecb.jrc.it/Documents/Biocides/TECHNICAL_NOTES_FOR_GUIDANCE/TNsG_DATA_REQUIREMENTS/chapter2.pdf

'Types of end-points to be investigated should parallel data requirements in the directive 91/414 and the existing guidance papers on e.g. Aquatic Ecotoxicology and Terrestrial Ecotoxicology'.

'For NTP, usually ED50 for fresh/dry weight or plant height are used as end-points. I have no data on hand regarding repeated assays under completely repeatable conditions (e.g. climatic chambers). Under glasshouse conditions ED50-values for plants may vary by up to the factor of 50, but is usually between 5 and 10. For end-points regarding other organisms, I would have to look up EU monographs, which you probably would have done by now yourself. Colleagues told me however, that the variation is in the same order of magnitude as for NTP'.

'I am not an expert on whom species are the most representative and sensitive species. It depends on the type of end-point you wish to address'.

'As far as regards the type of end-points to be considered there is a clear higher interest for long term toxicity ones because they cost much more'.

(2) How accurate are current registration data for these end-points (e.g. ± a factor of 2, 5 and 10)?

'Factor of about 2–5'.

'For NTP, usually ED50 for fresh/dry weight or plant height are used as end-points. I have no data on hand regarding repeated assays under completely repeatable conditions (e.g. climatic chambers). Under glasshouse conditions ED50-values for plants may vary by up to the factor of 50, but is usually between 5 and 10. For end-points regarding other organisms, I would have to look up EU monographs, which you probably would have done by now yourself. Colleagues told me however, that the variation is in the same order of magnitude as for NTP'.

'The criteria I use is that the uncertainty on the individual end-points has to be significant less than the difference in end-point value between chemicals'.

'About accuracy, acute tests generally fall in the range of factor 2–5, whereas for chronic ones the variability can be definitely more'.

(3) How accurately should QSARs predict actual toxicity (e.g. ± a factor of 2, 5 and 10)?

'Clearly, a given QSAR cannot be expected to make predictions that are more accurate than are repeat measurements of the experimental end-point it aims to predict. For this reason, the validity of a given QSAR should only be judged in the context of the experimental variability of the underlying data, and not according to fixed predictivity criteria'.

'Factor of 5'.

'I would expect the QSAR not to predict a certain ED50-value or the like but rather a certain probability for the substances ED50 to exceed a certain trigger value'.

'As precise and accurate as the input data, used for calibrating and validating the models'.

'Accuracy to be required to the model: that needed for acceptance seems to be the first answer. It's also clear to be pivotal that the attribution to either side of a cut off value is correct (i.e. that a compound predicted to be toxic with respect to an established threshold is really toxic and vice versa that a compound predicted not to be toxic is really not toxic)'.

(4) What reassurance would you require about the quality and quantity of the data used to produce the QSAR?

a) Use of homogeneous data for the development of the QSAR, i.e. all data should have been obtained by the same experimental protocol, but not necessarily in the same laboratory or by the same worker.

b) At least 10 data points per descriptor in the final QSAR model.
d) Full details of the training set, including descriptor and response data, and information on any manipulations of the training set carried out in QSAR development.
d) A clear definition of the domain of applicability of the QSAR.
e) If available, an indication of the experimental variability inherent in the response variable'.

'Studies according to existing Guidelines or well-described studies from the open literature'.

'The data used for development and validation of a QSAR-system have to meet requirements regarding laboratory and report standards, e.g. GLP. There should be several datasets for each chemical/test organism/end-point combination'.

'cf. 2) the data has to be validated according to testing of the significance of differences between individual pesticide end-points. This is the minimum criteria for developing of QSAR models that have to estimate the differences in toxicity end-points for the individual pesticides. Quantity depends on the domain of the models, but in general two third of the data for calibration and one third for validating the models. The requirements to the two third of the data used for calibrating the models is that these testing compounds span the property space of the model domain well (robustness)'.

'Reassurance about data to be used for the QSAR: deemed to be a question for experts, but someone in the group has suggested that the data set to be modeled has to include good numerical presence of compounds for the different mechanism of actions considered'.

(5) What type of validation would you expect to have been conducted?

'Minimum – assessment of internal goodness-of-fit (strictly, this is not "validation")

Better – additional use of cross-validation to demonstrate robustness of model parameters and predictions

Ideal – validation using a test set that is independent of the training set. This type of validation could be organised in the context of a formal framework by the Joint Research Centre (JRC), provided that the QSAR is sufficiently well-developed and documented to start with'.

'The substances in the training set should be listed and external predictivity should be established by external validations, i.e. 50% random out statistical cross-validations, true external validations etc. In case of true external validations it is important to show how well the substances used cover the domain of the model.

The models should give clear indications of whether a prediction is within the domain of the model'.

'The QSAR results should be compared with a great number of results of eco-toxicological assays e.g. from the EU-regulatory process. It may be necessary to conduct experiments specifically designed for validation, since often input parameters are lacking in older documents as we found when comparing pesticide leaching models'.

'Depends of the nature of the raw data, parametric tests, e.g. basic statistics such as F and t-tests. Probabilistic methods verifying the existence of significant difference in toxicity between pesticides. Alternatively a number of non-parametric tests may be used for securing the significance of differences in end-point values used when developing QSARs'.

'Similar is the reaction for the question on the type of validation, someone saying that has to be representative of the data set considered for the modelling, others instead stressing the importance of validating on the most critical cases (i.e. metabolites more toxic than the parent compound)'.

(6) Are there any other issues you would require to be addressed before you could accept the use of QSARs for regulatory purposes?

'This is a question for regulators, who would need to see the relevance of the QSAR in the context of their regulatory programme, and to be convinced that the predictions it makes are "good enough" (according to their criteria) for use as an alternative to the traditional test'.

'Good and solid validation'.

'At the moment, I cannot think of any, but this may be due to the fact that my experience in this field is yet restricted'.

'I think that analysis of the critical pathways in relation to exposure are crucial for selecting the most proper combination of end-points, matrices and species'.

'The basic interest is towards the implementation of a project useful for registration purposes at the EU level and therefore to models accepted at the EU level'.

Additional general comments

A short reaction to the DEMETRA project proposal:

- Success of the DEMETRA project stands and falls with the quality of the data in the database used for the QSAR calculations. As some important chemical companies are involved in this project, this may give the opportunity to retrieve high-quality (but protected) information, with tests using OECD guidelines derived form GLP labs. However some other important industries are not involved. It is strongly recommended to find a way also to obtain this information.

- The project proposal refers to "5 ecotoxicological end-points", does this include terrestrial toxicity, freshwater and marine aquatic toxicity, mammal and bird toxicity, bioaccumulation potential, LC50, NOEC values for different trophic levels (plants, algae, crustaceans, insects, fish, molluscs, bacteria etc.)?

- Several QSAR software programs are already existing, such as ECOSAR (USA) and TOPCAT-CASE (Denmark). Especially the latter one is extensive (but costly). It is recommended to find ways to use these programs as a starting point for DEMETRA, so that DEMETRA may target on the reliability of the QSARs.

- We find it of great importance that the programme has a feature that indicates the reliability of the QSAR results.

- Fancy techniques to prepare a QSAR may still give previously results. So keep it simple and transparent.

Edited by Emilio Benfenati

Appendix B

Toxicity values for five ECOTOX data sets for pesticide

Nadège Piclin[1]

[1]BioChemics Consulting SAS, Orléans, France

LIST OF THE 282 COMPOUNDS INCLUDED IN THE TROUT DATA SET

Three toxicity classes were generated according to the intervals established by the EC regulation, after regrouping the toxicity classes as too weakly represented:

LC_{50} (class 1) $< 1\,mg/l$; $1\,mg/l \leq LC_{50}$ (class 2) $< 10\,mg/l$; LC_{50} (class 3) $\geq 10\,mg/l$.

The symbol '*' in the ID column indicates the test set compounds.

ID	Structure	Name	CAS NB	Toxicity LC_{50}		
				mg/l	Class	–Log mmol/l
1	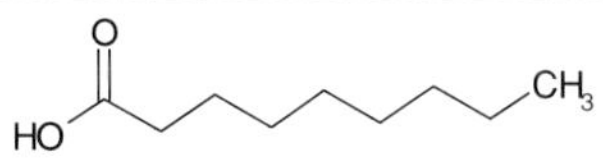	Pelargonic acid	112-05-0	91	3	0.24
2	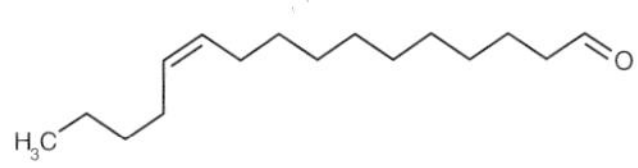	(Z)-11-Hexadecenal	53939-28-9	320	3	–0.13
3	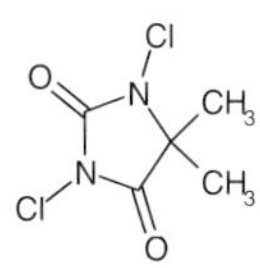	1,3-Dichloro-5,5-dimethyl-hydantoin (DCDMH)	118-52-5	0.91	1	2.34
4	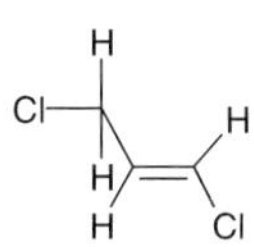	1,3-Dichloro-propene	542-75-6	3.94	2	1.45
5	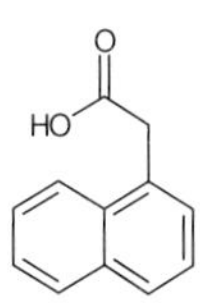	1-Naphthyl-acetic acid	86-87-3	28	3	0.82

ID	Structure	Name	CAS NB	Toxicity LC_{50}		
				mg/l	Class	–Log mmol/l
6	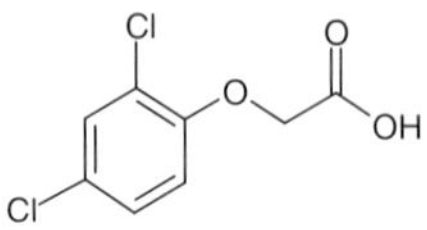	2,4-D	94-75-7	110	3	0.30
8*	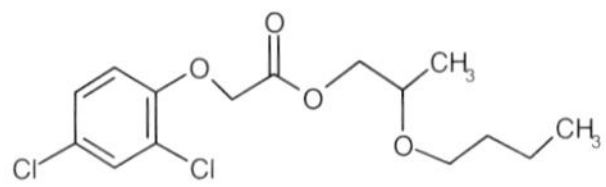	2,4-D 2-butoxymethylethyl ester	1320-18-9	0.95	1	2.55
10	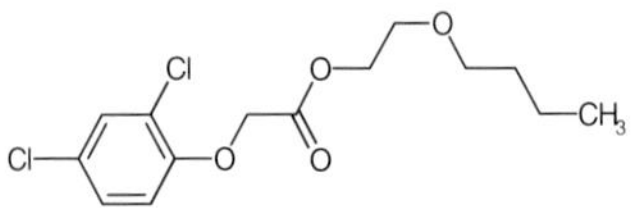	2,4-D butoxyethyl ester	1929-73-3	2.09	2	2.19
11	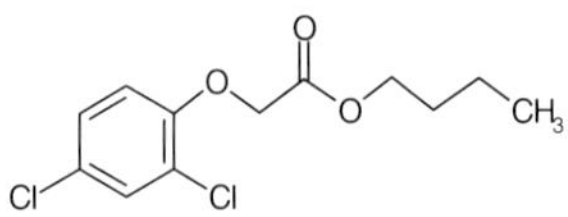	2,4-D butyl ester	94-80-4	0.4	1	2.84
12	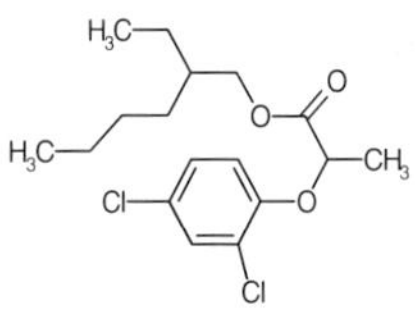	2-Ethylhexyl 2-(2,4-dichlorophenoxy) propionate	79270-78-3	7.9	2	1.64
13	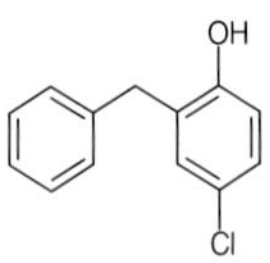	Clorophene	120-32-1	0.72	1	2.48
14	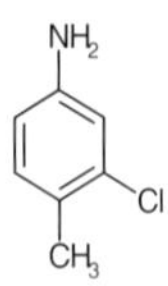	3-Chloro-*p*-toluidine	7745-89-3	7.71	2	1.26
15	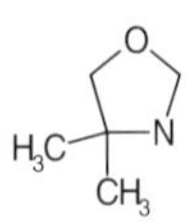	Oxadine A	51200-87-4	95.2	3	0.03
16	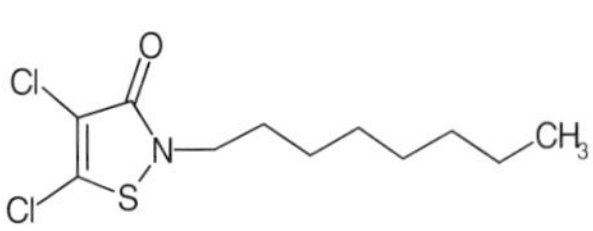	Kathon 930	64359-81-5	0.003	1	5.02

ID	Structure	Name	CAS NB	Toxicity LC_{50}		
				mg/l	Class	–Log mmol/l
17	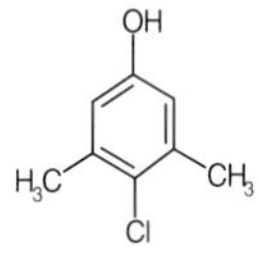	Chloroxylenol	88-04-0	0.36	1	2.64
19	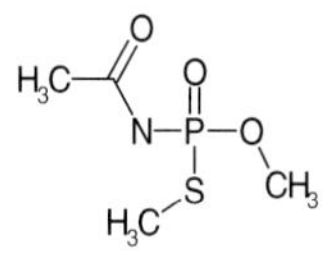	Acephate	30560-19-1	110	3	0.22
20	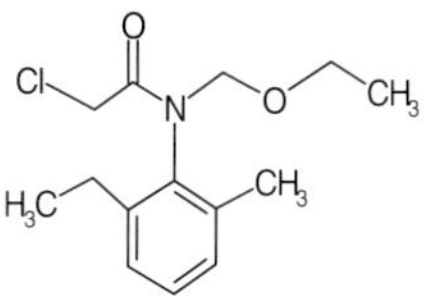	Acetochlor	34256-82-1	0.38	1	2.85
21	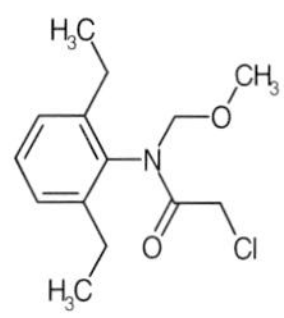	Alachlor	15972-60-8	1	2	2.43
22	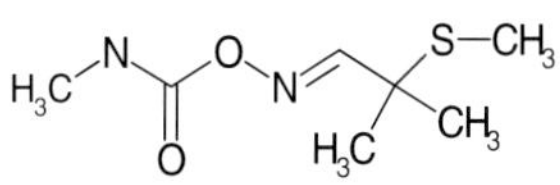	Aldicarb	116-06-3	0.56	1	2.53
23	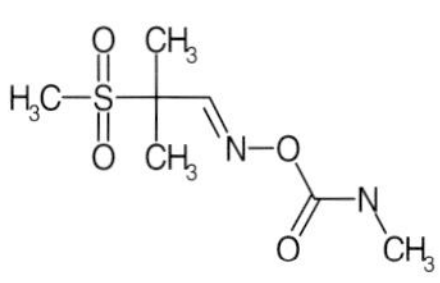	Aldicarb sulfone	1646-88-4	42	3	0.72
25	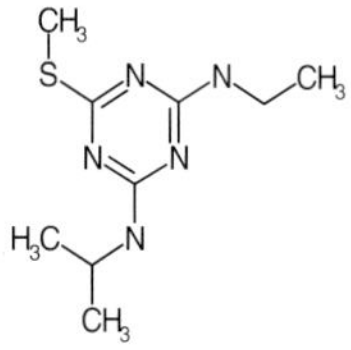	Ametryne	834-12-8	3.2	2	1.85
26	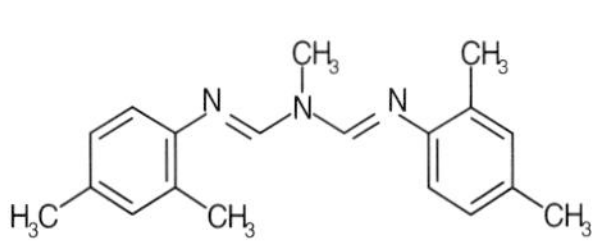	Amitraz	33089-61-1	0.74	1	2.60

ID	Structure	Name	CAS NB	Toxicity LC_{50}		
				mg/l	Class	–Log mmol/l
27	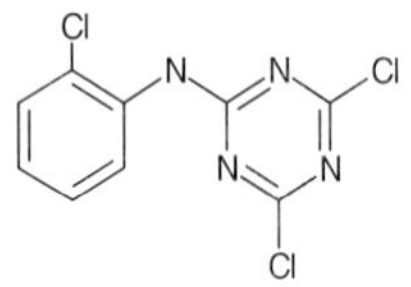	Anilazine	101-05-3	0.14	1	3.29
28	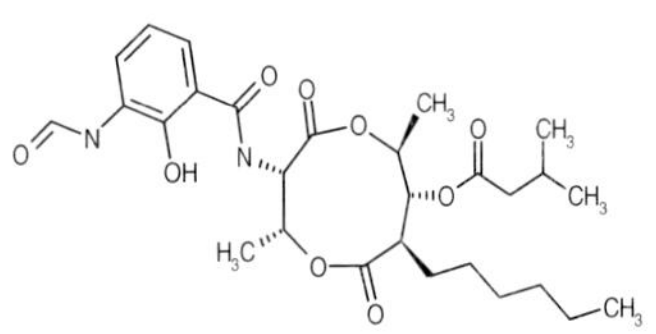	Antimycin A	1397-94-0	0.00001	1	7.74
29	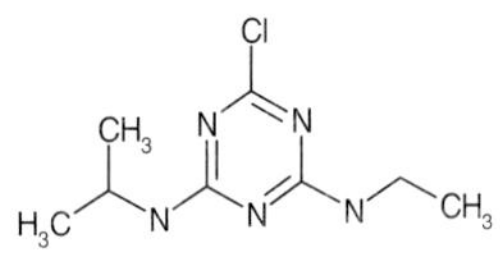	Atrazine	1912-24-9	4.5	2	1.68
30*	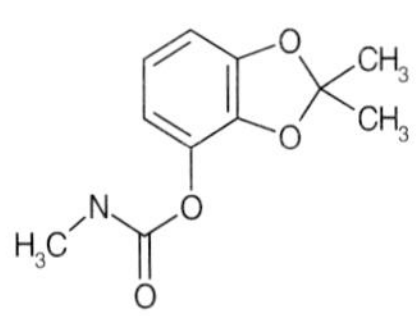	Bendiocarb	22781-23-3	1.2	2	2.27
31	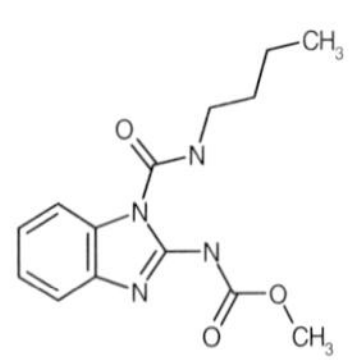	Benomyl	17804-35-2	0.17	1	3.23
32	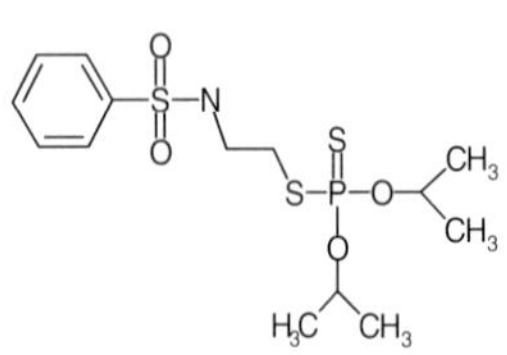	Bensulide	741-58-2	0.72	1	2.74

ID	Structure	Name	CAS NB	Toxicity LC_{50}		
				mg/l	Class	–Log mmol/l
33	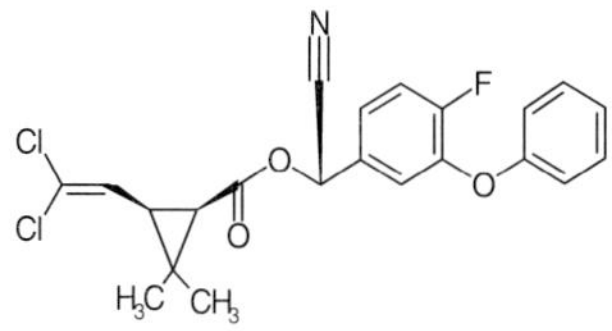	Beta-cyfluthrin	68359-37-5	0.0007	1	5.81
34	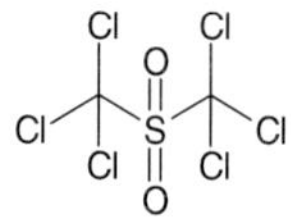	Bis(trichloro-methyl)sulfone	3064-70-8	0.029	1	4.02
35*	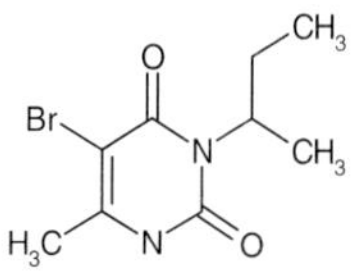	Bromacil	314-40-9	36	3	0.86
36	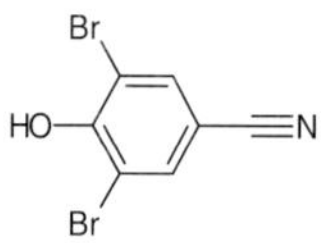	Bromoxynil	1689-84-5	2.09	2	2.12
37*	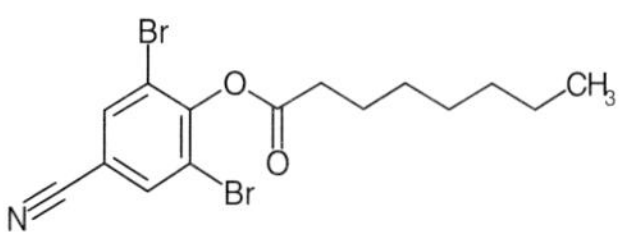	Bromoxynil octanoate	1689-99-2	0.1	1	3.59
38	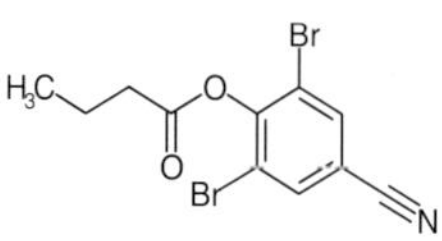	Bromoxynil butyrate	3861-41-4	0.029	1	4.08
39	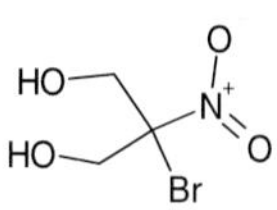	Bronopol	52-51-7	42	3	0.68

ID	Structure	Name	CAS NB	Toxicity LC_{50}		
				mg/l	Class	–Log mmol/l
40	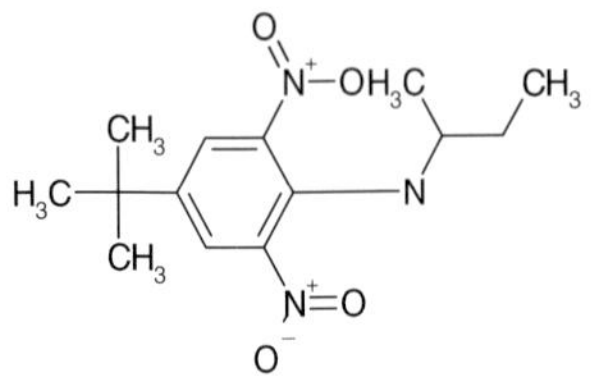	Butralin	33629-47-9	0.37	1	2.90
43	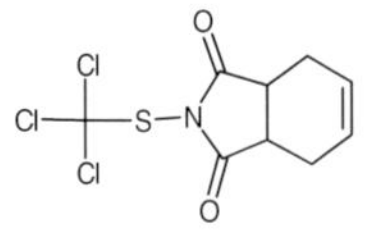	Captan	133-06-2	0.073	1	3.61
44	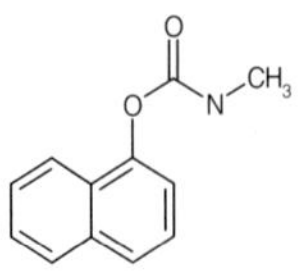	Carbaril	63-25-2	1.2	2	2.22
46	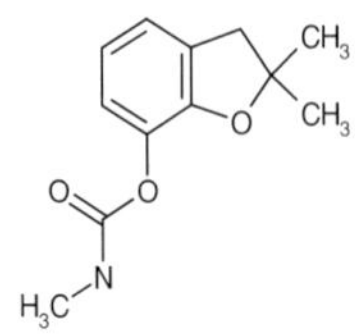	Carbofuran	1563-66-2	0.38	1	2.77
47	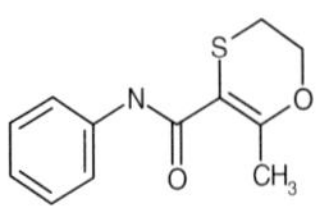	Carboxin	5234-68-4	2	2	2.07
48	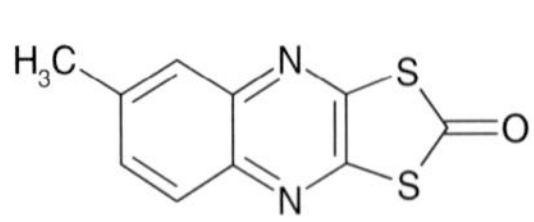	Chinomethio-nat	2439-01-2	0.128	1	3.26
49	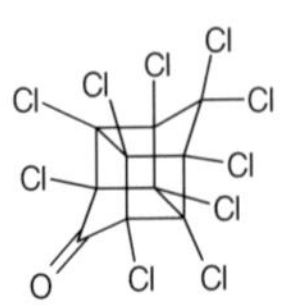	Chlordecone	143-50-0	0.029	1	4.23

ID	Structure	Name	CAS NB	Toxicity LC_{50}		
				mg/l	Class	–Log mmol/l
50	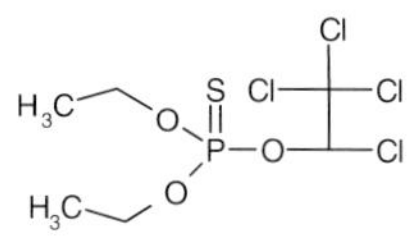	Chloroetho-xyfos	54593-83-8	0.089	1	3.58
51	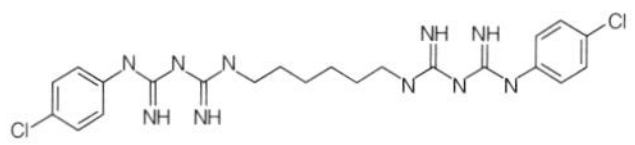	Chlorhexidine	55-56-1	1.512	2	2.52
52	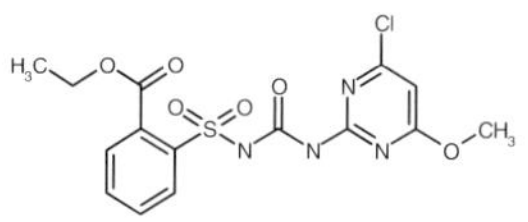	Chlorimuron ethyl	90982-32-4	8.4	2	1.69
53	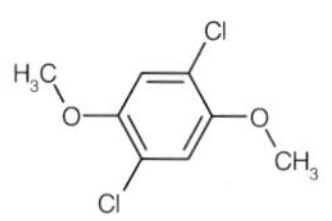	Chloroneb	2675-77-6	3.7	2	1.75
54*	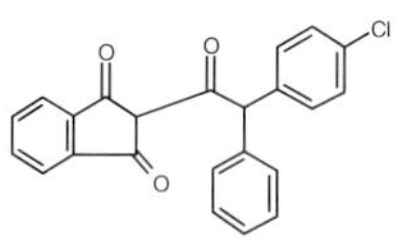	Chlorophaci-none	3691-35-8	0.252	1	3.17
55	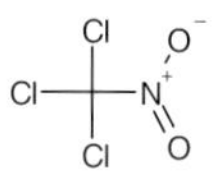	Chloropicrin	76-06-2	0.017	1	4.00
57	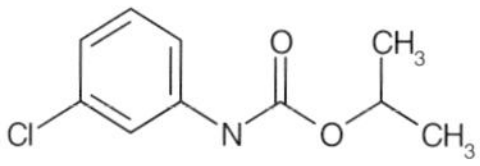	Chlorpropham	101-21-3	5.7	2	1.57
58	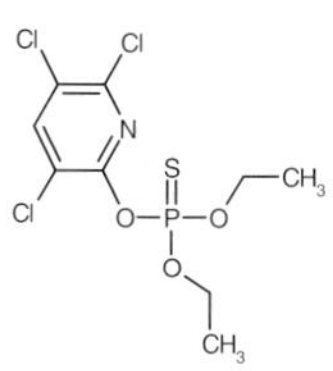	Chlorpyrifos	2921-88-2	0.027	1	4.11

ID	Structure	Name	CAS NB	Toxicity LC_{50}		
				mg/l	Class	–Log mmol/l
59	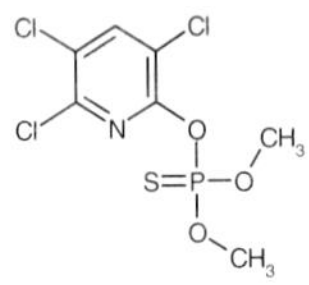	Chlorpyrifos-methyl	5598-13-0	0.013	1	4.41
60*	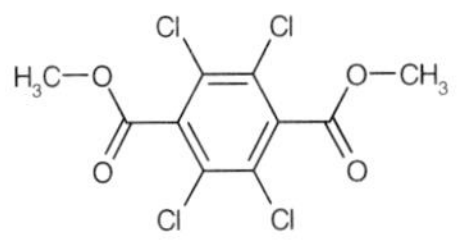	Chlorthal-dimethyl	1861-32-1	6.6	2	1.70
62	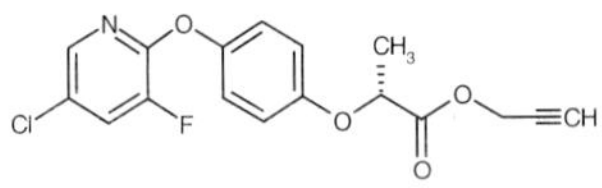	Clodinafop-propargyl	105512-06-9	0.3	1	3.07
63	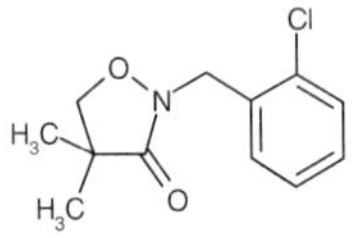	Clomazone	81777-89-1	19	3	1.10
67	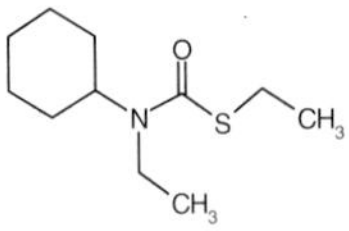	Cycloate	1134-23-2	4.5	2	1.68
68	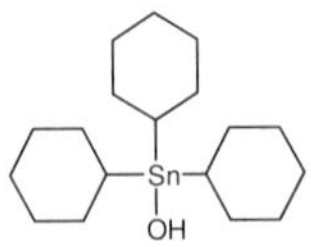	Cyhexatin	13121-70-5	0.0013	1	5.48
69	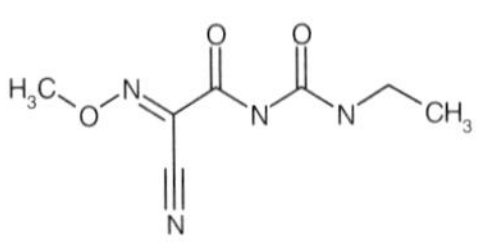	Cymoxanil	57966-95-7	61	3	0.51

ID	Structure	Name	CAS NB	Toxicity LC_{50}		
				mg/l	Class	–Log mmol/l
72	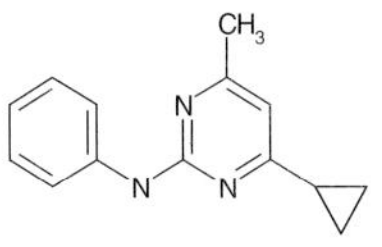	Cyprodinil	121552-61-2	2.41	2	1.97
73	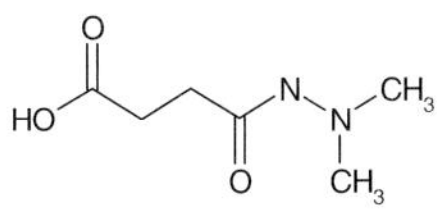	Daminozide	1596-84-5	149.3	3	0.03
75	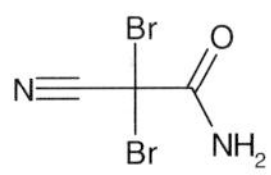	DBNPA	10222-01-2	1	2	2.38
76	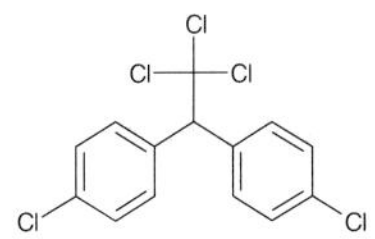	DDT	50-29-3	0.004	1	4.94
77*	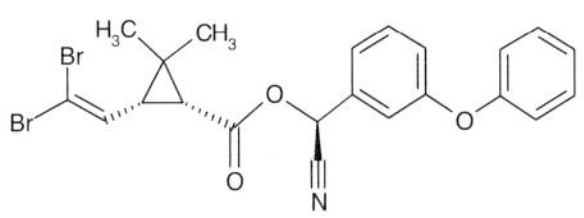	Deltamethrin	52918-63-5	0.001	1	5.74
81	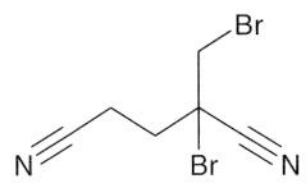	1,2-Dibromo-2,4-dicyano-butane	35691-65-7	1.75	2	2.18
83	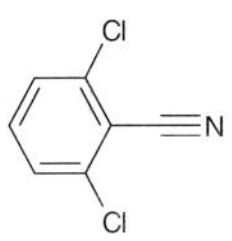	Dichlobenil	1194-65-6	4.93	2	1.54
84	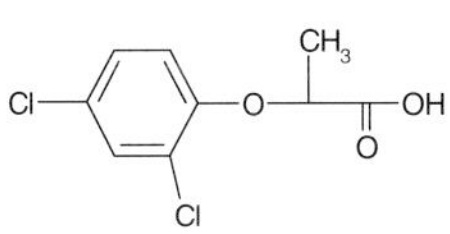	Dichlorprop	120-36-5	0.5	1	2.67

ID	Structure	Name	CAS NB	Toxicity LC_{50}		
				mg/l	Class	–Log mmol/l
85	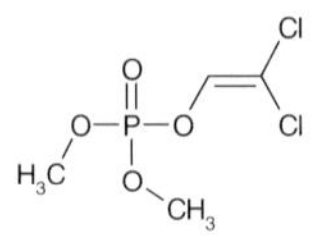	Dichlorvos	62-73-7	0.1	1	3.34
87	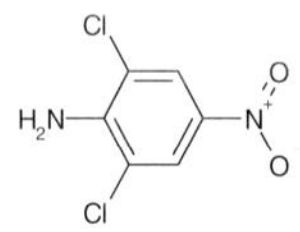	Dicloran	99-30-9	0.56	1	2.57
88*	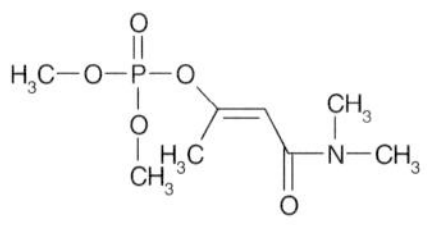	Dicrotophos	141-66-2	6.3	2	1.58
89*	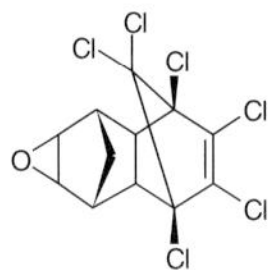	Dieldrin	60-57-1	0.001	1	5.50
90*	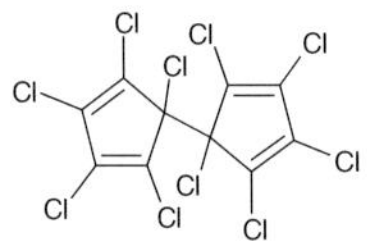	Dienochlor	2227-17-0	0.05	1	3.98
91	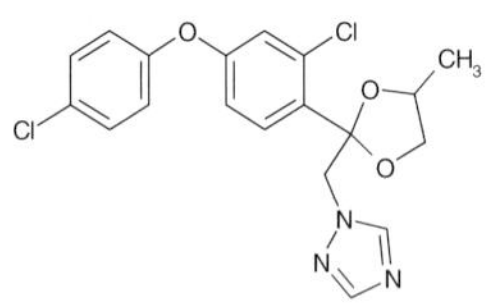	Difenocona-zole	119446-68-3	0.81	1	2.70
92	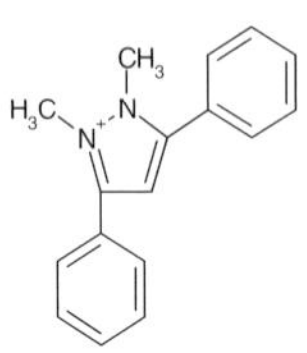	Difenzoquat	49866-87-7	491.9	3	–0.30

ID	Structure	Name	CAS NB	Toxicity LC_{50}		
				mg/l	Class	–Log mmol/l
93	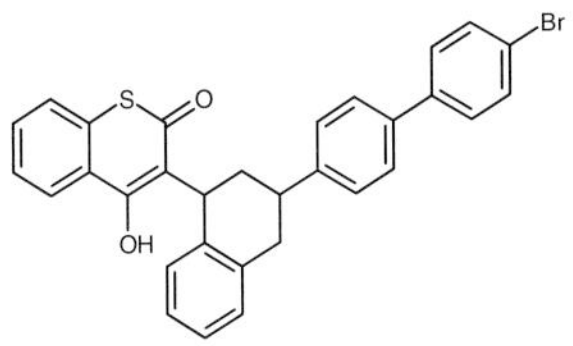	Difethialone	104653-34-1	0.051	1	4.02
94	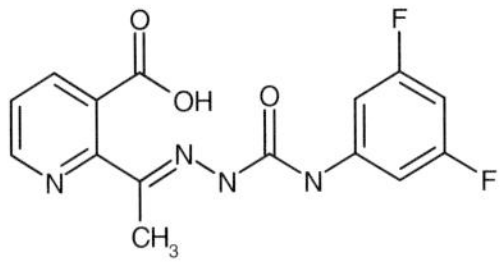	Diflufenzopyr	109293-97-2	99.47	3	0.53
95	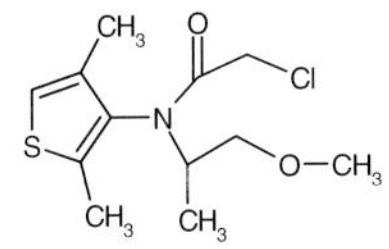	Dimethenamid	87674-68-8	2.6	2	2.03
96	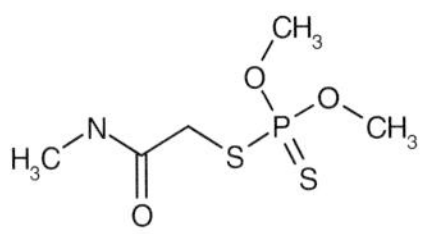	Dimethoate	60-51-5	6.2	2	1.57
98*	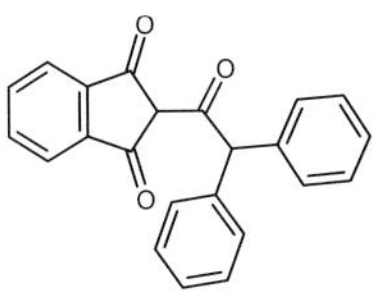	Diphacinone	82-66-6	2.82	2	2.08
101	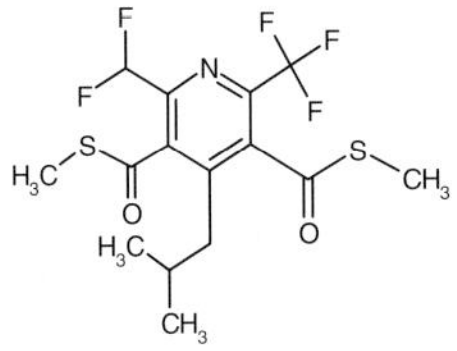	Dithiopyr	97886-45-8	0.46	1	2.94

ID	Structure	Name	CAS NB	Toxicity LC_{50}		
				mg/l	Class	–Log mmol/l
102	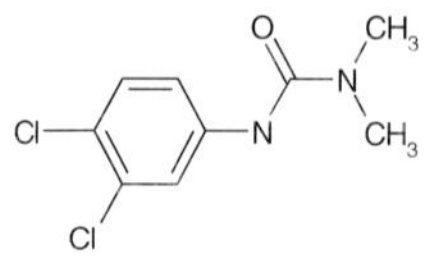	Diuron	330-54-1	1.95	2	2.08
103	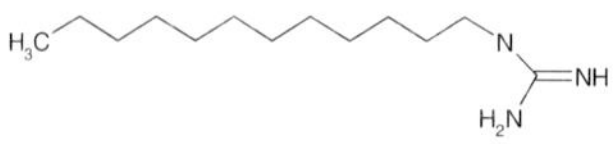	Dodine	2439-10-3	451.04	3	–0.30
104	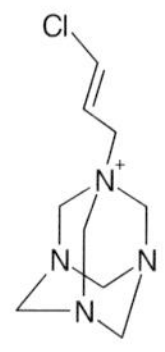	Dowicil	4080-31-3	17.61	3	1.09
107	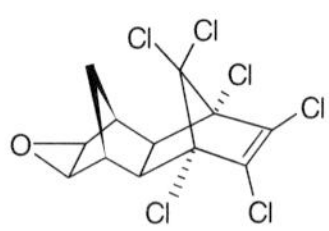	Endrin	72-20-8	0.0007	1	5.71
108	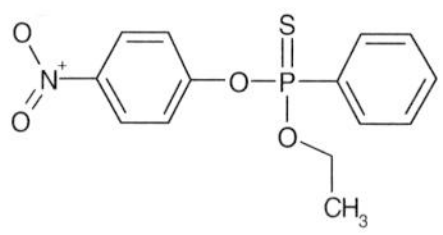	EPN	2104-64-5	0.21	1	3.19
109	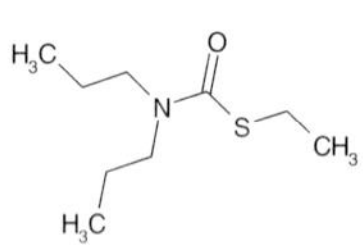	EPTC	759-94-4	19.96	3	0.98
110	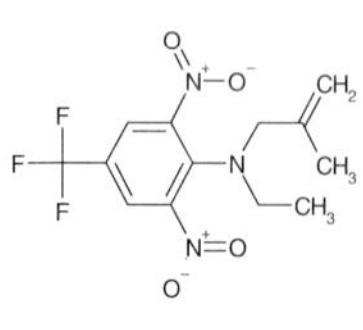	Ethalfluralin	55283-68-6	0.037	1	3.95

ID	Structure	Name	CAS NB	Toxicity LC_{50}		
				mg/l	Class	–Log mmol/l
111	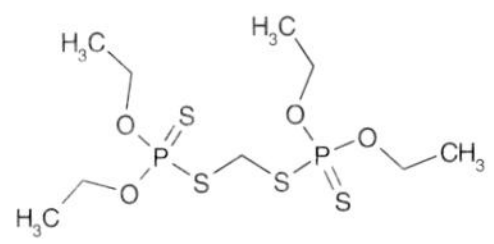	Ethion	563-12-2	0.5	1	2.89
115*		Ethylene dichloride	107-06-2	225	3	–0.36
116	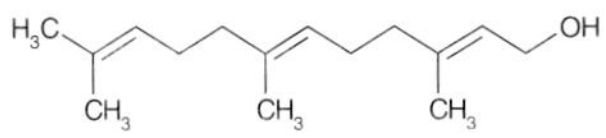	Farnesol	4602-84-0	1.8	2	2.09
119	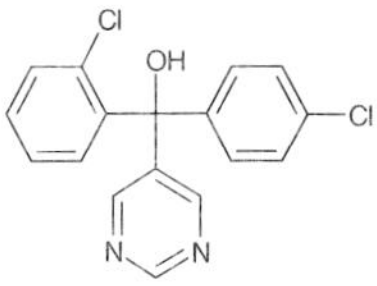	Fenarimol	60168-88-9	2.1	2	2.20
120	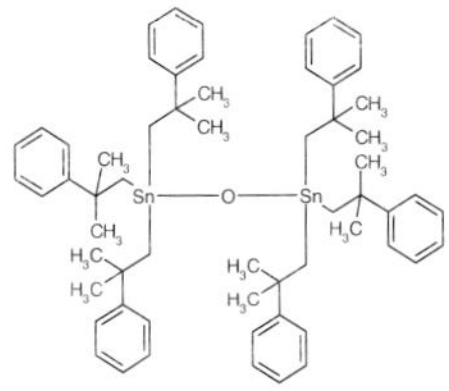	Fenbutatin oxide	13356-08-6	0.002	1	5.79
122*	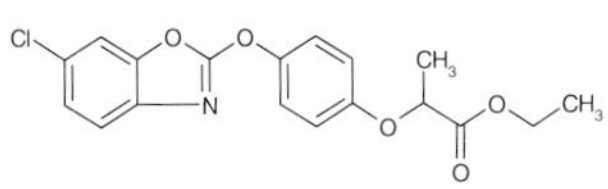	Fenoxaprop-ethyl	66441-23-4	0.46	1	2.90
123	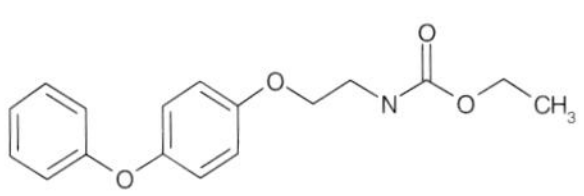	Fenoxycarb	79127-80-3	1.6	2	2.27

ID	Structure	Name	CAS NB	Toxicity LC_{50}		
				mg/l	Class	–Log mmol/l
124	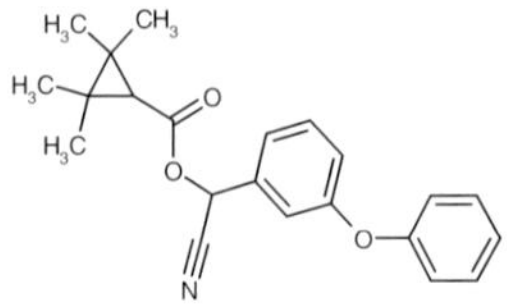	Fenpropathrin	39515-41-8	0.002	1	5.18
125*	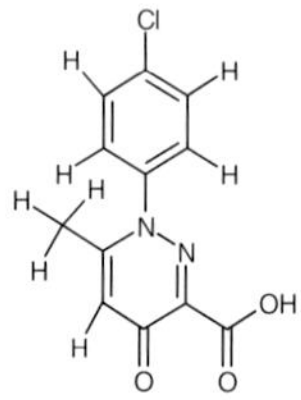	Fenridazone	68254-10-4	246.07	3	0.03
126*	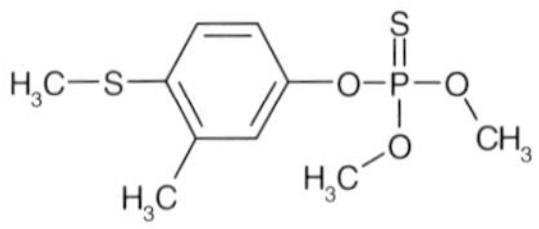	Fenthion	55-38-9	0.55	1	2.70
127	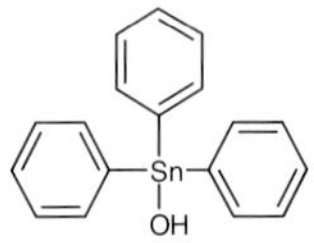	Fentin hydroxide	76-87-9	0.022	1	4.22
131		Fludioxonil	131341-86-1	0.47	1	2.72
133*	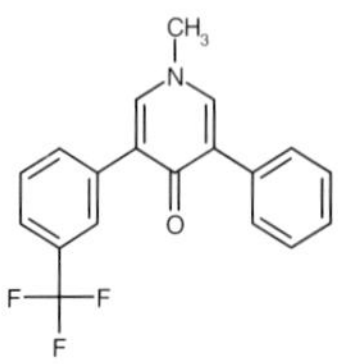	Fluridone	59756-60-4	4.25	2	1.89

ID	Structure	Name	CAS NB	Toxicity LC_{50} mg/l	Toxicity LC_{50} Class	Toxicity LC_{50} –Log mmol/l
134	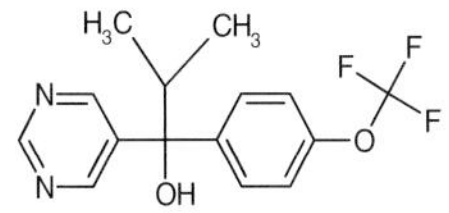	Flurprimidol	56425-91-3	18.6	3	1.23
135*	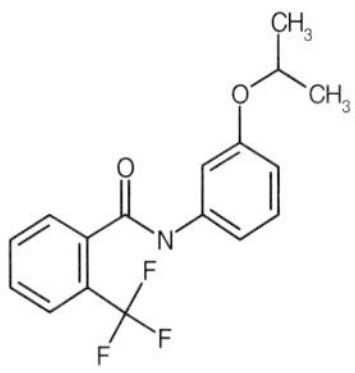	Flutolanil	66332-96-5	5.4	2	1.78
136	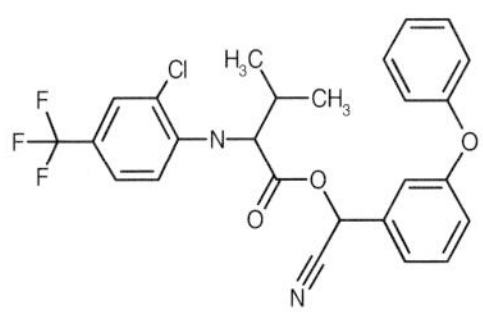	Fluvalinate	69409-94-5	0.003	1	5.24
137	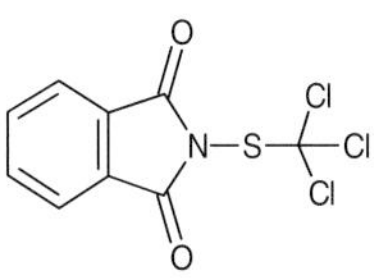	Folpet	133-07-3	0.015	1	4.30
138	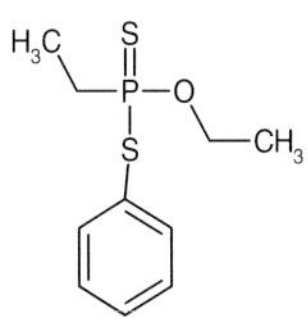	Fonofos	944-22-9	0.02	1	4.09
139	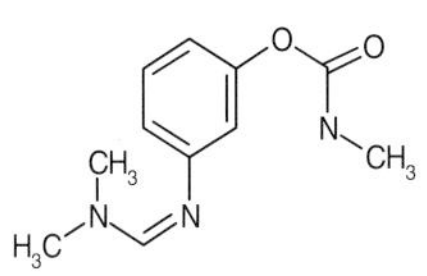	Formetanate	22259-30-9	3.78	2	1.77

ID	Structure	Name	CAS NB	Toxicity LC_{50}		
				mg/l	Class	–Log mmol/l
140	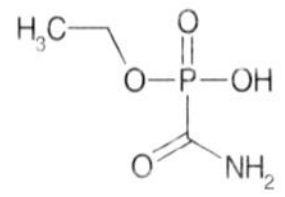	Fosamine	59682-52-9	339.3	3	–0.35
142	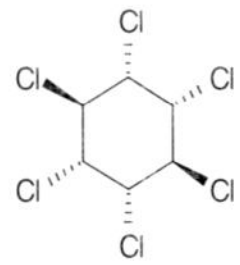	Gamma-HCH	58-89-9	0.018	1	4.21
143	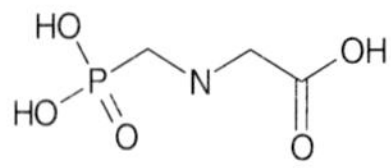	Glyphosate	1071-83-6	134	3	0.10
144	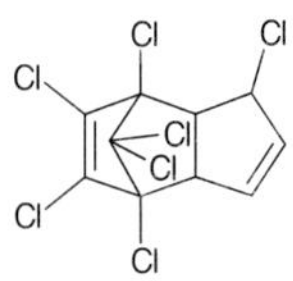	Heptachlor	76-44-8	0.007	1	4.73
145*	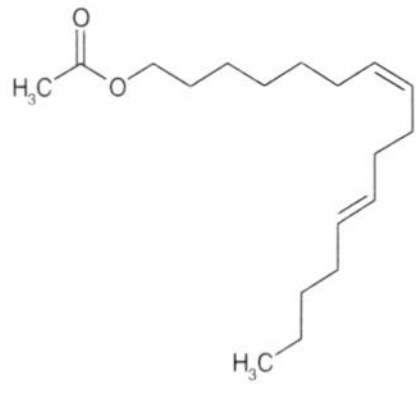	Hexadeca-dienol acetate	53042-79-8	94.9	3	0.47
146	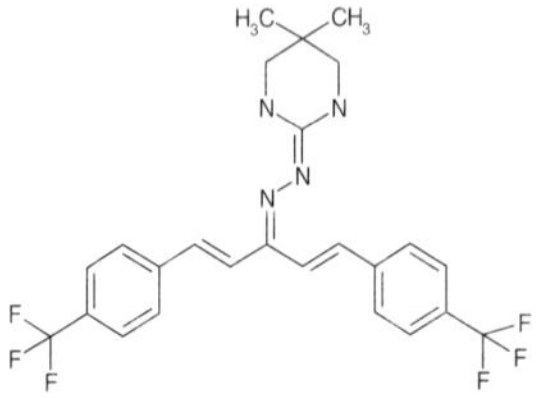	Hydra-methylnon	67485-29-4	0.15	1	3.52

ID	Structure	Name	CAS NB	Toxicity LC_{50}		
				mg/l	Class	–Log mmol/l
147	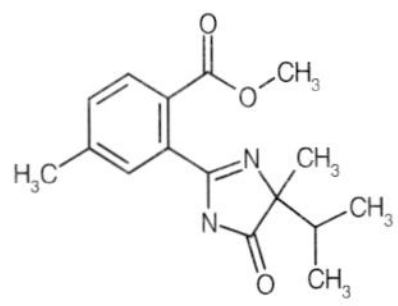	Imazethabenz	81405-85-8	280	3	0.01
148	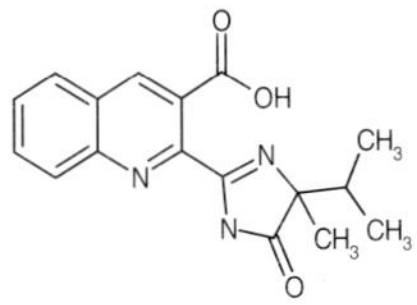	Imazaquin	81335-37-7	280	3	0.05
149	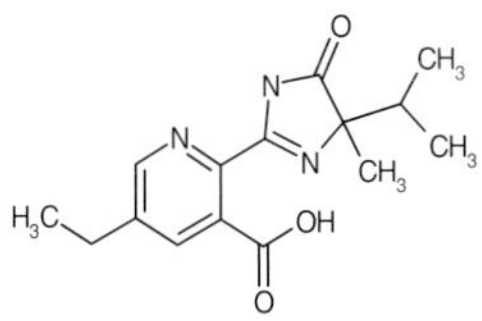	Imazethapyr	81335-77-5	344	3	–0.08
150	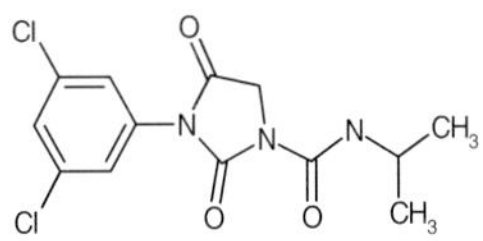	Iprodione	36734-19-7	4.2	2	1.90
151	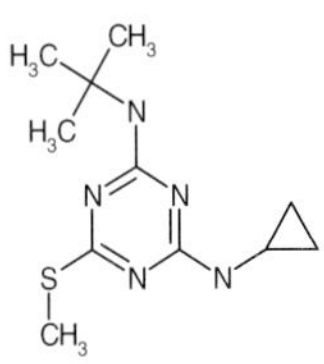	Irgarol	28159-98-0	0.75	1	2.53
152	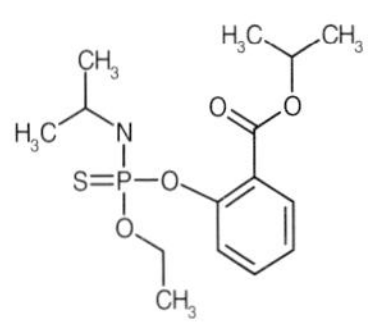	Isofenphos	25311-71-1	1.8	2	2.28

ID	Structure	Name	CAS NB	Toxicity LC_{50}		
				mg/l	Class	–Log mmol/l
153	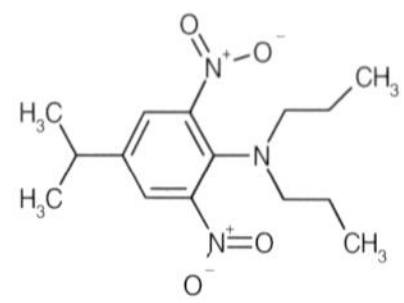	Isopropalin	33820-53-0	0.80	1	2.59
155	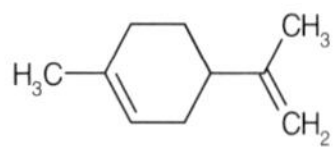	Limonene	138-86-3	80	3	0.23
156	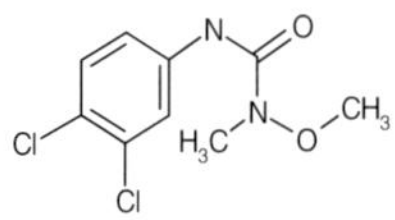	Linuron	330-55-2	3	2	1.92
157	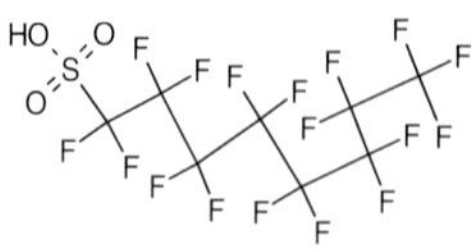	Perfluorooc-tane sulfonate	29457-72-5	4.15	2	2.08
158*	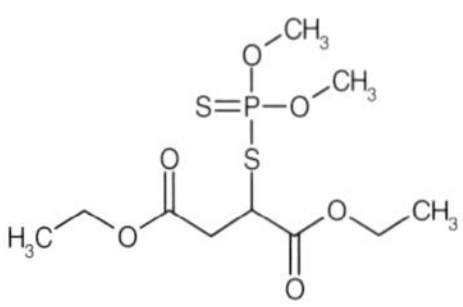	Malathion	121-75-5	0.004	1	4.92
160	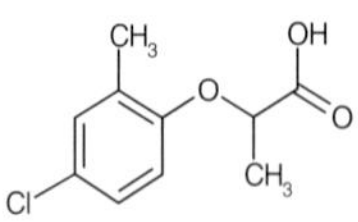	Mecoprop	7085-19-0	124.8	3	0.24
161*	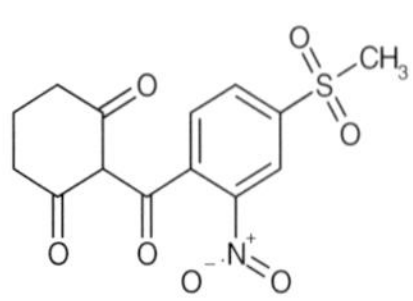	Mesotrione (AMBA)	104206-82-8	150	3	0.35

ID	Structure	Name	CAS NB	Toxicity LC_{50}		
				mg/l	Class	–Log mmol/l
162	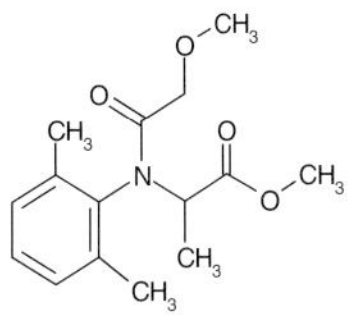	Metalaxyl	57837-19-1	130	3	0.33
163	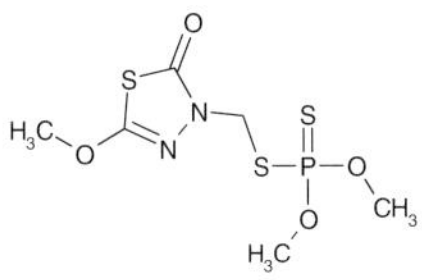	Methidathion	950-37-8	0.014	1	4.33
165	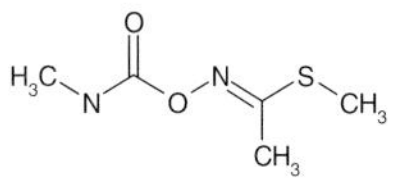	Methomyl	16752-77-5	0.86	1	2.28
167	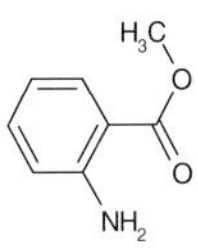	Methyl anthralinate	134-20-3	25.4	3	0.77
168	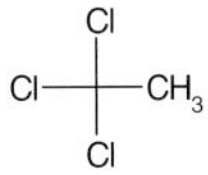	Methyl chloroform	71-55-6	52	3	0.41
169	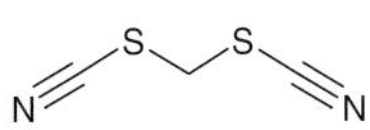	Methylene bis(thio-cyanate)	6317-18-6	0.089	1	3.17
171	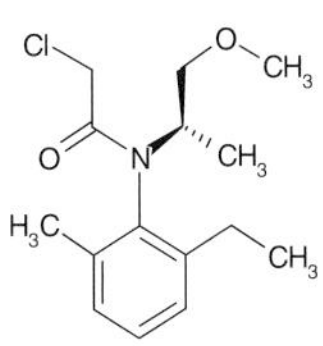	Metolachlor-*S*-isomer	87392-12-9	3.9	2	1.86

ID	Structure	Name	CAS NB	Toxicity LC_{50}		
				mg/l	Class	–Log mmol/l
172	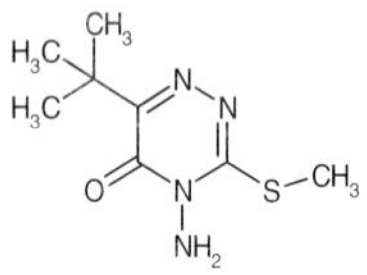	Metribuzin	21087-64-9	42	3	0.71
174	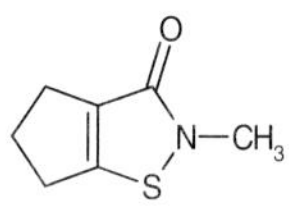	MTI	82633-79-2	0.89	1	2.24
176	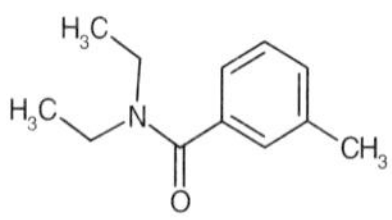	Diethyltolu-amide	134-62-3	71.25	3	0.43
177	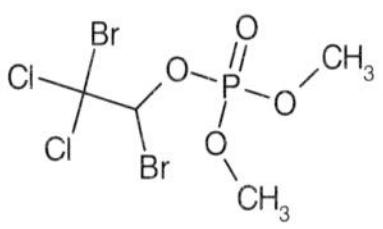	Naled	300-76-5	0.16	1	3.38
178	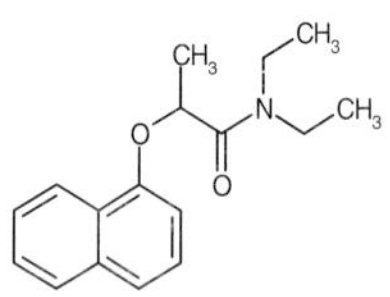	Napropamide	15299-99-7	9.4	2	1.46
179	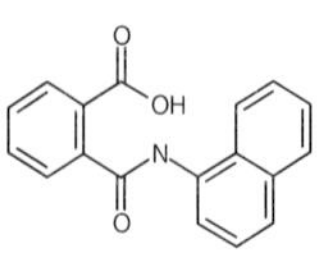	Naptalam	132-66-1	70.77	3	0.61
180	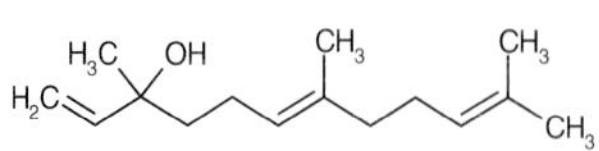	Nerolidol	7212-44-4	1.8	2	2.09

ID	Structure	Name	CAS NB	Toxicity LC_{50}		
				mg/l	Class	–Log mmol/l
182	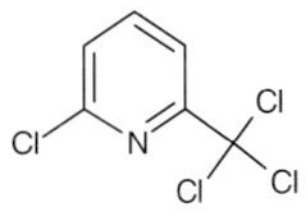	Nitrapyrin	1929-82-4	6.5	2	1.55
183	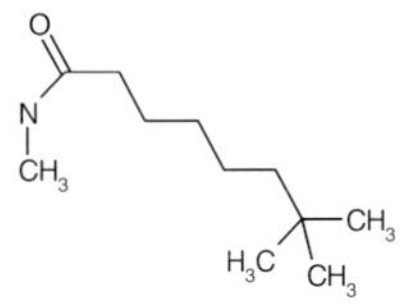	*N*-methylneo decanamide	105726-67-8	72	3	0.41
184	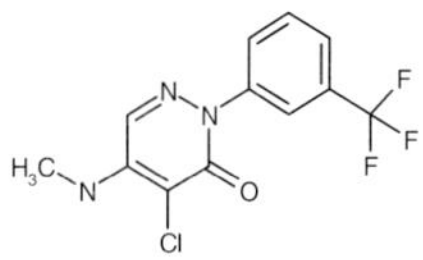	Norflurazon	27314-13-2	8.1	2	1.57
185	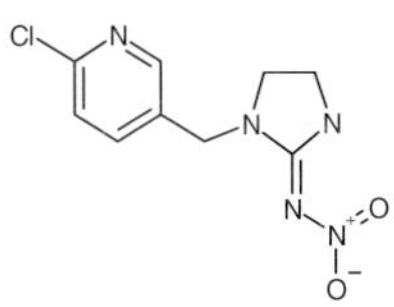	Imidacloprid	105827-78-9	229.1	3	0.05
186	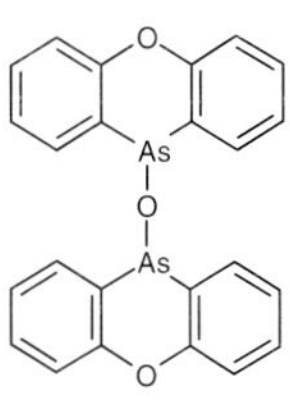	OBPA	58-36-6	0.0035	1	5.16
187	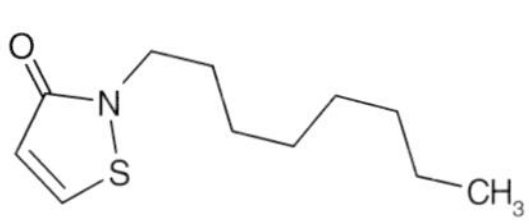	Octhilinone	26530-20-1	0.047	1	3.66

ID	Structure	Name	CAS NB	Toxicity LC_{50}		
				mg/l	Class	–Log mmol/l
188	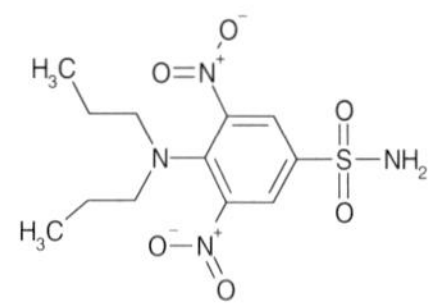	Oryzalin	19044-88-3	3.26	2	2.03
189	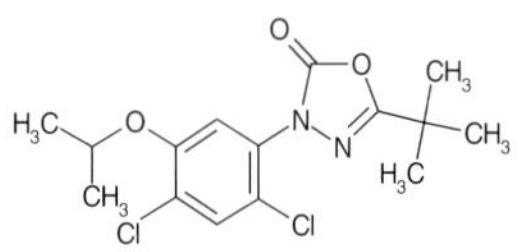	Oxadiazon	19666-30-9	1.2	2	2.46
190	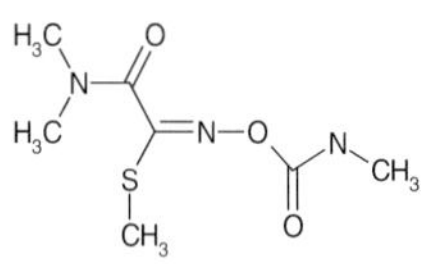	Oxamyl	23135-22-0	4.2	2	1.72
191	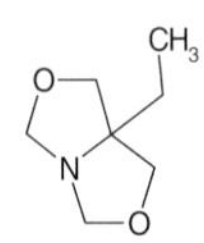	Oxazolidine E	7747-35-5	221	3	–0.19
192	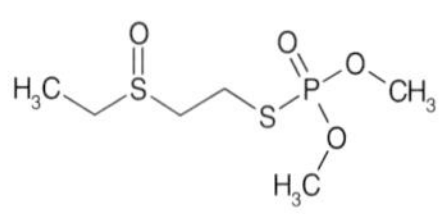	Oxydemeton-methyl	301-12-2	0.73	1	2.53
193	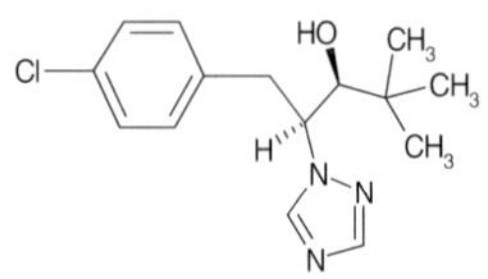	Paclobutrazol	76738-62-0	27.8	3	1.02
194		Paradichloro benzene	106-46-7	0.88	1	2.22

ID	Structure	Name	CAS NB	Toxicity LC_{50}		
				mg/l	Class	–Log mmol/l
195	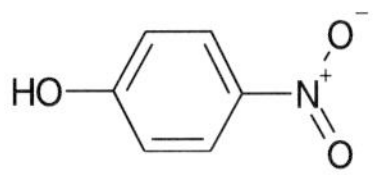	Paranitro-phenol	100-02-7	4.5	2	1.49
196	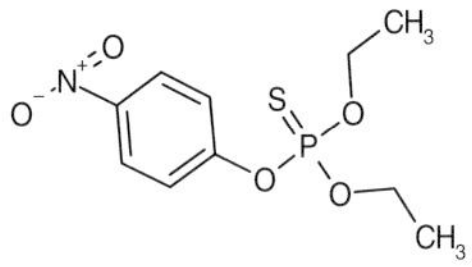	Parathion	56-38-2	0.78	1	2.57
197	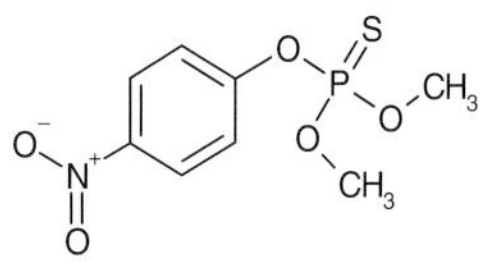	Parathion-methyl	298-00-0	2.75	2	1.98
198	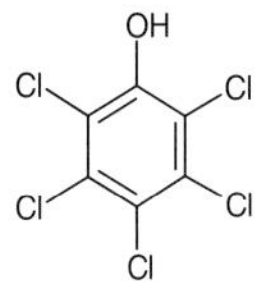	PCP	87-86-5	0.034	1	3.89
199	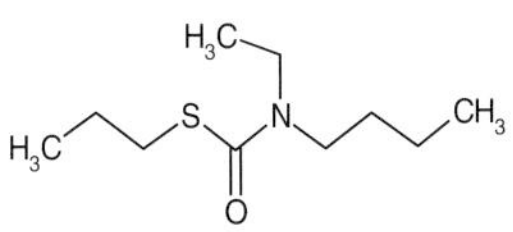	Pebulate	1114-71-2	7.4	2	1.44
200	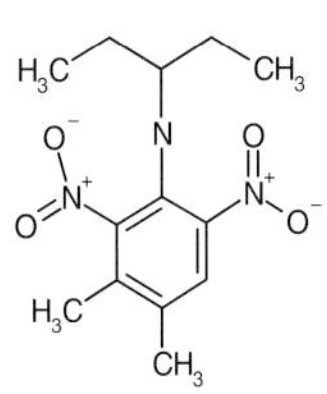	Pendimethalin	40487-42-1	0.138	1	3.31
202	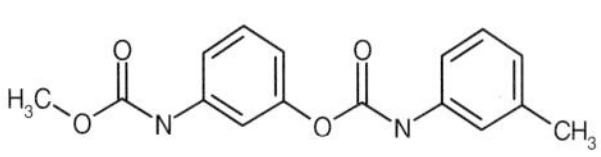	Phenmedipham	13684-63-4	1.41	2	2.33

ID	Structure	Name	CAS NB	Toxicity LC_{50}		
				mg/l	Class	–Log mmol/l
204	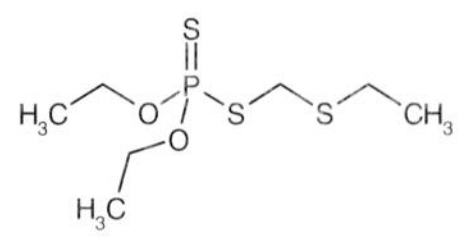	Phorate	298-02-2	0.013	1	4.30
205	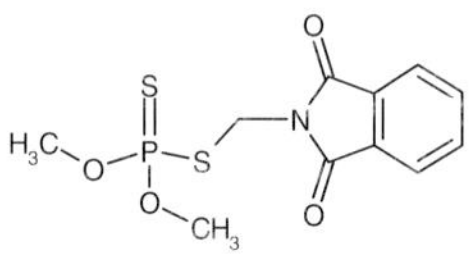	Phosmet	732-11-6	0.105	1	3.48
208	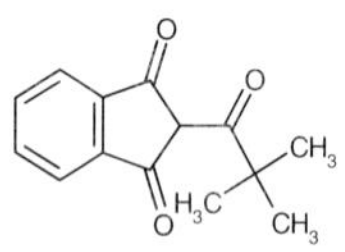	Pindone	83-26-1	0.21	1	3.04
209	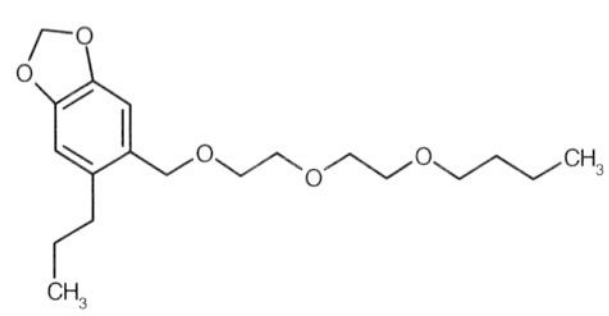	Piperonyl butoxide	51-03-6	1.8	2	2.27
212	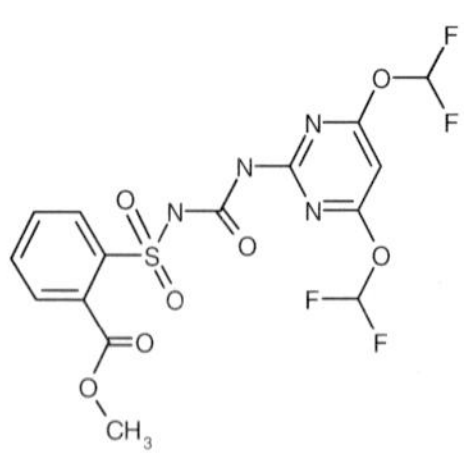	Primisulfuron-methyl	86209-51-0	210	3	0.35
213	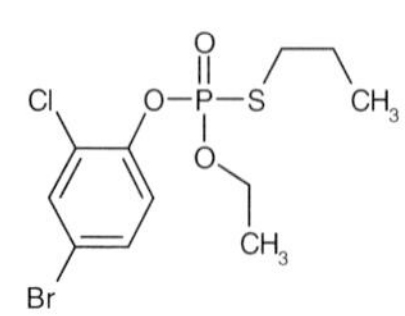	Profenofos	41198-08-7	0.021	1	4.25

ID	Structure	Name	CAS NB	Toxicity LC_{50} mg/l	Class	–Log mmol/l
214*	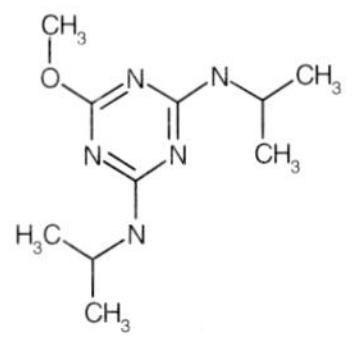	Prometon	1610-18-0	19.6	3	1.06
215	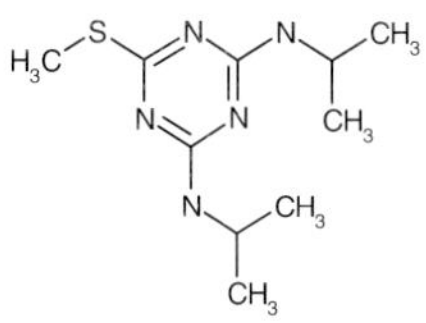	Prometryn	7287-19-6	2.9	2	1.92
216	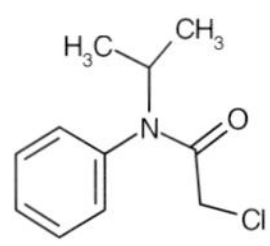	Propachlor	1918-16-7	0.17	1	3.10
217*	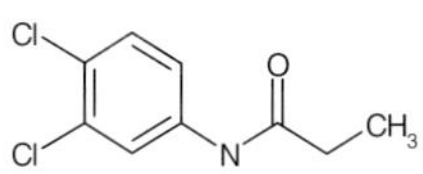	Propanil	709-98-8	2.3	2	1.98
218*	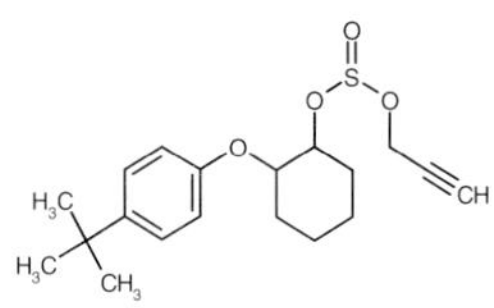	Propargite	2312-35-8	0.118	1	3.47
219	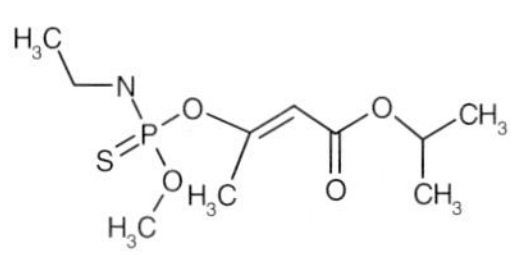	Propetamphos	31218-83-4	2.6	2	2.03

ID	Structure	Name	CAS NB	Toxicity LC_{50}		
				mg/l	Class	–Log mmol/l
223	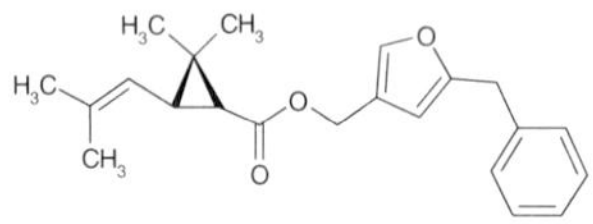	Resmethrin	10453-86-8	0.0008	1	5.62
224	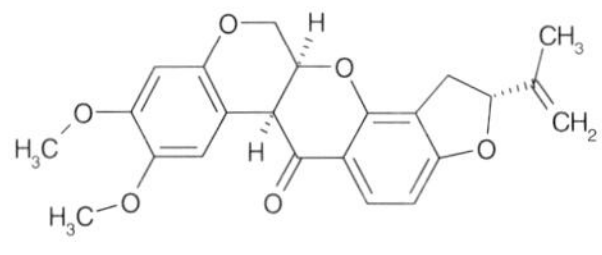	Rotenone	83-79-4	0.002	1	5.32
225*	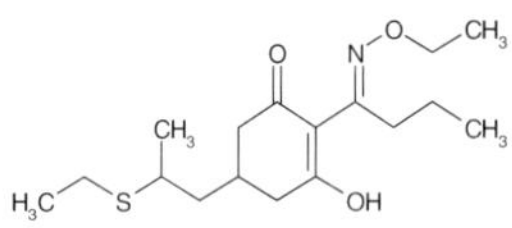	Sethoxydim	74051-80-2	170	3	0.28
226	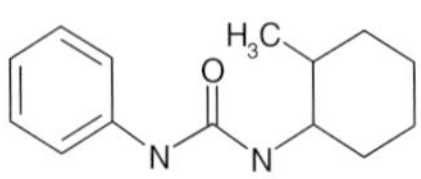	Siduron	1982-49-6	8.1	2	1.46
227	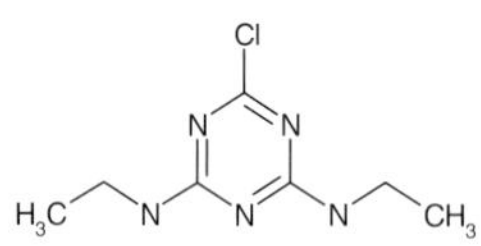	Simazine	122-34-9	70.5	3	0.46
228	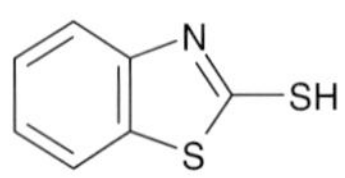	2-Mercapto benzothiazole	149-30-4	0.645	1	2.41
230	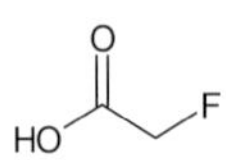	Fluoroacetic acid	144-49-0	42.14	3	0.27

ID	Structure	Name	CAS NB	Toxicity LC_{50}		
				mg/l	Class	–Log mmol/l
231	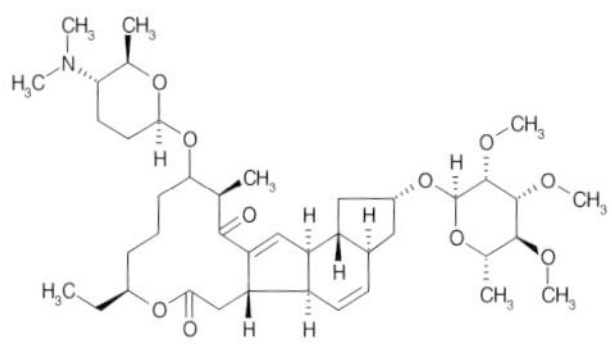	Spinosad	131929-60-7	30	3	1.39
232*	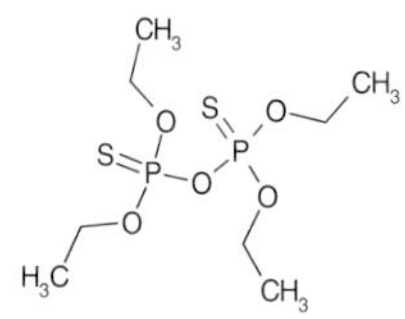	Sulfotep	3689-24-5	1	2	2.51
233	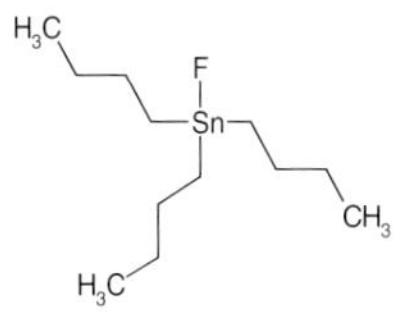	TBTF	1983-10-4	0.003	1	4.96
234	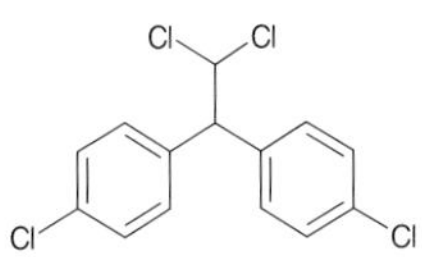	TDE	72-54-8	0.07	1	3.66
236	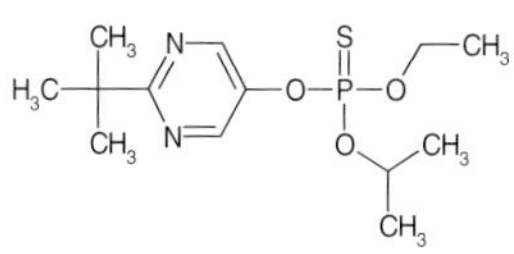	Tebupirimfos	96182-53-5	2.22	2	2.16
237*	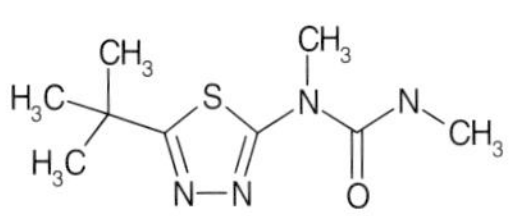	Tebuthiuron	34014-18-1	143	3	0.20

ID	Structure	Name	CAS NB	Toxicity LC_{50}		
				mg/l	Class	–Log mmol/l
238	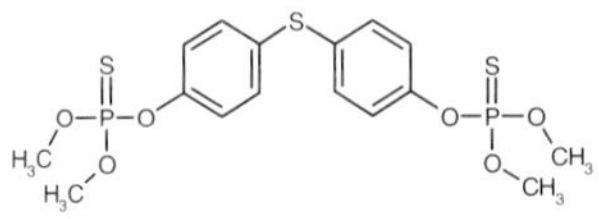	Temephos	3383-96-8	3.49	2	2.13
239*	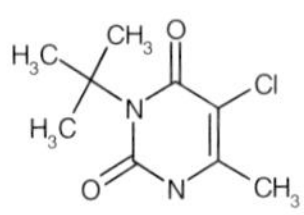	Terbacil	5902-51-2	46.2	3	0.67
240	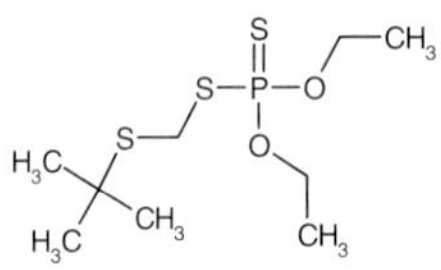	Terbufos	13071-79-9	0.008	1	4.58
241	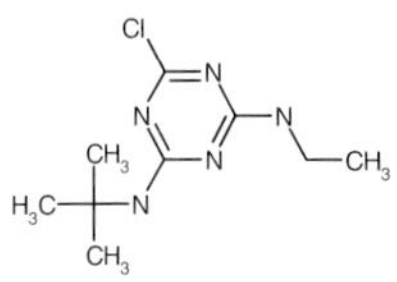	Terbuthylazine	5915-41-3	3.4	2	1.83
244	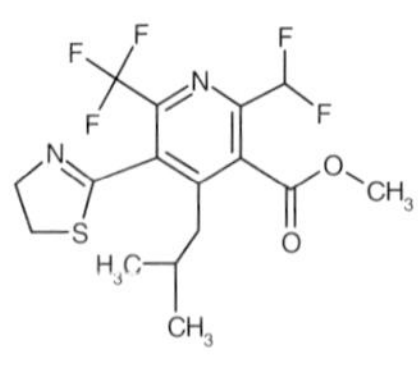	Thiazopyr	117718-60-2	3.4	2	2.07
245	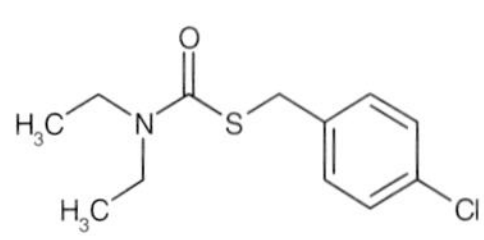	Thiobencarb	28249-77-6	1.05	2	2.39

ID	Structure	Name	CAS NB	Toxicity LC_{50}		
				mg/l	Class	–Log mmol/l
247	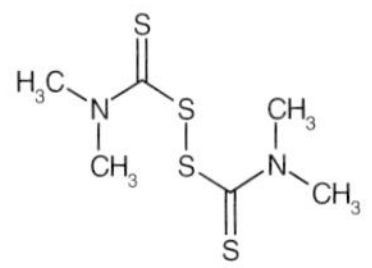	Thiram	137-26-8	0.13	1	3.27
248	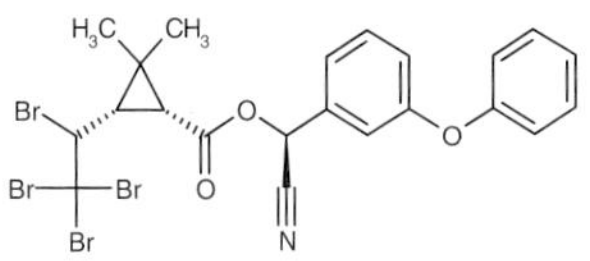	Tralomethrin	66841-25-6	0.002	1	5.62
249	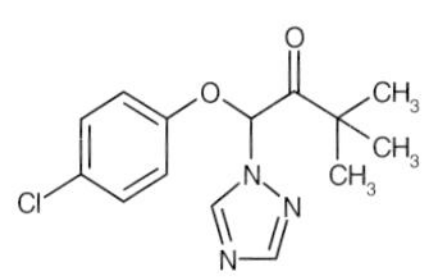	Triadimefon	43121-43-3	4.1	2	1.86
250*	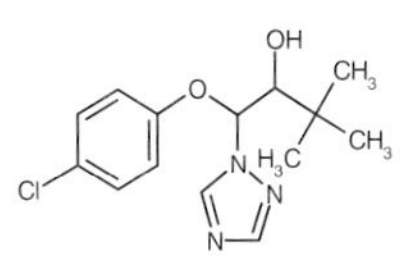	Triadimenol	55219-65-3	14	3	1.32
251	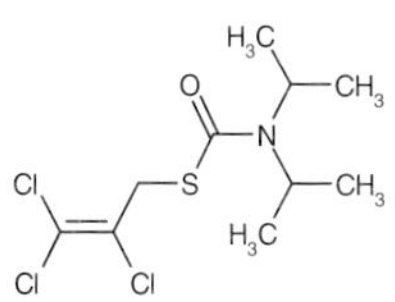	Triallate	2303-17-5	1.2	2	2.40
252	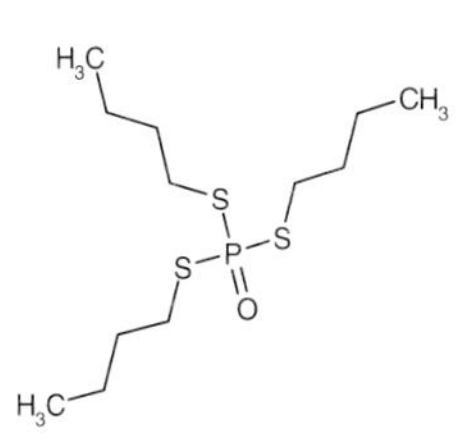	Tribufos	78-48-8	0.66	1	2.68

ID	Structure	Name	CAS NB	Toxicity LC_{50}		
				mg/l	Class	–Log mmol/l
254	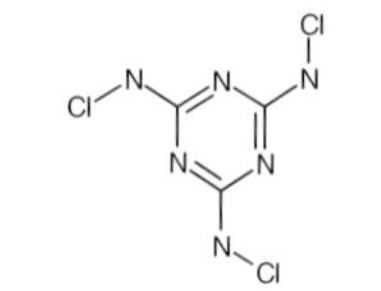	Trichloro melamine	7673-09-8	2	2	2.06
256	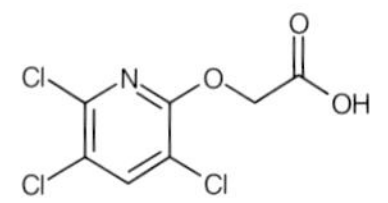	Triclopyr	55335-06-3	117	3	0.34
257	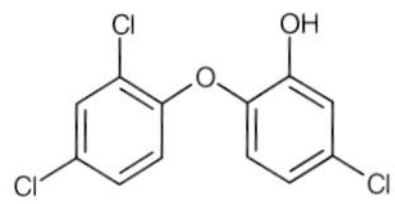	Triclosan	3380-34-5	0.288	1	3.00
258*	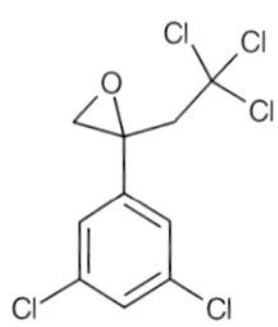	Tridiphane	58138-08-2	0.53	1	2.78
259	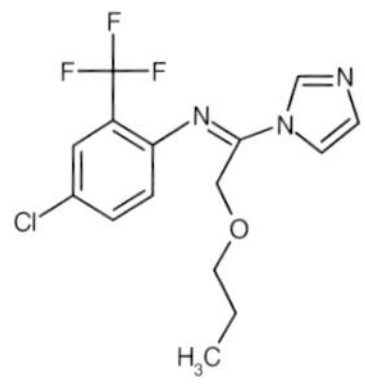	Triflumizole	68694-11-1	0.58	1	2.78
260	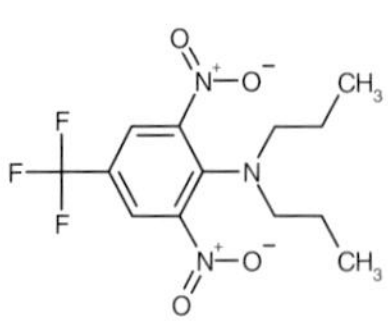	Trifluralin	1582-09-8	0.022	1	4.18

ID	Structure	Name	CAS NB	Toxicity LC_{50}		
				mg/l	Class	–Log mmol/l
261*	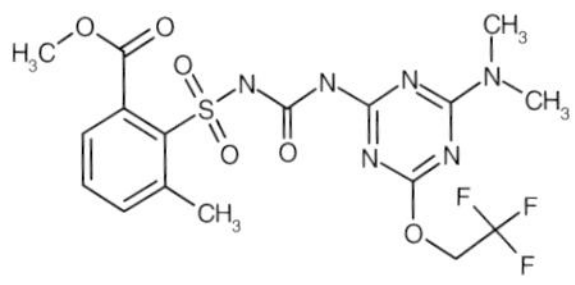	Triflusulfuron-methyl	126535-15-7	648	3	–0.12
262	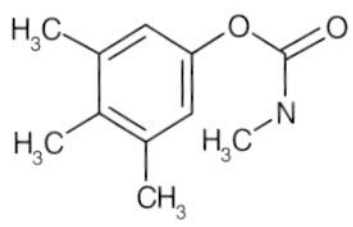	Trimethacarb	2686-99-9	4.7	2	1.61
263	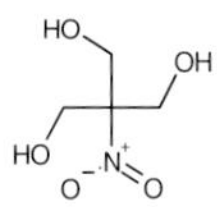	Tris-nitro	126-11-4	410	3	–0.43
264	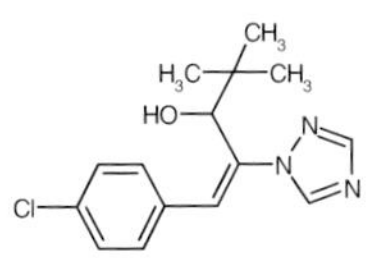	Uniconazole	83657-17-4	14.8	3	1.29
265	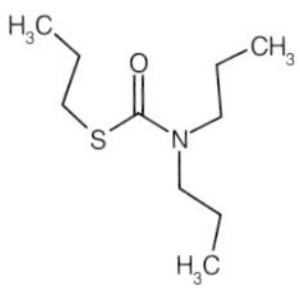	Vernolate	1929-77-7	3.5	2	1.76
267	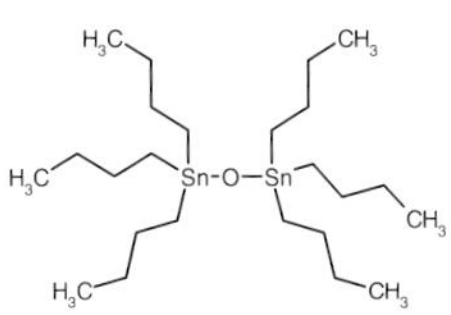	Tributyltin oxide	56-35-9	0.006	1	5.03
268	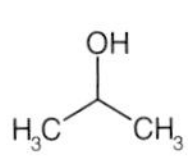	Isopropanol	67-63-0	13000	3	–2.33

ID	Structure	Name	CAS NB	Toxicity LC_{50}		
				mg/l	Class	–Log mmol/l
269	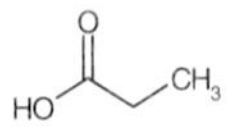	Propionic acid	79-09-4	51	3	0.16
270*	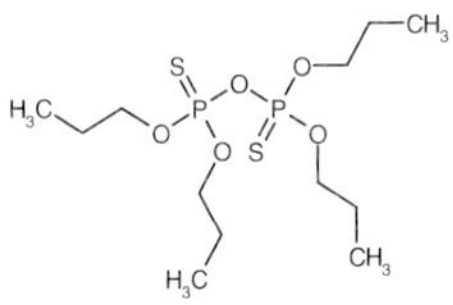	Tetrapropyl thiodiphos-phate	3244-90-4	0.076	1	3.70
271	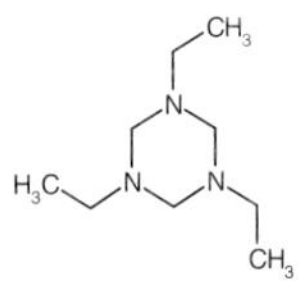	1,3,5-Triethyl hexahydro-*s*-triazine	7779-27-3	23.3	3	0.87
272*	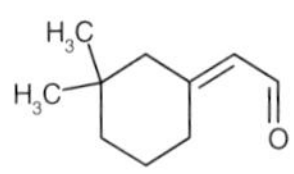	(E)-(3,3Di-methyl cyclo-hexylidene) acetaldehyde	26532-25-2	2.2	2	1.84
273	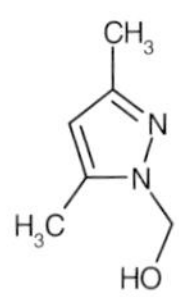	Dimethyl hydroxy methyl pyrazole	85264-33-1	123	3	0.01
274*	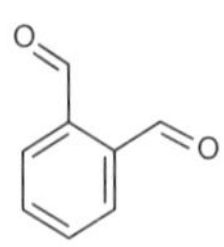	1,2-Benzene dicarboxalde-hyde	643-79-8	0.072	1	3.27
275	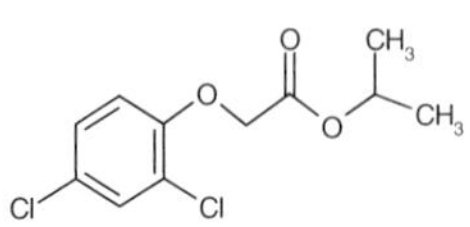	2,4-D Isopropyl ester	94-11-1	0.69	1	2.58

ID	Structure	Name	CAS NB	Toxicity LC_{50}		
				mg/l	Class	–Log mmol/l
276	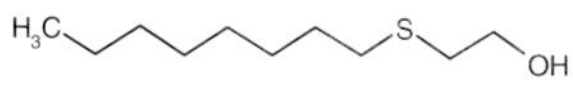	2-Hydroxyethyl octyl sulfide	3547-33-9	2.9	2	1.82
277	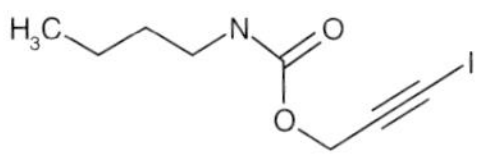	3-Iodo-2 propynyl butyl carbamate	55406-53-6	0.067	1	3.62
278	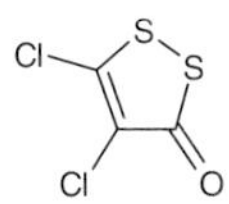	4,5-Dichloro-1,2-dithiol-3-one	1192-52-5	0.014	1	4.13
279*	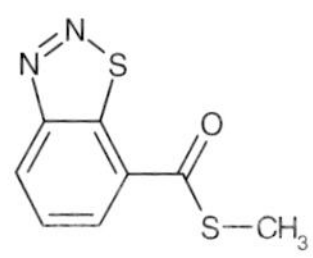	Acibenzolar-*s*-methyl	135158-54-2	0.88	1	2.38
280	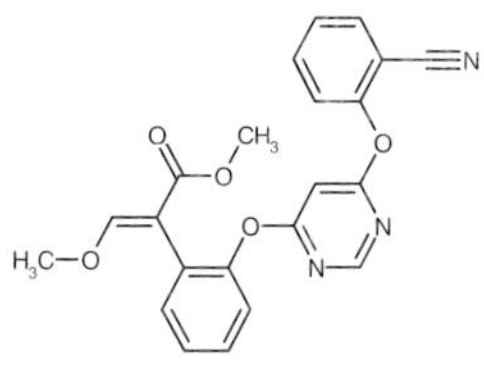	Azoxystrobin	131860-33-8	0.47	1	2.93
281	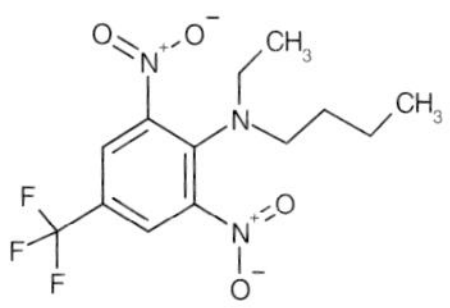	Benfluralin	1861-40-1	0.081	1	3.62
282	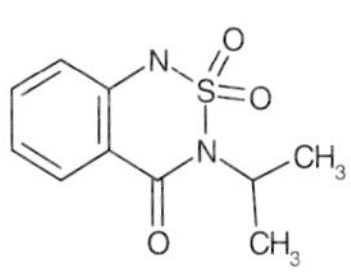	Bentazone	25057-89-0	174.1	3	0.14

ID	Structure	Name	CAS NB	Toxicity LC_{50}		
				mg/l	Class	–Log mmol/l
283	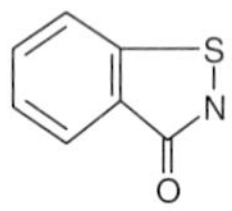	Benzisothiazo-lin-3-one	2634-33-5	1.6	2	1.98
284	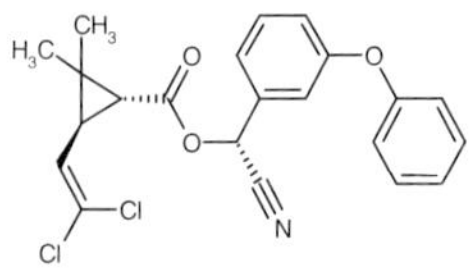	Beta cypermethrin	66841-24-5	0.0004	1	6.03
285*	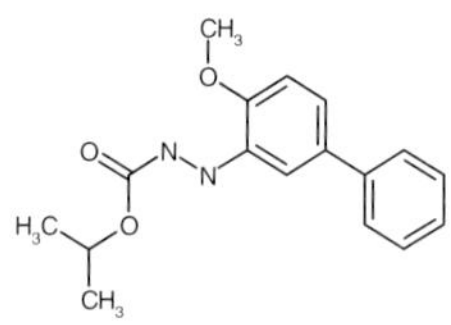	Bifenazate	149877-41-8	0.76	1	2.60
286	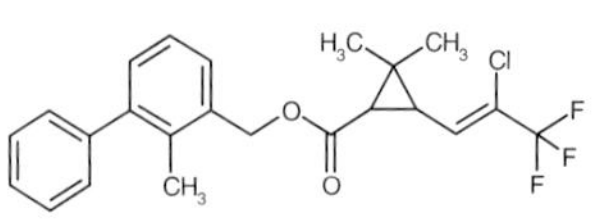	Bifenthrin	82657-04-3	0.00015	1	6.45
287	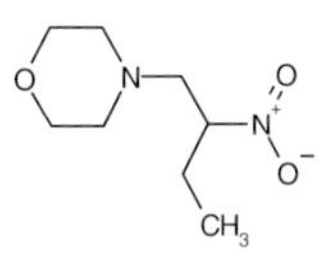	4-(2-Nitrobutyl) morpholine	2224-44-4	0.75	1	2.40
288	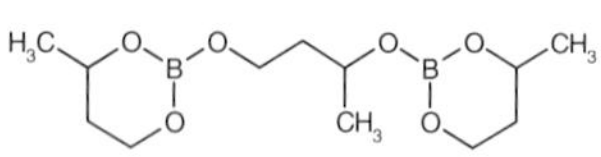	Biobor	2665-13-6	71	3	0.61
289	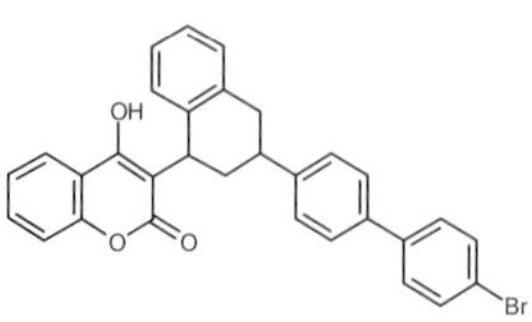	Brodifacoum	56073-10-0	0.025	1	4.32

ID	Structure	Name	CAS NB	Toxicity LC_{50}		
				mg/l	Class	–Log mmol/l
290	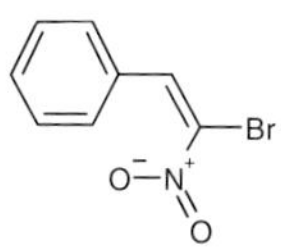	Bromonitro styrene	7166-19-0	0.027	1	3.93
291	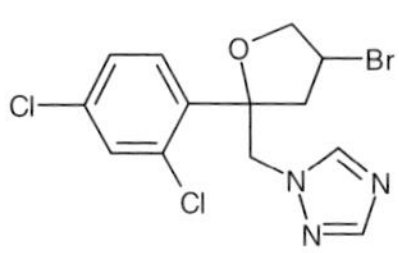	Bromuconazole	116255-48-2	1.7	2	2.35
292*	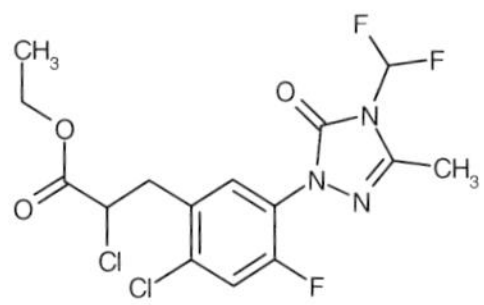	Carfentrazone-ethyl (F8246)	128639-02-1	1.6	2	2.41
293	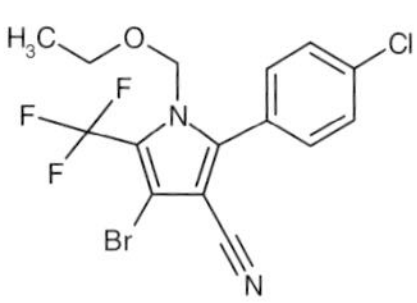	Chlorfenapyr (Pirate)	122453-73-0	0.007	1	4.74
294	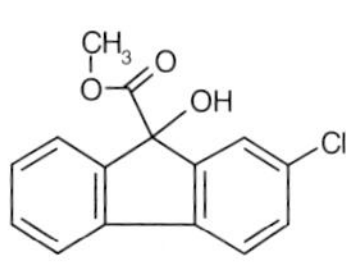	Chlorflurenol methyl	2536-31-4	2.83	2	1.99
295	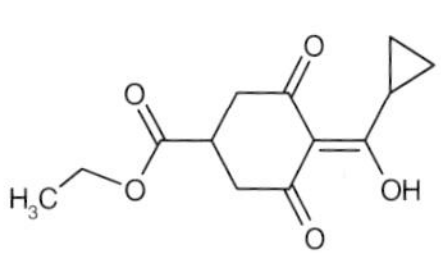	Cimecticarb	95266-40-3	65.7	3	0.58

ID	Structure	Name	CAS NB	Toxicity LC_{50}		
				mg/l	Class	–Log mmol/l
296	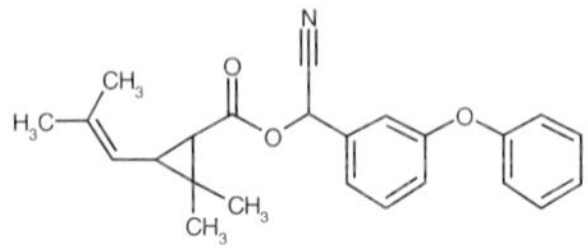	Cyphenothrin	39515-40-7	0.0003	1	6.04
297	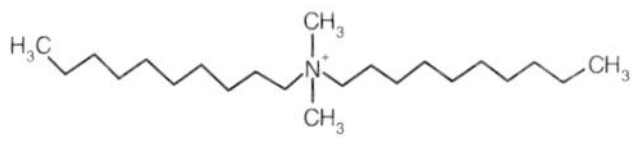	DDAC	7173-51-5	2.53	2	2.11
298	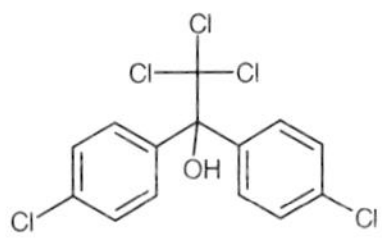	Dicofol	115-32-2	0.124	1	3.48
299*	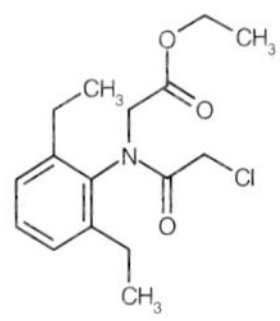	Diethatyl ethyl	38727-55-8	2.85	2	2.04
300	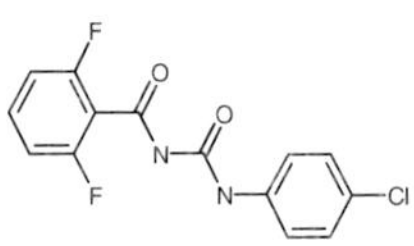	Diflubenzuron	35367-38-5	140	3	0.35
301	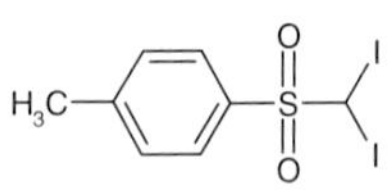	Diiodomethyl *p*-tolyl sulfone	20018-09-1	0.13	1	3.51
302	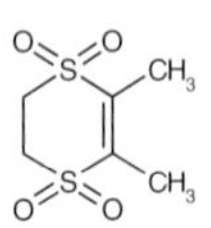	Dimethipin	55290-64-7	56	3	0.57
303	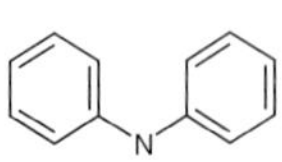	Diphenylamine	122-39-4	2.2	2	1.89

ID	Structure	Name	CAS NB	Toxicity LC_{50} mg/l	Class	–Log mmol/l
304	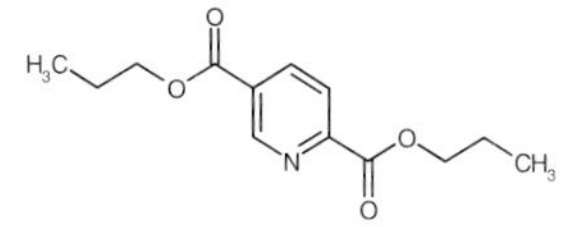	Dipropyl isocincho meronate	136-45-8	1	2	2.40
305	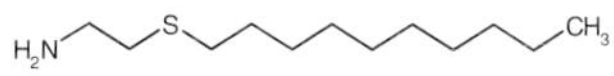	DTEA	29873-30-1	0.66	1	2.52
306	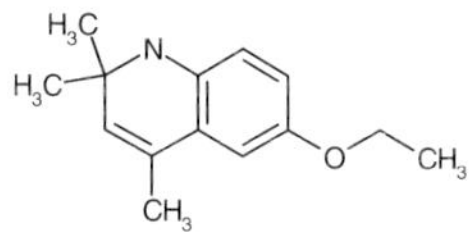	Ethoxyquin	91-53-2	18	3	1.05
307	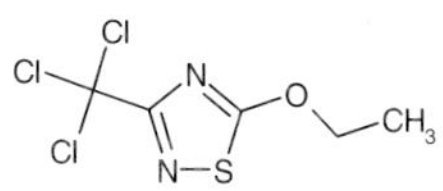	Etridiazole	2593-15-9	0.77	1	2.51
308	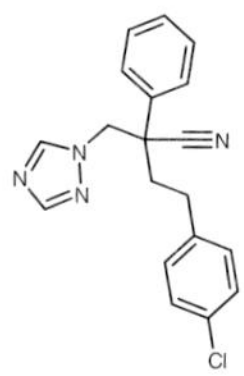	Fenbuconazole	114369-43-6	1.5	2	2.35
309	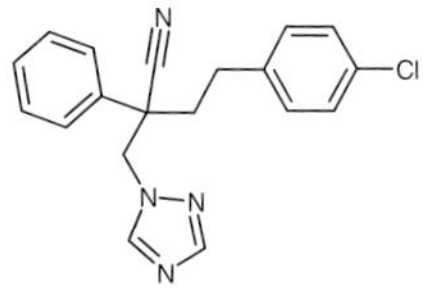	Fenhexamid	126833-17-8	1.34	2	2.35

ID	Structure	Name	CAS NB	Toxicity LC_{50}		
				mg/l	Class	–Log mmol/l
310	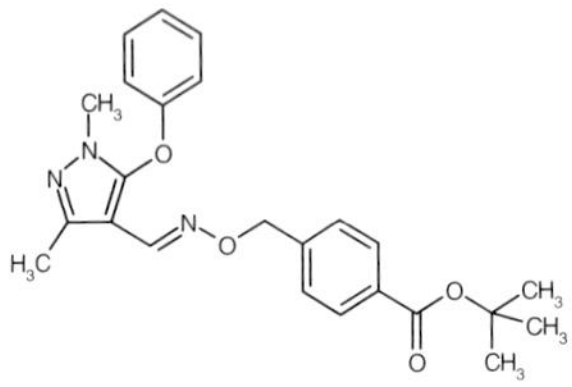	Fenpyroximate	134098-61-6	0.001	1	5.58
311*	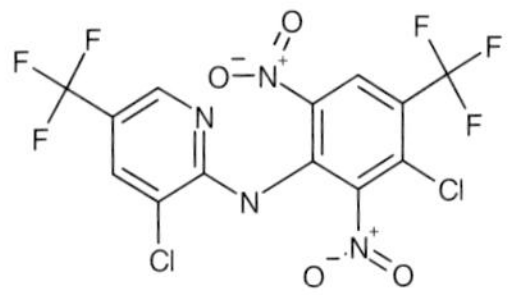	Fluazinam	79622-59-6	0.036	1	4.11
312	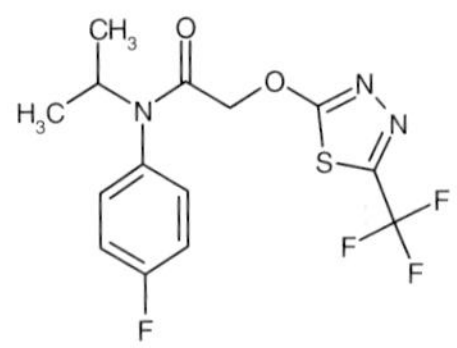	Flufenacet	142459-58-3	5.84	2	1.79
313*	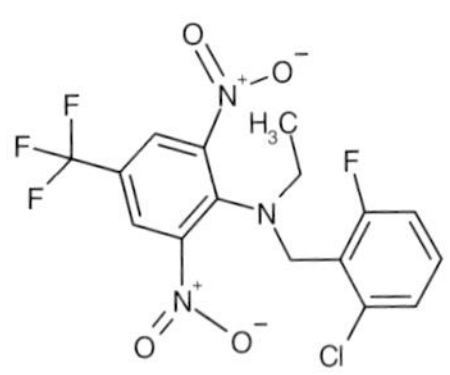	Flumetralin	62924-70-3	0.024	1	4.24
314	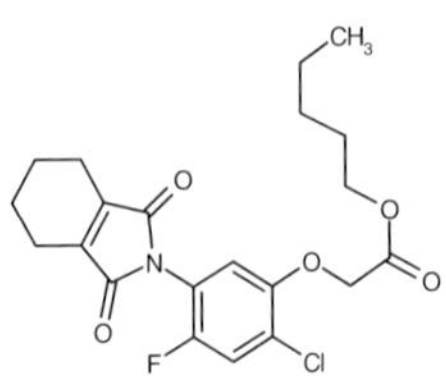	Flumiclorac pentyl	87546-18-7	1.1	2	2.59

ID	Structure	Name	CAS NB	Toxicity LC_{50}		
				mg/l	Class	–Log mmol/l
315	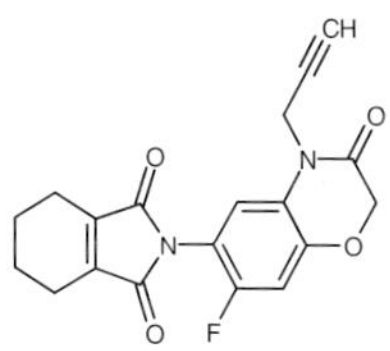	Flumioxazin (V-53482)	103361-09-7	2.3	2	2.19
316	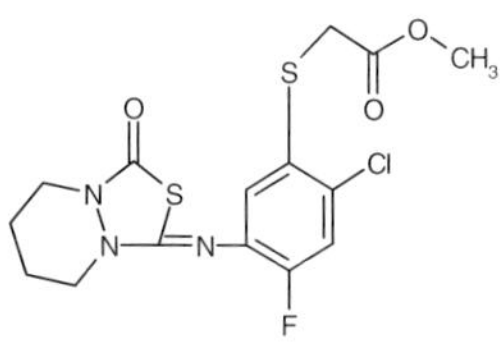	Fluthiacet methyl	117337-19-6	0.043	1	3.97
317*	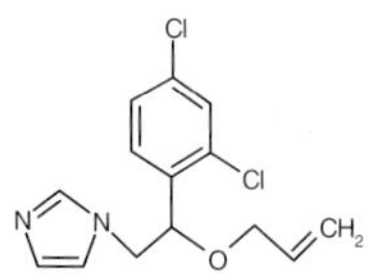	Imazalil	35554-44-0	1.48	2	2.30
318	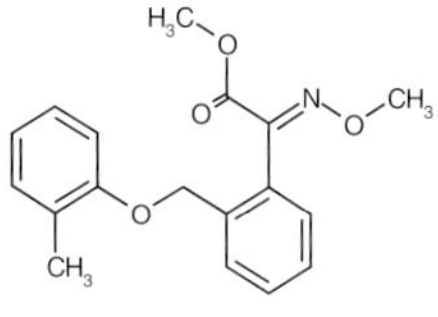	Kresoxim methyl	143390-89-0	0.19	1	3.22
319	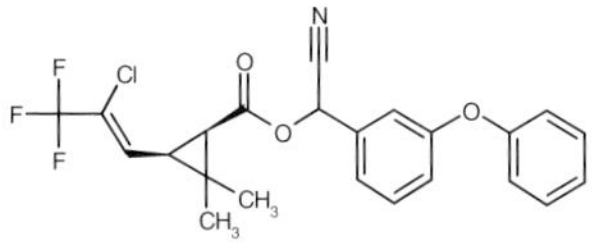	Lambda-cyhalothrin	91465-08-6	0.0002	1	6.27
320	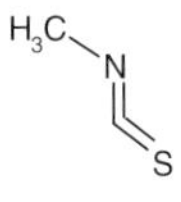	Methyl isothiocyanate	556-61-6	0.094	1	2.89

ID	Structure	Name	CAS NB	Toxicity LC_{50}		
				mg/l	Class	–Log mmol/l
321	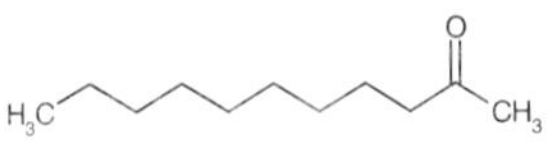	Methyl nonyl ketone	112-12-9	3	2	1.75
322	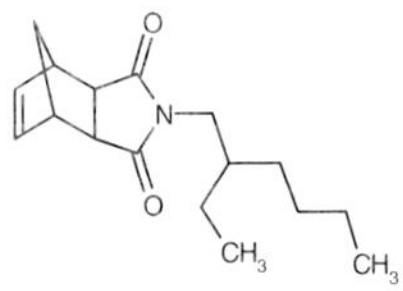	MGK 264	113-48-4	1.4	2	2.29
323	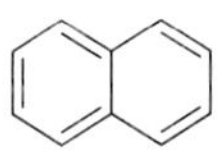	Naphthalene	91-20-3	2	2	1.81
324	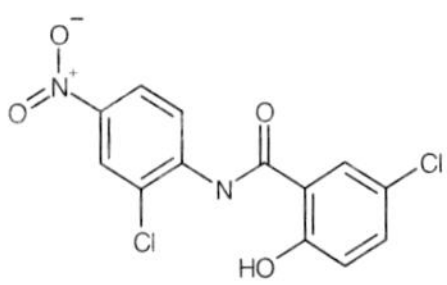	Niclosamide	50-65-7	0.04	1	3.89
325	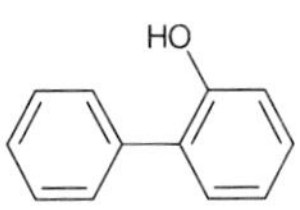	*O*-Phenylphenol	90-43-7	2.75	2	1.79
327	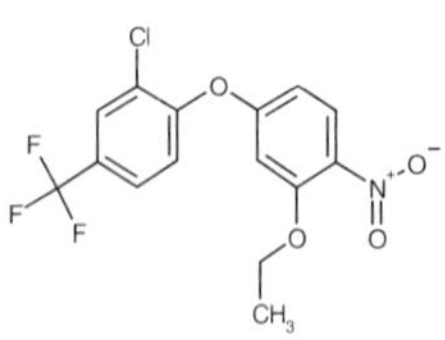	Oxyfluorfen	42874-03-3	0.41	1	2.95
328	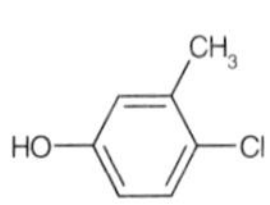	Parachlorometacresol	59-50-7	0.917	1	2.19

ID	Structure	Name	CAS NB	Toxicity LC_{50}		
				mg/l	Class	–Log mmol/l
329*	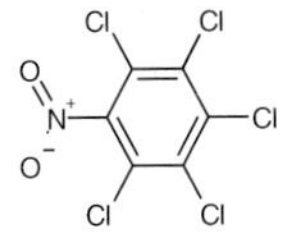	PCNB	82-68-8	0.32	1	2.97
330	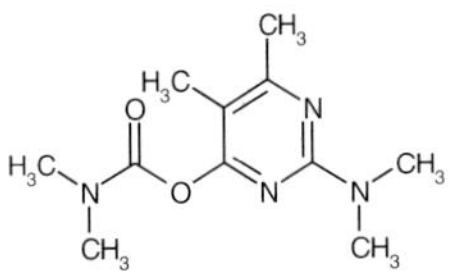	Pirimicarb	23103-98-2	29	3	0.91
331	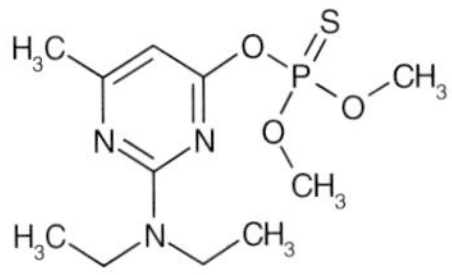	Pirimiphos methyl	29232-93-7	0.404	1	2.88
332*	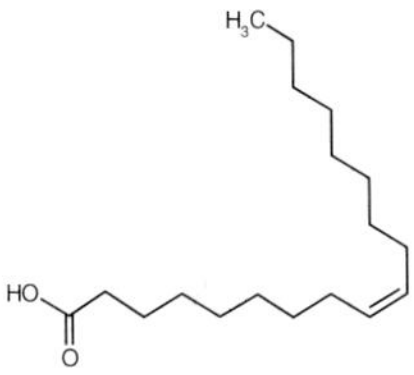	Oleic acid	112-80-1	8.1	2	1.54
333*	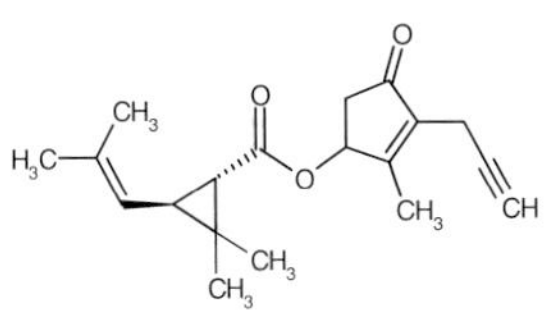	Prallethrin	23031-36-9	0.012	1	4.40
334	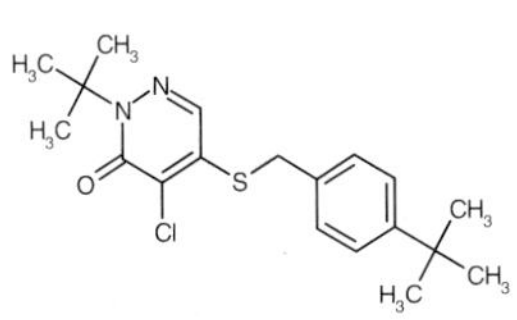	Pyridaben	96489-71-3	0.0007	1	5.70

ID	Structure	Name	CAS NB	Toxicity LC_{50} mg/l	Toxicity LC_{50} Class	Toxicity LC_{50} –Log mmol/l
335*	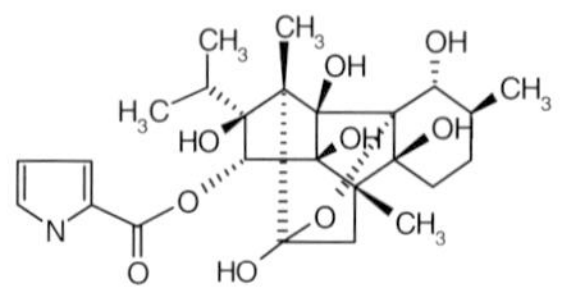	Ryanodine	15662-33-6	3.2	2	2.19
337	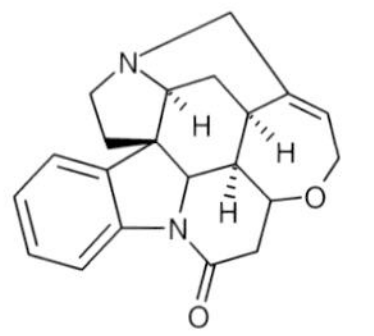	Strychnine	57-24-9	2.3	2	2.16
338	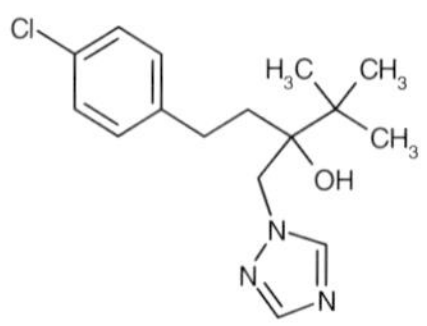	Tebuconazole	107534-96-3	4.4	2	1.84
339*	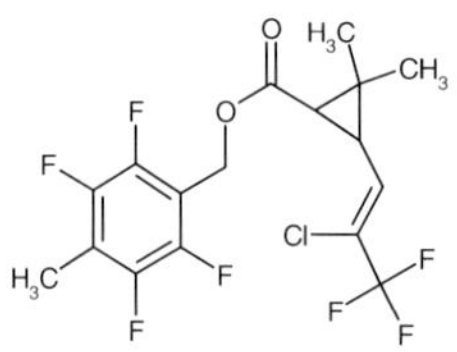	Tefluthrin	79538-32-2	0.00006	1	6.84
340*	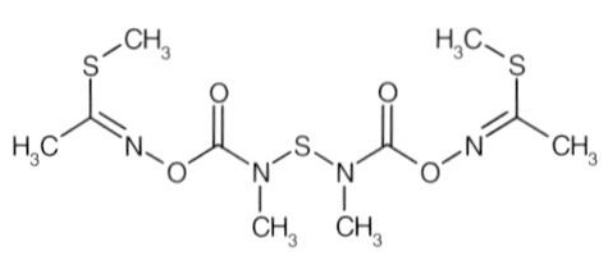	Thiodicarb	59669-26-0	2.65	2	2.13
341	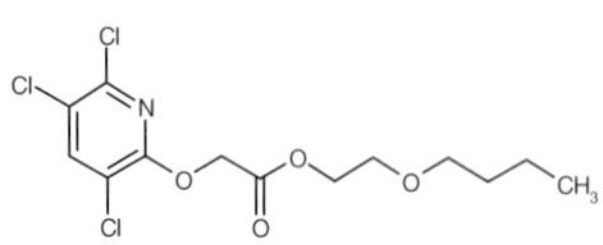	Triclopyr BEE	64700-56-7	0.65	1	2.74

ID	Structure	Name	CAS NB	Toxicity LC_{50}		
				mg/l	Class	–Log mmol/l
342	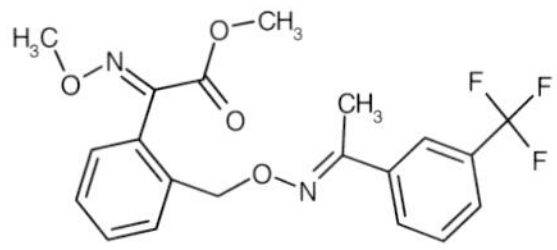	Trifloxystrobin	141517-21-7	0.014	1	4.46
344	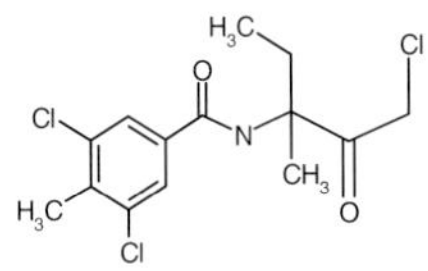	Zoxamide	156052-68-5	0.156	1	3.33
345	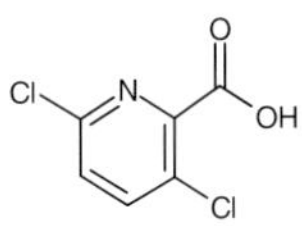	Clopyralid	1702-17-6	78.52	3	0.39
346	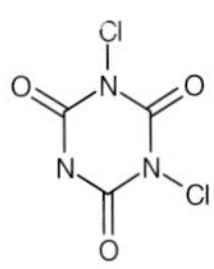	Dichloroisocyanuric acid	2782-57-2	0.22	1	2.94

LIST OF THE 264 COMPOUNDS INCLUDING IN THE DAPHNIA DATA SET

Three toxicity classes were generated according to the intervals established by the EC regulation, after regrouping the toxicity classes as too weakly represented:

LC_{50} (class 1) < 1 mg/l; 1 mg/l ≤ LC_{50} (class 2) < 10 mg/l; LC_{50} (class 3) ≥ 10 mg/l.

The symbol '*' in the ID column indicates the test set compounds.

ID	Structure	Name	CAS NB	Toxicity LC_{50}		
				mg/l	Class	–Log mmol/l
1*	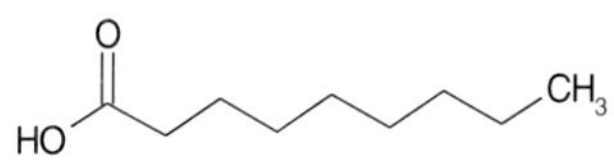	Pelargonic acid	112-05-0	96	3	0.22
2	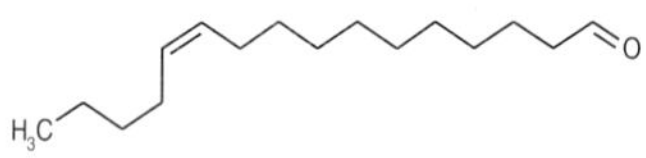	(Z)-11-Hexadecenal	53939-28-9	1	2	2.38
3	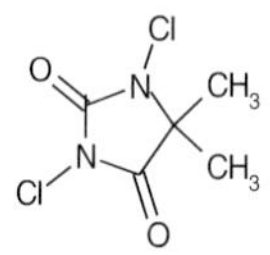	1,3-Dichloro-5,5-dimethylhydantoin (DCDMH)	118-52-5	0.5	1	2.60
4	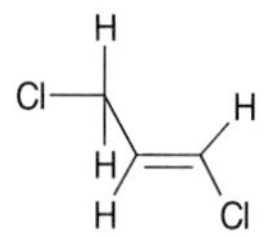	1,3-Dichloro propene	542-75-6	0.09	1	3.09
5	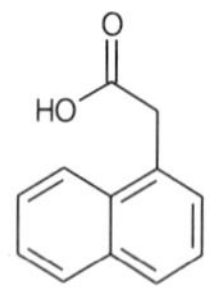	1-Naphthylacetic acid	86-87-3	180	3	0.01
6	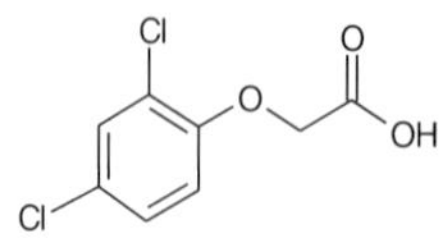	2,4-D	94-75-7	25	3	0.95
8*	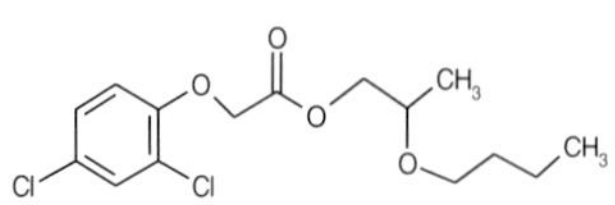	2,4-D 2-butoxymethyl ethyl ester	1320-18-9	1.2	2	2.45

ID	Structure	Name	CAS NB	Toxicity LC_{50}		
				mg/l	Class	–Log mmol/l
10	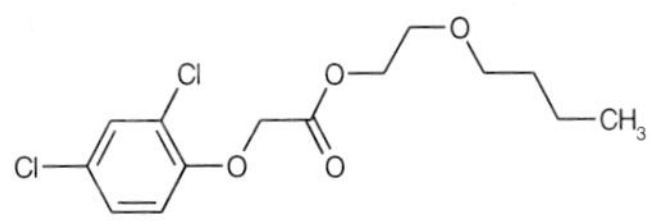	2,4-D butoxyethyl ester	1929-73-3	7.2	2	1.65
13	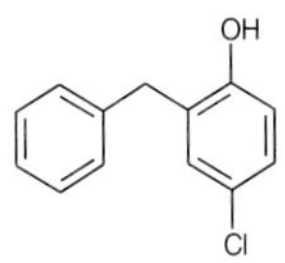	Clorophene	120-32-1	0.59	1	2.57
14	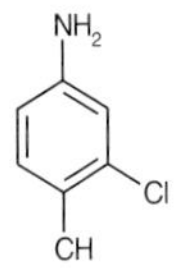	3-Chloro-*p*-toluidine	7745-89-3	0.056	1	3.41
16*	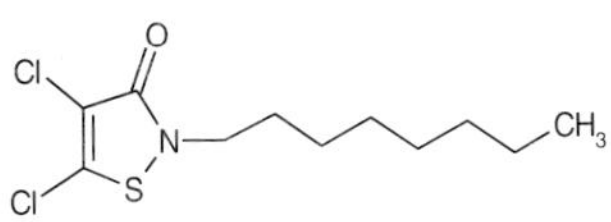	Kathon 930	64359-81-5	0.005	1	4.73
17	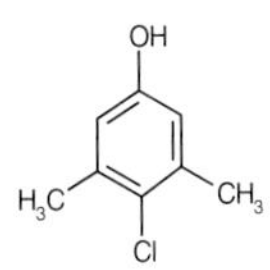	Chloroxylenol	88-04-0	7.7	2	1.31
20	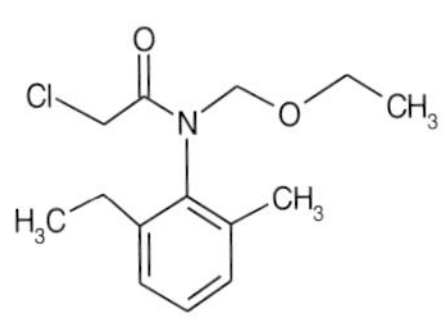	Acetochlor	34256-82-1	8.2	2	1.52
21	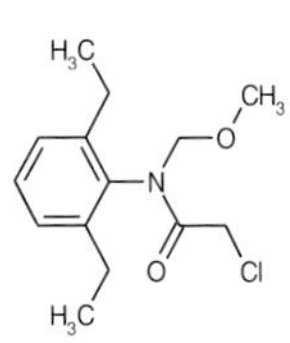	Alachlor	15972-60-8	21	3	1.11

ID	Structure	Name	CAS NB	Toxicity LC_{50}		
				mg/l	Class	–Log mmol/l
22	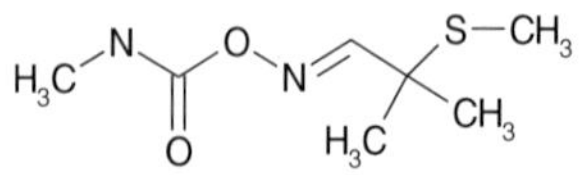	Aldicarb	116-06-3	0.41	1	2.67
23	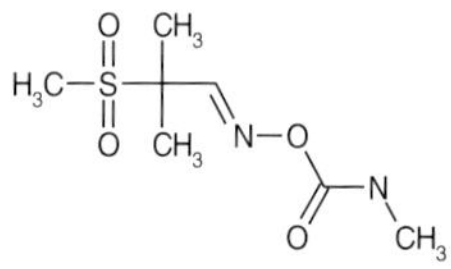	Aldicarb sulfone	1646-88-4	0.28	1	2.90
25	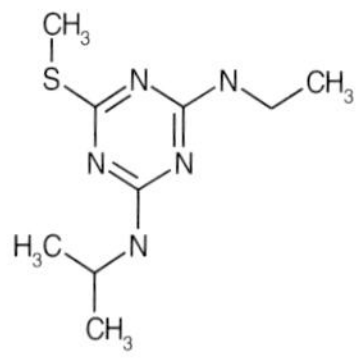	Ametryne	834-12-8	28	3	0.91
26	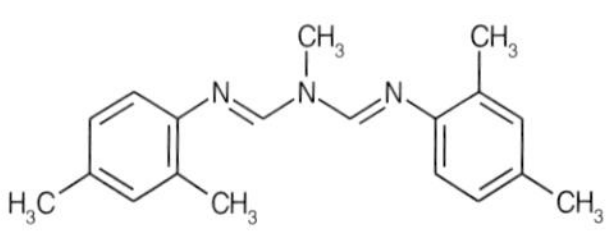	Amitraz	33089-61-1	0.035	1	3.92
27	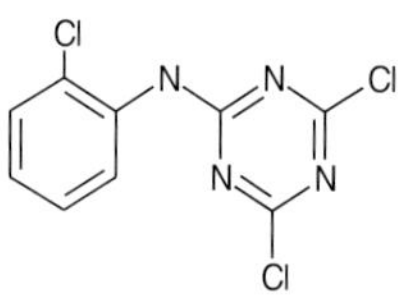	Anilazine	101-05-3	1.1	2	2.40
30	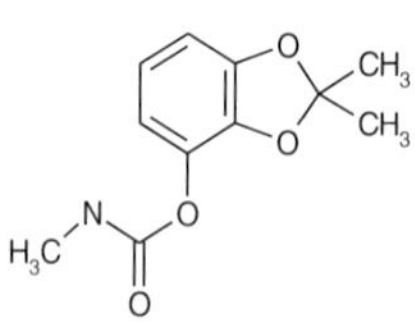	Bendiocarb	22781-23-3	0.03	1	3.88
32	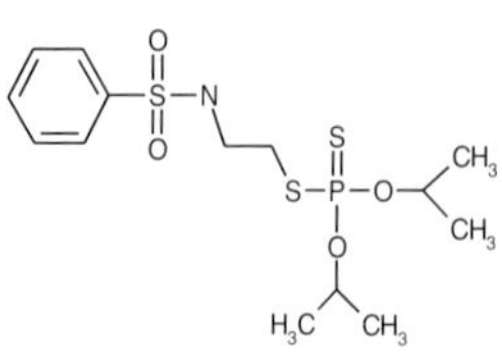	Bensulide	741-58-2	0.58	1	2.84

ID	Structure	Name	CAS NB	Toxicity LC_{50}		
				mg/l	Class	–Log mmol/l
33	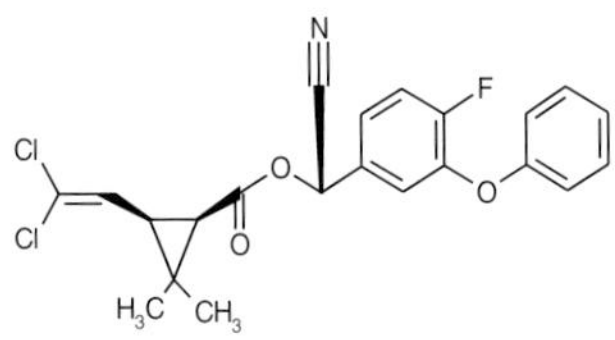	Beta-cyfluthrin	68359-37-5	0.0001	1	6.49
34*	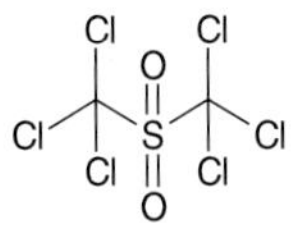	Bis(trichlorometh yl) sulfone	3064-70-8	0.17	1	3.24
35	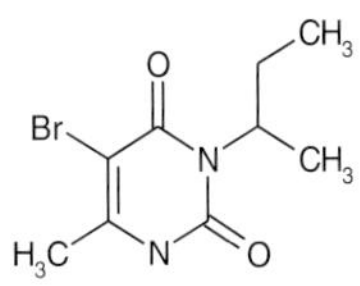	Bromacil	314-40-9	121	3	0.33
36	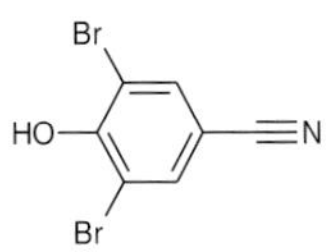	Bromoxynil	1689-84-5	19.2	3	1.16
37*	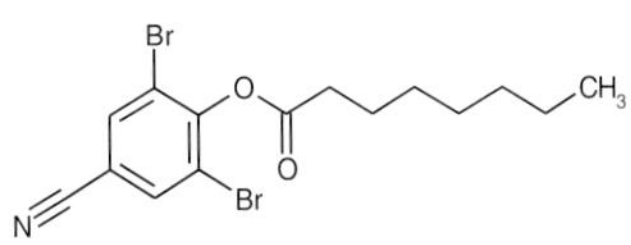	Bromoxynil octanoate	1689-99-2	0.11	1	3.56
38	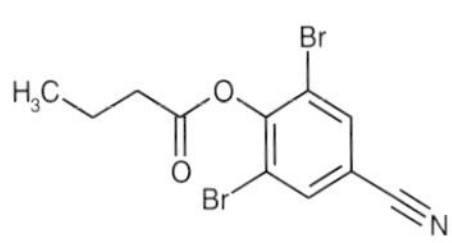	Bromoxynil butyrate	3861-41-4	0.2	1	3.24
39	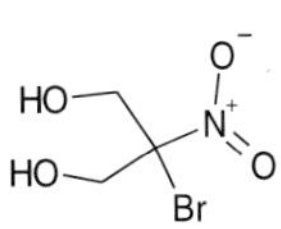	Bronopol	52-51-7	1.6	2	2.10

ID	Structure	Name	CAS NB	Toxicity LC_{50}		
				mg/l	Class	–Log mmol/l
43	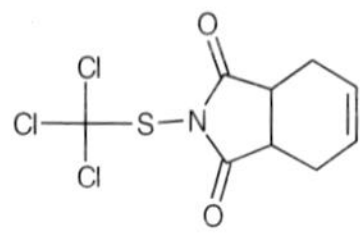	Captan	133-06-2	8.4	2	1.55
44	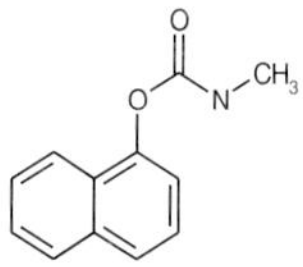	Carbaril	63-25-2	0.006	1	4.56
46*	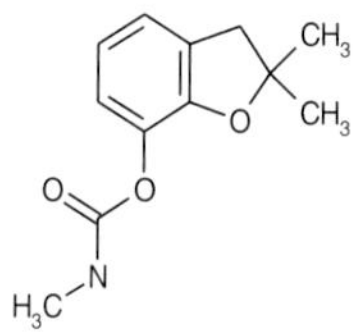	Carbofuran	1563-66-2	0.03	1	3.88
47*	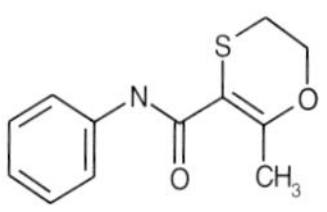	Carboxin	5234-68-4	84.4	3	0.45
48	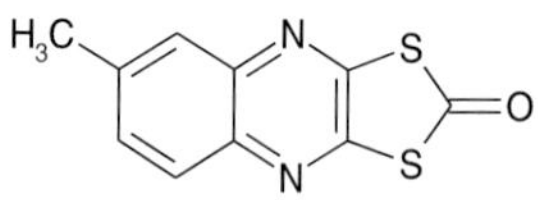	Chinomethionat	2439-01-2	0.12	1	3.29
49	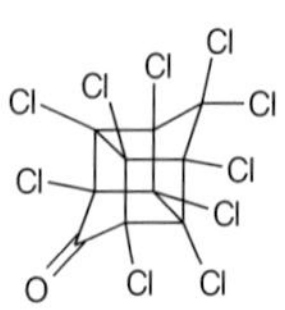	Chlordecone	143-50-0	0.26	1	3.28
51	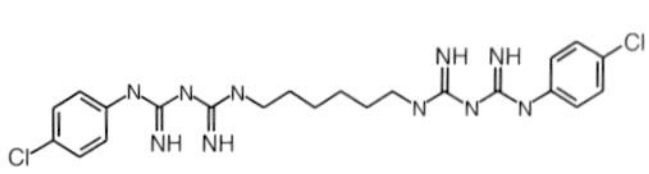	Chlorhexidine	55-56-1	0.051	1	4.00

ID	Structure	Name	CAS NB	Toxicity LC_{50}		
				mg/l	Class	–Log mmol/l
53*	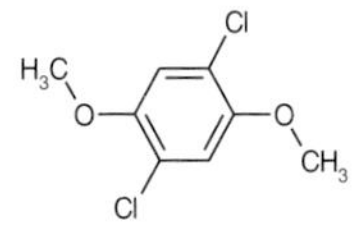	Chloroneb	2675-77-6	6.19	2	1.52
54	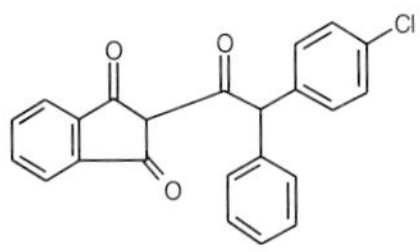	Chloro phacinone	3691-35-8	0.43	1	2.94
57	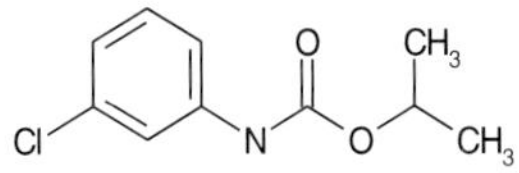	Chlorpropham	101-21-3	3.7	2	1.76
59*	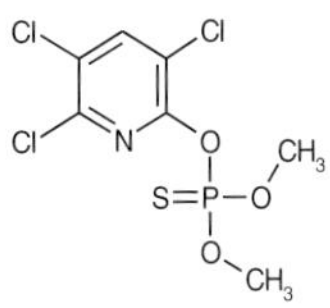	Chlorpyrifos-methyl	5598-13-0	0.001	1	5.46
63	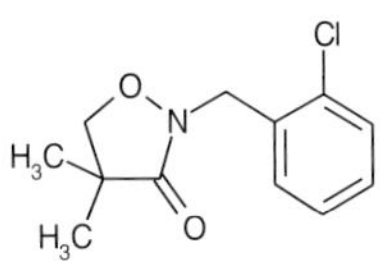	Clomazone	81777-89-1	5.2	2	1.66
67	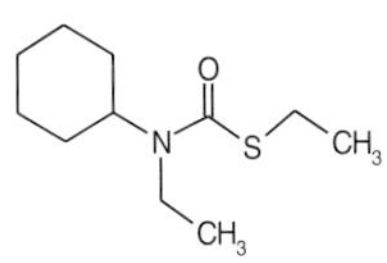	Cycloate	1134-23-2	24	3	0.95
68	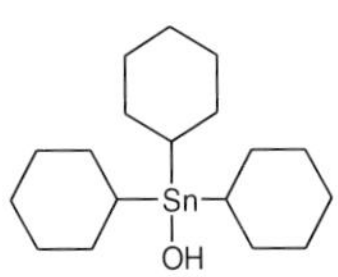	Cyhexatin	13121-70-5	0.0002	1	6.36

ID	Structure	Name	CAS NB	Toxicity LC_{50}		
				mg/l	Class	–Log mmol/l
69	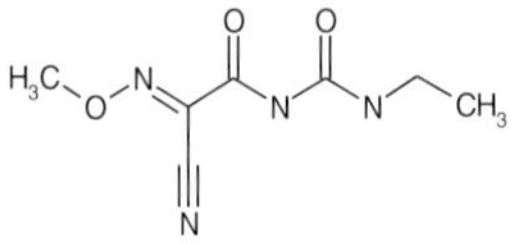	Cymoxanil	57966-95-7	28	3	0.85
72	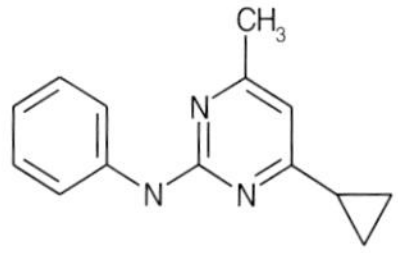	Cyprodinil	121552-61-2	0.03	1	3.85
73	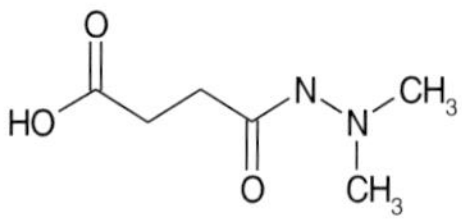	Daminozide	1596-84-5	98.5	3	0.21
75	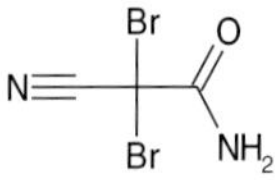	DBNPA	10222-01-2	0.86	1	2.45
76	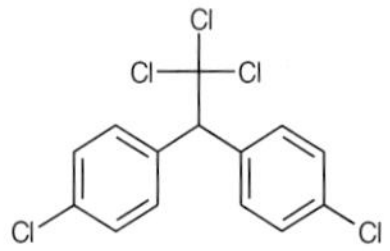	DDT	50-29-3	0.005	1	4.88
81	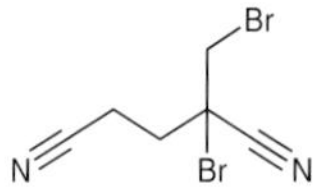	1,2-Dibromo-2,4-dicyano butane	35691-65-7	2.2	2	2.08
83*	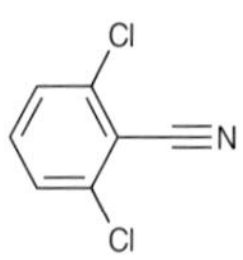	Dichlobenil	1194-65-6	6.2	2	1.44
87	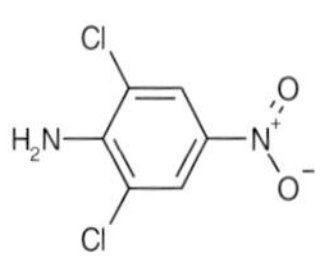	Dicloran	99-30-9	2.07	2	2.00

ID	Structure	Name	CAS NB	Toxicity LC_{50} mg/l	Class	–Log mmol/l
88*	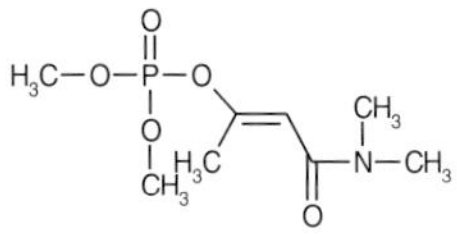	Dicrotophos	141-66-2	0.01	1	4.27
90	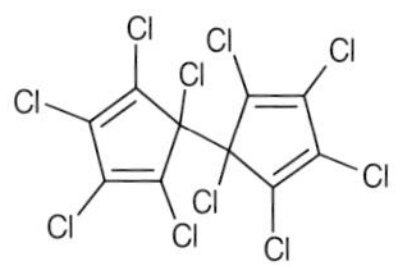	Dienochlor	2227-17-0	1.2	2	2.60
91	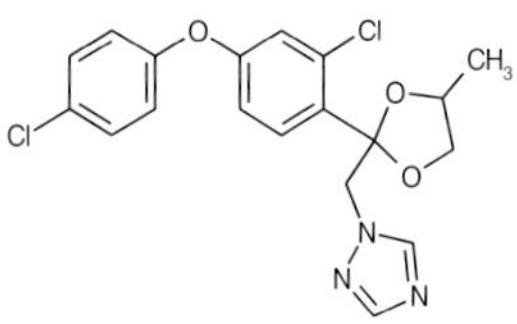	Difenoconazole	119446-68-3	0.77	1	2.72
92*	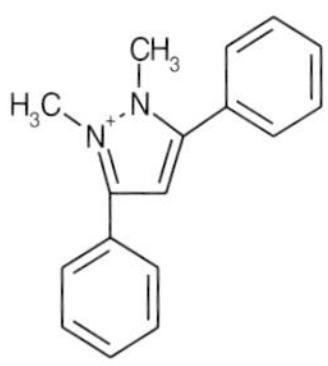	Difenzoquat	49866-87-7	1.75	2	2.15
93*	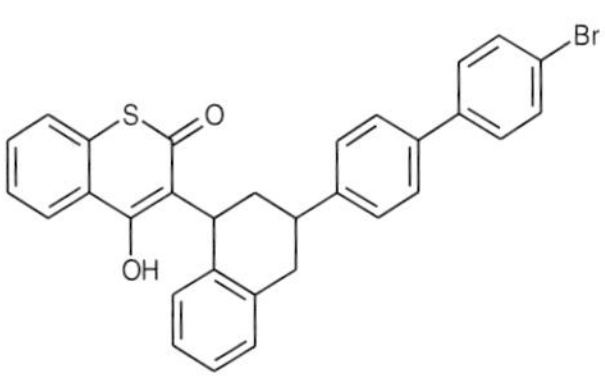	Difethialone	104653-34-1	0.004	1	5.09
94	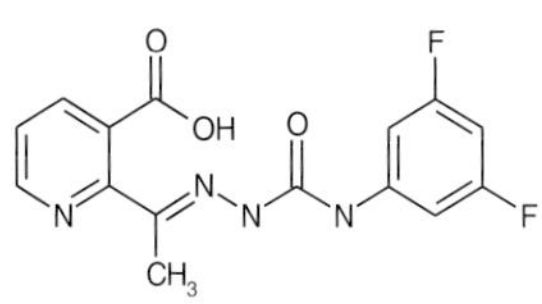	Diflufenzopyr	109293-97-2	14.08	3	1.38

ID	Structure	Name	CAS NB	Toxicity LC_{50}		
				mg/l	Class	–Log mmol/l
95	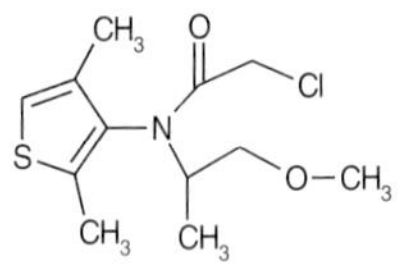	Dimethenamid	87674-68-8	12	3	1.36
98*	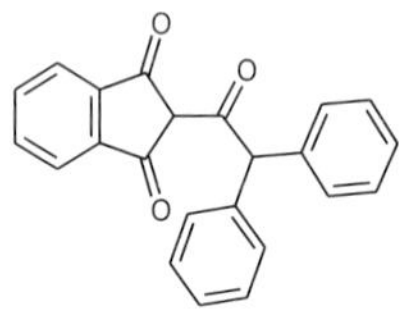	Diphacinone	82-66-6	1.8	2	2.28
101*	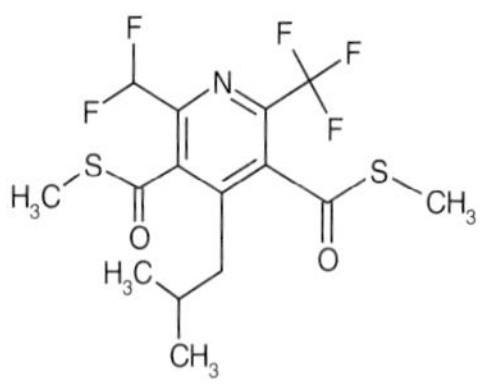	Dithiopyr	97886-45-8	17	3	1.37
104	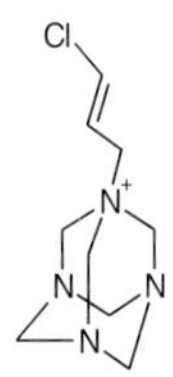	Dowicil	4080-31-3	34.36	3	0.80
110	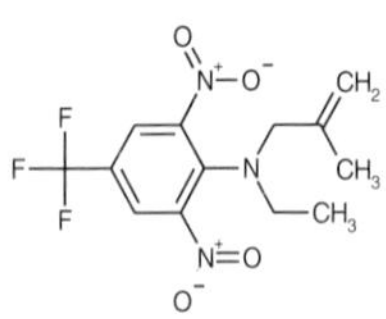	Ethalfluralin	55283-68-6	0.06	1	3.74
111	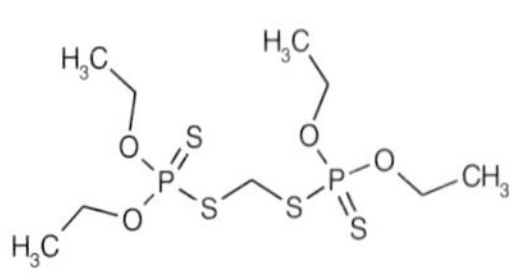	Ethion	563-12-2	0.0001	1	6.84

ID	Structure	Name	CAS NB	Toxicity LC_{50} mg/l	Class	–Log mmol/l
116	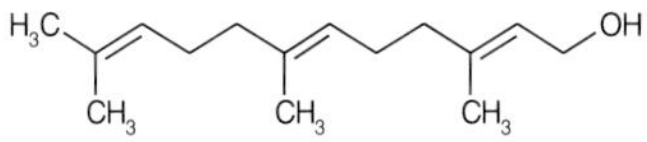	Farnesol	4602-84-0	2.2	2	2.00
119	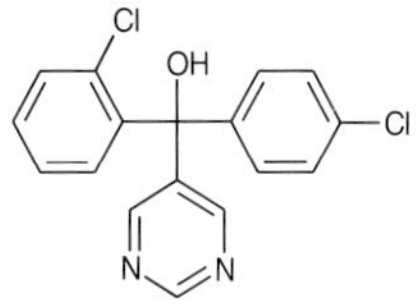	Fenarimol	60168-88-9	6.8	2	1.69
120	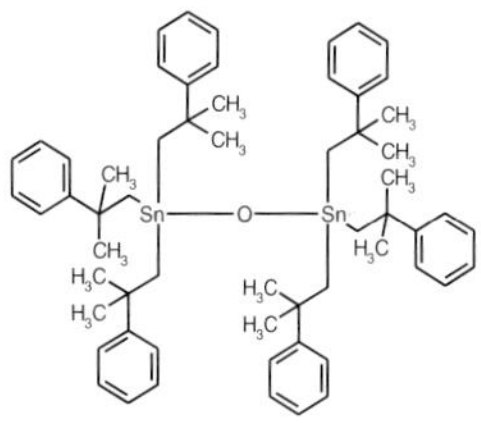	Fenbutatin oxide	13356-08-6	0.03	1	4.53
122	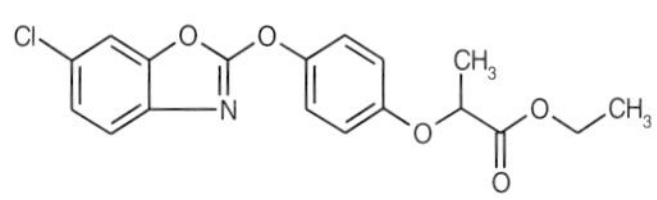	Fenoxaprop-ethyl	66441-23-4	3.18	2	2.06
123*	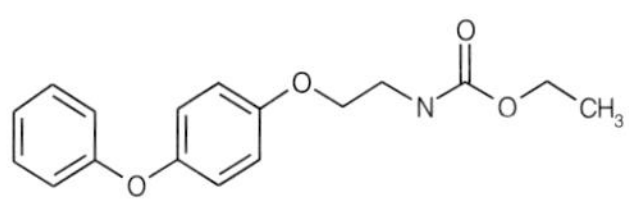	Fenoxycarb	79127-80-3	0.4	1	2.88
124	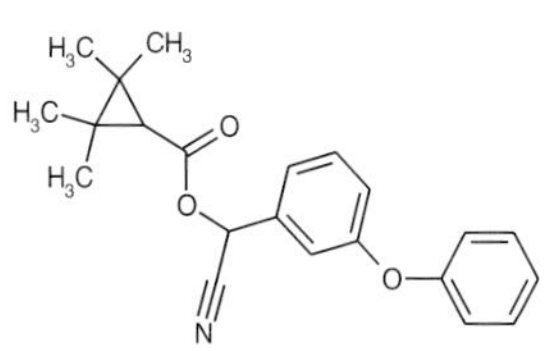	Fenpropathrin	39515-41-8	0.0005	1	5.82

ID	Structure	Name	CAS NB	Toxicity LC_{50}		
				mg/l	Class	–Log mmol/l
125	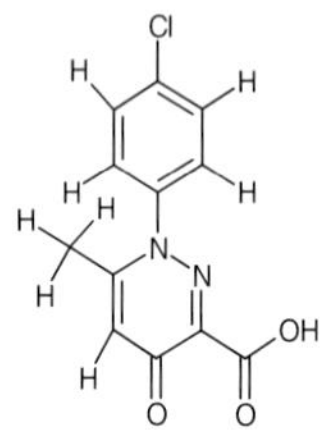	Fenridazone	68254-10-4	173.6	3	0.18
126	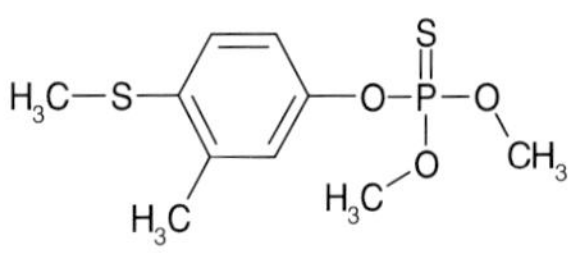	Fenthion	55-38-9	0.005	1	4.73
127	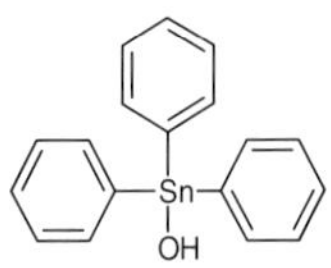	Fentin hydroxide	76-87-9	0.01	1	4.56
131	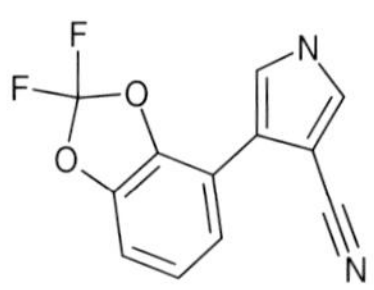	Fludioxonil	131341-86-1	0.9	1	2.44
133	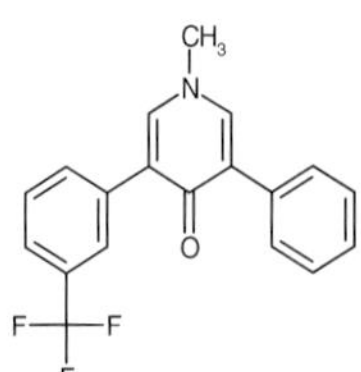	Fluridone	59756-60-4	4.4	2	1.87
134	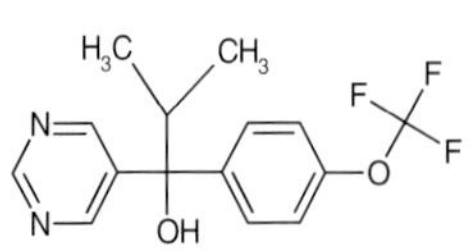	Flurprimidol	56425-91-3	11.8	3	1.42

ID	Structure	Name	CAS NB	Toxicity LC_{50}		
				mg/l	Class	–Log mmol/l
137	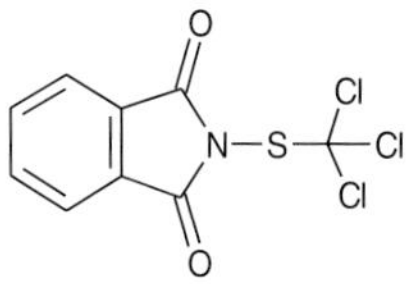	Folpet	133-07-3	0.02	1	4.17
138	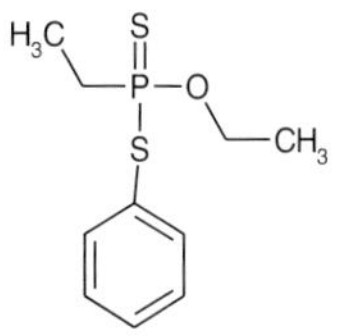	Fonofos	944-22-9	0.002	1	5.09
139	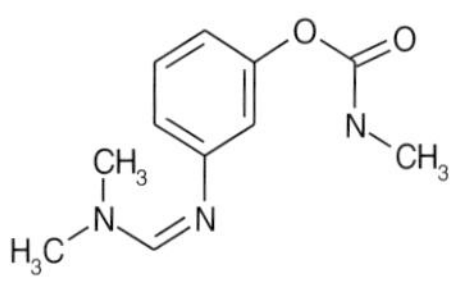	Formetanate	22259-30-9	0.075	1	3.47
140	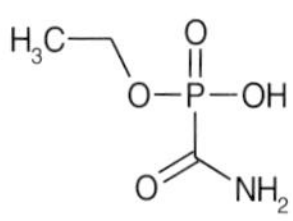	Fosamine	59682-52-9	1371.6	3	–0.95
143	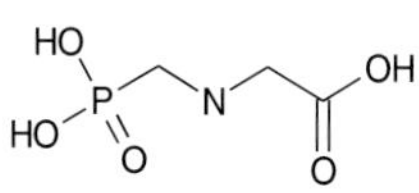	Glyphosate	1071-83-6	134	3	0.10
145	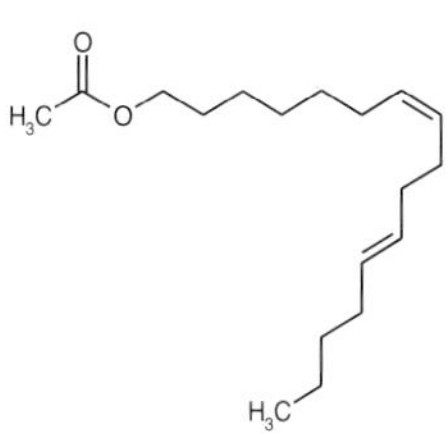	Hexadecadienol acetate	53042-79-8	0.7	1	2.60

ID	Structure	Name	CAS NB	Toxicity LC_{50}		
				mg/l	Class	–Log mmol/l
146*	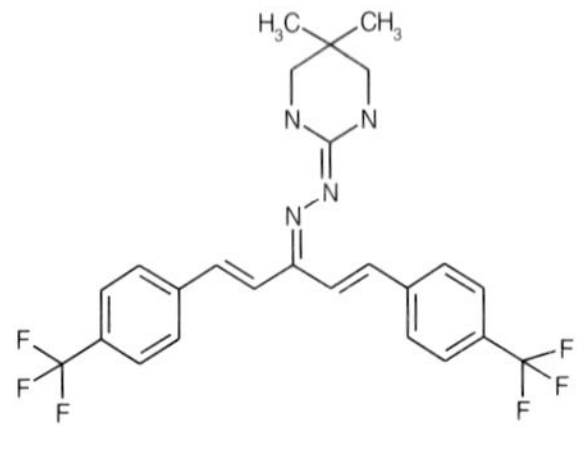	Hydramethylnon	67485-29-4	1.14	2	2.64
147	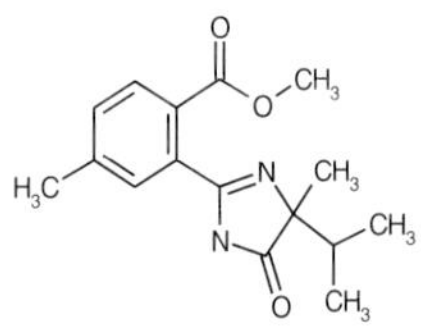	Imazamethabenz-methyl	81405-85-8	220	3	0.12
148	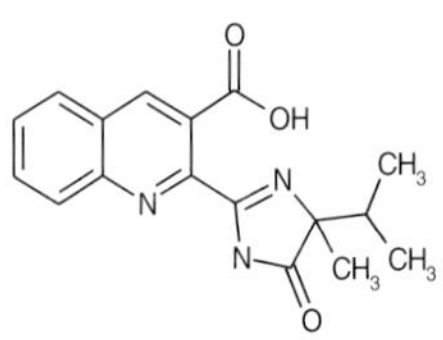	Imazaquin	81335-37-7	280	3	0.05
151*	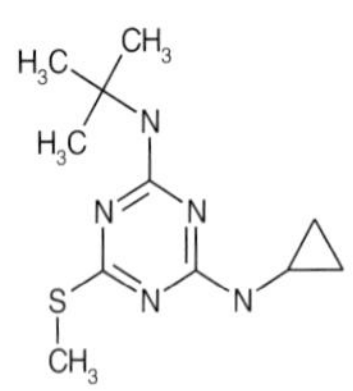	Irgarol	28159-98-0	5.3	2	1.68
152	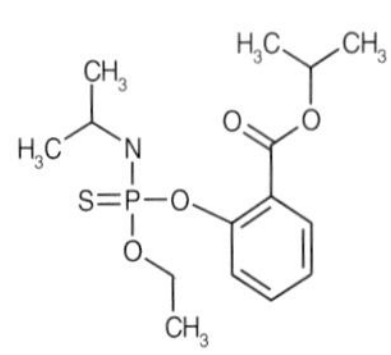	Isofenphos	25311-71-1	0.004	1	4.95

ID	Structure	Name	CAS NB	Toxicity LC_{50}		
				mg/l	Class	–Log mmol/l
153*	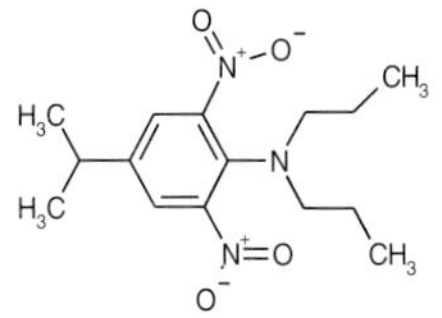	Isopropalin	33820-53-0	0.27	1	3.06
157	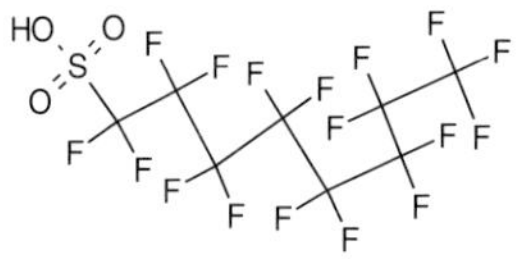	Perfluorooctane sulfonate	29457-72-5	66	3	0.88
158	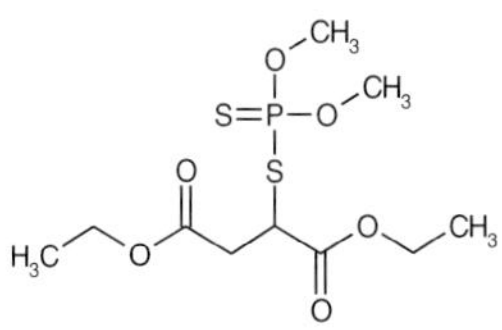	Malathion	121-75-5	0.001	1	5.52
163	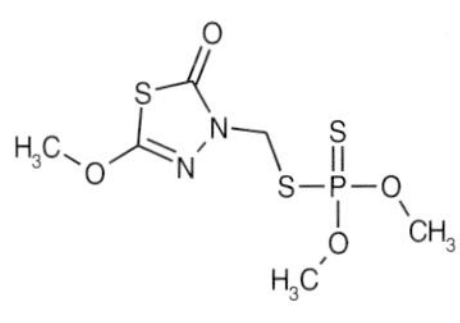	Methidathion	950-37-8	0.006	1	4.67
165	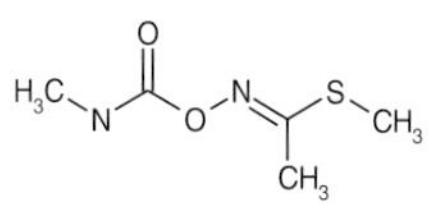	Methomyl	16752-77-5	0.009	1	4.27
168	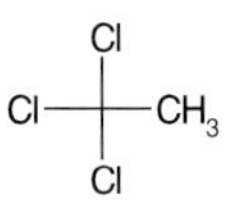	Methyl chloroform	71-55-6	11.2	3	1.08
169	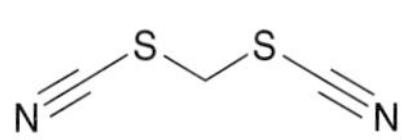	Methylene bis(thiocyanate)	6317-18-6	0.04	1	3.52

ID	Structure	Name	CAS NB	Toxicity LC_{50}		
				mg/l	Class	–Log mmol/l
171	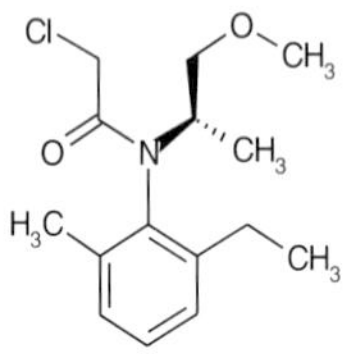	Metolachlor-S-isomer	87392-12-9	23.5	3	1.08
172	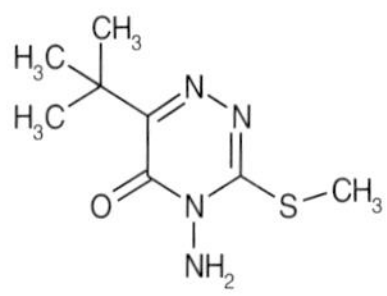	Metribuzin	21087-64-9	4.18	2	1.71
174	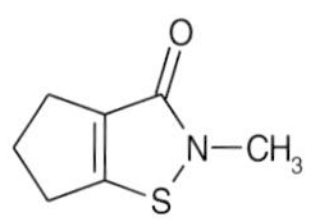	MTI	82633-79-2	1.3	2	2.08
176	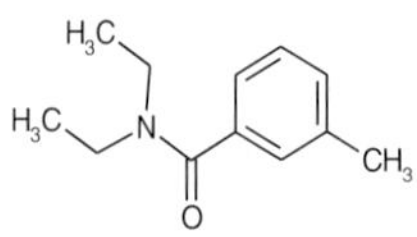	Diethyltoluamide	134-62-3	75	3	0.41
177	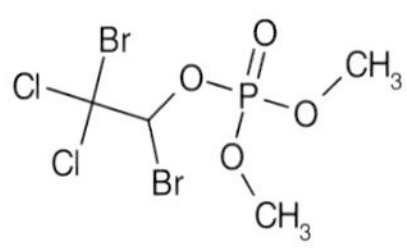	Naled	300-76-5	0.0004	1	5.98
178	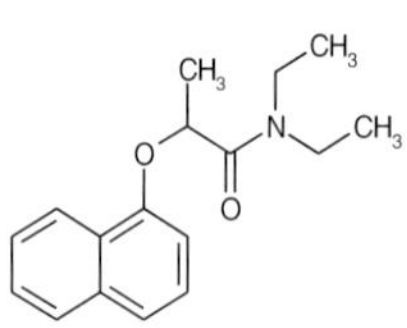	Napropamide	15299-99-7	14.3	3	1.28
179	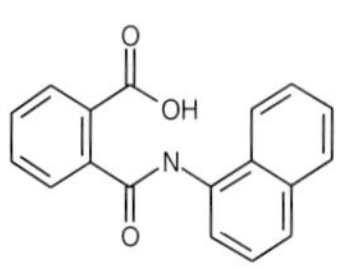	Naptalam	132-66-1	110.2	3	0.42

ID	Structure	Name	CAS NB	Toxicity LC_{50}		
				mg/l	Class	–Log mmol/l
180*	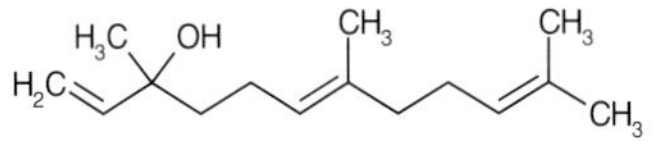	Nerolidol	7212-44-4	2.2	2	2.00
182	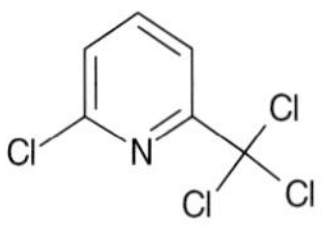	Nitrapyrin	1929-82-4	2.2	2	2.02
183	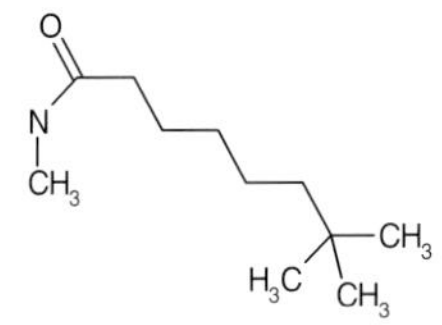	N-methylneo decanamide	105726-67-8	130	3	0.15
185	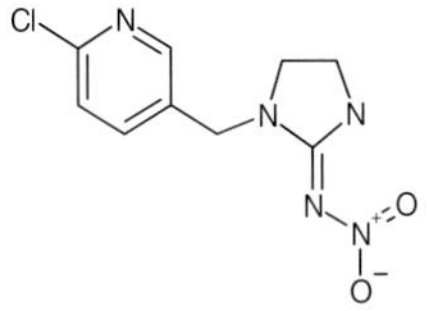	Imidacloprid	105827-78-9	85.2	3	0.48
186	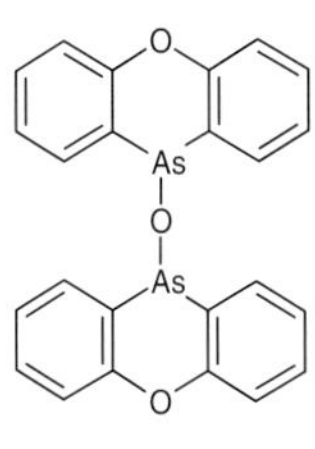	OBPA	58-36-6	0.005	1	5.02
187		Octhilinone	26530-20-1	0.18	1	3.07

ID	Structure	Name	CAS NB	Toxicity LC_{50}		
				mg/l	Class	–Log mmol/l
188	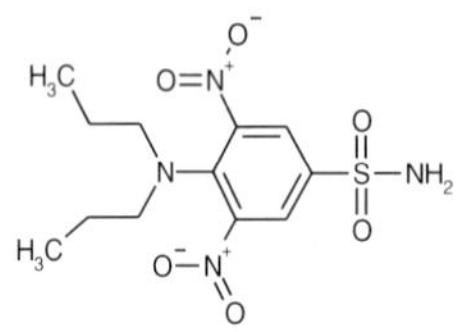	Oryzalin	19044-88-3	1.5	2	2.36
189	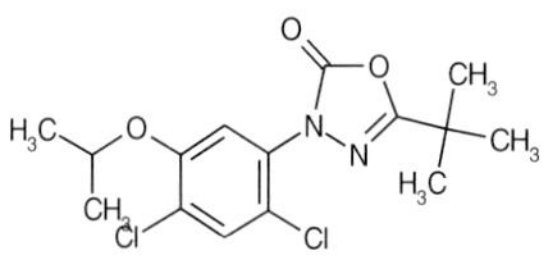	Oxadiazon	19666-30-9	2.18	2	2.20
191	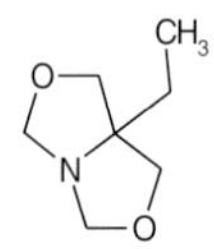	Oxazolidine E	7747-35-5	42	3	0.53
192	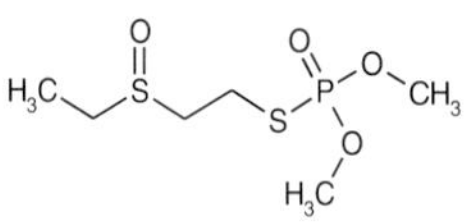	Oxydemeton-methyl	301-12-2	0.24	1	3.01
193*	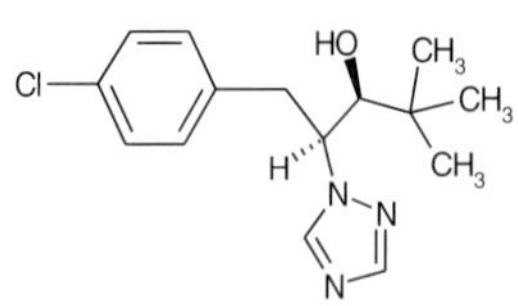	Paclobutrazol	76738-62-0	33.2	3	0.95
197	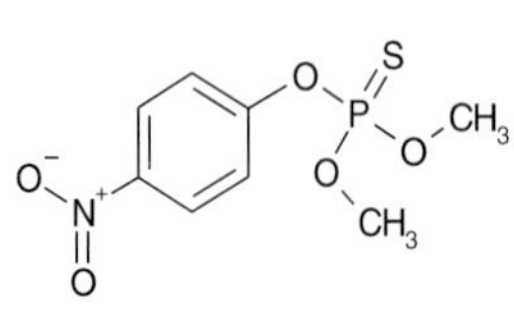	Parathion-methyl	298-00-0	0.0001	1	6.27

ID	Structure	Name	CAS NB	Toxicity LC_{50}		
				mg/l	Class	–Log mmol/l
198	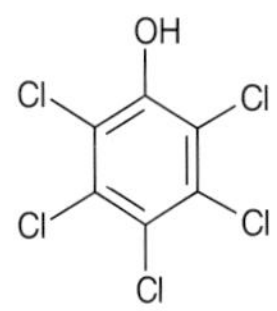	PCP	87-86-5	0.24	1	3.05
199	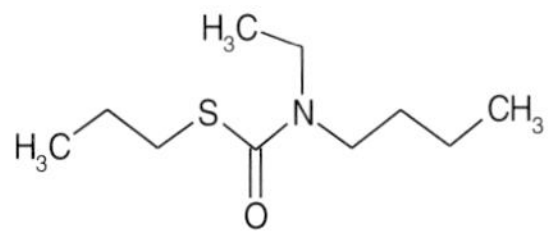	Pebulate	1114-71-2	6.83	2	1.47
200	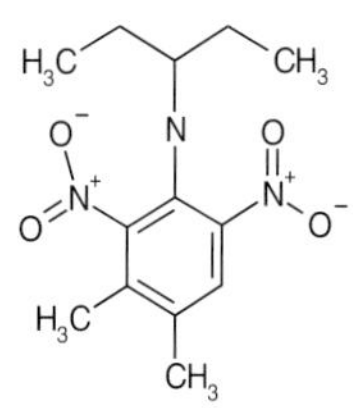	Pendimethalin	40487-42-1	0.28	1	3.00
205	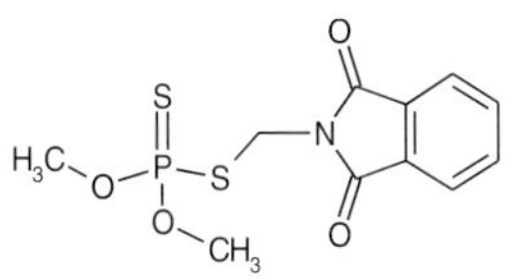	Phosmet	732-11-6	0.006	1	4.75
209	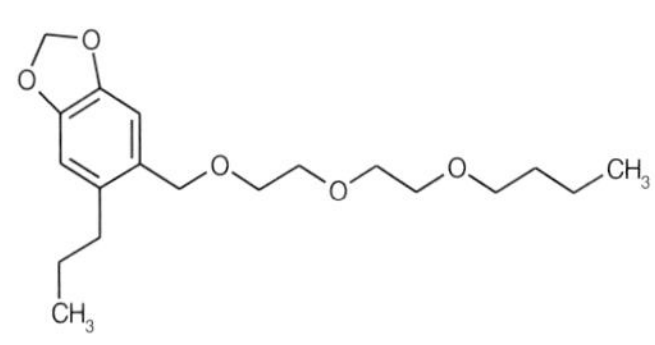	Piperonyl butoxide	51-03-6	0.51	1	2.82
213	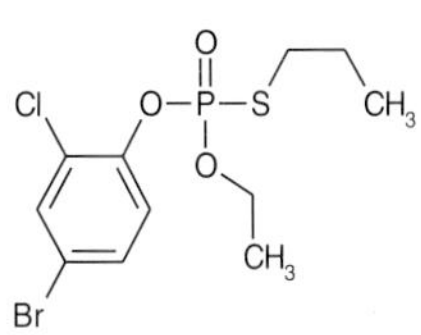	Profenofos	41198-08-7	0.001	1	5.60

ID	Structure	Name	CAS NB	Toxicity LC_{50}		
				mg/l	Class	–Log mmol/l
214	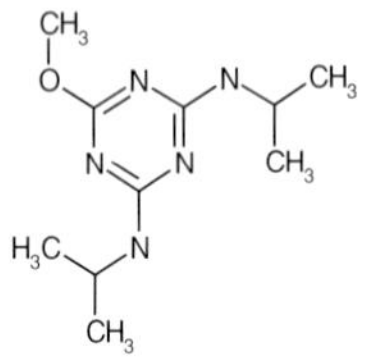	Prometon	1610-18-0	25.7	3	0.94
215*	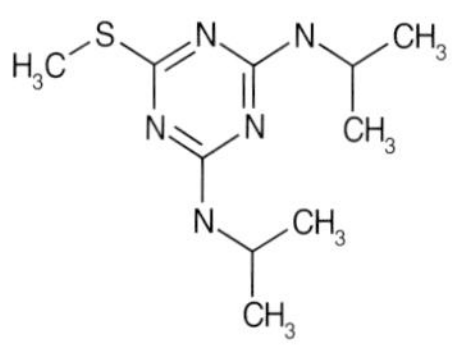	Prometryn	7287-19-6	18.6	3	1.11
216	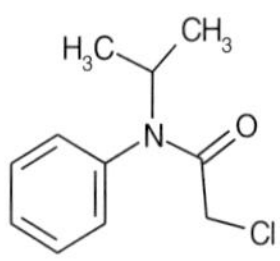	Propachlor	1918-16-7	6.9	2	1.49
217	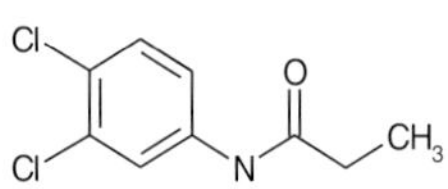	Propanil	709-98-8	6.7	2	1.51
224	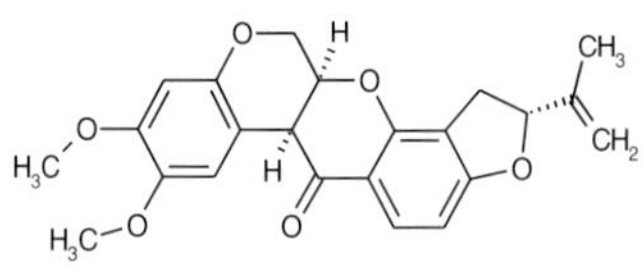	Rotenone	83-79-4	0.004	1	5.03
225*	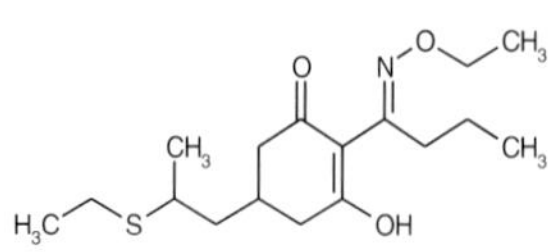	Sethoxydim	74051-80-2	78.1	3	0.62
227	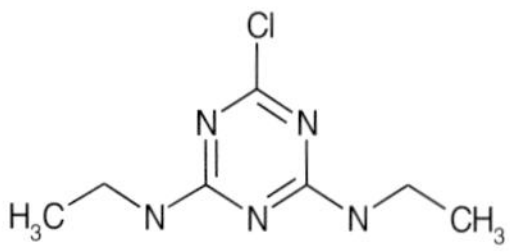	Simazine	122-34-9	1.1	2	2.26

ID	Structure	Name	CAS NB	Toxicity LC_{50}		
				mg/l	Class	–Log mmol/l
228*	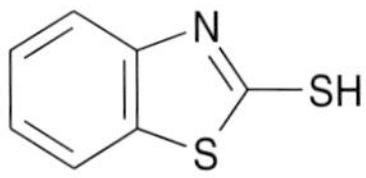	2-Mercapto benzothiazole	149-30-4	2.6	2	1.81
230*	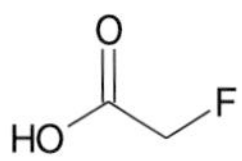	Fluoroacetic acid	144-49-0	273.1	3	–0.54
233	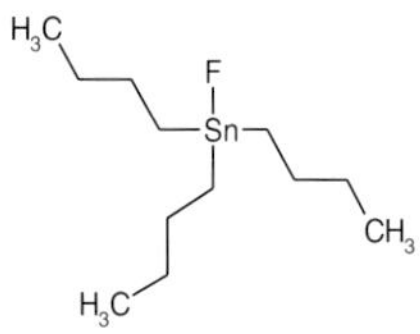	TBTF	1983-10-4	0.0002	1	6.09
236	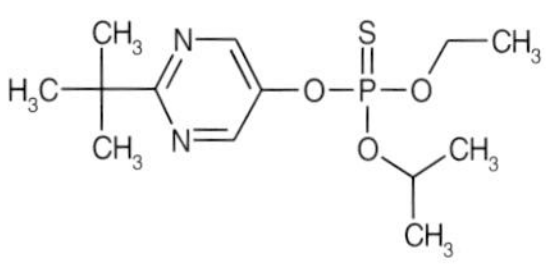	Tebupirimfos	96182-53-5	0.0001	1	6.61
237	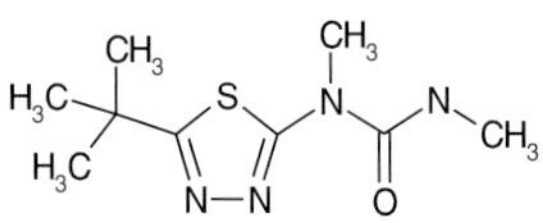	Tebuthiuron	34014-18-1	297	3	–0.11
238	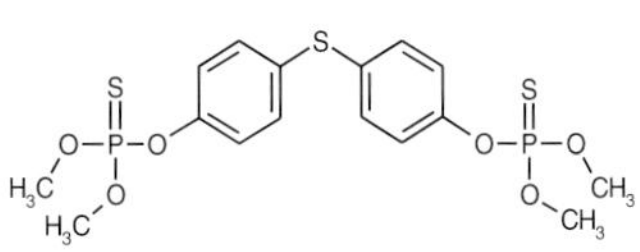	Temephos	3383-96-8	0.0001	1	7.63
239	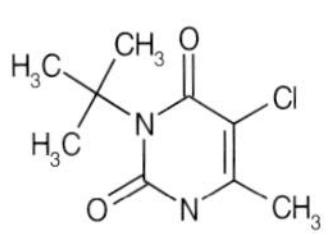	Terbacil	5902-51-2	65	3	0.52

ID	Structure	Name	CAS NB	Toxicity LC_{50}		
				mg/l	Class	–Log mmol/l
240	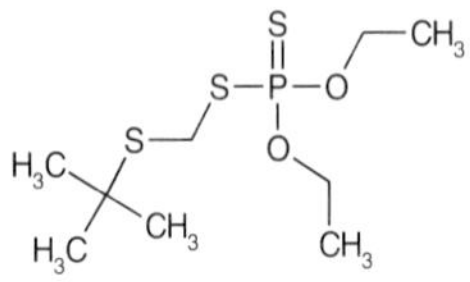	Terbufos	13071-79-9	0.0003	1	5.97
241	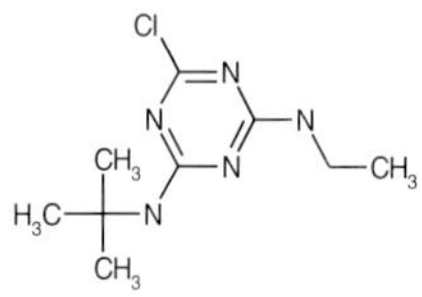	Terbuthylazine	5915-41-3	21.2	3	1.03
244	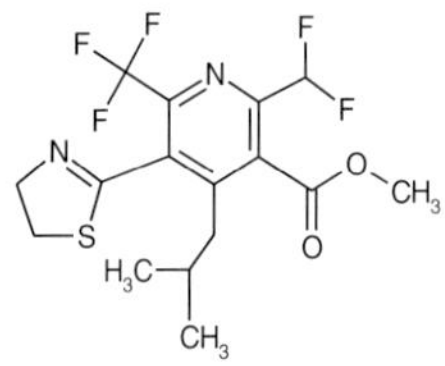	Thiazopyr	117718-60-2	6.1	2	1.81
245*	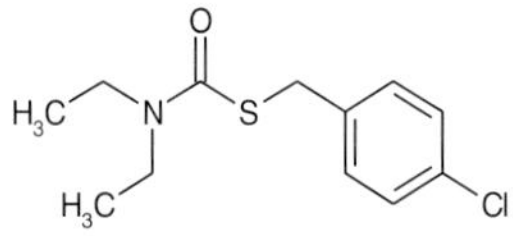	Thiobencarb	28249-77-6	0.1	1	3.41
248	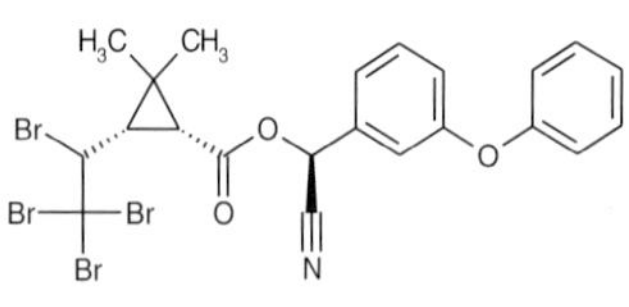	Tralomethrin	66841-25-6	0.0004	1	7.23
254	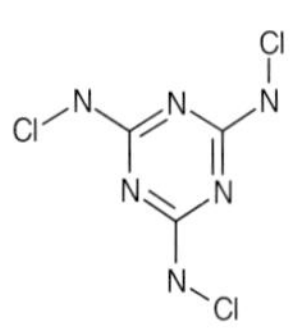	Trichloro melamine	7673-09-8	1.2	2	2.28

ID	Structure	Name	CAS NB	Toxicity LC_{50}		
				mg/l	Class	–Log mmol/l
256	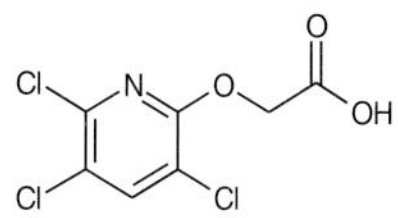	Triclopyr	55335-06-3	132.9	3	0.29
257	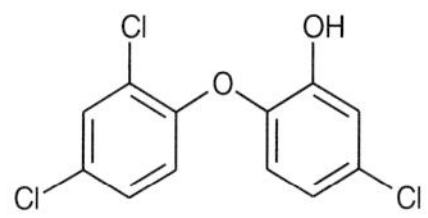	Triclosan	3380-34-5	0.39	1	2.87
258	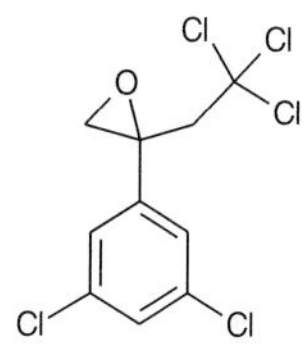	Tridiphane	58138-08-2	0.25	1	3.11
262	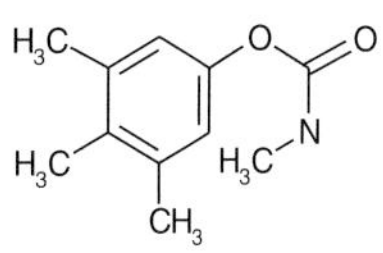	Trimethacarb	2686-99-9	0.02	1	4.03
263	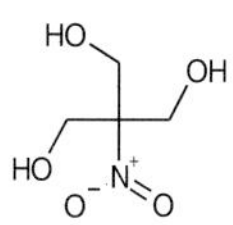	Tris nitro	126-11-4	80	3	0.28
269	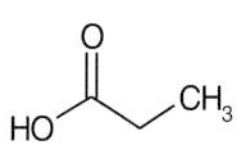	Propionic acid	79-09-4	22.7	3	0.51
271*	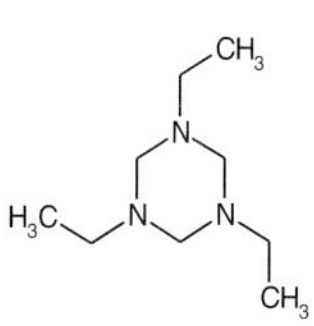	1,3,5-Triethyl hexahydro-s-triazine	7779-27-3	15.3	3	1.05

ID	Structure	Name	CAS NB	Toxicity LC_{50}		
				mg/l	Class	–Log mmol/l
272	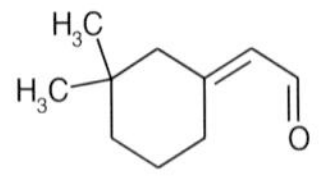	(E)-(3,3-Dimethyl cyclohexylidene) acetaldehyde	26532-25-2	4.7	2	1.51
273	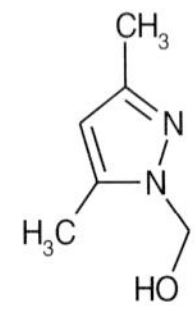	Dimethyl hydroxy methyl pyrazole	85264-33-1	31.7	3	0.60
274*	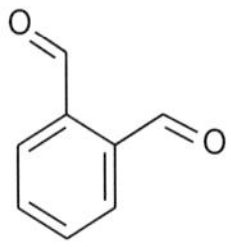	1,2-Benzene dicarbox aldehyde	643-79-8	0.09	1	3.17
275	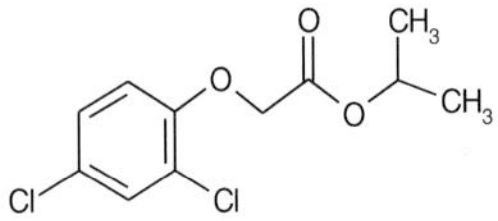	2,4-D isopropyl ester	94-11-1	2.6	2	2.01
276	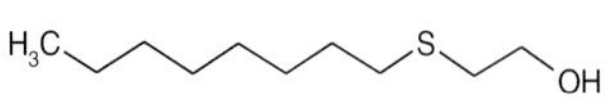	2-Hydroxyethyl octyl sulfide	3547-33-9	0.37	1	2.71
278	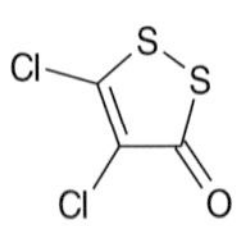	4,5-Dichloro-1,2-dithiol-3-one	1192-52-5	0.01	1	4.23
279	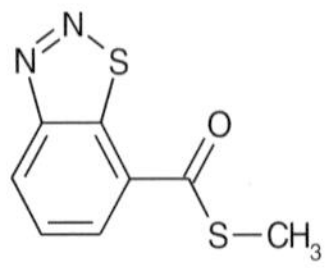	Acibenzolar-*S*-methyl	135158-54-2	2.9	2	1.86

ID	Structure	Name	CAS NB	Toxicity LC_{50}		
				mg/l	Class	–Log mmol/l
280	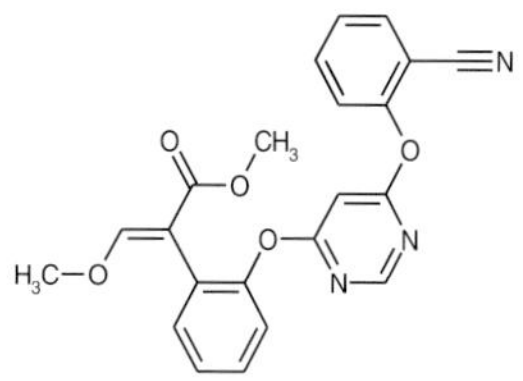	Azoxystrobin	131860-33-8	0.26	1	3.19
285	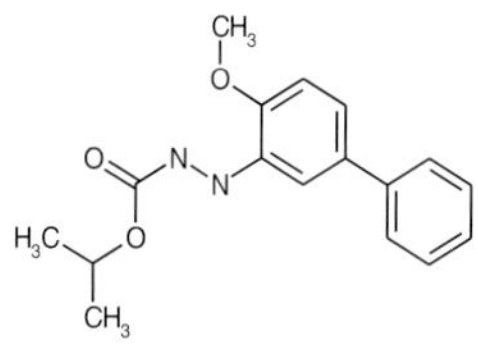	Bifenazate	149877-41-8	0.5	1	2.78
286	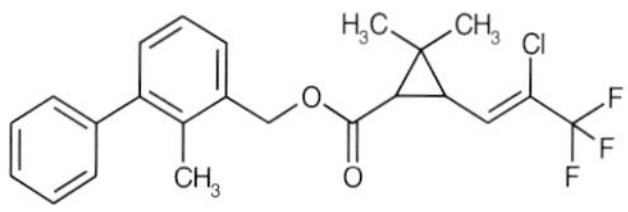	Bifenthrin	82657-04-3	0.002	1	5.42
288	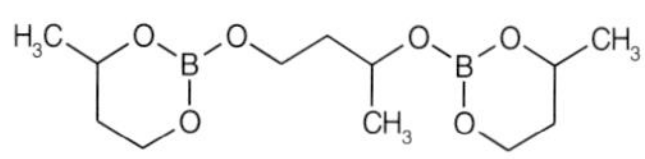	Biobor	2665-13-6	730	3	–0.41
289*	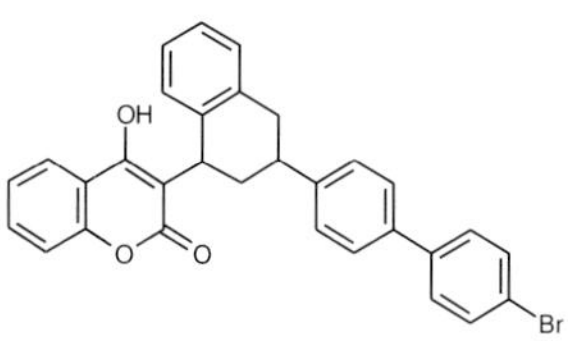	Brodifacoum	56073-10-0	0.98	1	2.73
290	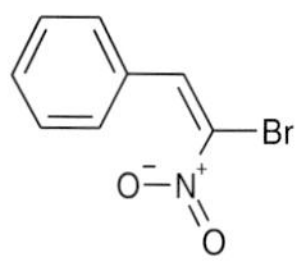	Bromonitro styrene	7166-19-0	0.02	1	3.98
297	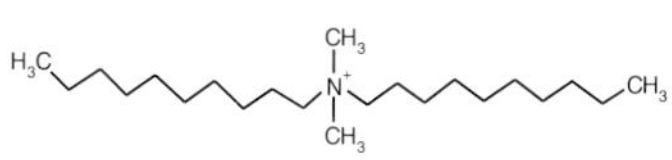	DDAC	7173-51-5	0.02	1	4.30

ID	Structure	Name	CAS NB	Toxicity LC_{50}		
				mg/l	Class	–Log mmol/l
300	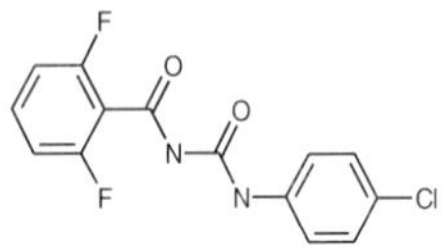	Diflubenzuron	35367-38-5	0.004	1	4.92
302	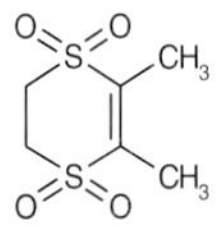	Dimethipin	55290-64-7	21.3	3	0.99
303	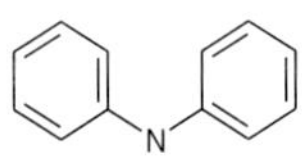	Diphenylamine	122-39-4	1.2	2	2.15
304	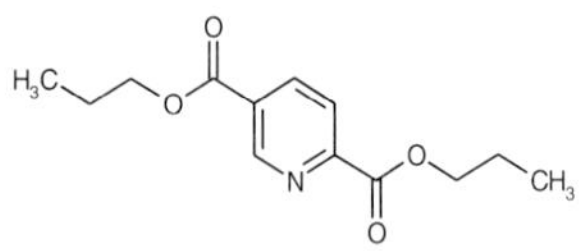	Dipropyl isocincho meronate	136-45-8	18	3	1.14
305	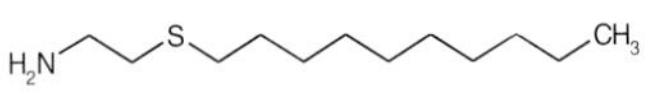	DTEA	29873-30-1	0.031	1	3.85
306	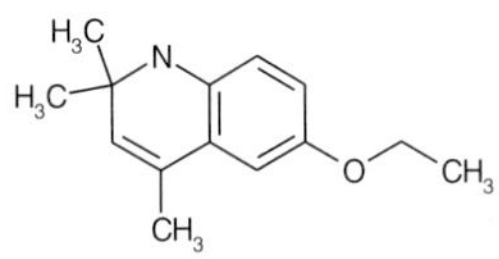	Ethoxyquin	91-53-2	2	2	2.04
307	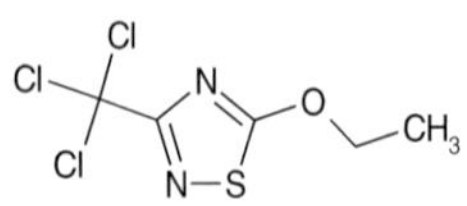	Etridiazole	2593-15-9	4.9	2	1.70
308	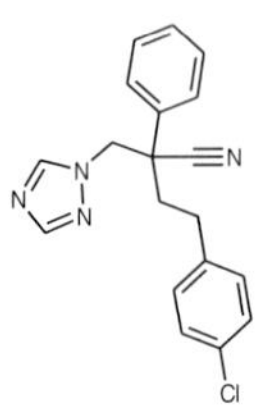	Fenbuconazole	114369-43-6	2.3	2	2.17

ID	Structure	Name	CAS NB	Toxicity LC_{50} mg/l	Class	–Log mmol/l
310	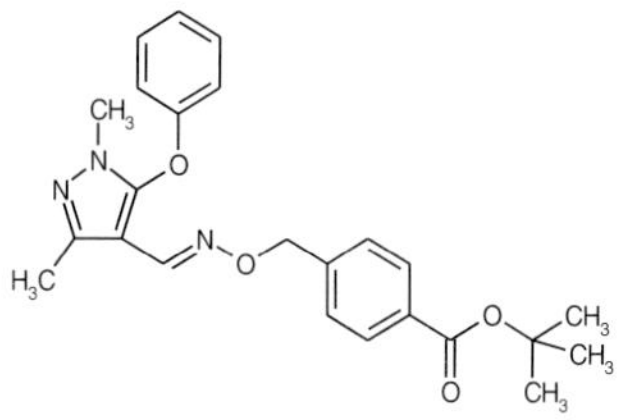	Fenpyroximate	134098-61-6	0.004	1	5.07
311	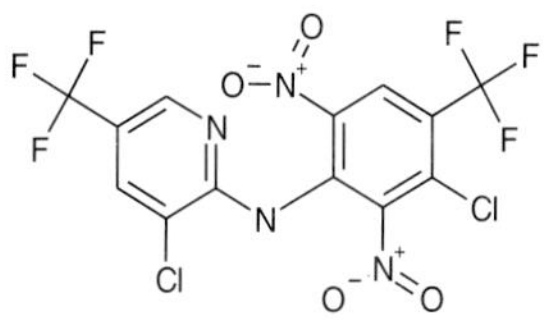	Fluazinam	79622-59-6	0.18	1	3.41
315	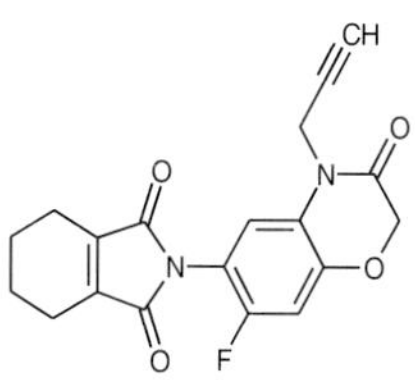	Flumioxazin (V-53482)	103361-09-7	5.5	2	1.81
317	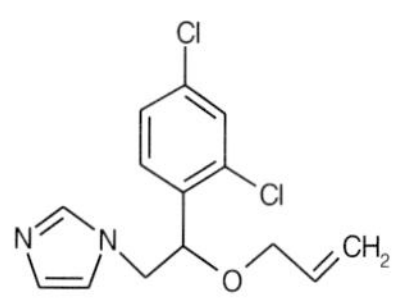	Imazalil	35554-44-0	3.16	2	1.97
318	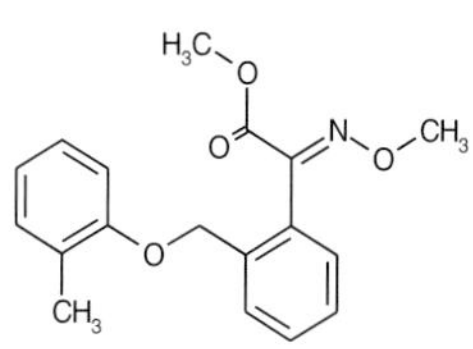	Kresoxim methyl	143390-89-0	0.33	1	2.97

ID	Structure	Name	CAS NB	Toxicity LC_{50}		
				mg/l	Class	–Log mmol/l
319	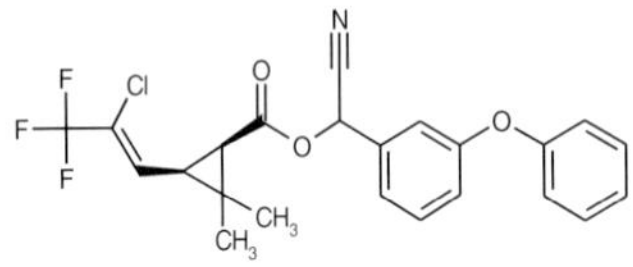	Lambda-cyhalothrin	91465-08-6	0.0002	1	6.29
320	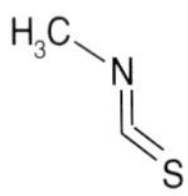	Methyl isothiocyanate	556-61-6	0.055	1	3.12
321	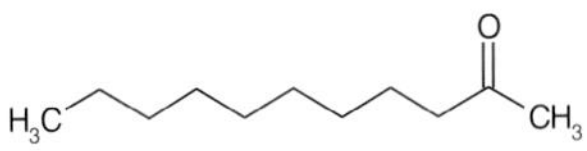	Methyl nonyl ketone	112-12-9	0.54	1	2.50
322*	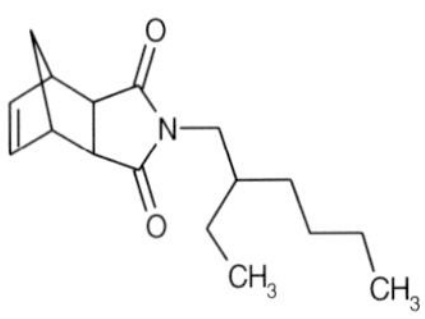	MGK 264	113-48-4	2.3	2	2.08
323	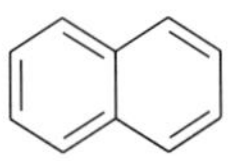	Naphthalene	91-20-3	1.6	2	1.90
328	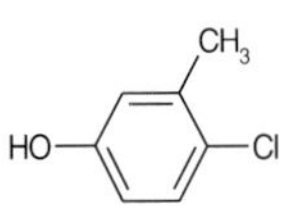	Parachloro metacresol	59-50-7	2.3	2	1.79
329	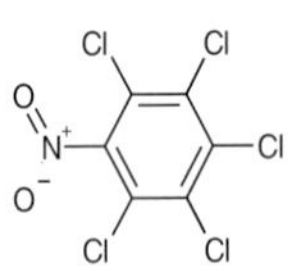	PCNB	82-68-8	0.77	1	2.58
330*	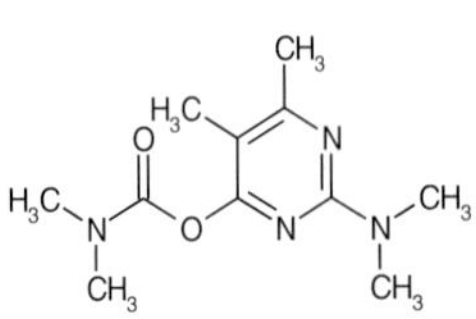	Pirimicarb	23103-98-2	0.0065	1	4.56

ID	Structure	Name	CAS NB	Toxicity LC_{50}		
				mg/l	Class	–Log mmol/l
331	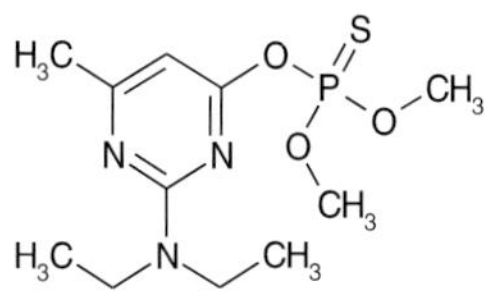	Pirimiphos methyl	29232-93-7	0.0002	1	6.16
332	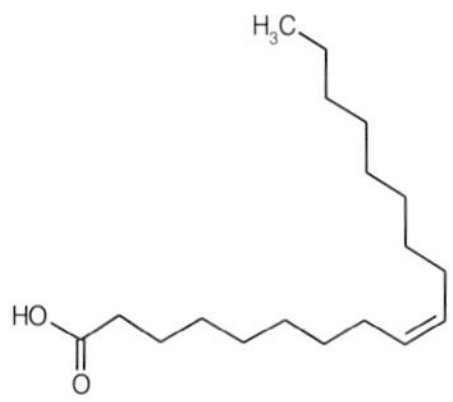	Oleic acid	112-80-1	0.5	1	2.75
333	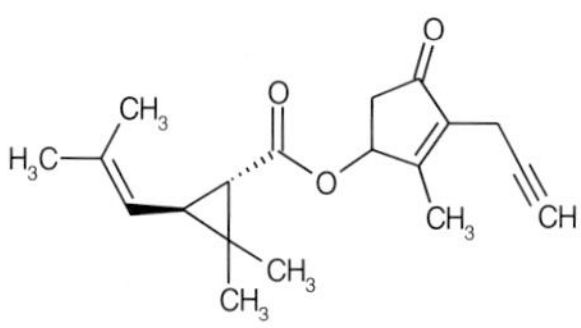	Prallethrin	23031-36-9	0.006	1	4.69
334	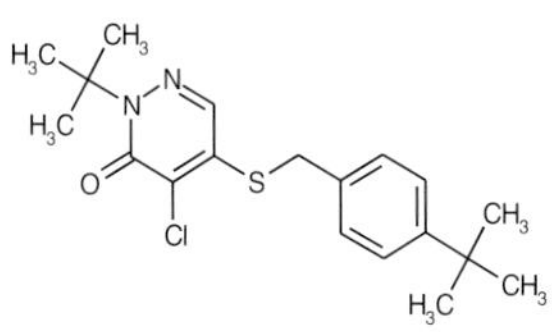	Pyridaben	96489-71-3	0.0005	1	5.84
337	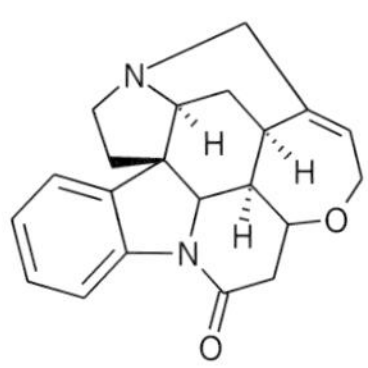	Strychnine	57-24-9	10	2	1.52

ID	Structure	Name	CAS NB	Toxicity LC$_{50}$		
				mg/l	Class	–Log mmol/l
338	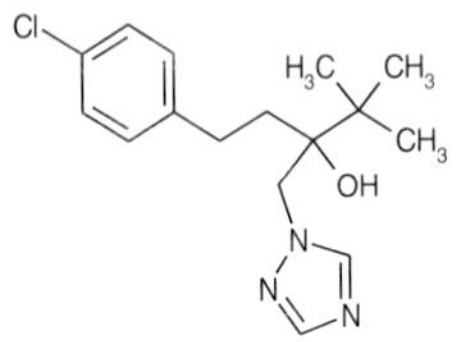	Tebuconazole	107534-96-3	4	2	1.89
339	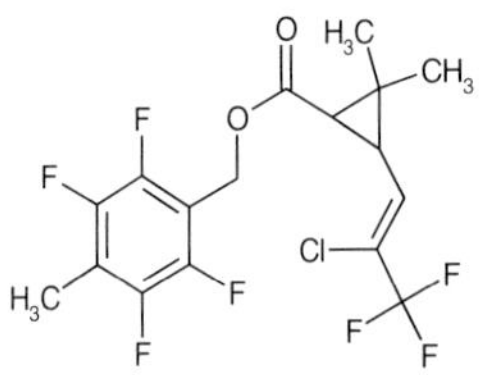	Tefluthrin	79538-32-2	0.0001	1	6.78
340	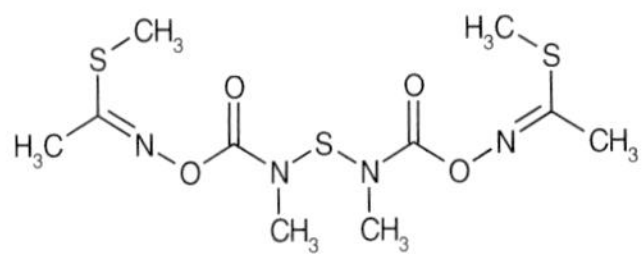	Thiodicarb	59669-26-0	0.03	1	4.12
342	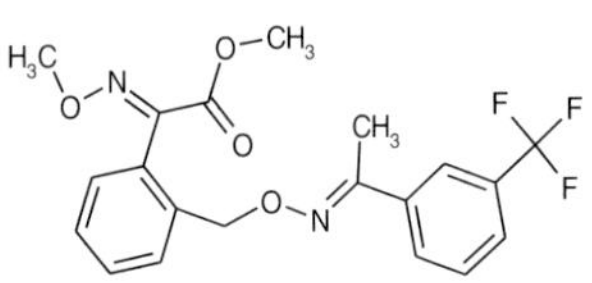	Trifloxystrobin	141517-21-7	0.025	1	4.21
345	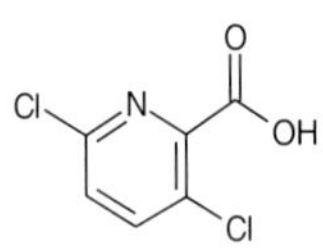	Clopyralid	1702-17-6	170.7	3	0.05
346	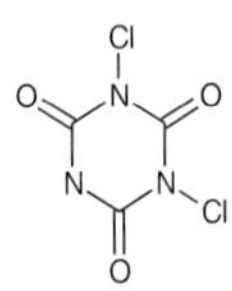	Dichloroiso Cyanuric acid	2782-57-2	0.1	1	3.30
347*	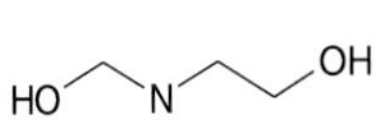	2-(Hydroxy methylamino) ethanol	34375-28-5	25.2	3	0.56

ID	Structure	Name	CAS NB	Toxicity LC_{50}		
				mg/l	Class	–Log mmol/l
348	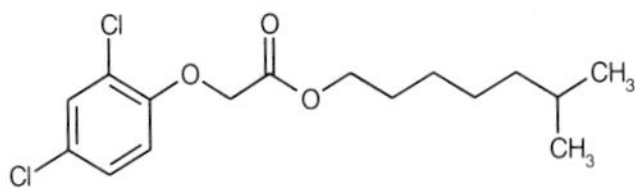	2,4-D isooctyl ester	25168-26-7	0.02	1	4.25
349	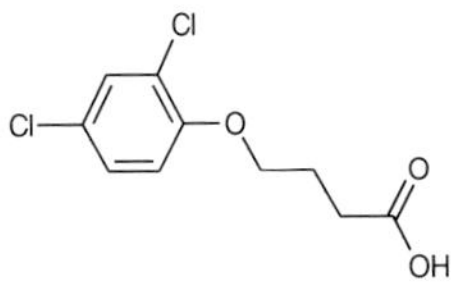	2,4-DB	94-82-6	25	3	1.00
350	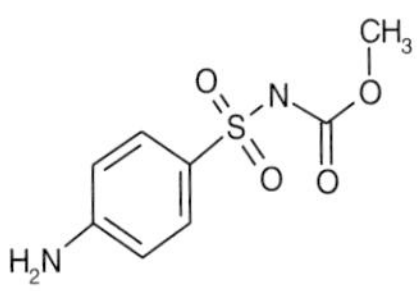	Asulam	3337-71-1	27	3	0.93
351	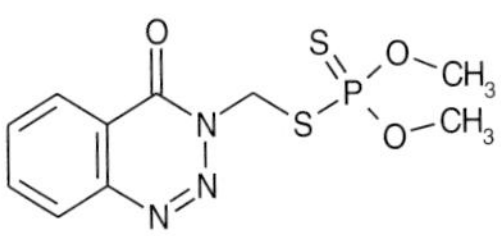	Azinphos-methyl	86-50-0	0.001	1	5.46
352	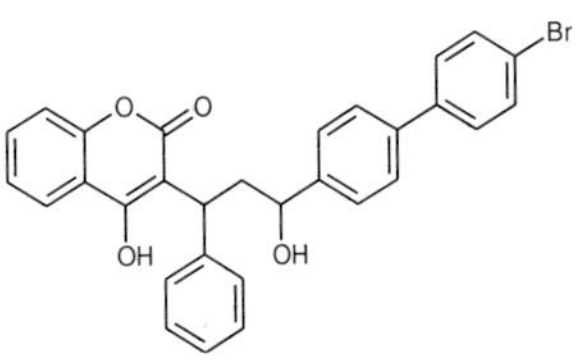	Bromadiolone	28772-56-7	2	2	2.42
353	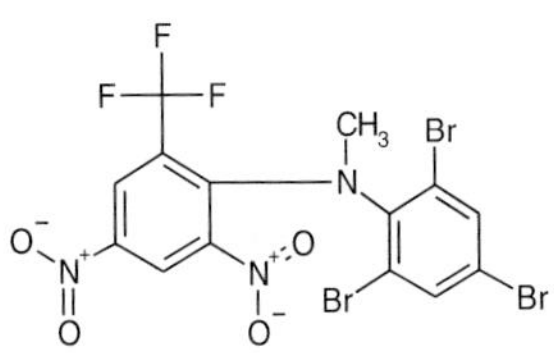	Bromethalin	63333-35-7	0.002	1	5.47

ID	Structure	Name	CAS NB	Toxicity LC_{50}		
				mg/l	Class	–Log mmol/l
354	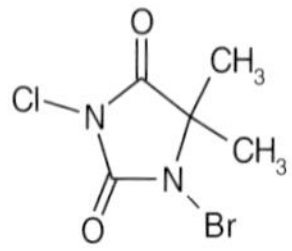	Bromo-3-chloro-5,5-dimethyl hydantoin (BCDMH)	16079-88-2	0.75	1	2.51
355	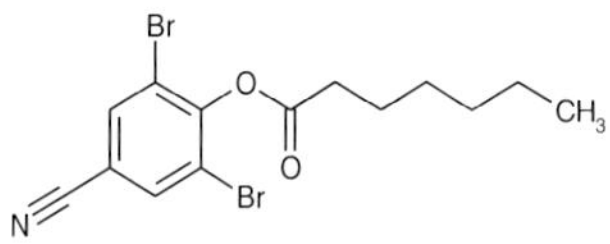	Bromoxynil heptanoate	56634-95-8	0.03	1	4.10
356	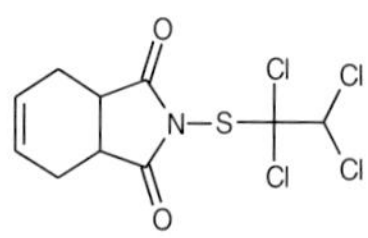	Captafol	2425-06-1	3.4	2	2.01
357	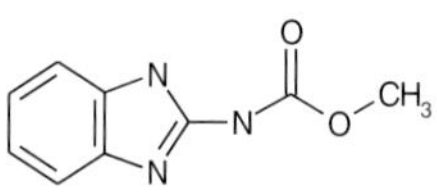	Carbendazim	10605-21-7	0.35	1	2.74
358	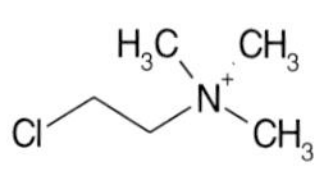	Chloroethyl trimethyl ammonium	7003-89-6	12.96	3	0.98
359	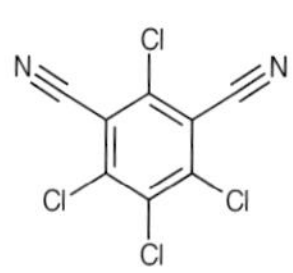	Chlorothalonil	1897-45-6	0.07	1	3.58
360	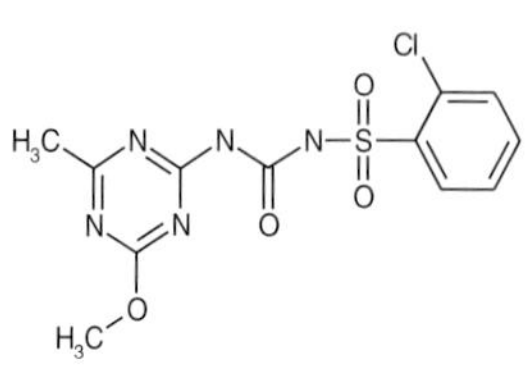	Chlorsulfuron	64902-72-3	370	3	–0.01

ID	Structure	Name	CAS NB	Toxicity LC_{50}		
				mg/l	Class	–Log mmol/l
361*	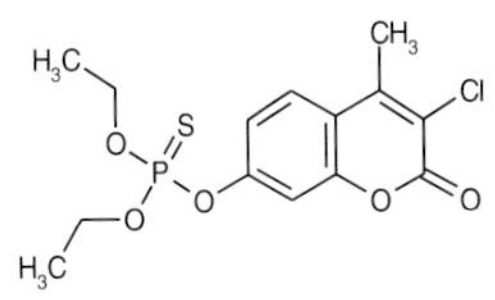	Coumaphos	56-72-4	0.0002	1	6.28
362	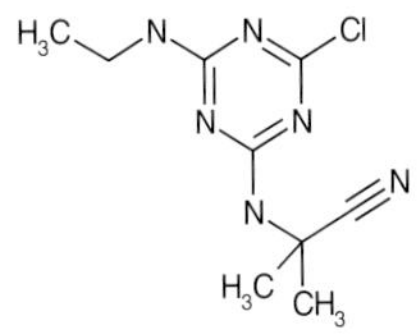	Cyanazine	21725-46-2	49	3	0.69
363	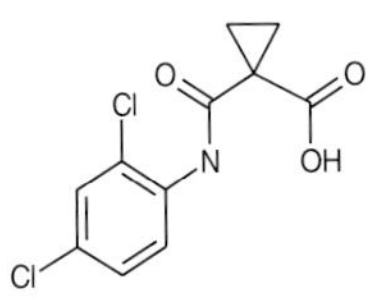	Cyclanilide	113136-77-9	5	2	1.74
364	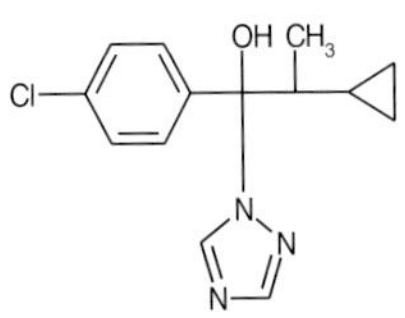	Cyproconazole		26	3	1.05
365		Decanol	112-30-1	6.5	2	1.39
366	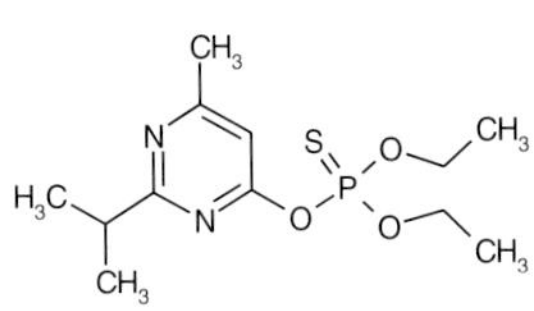	Diazinon	333-41-5	0.001	1	5.50
368	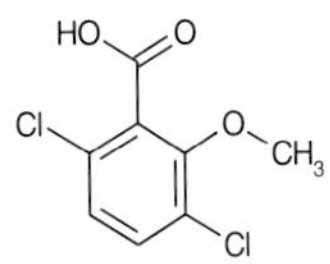	Dicamba	1918-00-9	111	3	0.30

ID	Structure	Name	CAS NB	Toxicity LC_{50}		
				mg/l	Class	–Log mmol/l
369	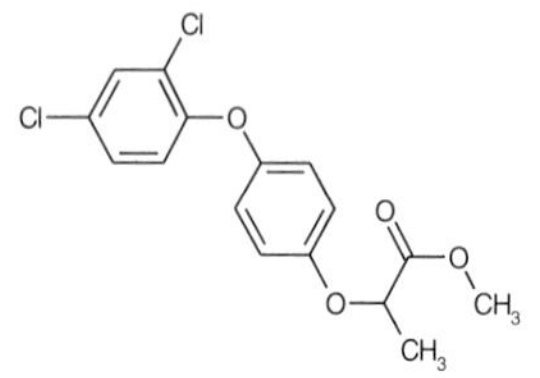	Diclofop-methyl	51338-27-3	0.23	1	3.17
370	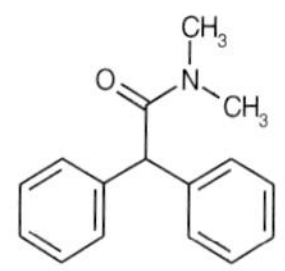	Diphenamid	957-51-7	58	3	0.62
371	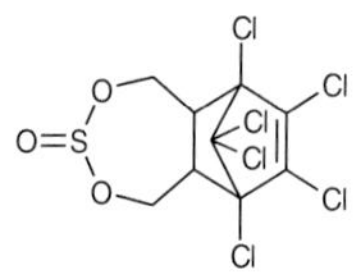	Endosulfan	115-29-7	0.17	1	3.39
372	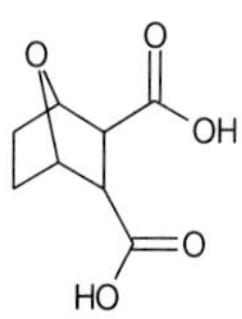	Endothall	145-73-3	32.5	3	0.76
373	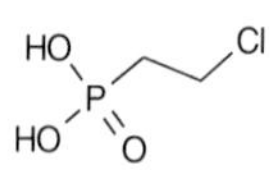	Ethephon	16672-87-0	31.7	3	0.66
374*	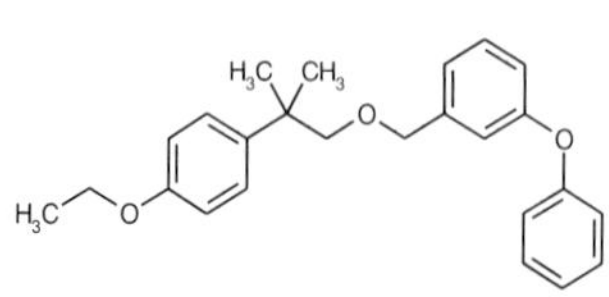	Ethofenprox	80844-07-1	0.0006	1	5.82

ID	Structure	Name	CAS NB	Toxicity LC_{50}		
				mg/l	Class	–Log mmol/l
375	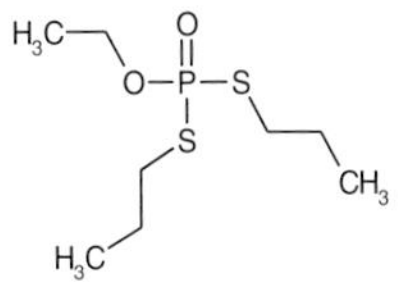	Ethoprophos	13194-48-4	0.09	1	3.42
376	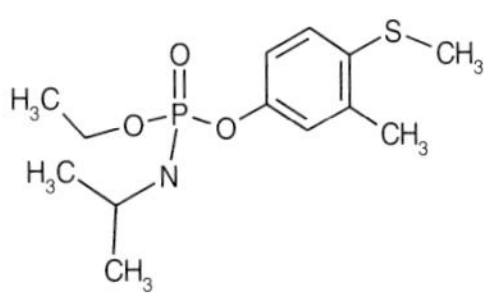	Fenamiphos	22224-92-6	0.002	1	5.20
377	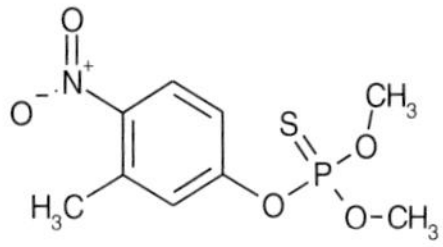	Fenitrothion	122-14-5	0.01	1	4.40
378	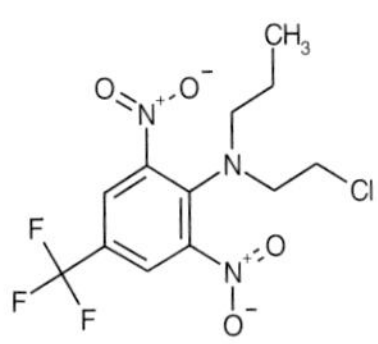	Fluchloralin	33245-39-5	0.6	1	2.80
379*	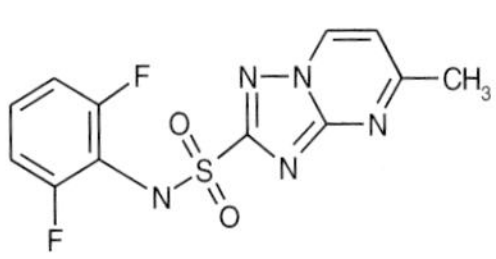	Flumetsulam	98967-40-9	254	3	0.11
380	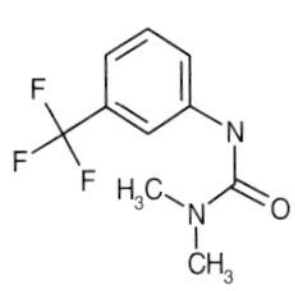	Fluometuron	2164-17-2	9.9	2	1.37

ID	Structure	Name	CAS NB	Toxicity LC_{50}		
				mg/l	Class	–Log mmol/l
381	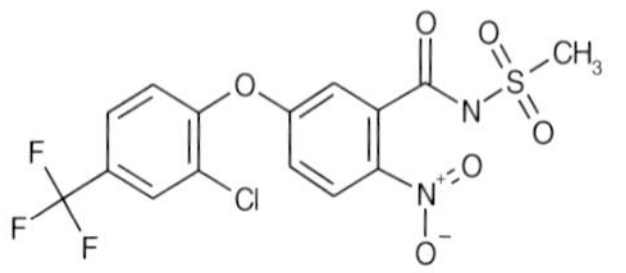	Fomesafen	72178-02-0	294	3	0.17
382	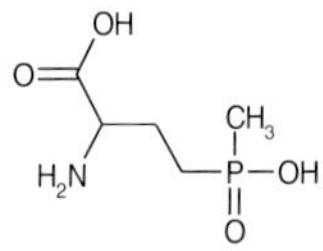	Glufosinate	51276-47-2	610.3	3	–0.53
383*	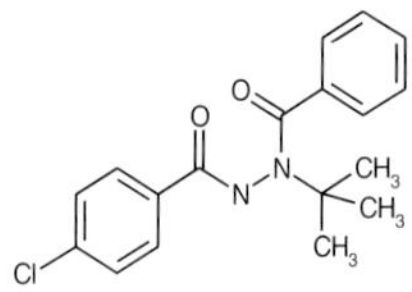	Halofenoxide	112226-61-6	3.6	2	1.96
384*	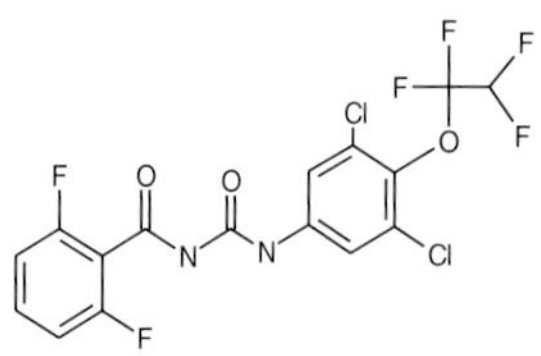	Hexaflumuron	86479-06-3	0.0001	1	6.62
385	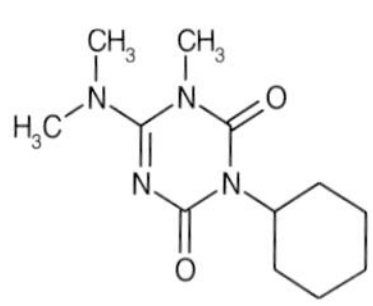	Hexazinone	51235-04-2	151.6	3	0.22
386	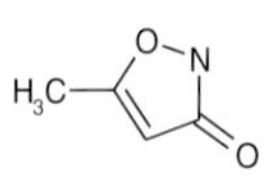	Hymexazol	10004-44-1	30.8	3	0.51
387	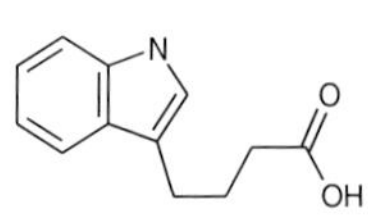	Indole-3-butyric acid	133-32-4	57	3	0.55

ID	Structure	Name	CAS NB	Toxicity LC_{50}		
				mg/l	Class	–Log mmol/l
389	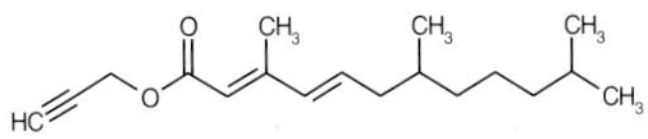	Kinoprene	42588-37-4	0.1	1	3.44
390	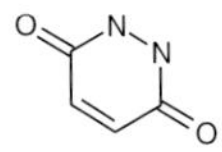	Maleic hydrazide	123-33-1	107.5	3	0.02
391*	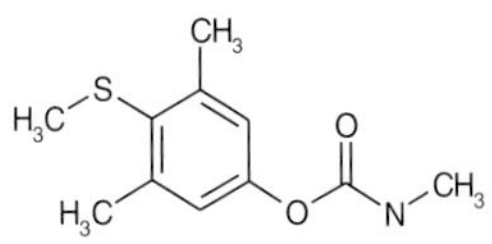	Methiocarb	2032-65-7	0.02	1	4.07
392	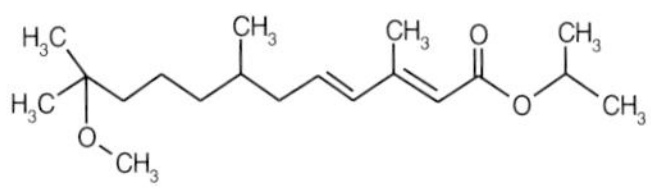	Methoprene	40596-69-8	0.89	1	2.54
393	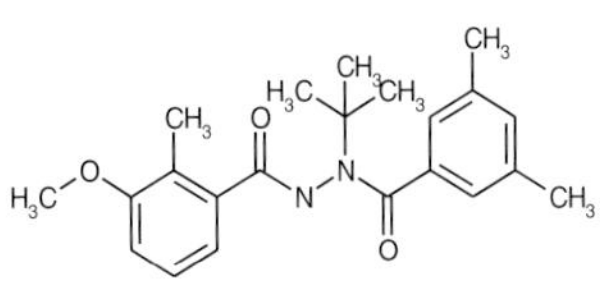	Methoxy fenozide	161050-58-4	3.7	2	2.00
394	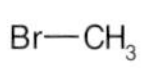	Methyl bromide	74-83-9	2.6	2	1.56
395	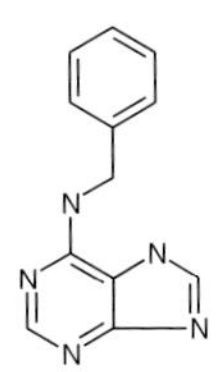	6-Benzyl adenine	1214-39-7	20.5	3	1.04
396	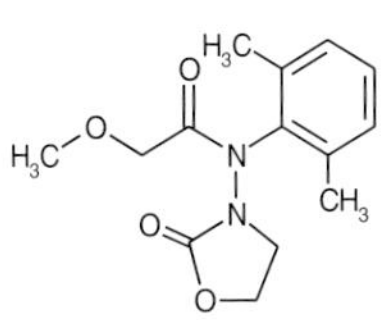	Oxadixyl	77732-09-3	530	3	–0.28

ID	Structure	Name	CAS NB	Toxicity LC_{50}		
				mg/l	Class	–Log mmol/l
397	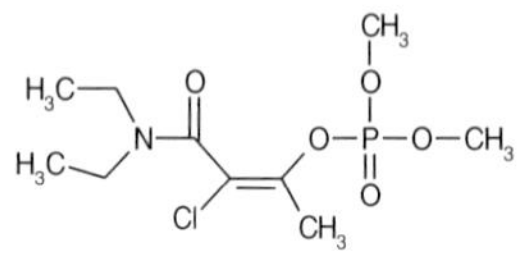	Phosphamidon	13171-21-6	0.013	1	4.37
398	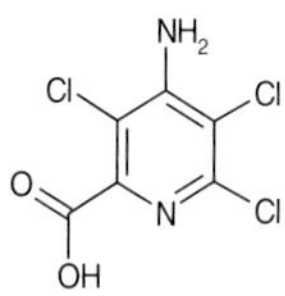	Picloram	1918-02-1	68.3	3	0.55
399	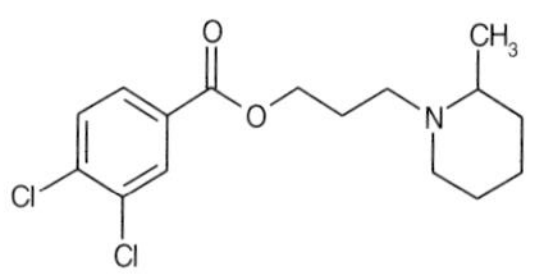	Piperalin	3478-94-2	1.89	2	2.24
400	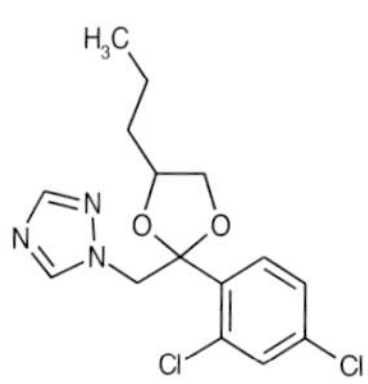	Propiconazole	60207-90-1	4.8	2	1.85
401	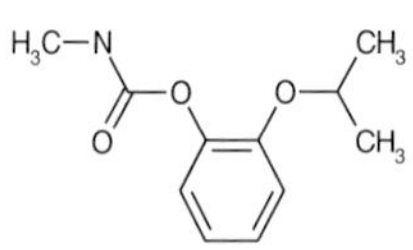	Propoxur	114-26-1	0.01	1	4.28
402	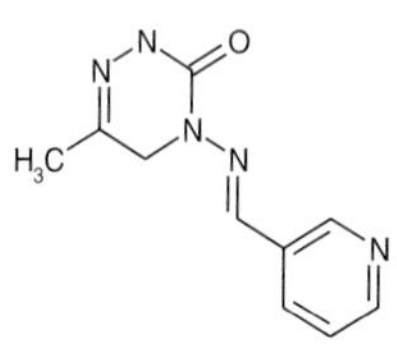	Pymetrozine	123312-89-0	87	3	0.40

ID	Structure	Name	CAS NB	Toxicity LC_{50}		
				mg/l	Class	–Log mmol/l
403	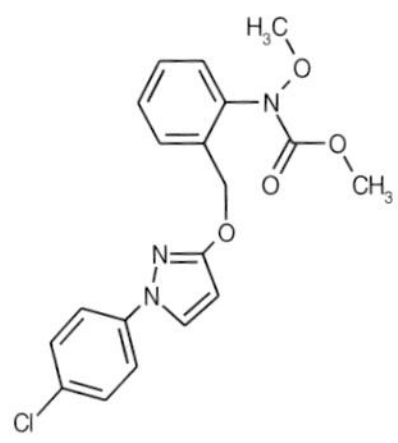	Pyraclostrobin	175013-18-0	0.016	1	4.39
404*	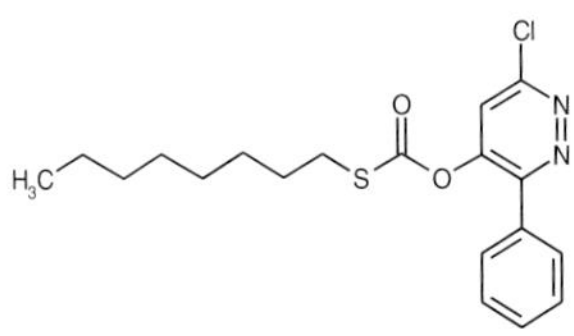	Pyridate	55512-33-9	1.08	2	2.55
405	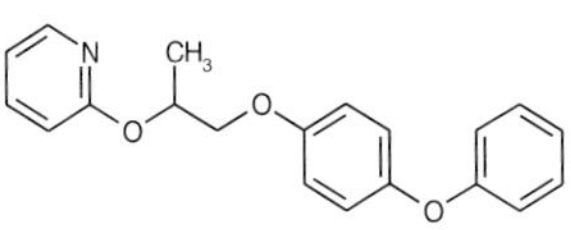	Pyriproxyfen	95737-68-1	0.4	1	2.90
406*	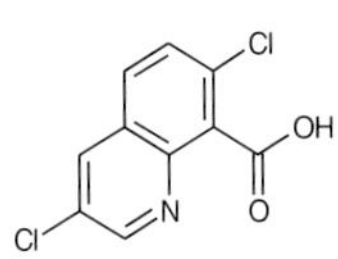	Quinclorac	84087-01-4	29.8	3	0.91
407	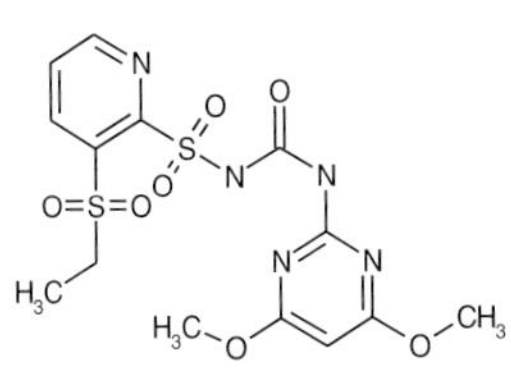	Rimsulfuron	122931-48-0	1000	3	–0.37
408*	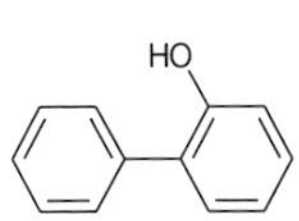	2-Phenylphenol	90-43-7	3.37	2	1.70

ID	Structure	Name	CAS NB	Toxicity LC_{50} mg/l	Toxicity LC_{50} Class	Toxicity LC_{50} –Log mmol/l
409*	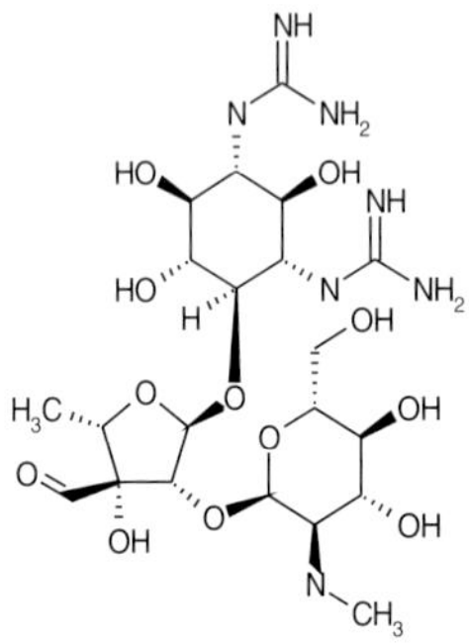	Streptomycin	57-92-1	556.3	3	0.02
410	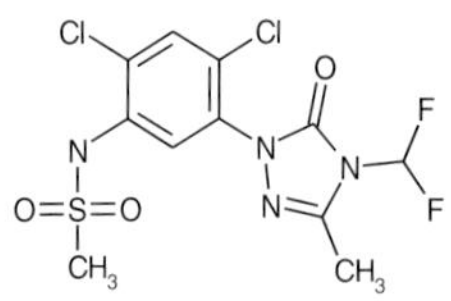	Sulfentrazone	122836-35-5	60.4	3	0.81
411	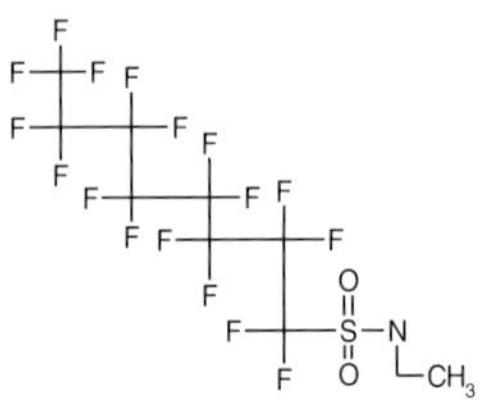	Sulfluramid	4151-50-2	0.37	1	3.15
412	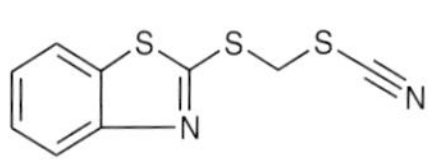	TCMTB	21564-17-0	0.02	1	4.02
413	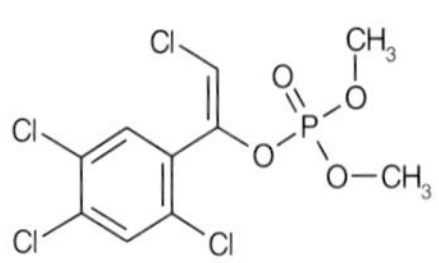	Tetrachlor vinphos	22248-79-9	0.002	1	5.28
414	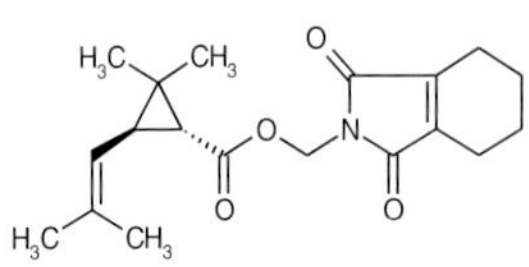	Tetramethrin	7696-12-0	0.045	1	3.87

ID	Structure	Name	CAS NB	Toxicity LC_{50}		
				mg/l	Class	–Log mmol/l
415	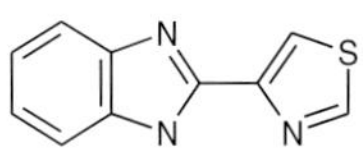	Thiabendazole	148-79-8	0.31	1	2.81
416	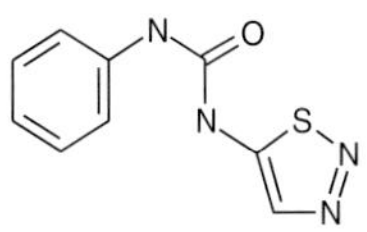	Thidiazuron	51707-55-2	10	2	1.34
417	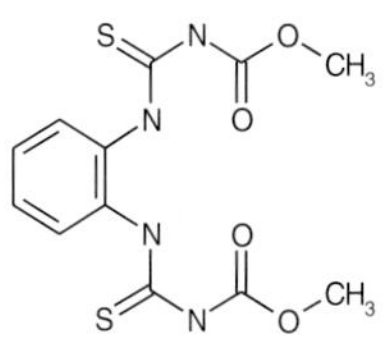	Thiophanate-methyl	23564-05-8	5.4	2	1.80
418	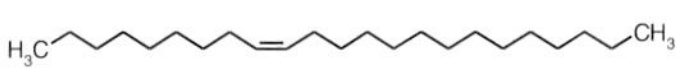	Tricosene	27519-02-4	1.08	2	2.48
419	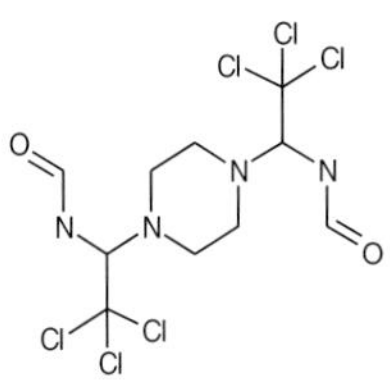	Triforine	26644-46-2	28	3	1.19
420	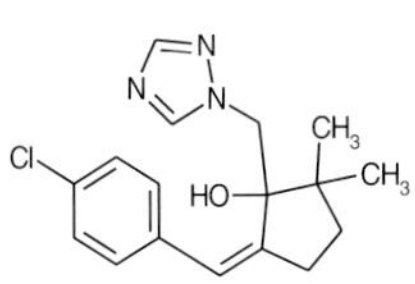	Triticonazole	131983-72-7	7.6	2	1.62
421	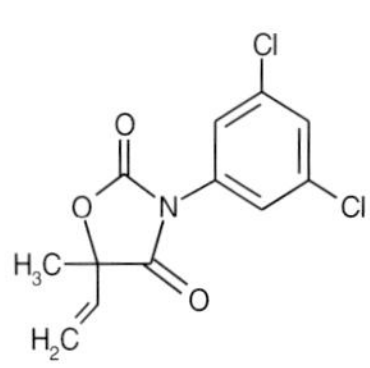	Vinclozolin	50471-44-8	3.65	2	1.89

ID	Structure	Name	CAS NB	Toxicity LC_{50}		
				mg/l	Class	–Log mmol/l
422*	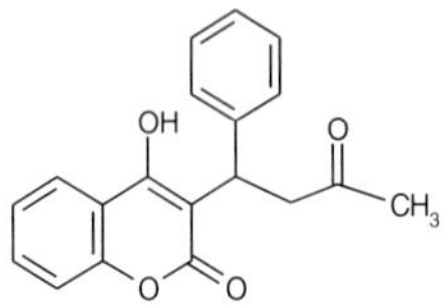	Warfarin	81-81-2	130	3	0.38

LIST OF THE 123 COMPOUNDS INCLUDED IN THE DIETARY QUAIL DATA SET

Three toxicity classes were generated according to the hazard criteria adopted by the USEPA and Office of Pollution Prevention and Toxics (OPPT):

LD_{50} (class 1) < 1000 mg/kg; 1000 mg/kg ≤ LD_{50} (class 2) < 5000 mg/kg; LD_{50} (class 3) ≥ 5000 mg/kg. The symbol '*' in the ID column indicated the test set compounds.

ID	Structure	Name	CAS NB	Toxicity LD_{50}		
				mg/kg	Class	–Log mmol/kg
11	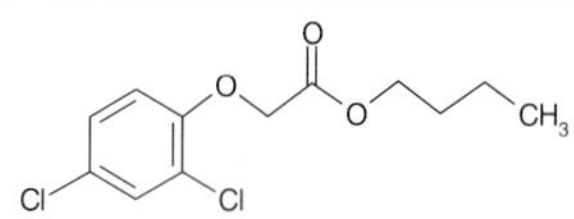	2,4-D butyl ester	94-80-4	10000	3	–1.56
19	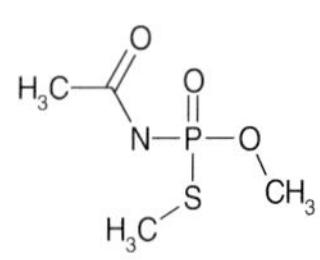	Acephate	30560-19-1	1280	2	–0.84
22*	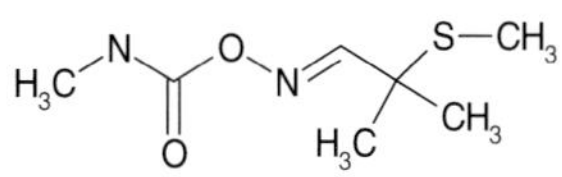	Aldicarb	116-06-3	71	1	0.43
23	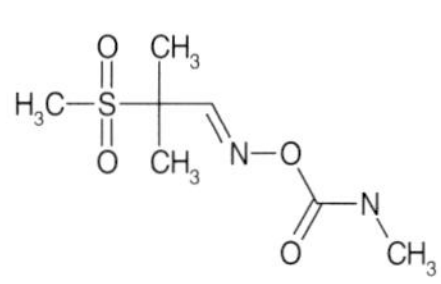	Aldicarb sulfone	1646-88-4	5706	3	–1.41
26	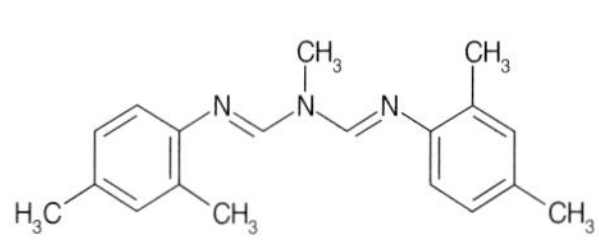	Amitraz	33089-61-1	3081	2	–1.02
30	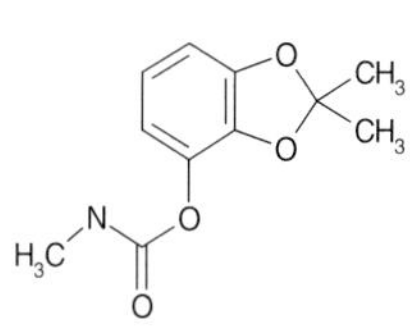	Bendiocarb	22781-23-3	1770	2	–0.90
36	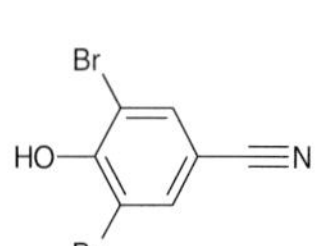	Bromoxynil	1689-84-5	2010	2	–0.86

ID	Structure	Name	CAS NB	Toxicity LD_{50}		
				mg/kg	Class	–Log mmol/kg
37	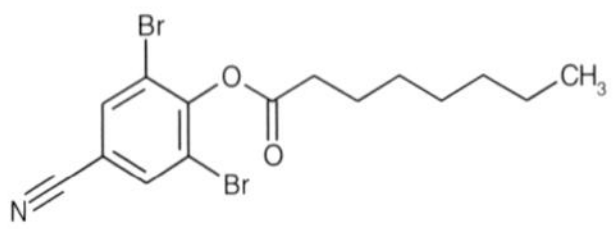	Bromoxynil octanoate	1689-99-2	1315	2	–0.51
38	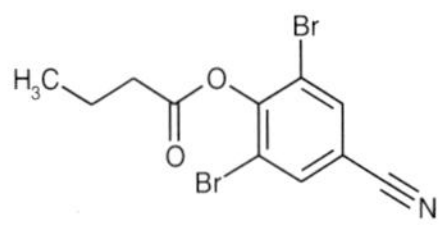	Bromoxynil butyrate	3861-41-4	3159	2	–0.96
39	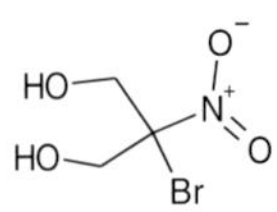	Bronopol	52-51-7	4487	2	–1.35
46	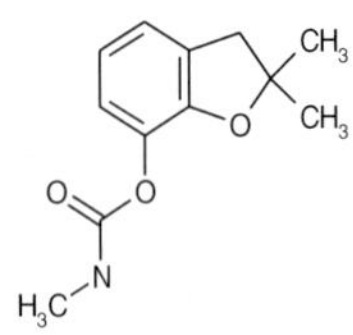	Carbofuran	1563-66-2	714	1	–0.51
48*	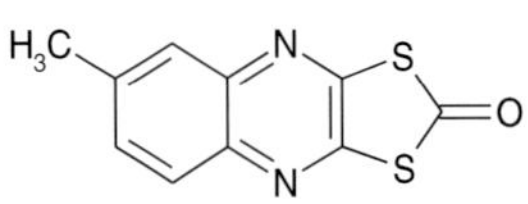	Chino methionat	2439-01-2	2170	2	–0.97
50	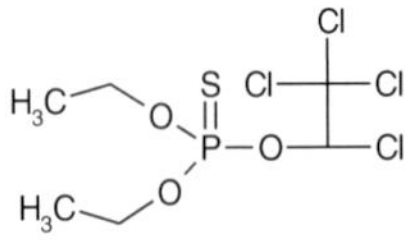	Chloro ethoxyfos	54593-83-8	181	1	0.27
54	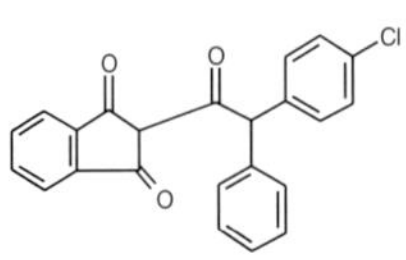	Chloro phacinone	3691-35-8	242	1	0.19

ID	Structure	Name	CAS NB	Toxicity LD_{50}		
				mg/kg	Class	–Log mmol/kg
58	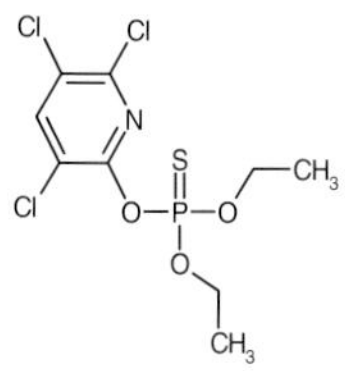	Chlorpyrifos	2921-88-2	423	1	–0.08
59*	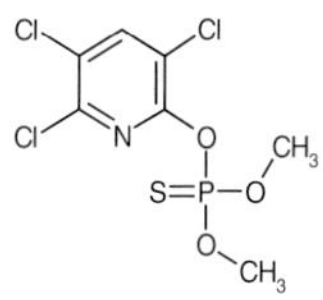	Chlorpyrifos-methyl	5598-13-0	2010	2	–0.79
76	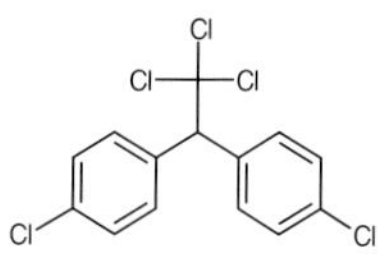	DDT	50-29-3	611	1	–0.24
81	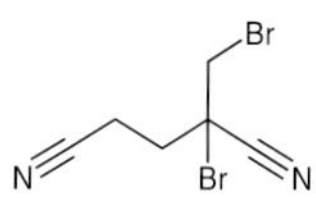	1,2-Dibromo-2,4-dicyano butane	35691-65-7	4042	2	–1.18
84	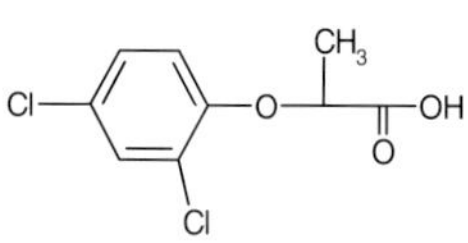	Dichlorprop	120-36-5	9907	3	–1.62
87	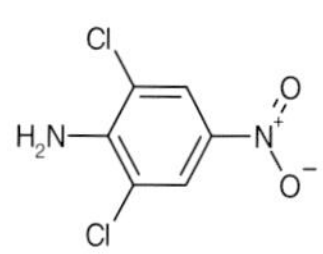	Dicloran	99-30-9	1219	2	–0.77
88*	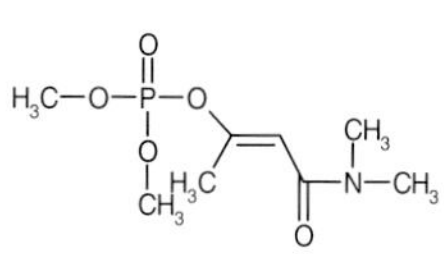	Dicrotophos	141-66-2	13	1	1.26

ID	Structure	Name	CAS NB	Toxicity LD_{50}		
				mg/kg	Class	–Log mmol/kg
89	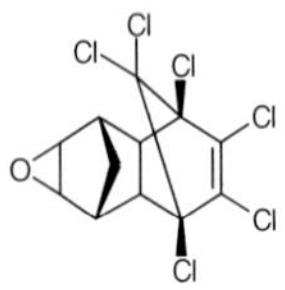	Dieldrin	60-57-1	37	1	1.01
91*	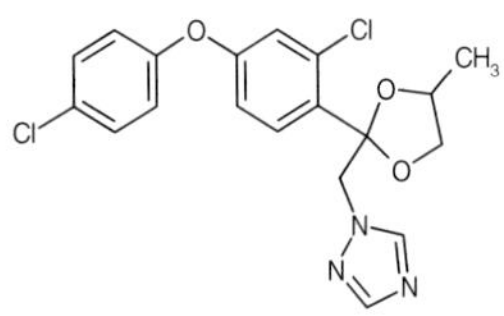	Difeno conazole	119446-68-3	4760	2	–1.07
98	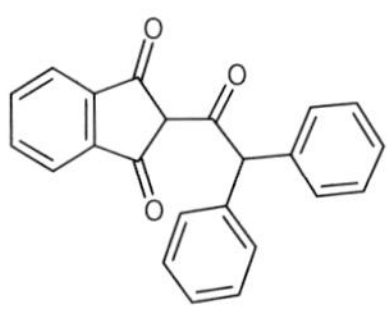	Diphacinone	82-66-6	4485	2	–1.12
102*	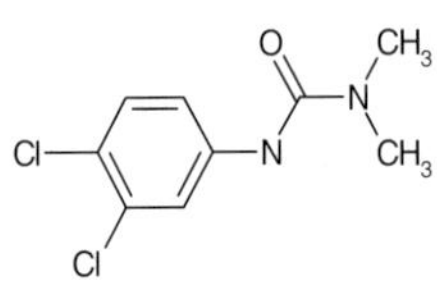	Diuron	330-54-1	1730	2	–0.87
103	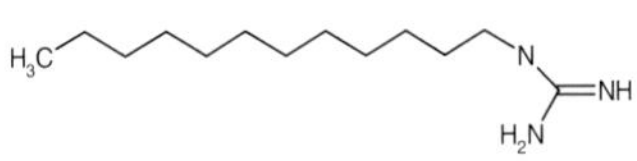	Dodine	2439-10-3	6657	3	–1.47
104	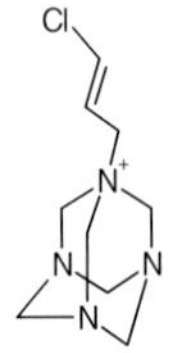	Dowicil	4080-31-3	2272	2	–1.02

ID	Structure	Name	CAS NB	Toxicity LD_{50}		
				mg/kg	Class	–Log mmol/kg
107	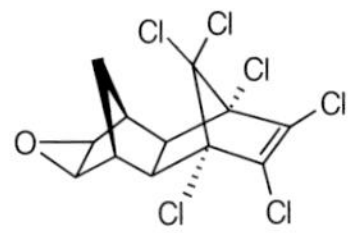	Endrin	72-20-8	14	1	1.43
108	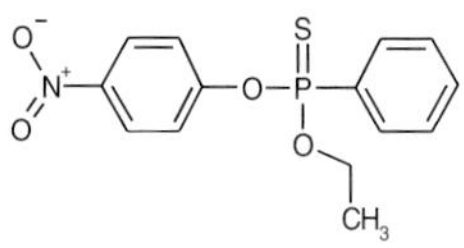	EPN	2104-64-5	349	1	–0.03
120	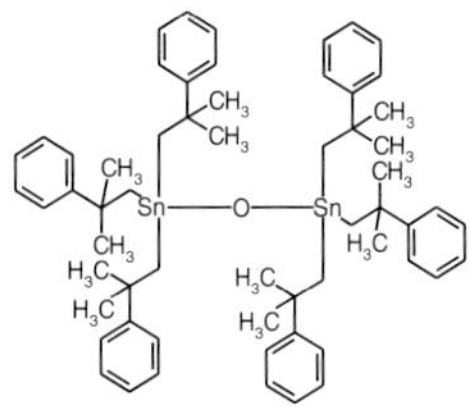	Fenbutatin oxide	13356-08-6	5065	3	–0.68
123*	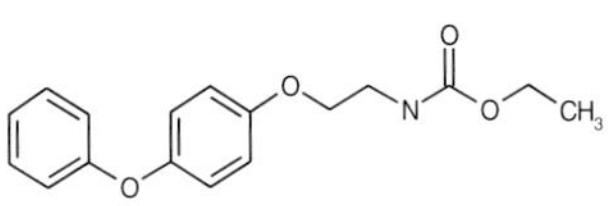	Fenoxycarb	79127-80-3	11574	3	–1.58
126	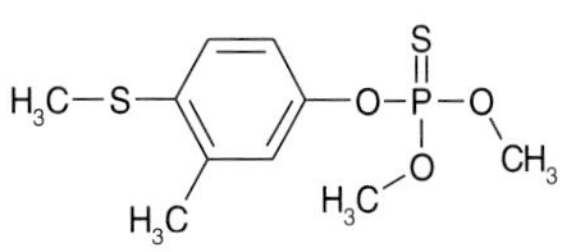	Fenthion	55-38-9	30	1	0.97
127	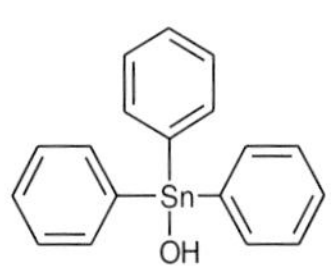	Fentin hydroxide	76-87-9	253	1	0.16

ID	Structure	Name	CAS NB	Toxicity LD_{50}		
				mg/kg	Class	–Log mmol/kg
136	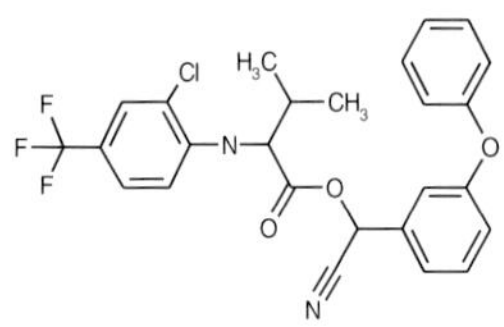	Fluvalinate	69409-94-5	5627	3	–1.05
138*	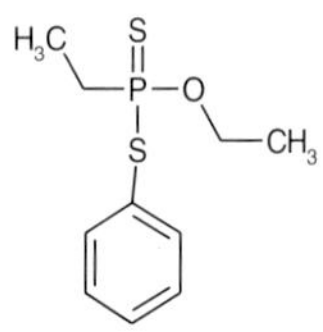	Fonofos	944-22-9	133	1	0.27
139	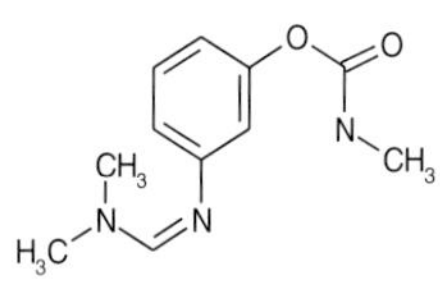	Formetanate	22259-30-9	3403	2	–1.19
142	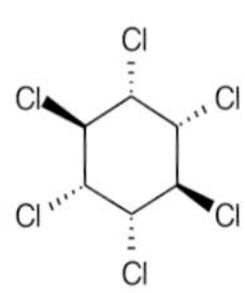	Gamma-HCH	58-89-9	882	1	–0.48
144	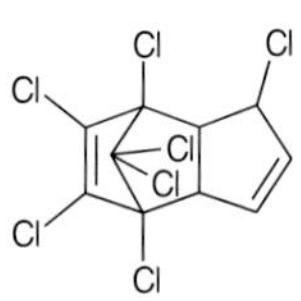	Heptachlor	76-44-8	92	1	0.61
146	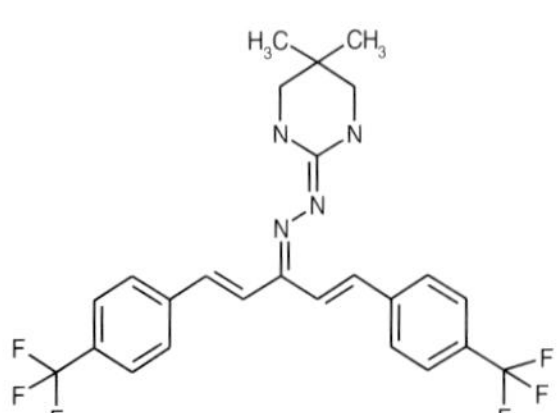	Hydra methylnon	67485-29-4	1136	2	–0.36

ID	Structure	Name	CAS NB	Toxicity LD_{50}		
				mg/kg	Class	–Log mmol/kg
150	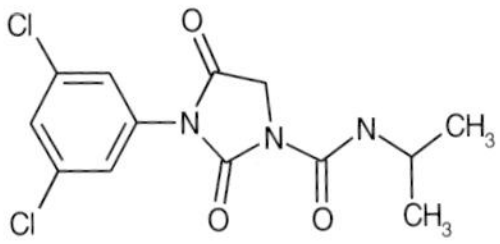	Iprodione	36734-19-7	9200	3	–1.45
152	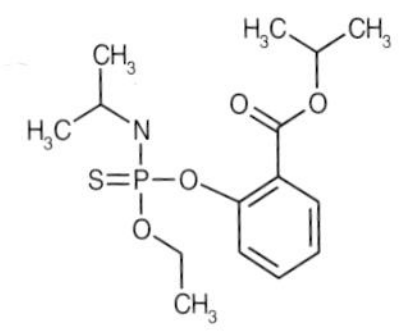	Isofenphos	25311-71-1	145	1	0.38
156	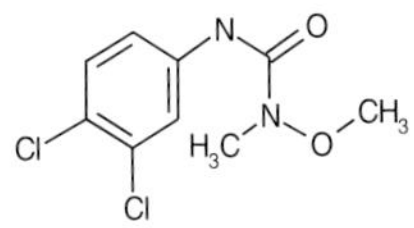	Linuron	330-55-2	1700	2	–0.83
157	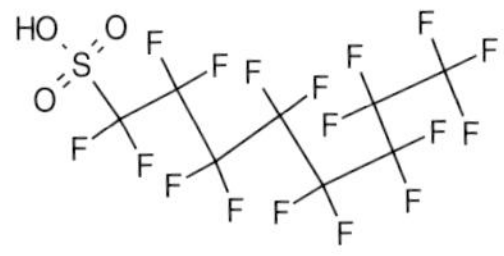	Perfluoro octane sulfonate	29457-72-5	217	1	0.36
158	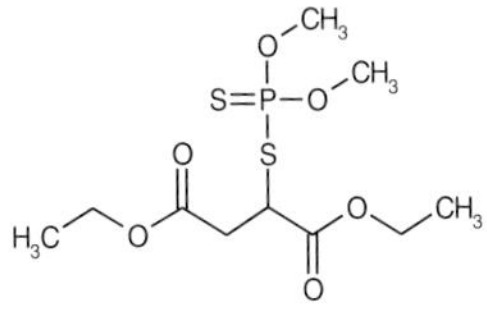	Malathion	121-75-5	3497	2	–1.02
160	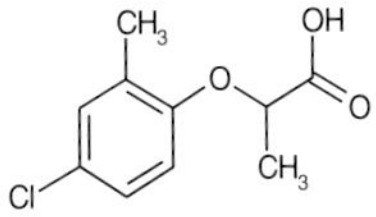	Mecoprop	7085-19-0	5000	3	–1.37

ID	Structure	Name	CAS NB	Toxicity LD_{50}		
				mg/kg	Class	–Log mmol/kg
163	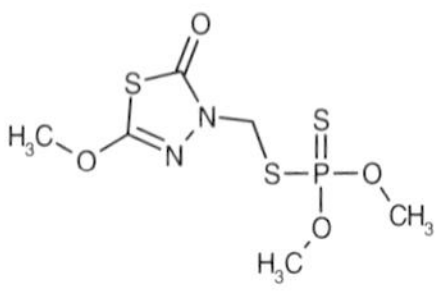	Methidathion	950-37-8	224	1	0.13
165	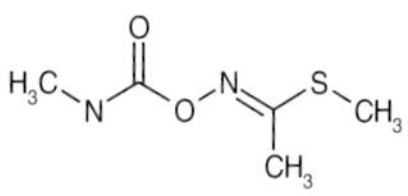	Methomyl	16752-77-5	1100	2	–0.83
169	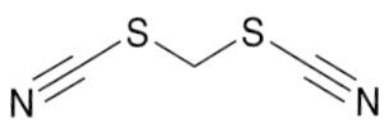	Methylene bis(thio cyanate)	6317-18-6	2230	2	–1.23
177	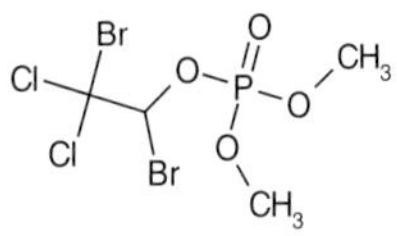	Naled	300-76-5	2117	2	–0.75
182	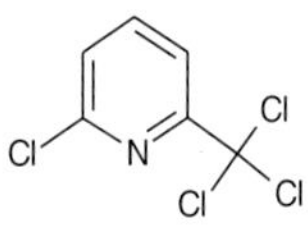	Nitrapyrin	1929-82-4	2135	2	–0.97
185	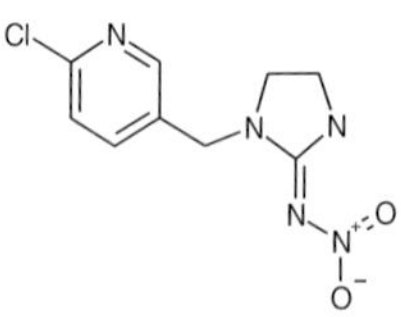	Imidacloprid	105827-78-9	1536	2	–0.78
196*	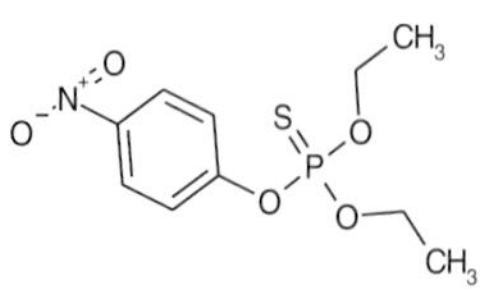	Parathion	56-38-2	194	1	0.18

ID	Structure	Name	CAS NB	Toxicity LD_{50}		
				mg/kg	Class	–Log mmol/kg
197	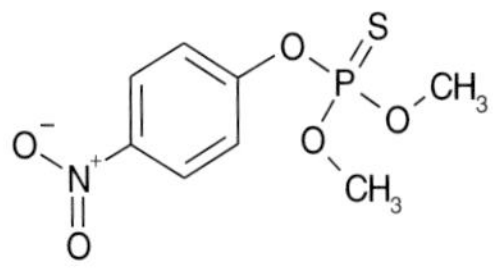	Parathion-methyl	298-00-0	90	1	0.47
198	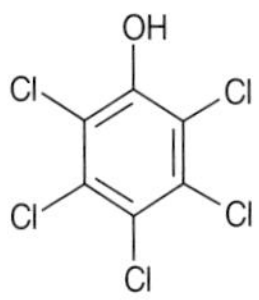	PCP	87-86-5	5581	3	–1.32
200*	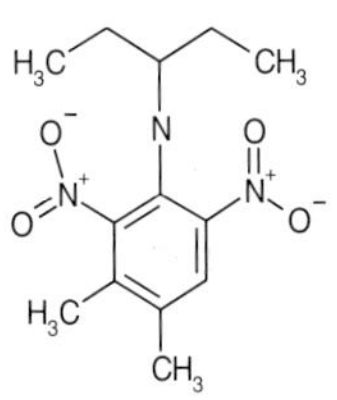	Pendimethalin	40487-42-1	4187	2	–1.17
204	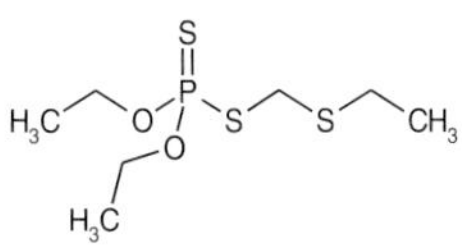	Phorate	298-02-2	373	1	–0.16
205	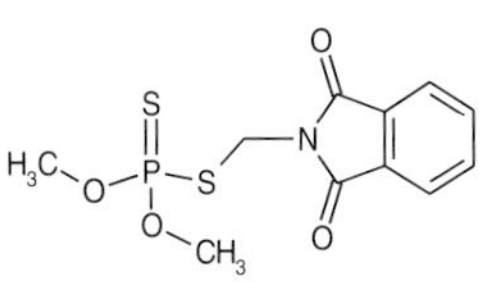	Phosmet	732-11-6	501	1	–0.20
208	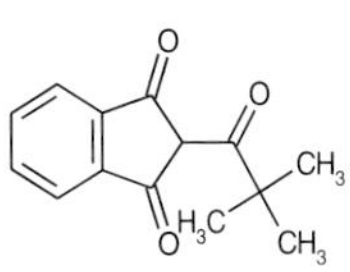	Pindone	83-26-1	1560	2	–0.83

ID	Structure	Name	CAS NB	Toxicity LD_{50}		
				mg/kg	Class	–Log mmol/kg
213	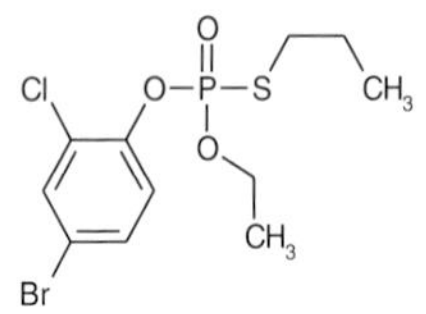	Profenofos	41198-08-7	201	1	0.27
215	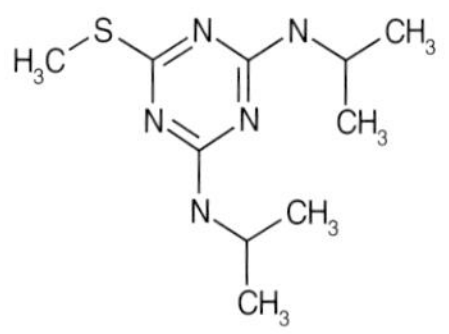	Prometryn	7287-19-6	18184	3	–1.88
217	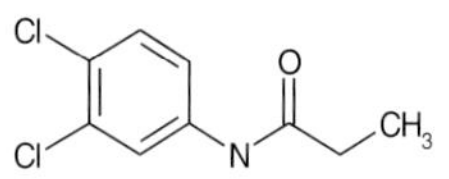	Propanil	709-98-8	2311	2	–1.03
218	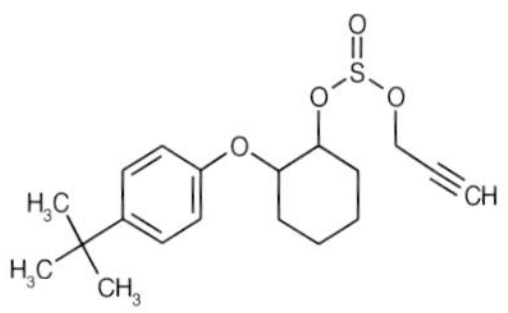	Propargite	2312-35-8	3401	2	–0.99
219	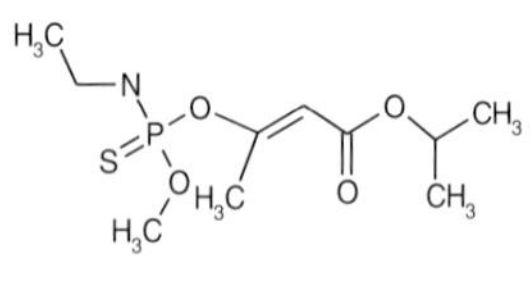	Propetamphos	31218-83-4	258	1	0.04
230	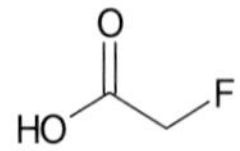	Fluoroacetic acid	144-49-0	379	1	–0.69

ID	Structure	Name	CAS NB	Toxicity LD_{50}		
				mg/kg	Class	–Log mmol/kg
233	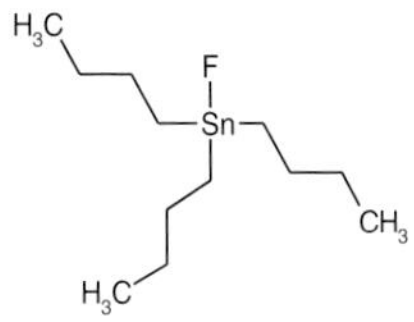	TBTF	1983-10-4	330	1	–0.03
234*	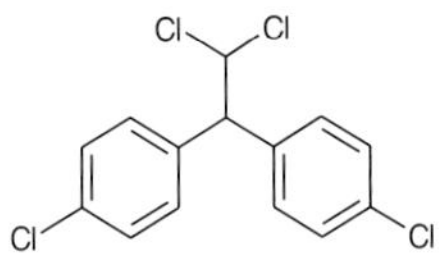	TDE	72-54-8	2178	2	–0.83
236	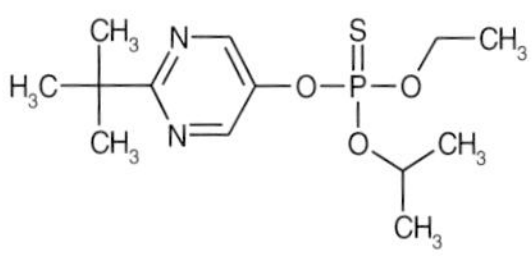	Tebupirimfos	96182-53-5	191	1	0.22
238	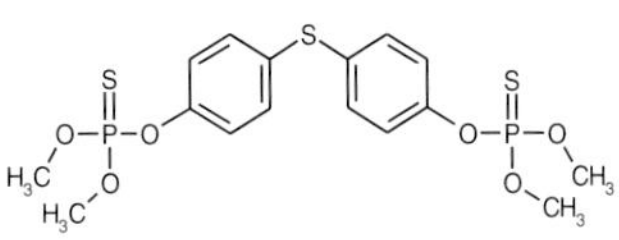	Temephos	3383-96-8	92	1	0.71
240	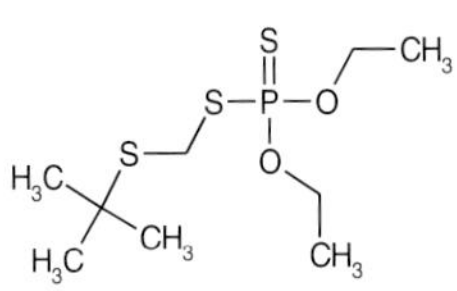	Terbufos	13071-79-9	143	1	0.30
247	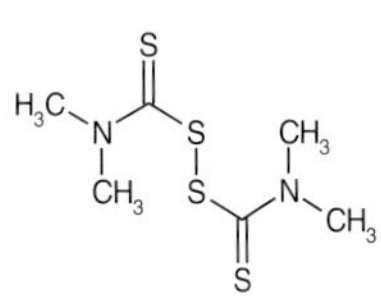	Thiram	137-26-8	3950	2	–1.22

ID	Structure	Name	CAS NB	Toxicity LD_{50}		
				mg/kg	Class	–Log mmol/kg
248	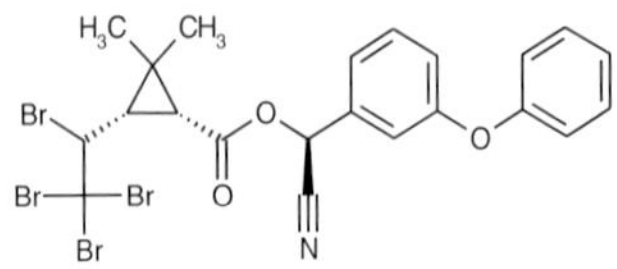	Tralomethrin	66841-25-6	4301	2	–0.81
249*	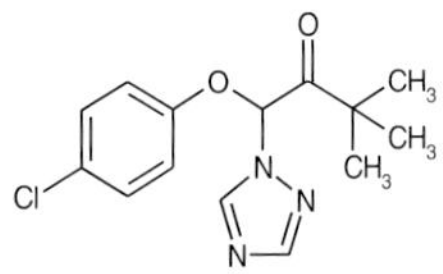	Triadimefon	43121-43-3	8392	3	–1.46
252	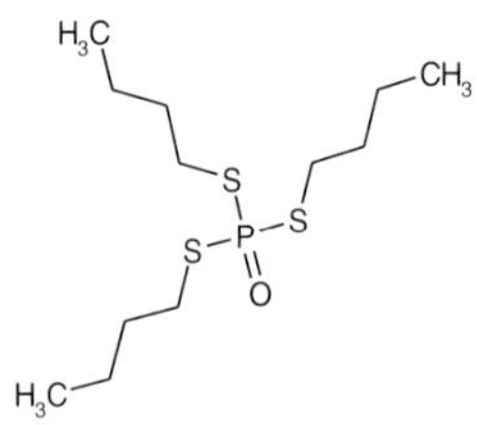	Tribufos	78-48-8	1519	2	–0.68
256	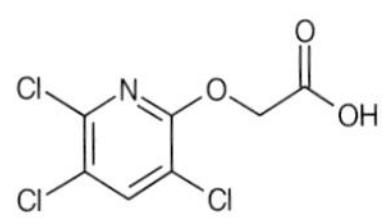	Triclopyr	55335-06-3	2934	2	–1.06
262	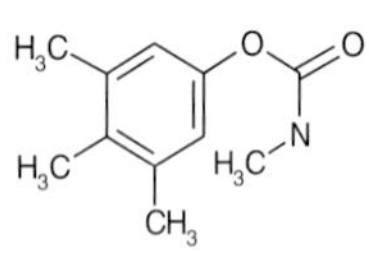	Trimethacarb	2686-99-9	2650	2	–1.14
267	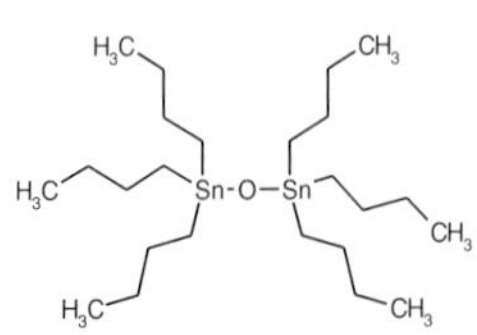	Tributyltin oxide	56-35-9	545	1	0.04

ID	Structure	Name	CAS NB	Toxicity LD_{50}		
				mg/kg	Class	–Log mmol/kg
272*	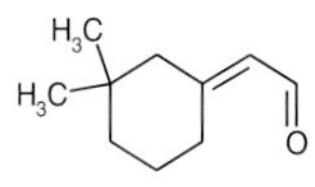	(E)-(3,3-Dimethyl cyclo hexylidene) acetaldehyde	26532-25-2	808	1	–0.72
285	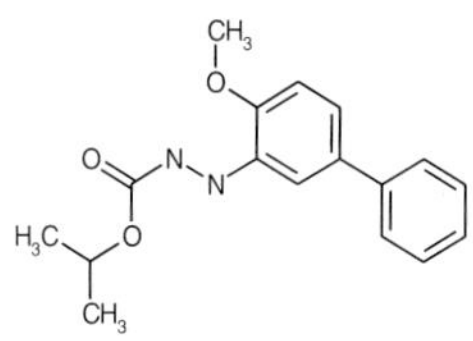	Bifenazate	149877-41-8	1862	2	–0.79
286	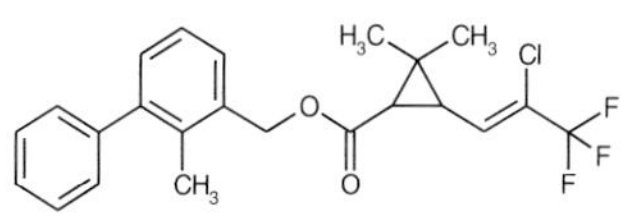	Bifenthrin	82657-04-3	4450	2	–1.02
293	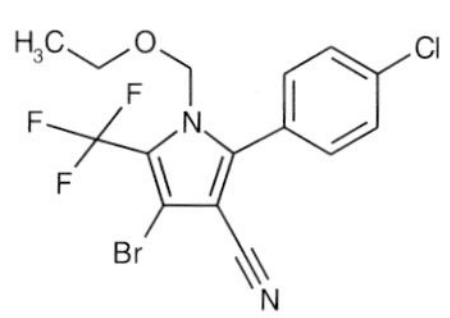	Chlorfenapyr (Pirate)	122453-73-0	132	1	0.49
298	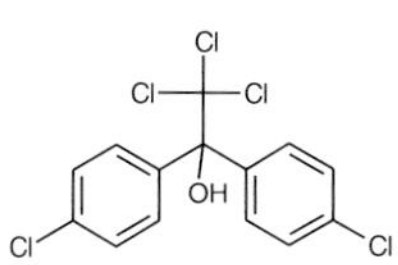	Dicofol	115-32-2	3010	2	–0.91
304	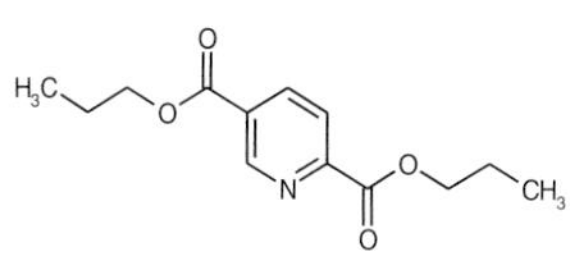	Dipropyl isocincho meronate	136-45-8	5620	3	–1.35

ID	Structure	Name	CAS NB	Toxicity LD_{50}		
				mg/kg	Class	–Log mmol/kg
308	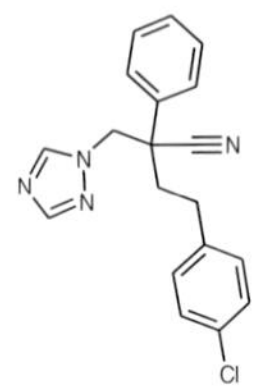	Fenbu conazole	114369-43-6	4050	2	–1.08
317	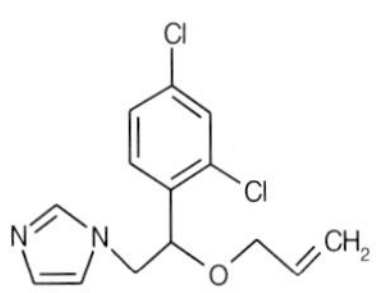	Imazalil	35554-44-0	6290	3	–1.33
319	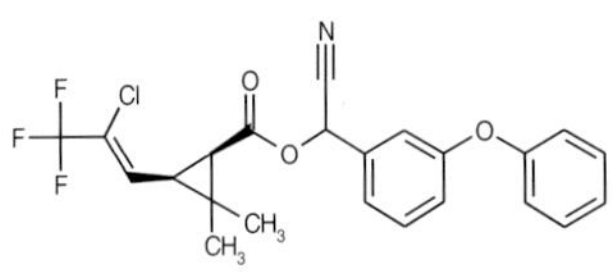	Lambda-cyhalothrin	91465-08-6	2354	2	–0.72
330	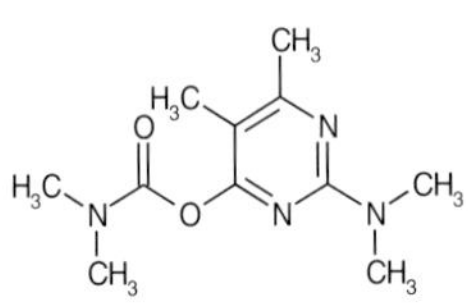	Pirimicarb	23103-98-2	3425	2	–1.16
331	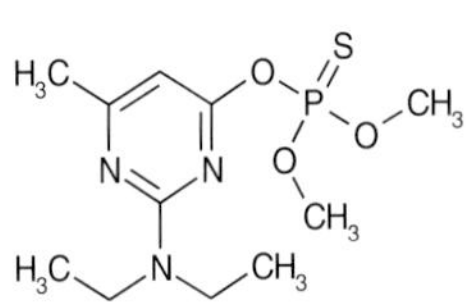	Pirimiphos methyl	29232-93-7	207	1	0.17
337*	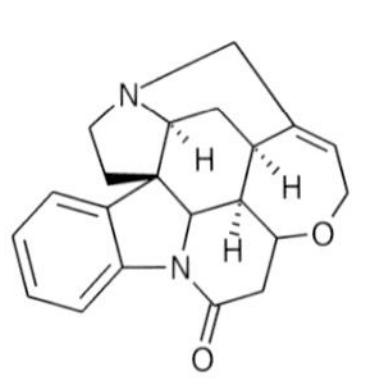	Strychnine	57-24-9	3536	2	–1.02

ID	Structure	Name	CAS NB	Toxicity LD_{50}		
				mg/kg	Class	–Log mmol/kg
339	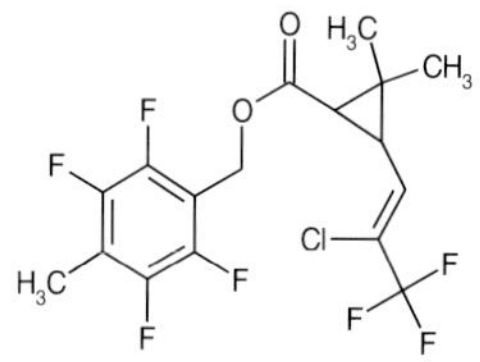	Tefluthrin	79538-32-2	15000	3	–1.55
341	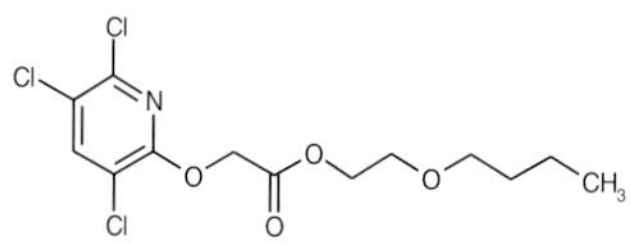	Triclopyr BEE	64700-56-7	9026	3	–1.40
348*	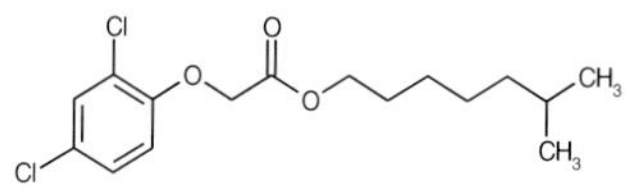	2,4-D isooctyl ester	25168-26-7	7187	3	–1.33
351*	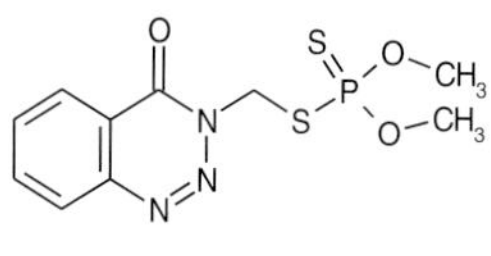	Azinphos-methyl	86-50-0	488	1	–0.19
353	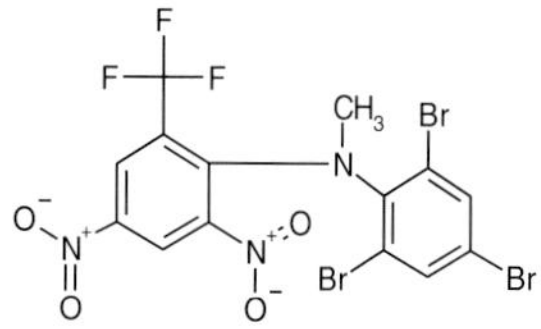	Bromethalin	63333-35-7	210	1	0.44
355	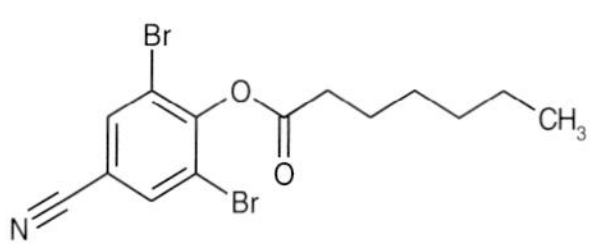	Bromoxynil heptanoate	56634-95-8	4525	2	–1.07

ID	Structure	Name	CAS NB	Toxicity LD_{50}		
				mg/kg	Class	–Log mmol/kg
359	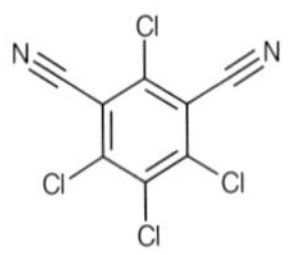	Chlorothalonil	1897-45-6	5200	3	–1.29
361*	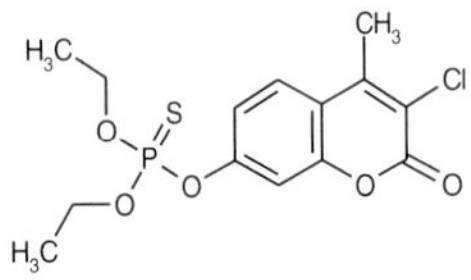	Coumaphos	56-72-4	85	1	0.63
362	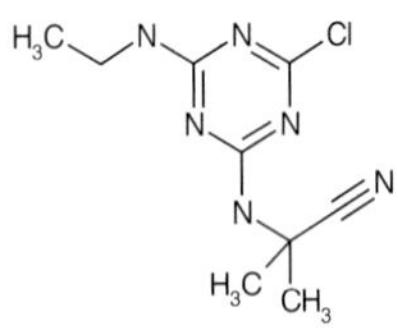	Cyanazine	21725-46-2	2654	2	–1.04
363	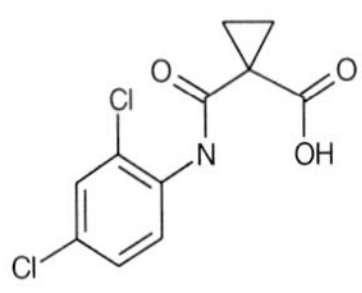	Cyclanilide	113136-77-9	2580	2	–0.97
364	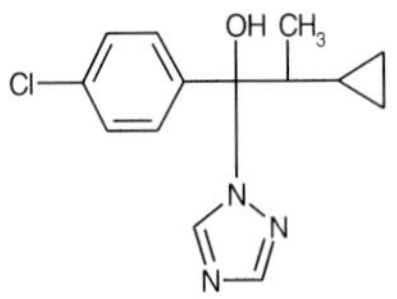	Cyproconazole		816	1	–0.45
366	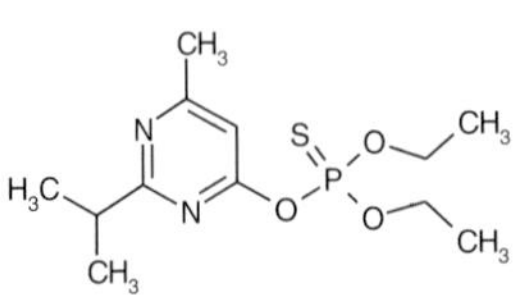	Diazinon	333-41-5	245	1	0.09

ID	Structure	Name	CAS NB	Toxicity LD_{50}		
				mg/kg	Class	–Log mmol/kg
369	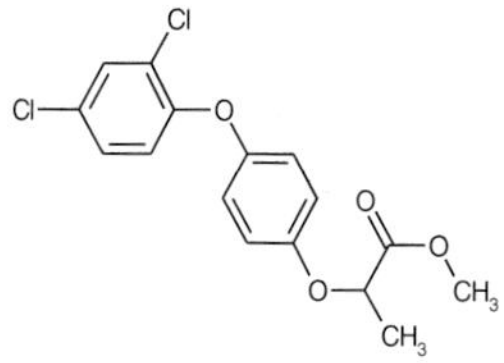	Diclofop-methyl	51338-27-3	13000	3	–1.58
371	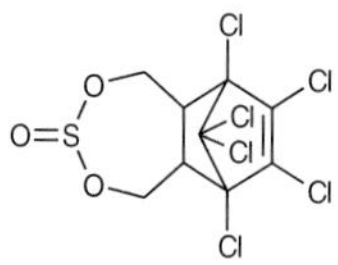	Endosulfan	115-29-7	805	1	–0.30
375	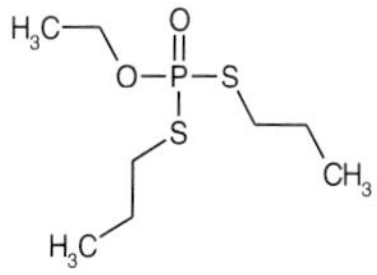	Ethoprophos	13194-48-4	33	1	0.87
376	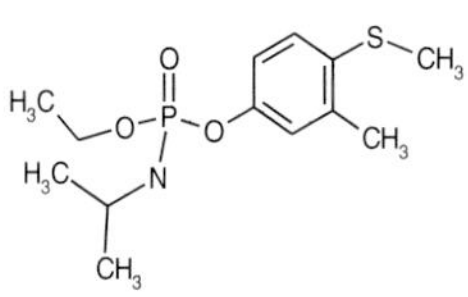	Fenamiphos	22224-92-6	38	1	0.90
377	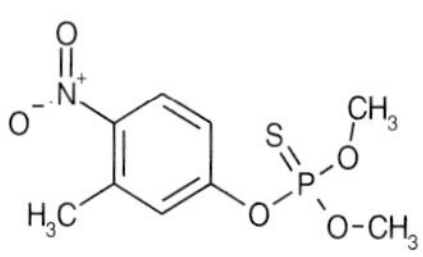	Fenitrothion	122-14-5	157	1	0.25
380*	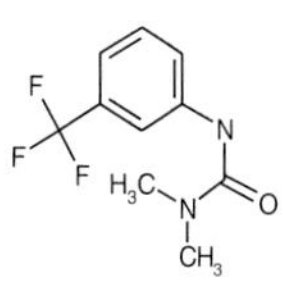	Fluometuron	2164-17-2	4620	2	–1.30

ID	Structure	Name	CAS NB	Toxicity LD_{50}		
				mg/kg	Class	–Log mmol/kg
384*	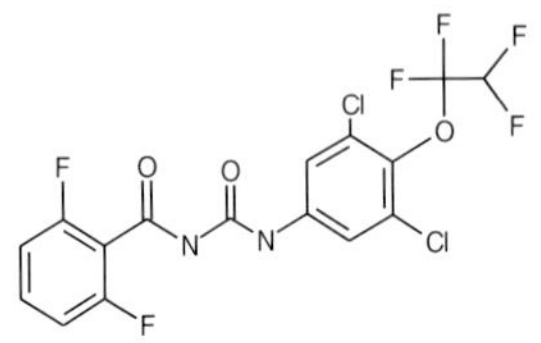	Hexaflumuron	86479-06-3	2201	2	–0.68
391	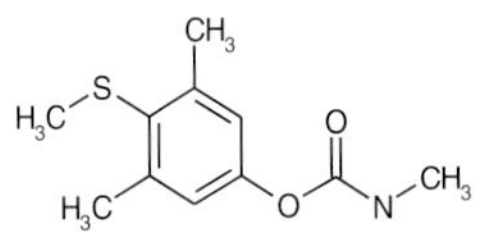	Methiocarb	2032-65-7	827	1	–0.56
411	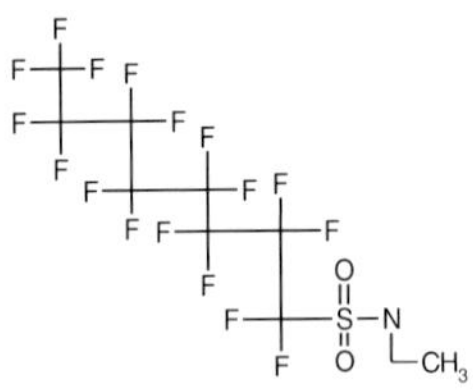	Sulfluramid	4151-50-2	286	1	0.27
422	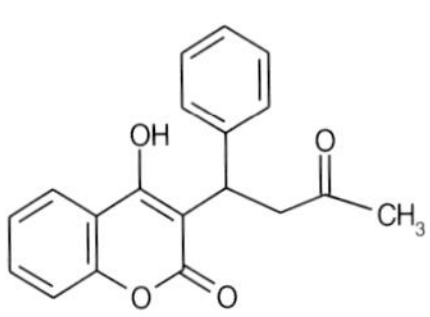	Warfarin	81-81-2	3900	2	–1.10
430	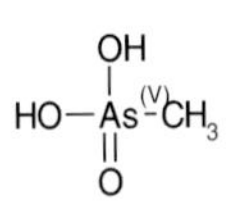	Methane arsonate	144-21-8	3573	2	–1.41
431*	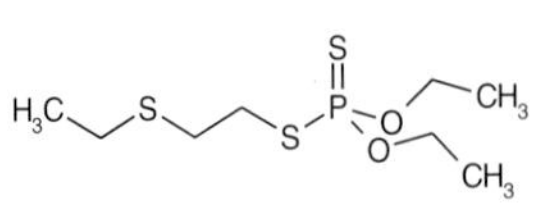	Disulfoton	298-04-4	544	1	–0.30

ID	Structure	Name	CAS NB	Toxicity LD_{50}		
				mg/kg	Class	–Log mmol/kg
440	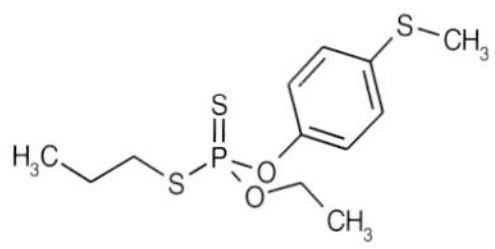	Sulprofos	35400-43-2	99	1	0.51
441	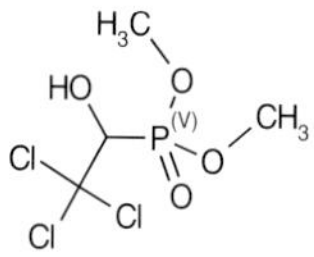	Trichlorfon	52-68-6	720	1	–0.45
442	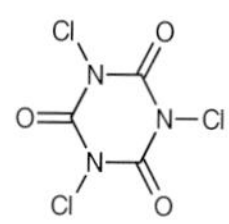	Trichloro-s-triazinetrione	87-90-1	7235	3	–1.49
443	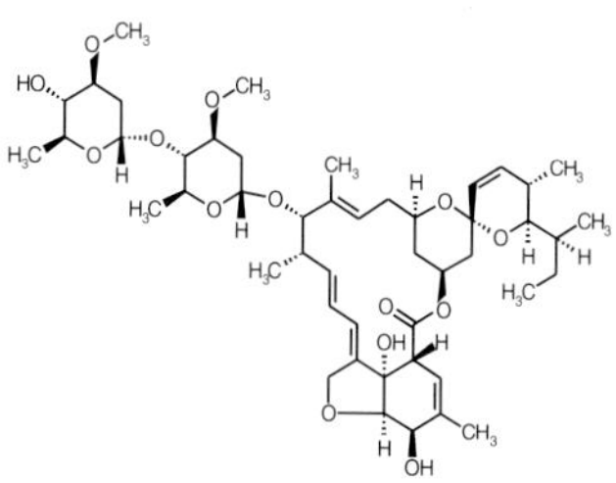	Avermectin	65195-55-3	3102	2	–0.55
444	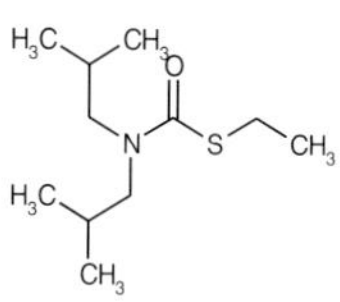	Butylate	2008-41-5	40000	3	–2.26
445	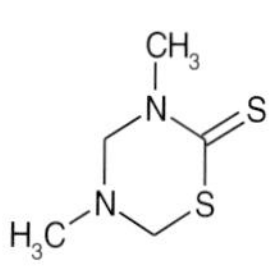	Dazomet	533-74-4	2300	2	–1.15

ID	Structure	Name	CAS NB	Toxicity LD_{50}		
				mg/kg	Class	–Log mmol/kg
446	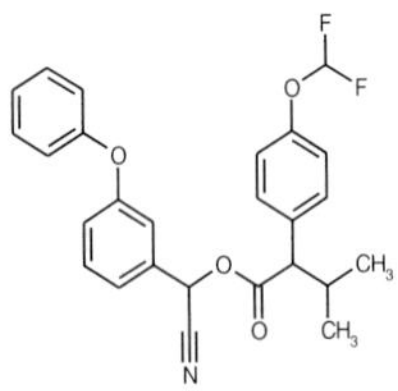	Flucythrinate	70124-77-5	3443	2	–0.88
447	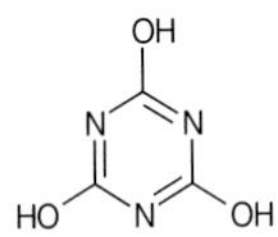	Isocyanuric acid	108-80-5	7235	3	–1.75
448	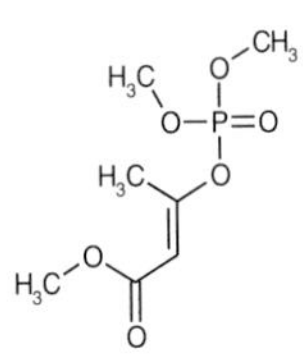	Mevinphos	7786-34-7	1000	2	–0.65
449	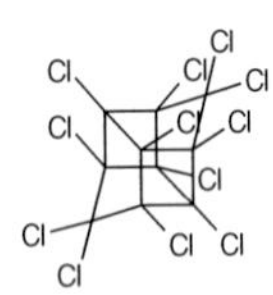	Mirex	2385-85-5	2511	2	–0.66
450	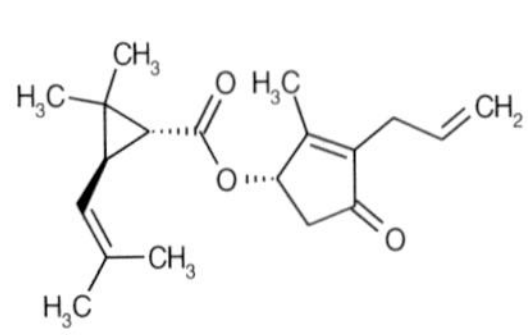	*s*-Bioallethrin	28434-00-6	2030	2	–0.83

LIST OF THE 116 COMPOUNDS INCLUDED IN THE ORAL QUAIL DATA SET

Three toxicity classes were generated according to the intervals established by the EC regulation, after regrouping the toxicity classes as too weakly represented:

LD_{50} (class 1) < 50 mg/kg; 50 mg/kg ≤ LD_{50} (class 2) < 500 mg/kg; LD_{50} (class 3) ≥ 500 mg/kg.

The symbol '*' in the ID column indicated the test set compounds.

ID	Structure	Name	CAS NB	Toxicity LD_{50}		
				mg/kg	Class	–Log mmol/kg
3	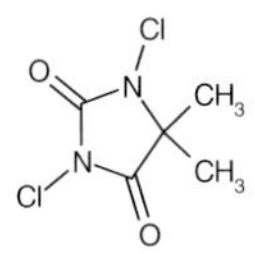	1,3-Dichloro -5,5-dimethyl hydantoin (DCDMH)	118-52-5	1715	3	–0.94
4	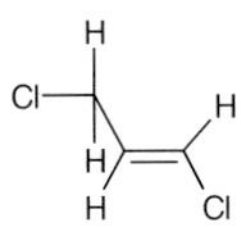	1,3-Dichloro propene	542-75-6	152	2	–0.14
21	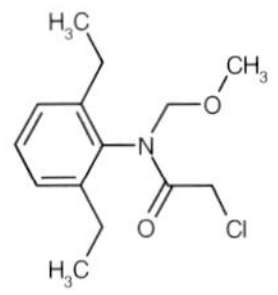	Alachlor	15972-60-8	1499	3	–0.74
22	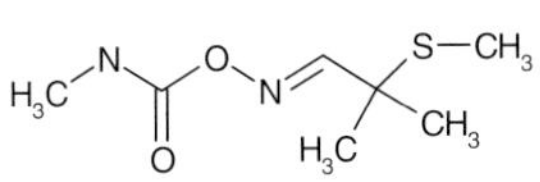	Aldicarb	116-06-3	2	1	1.98
25	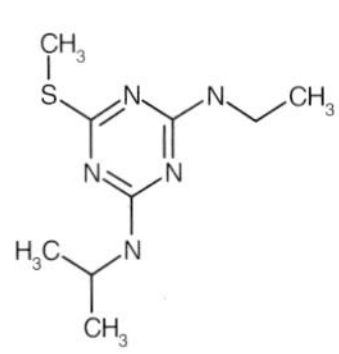	Ametryne	834-12-8	2250	3	–1.00
26	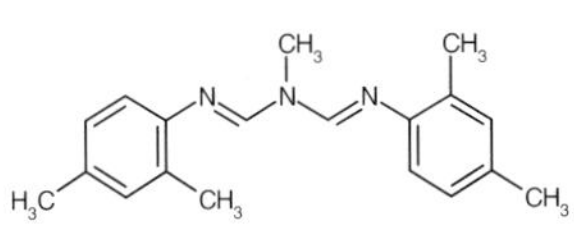	Amitraz	33089-61-1	788	3	–0.43

ID	Structure	Name	CAS NB	Toxicity LD_{50}		
				mg/kg	Class	–Log mmol/kg
29	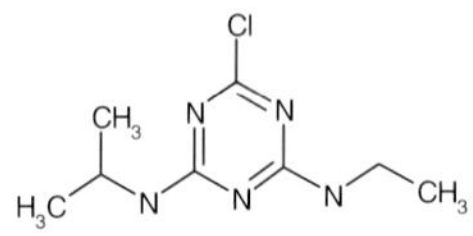	Atrazine	1912-24-9	940	3	–0.64
30	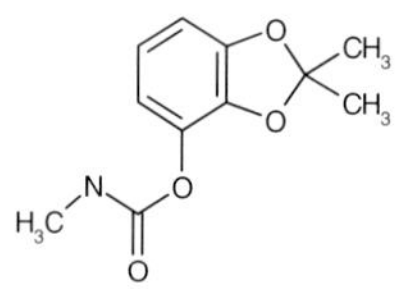	Bendiocarb	22781-23-3	19	1	1.07
32*	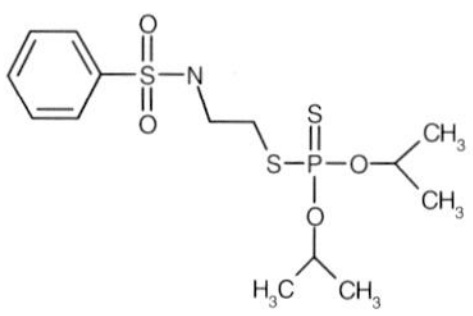	Bensulide	741-58-2	1386	3	–0.54
35	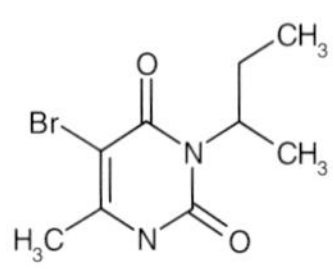	Bromacil	314-40-9	2250	1	–0.96
46*	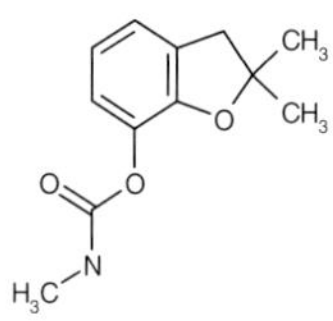	Carbofuran	1563-66-2	5	2	1.64
48*	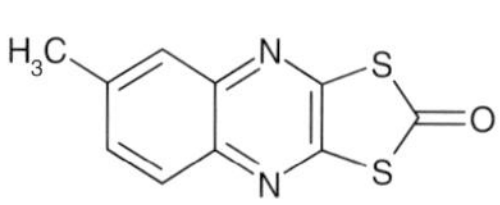	Chinomethionat	2439-01-2	196	1	0.08

ID	Structure	Name	CAS NB	Toxicity LD_{50}		
				mg/kg	Class	–Log mmol/kg
50	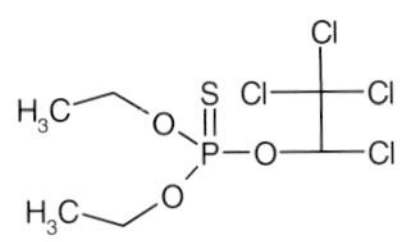	Chloroethoxyfos	54593-83-8	28	3	1.08
51	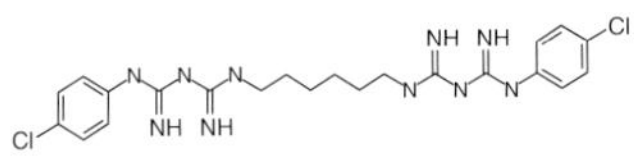	Chlorhexidine	55-56-1	1627	2	–0.51
54*	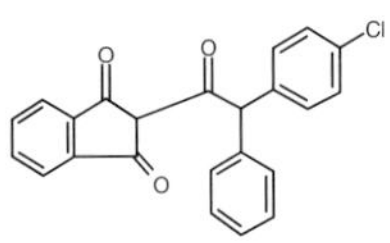	Chlorophacinone	3691-35-8	495	1	–0.12
58	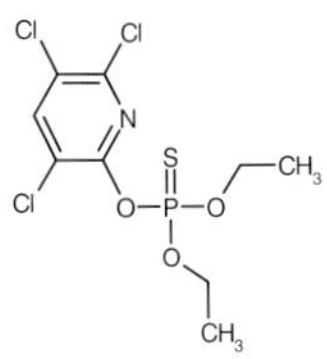	Chlorpyrifos	2921-88-2	32	3	1.04
62	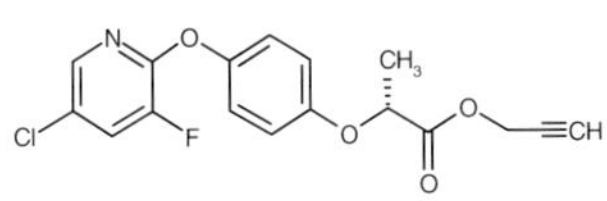	Clodinafop-propargyl	105512-06-9	1455	3	–0.62
63	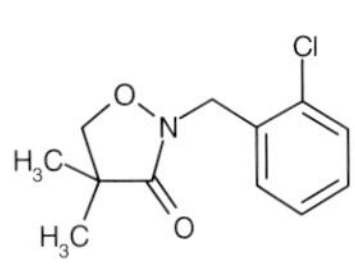	Clomazone	81777-89-1	2510	2	–1.02
68	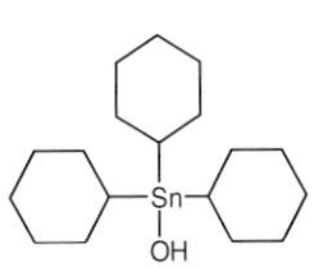	Cyhexatin	13121-70-5	300	2	0.11

ID	Structure	Name	CAS NB	Toxicity LD_{50}		
				mg/kg	Class	–Log mmol/kg
75*	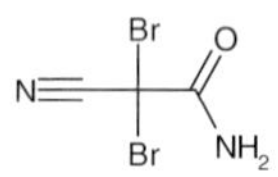	DBNPA	10222-01-2	354	3	–0.17
83	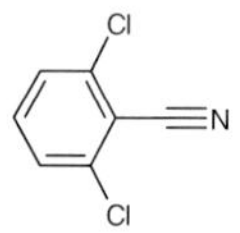	Dichlobenil	1194-65-6	683	2	–0.60
84	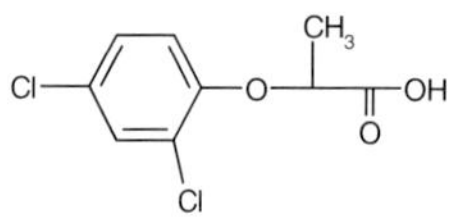	Dichlorprop	120-36-5	354	1	–0.18
85	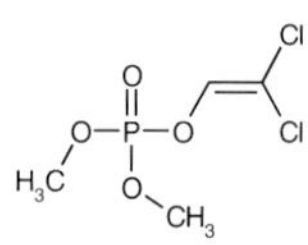	Dichlorvos	62-73-7	9	3	1.40
87	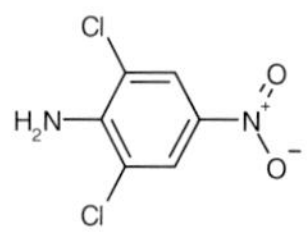	Dicloran	99-30-9	900	3	–0.64
90	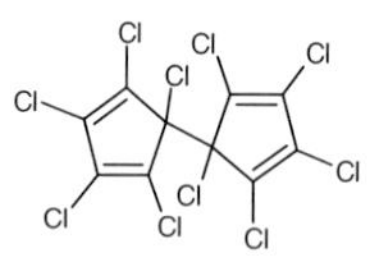	Dienochlor	2227-17-0	705	3	–0.17
95	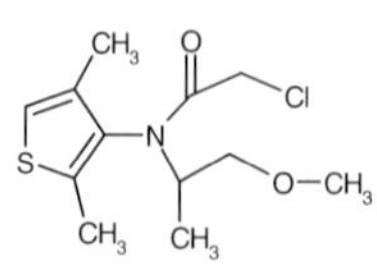	Dimethenamid	87674-68-8	1068	3	–0.59
103	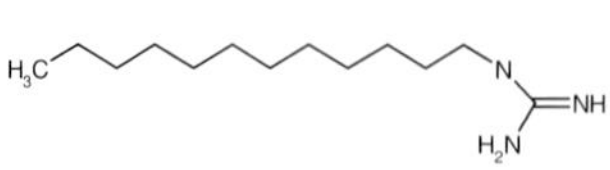	Dodine	2439-10-3	546	3	–0.38

ID	Structure	Name	CAS NB	Toxicity LD_{50}		
				mg/kg	Class	–Log mmol/kg
104	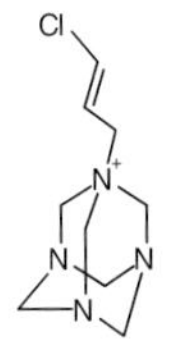	Dowicil	4080-31-3	1237	2	–0.76
111	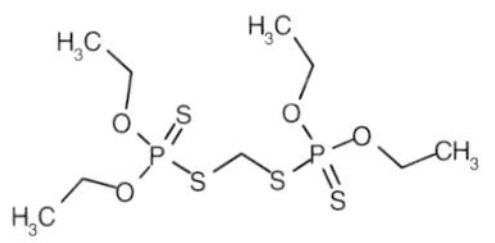	Ethion	563-12-2	128	3	0.48
125	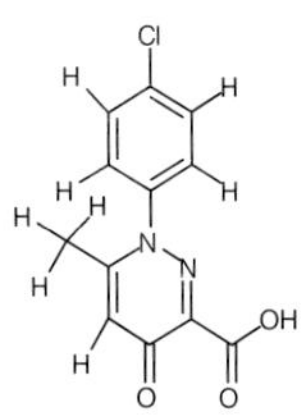	Fenridazone	68254-10-4	3941	1	–1.17
126	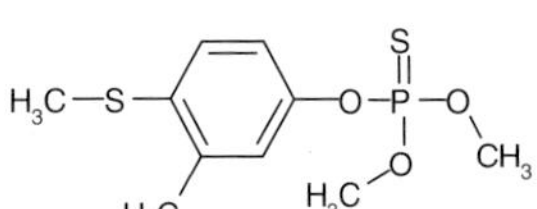	Fenthion	55-38-9	7	1	1.59
139	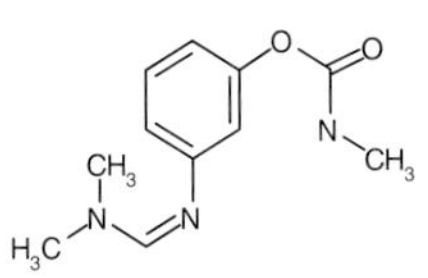	Formetanate	22259-30-9	36	3	0.79

ID	Structure	Name	CAS NB	Toxicity LD_{50}		
				mg/kg	Class	–Log mmol/kg
146	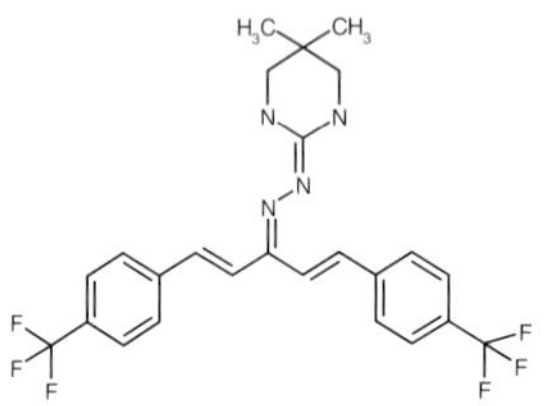	Hydramethylnon	67485-29-4	1828	3	–0.57
150	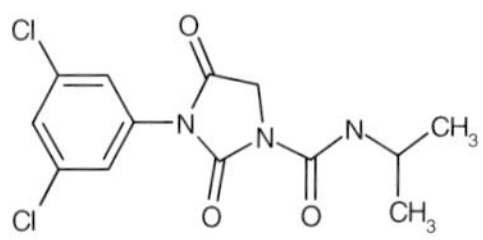	Iprodione	36734-19-7	930	1	–0.45
152	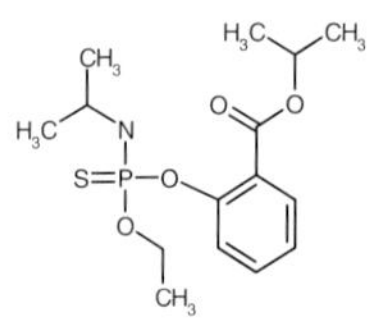	Isofenphos	25311-71-1	9	1	1.60
157	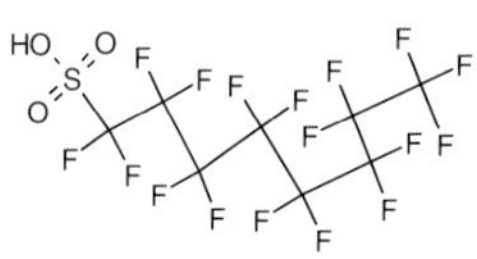	Perfluorooctane sulfonate	29457-72-5	4 2	3	1.08
160	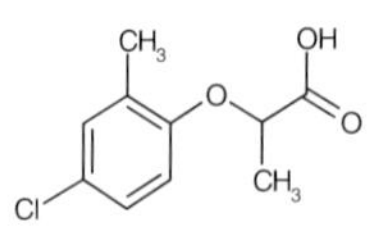	Mecoprop	7085-19-0	707	1	–0.52
165	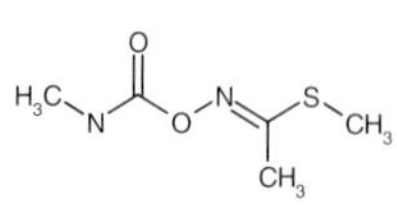	Methomyl	16752-77-5	24	3	0.83

ID	Structure	Name	CAS NB	Toxicity LD_{50}		
				mg/kg	Class	–Log mmol/kg
176*	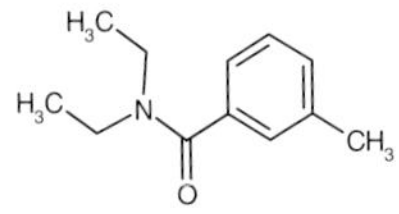	Diethyltoluamide	134-62-3	1375	2	–0.86
185	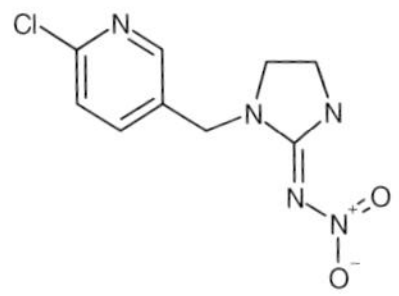	Imidacloprid	105827-78-9	152	2	0.23
187	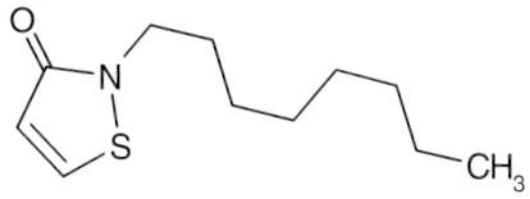	Octhilinone	26530-20-1	384	3	–0.26
188	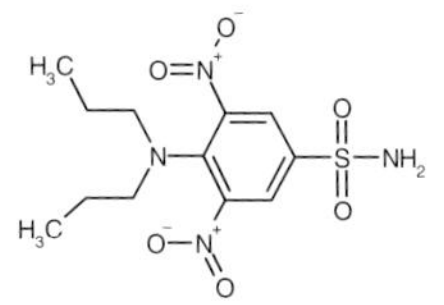	Oryzalin	19044-88-3	507	3	–0.17
194*		Paradichloro benzene	106-46-7	1608	3	–1.04
195	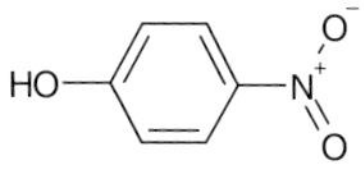	Paranitrophenol	100-02-7	577	3	–0.62
197	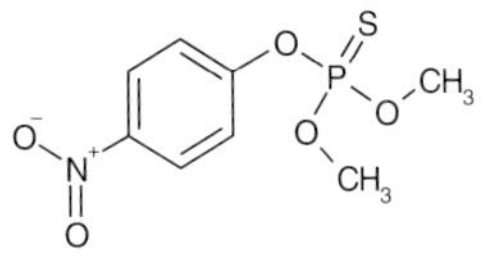	Parathion-methyl	298-00-0	8	1	1.54

ID	Structure	Name	CAS NB	Toxicity LD_{50}		
				mg/kg	Class	–Log mmol/kg
198	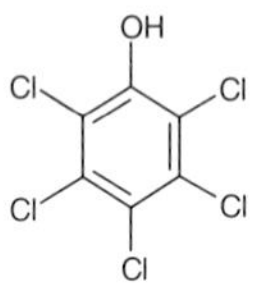	PCP	87-86-5	627	3	–0.37
204	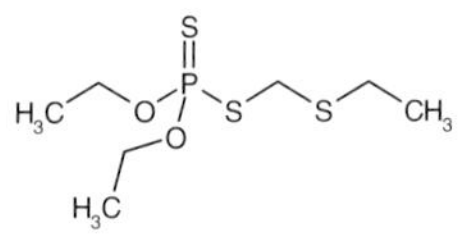	Phorate	298-02-2	7	1	1.57
216*	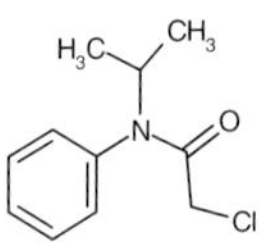	Propachlor	1918-16-7	88	2	0.38
217	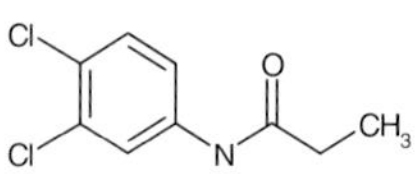	Propanil	709-98-8	201	2	0.04
236	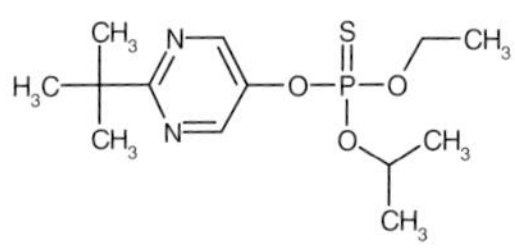	Tebupirimfos	96182-53-5	20	1	1.20
238	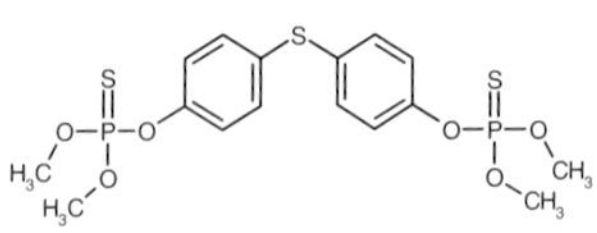	Temephos	3383-96-8	27	1	1.23
240	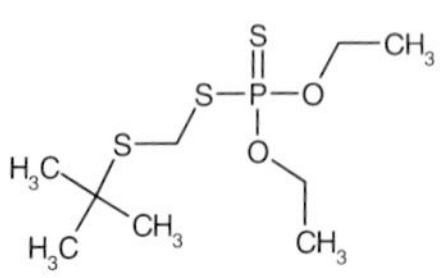	Terbufos	13071-79-9	29	1	1.00

ID	Structure	Name	CAS NB	Toxicity LD_{50}		
				mg/kg	Class	–Log mmol/kg
244	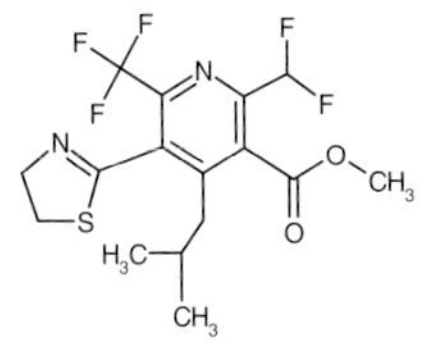	Thiazopyr	117718-60-2	1913	3	–0.68
252	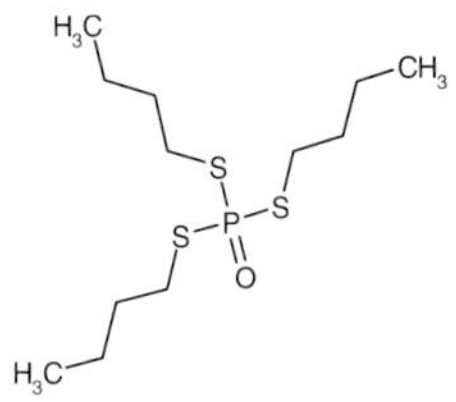	Tribufos	78-48-8	151	2	0.32
257	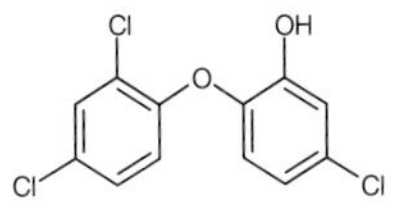	Triclosan	3380-34-5	825	3	–0.45
262	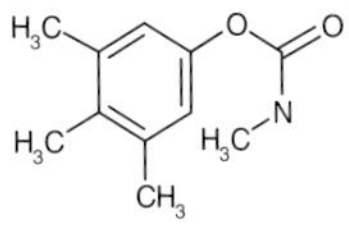	Trimethacarb	2686-99-9	238	2	–0.09
264	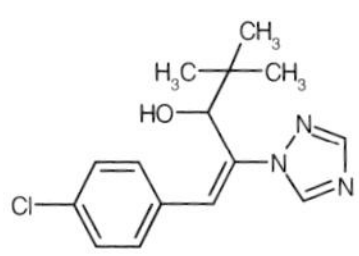	Uniconazole	83657-17-4	1461	3	–0.70
273	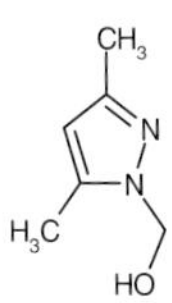	Dimethyl hydroxy methyl pyrazole	85264-33-1	759	3	–0.78

ID	Structure	Name	CAS NB	Toxicity LD_{50}		
				mg/kg	Class	–Log mmol/kg
275	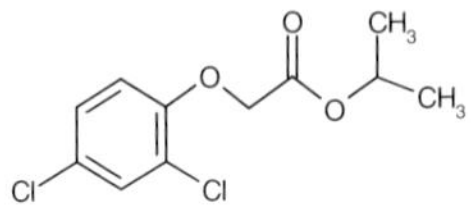	2,4-D isopropyl ester	94-11-1	1879	3	–0.85
277*	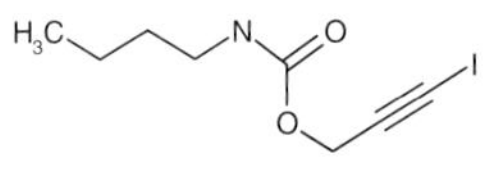	3-Iodo-2-propynyl butylcarbamate	55406-53-6	749	3	–0.43
278	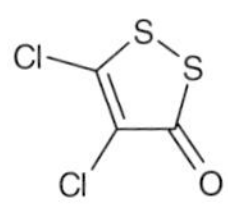	4,5-Dichloro-1,2-dithiol-3-one	1192-52-5	247	2	–0.12
282*	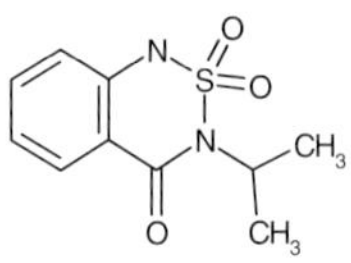	Bentazone	25057-89-0	1073	3	–0.65
285	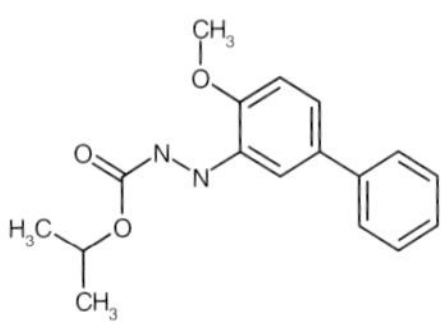	Bifenazate	149877-41-8	1032	3	–0.54
297	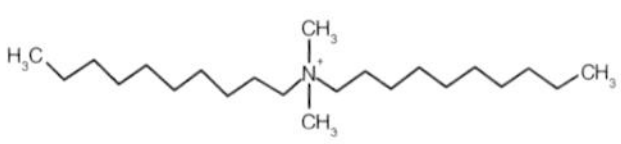	DDAC	7173-51-5	196	2	0.22
304	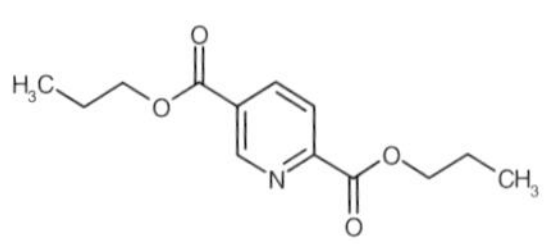	Dipropyl isocincho meronate	136-45-8	1350	3	–0.73
307	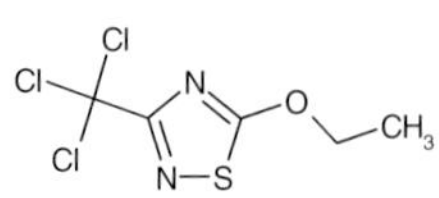	Etridiazole	2593-15-9	560	3	–0.35

ID	Structure	Name	CAS NB	Toxicity LD$_{50}$		
				mg/kg	Class	–Log mmol/kg
311	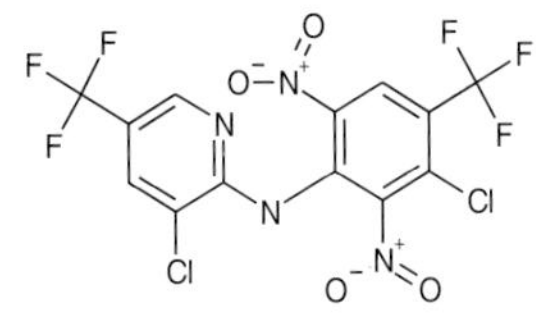	Fluazinam	79622-59-6	1782	3	–0.58
323	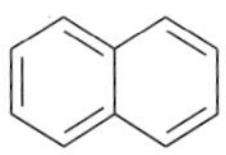	Naphthalene	91-20-3	2690	3	–1.32
328	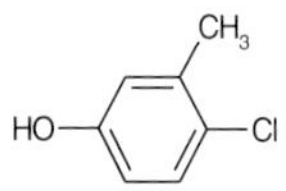	Parachloro metacresol	59-50-7	1540	3	–1.03
331	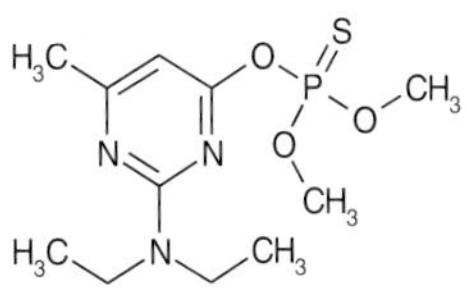	Pirimiphos methyl	29232-93-7	40	1	0.88
333	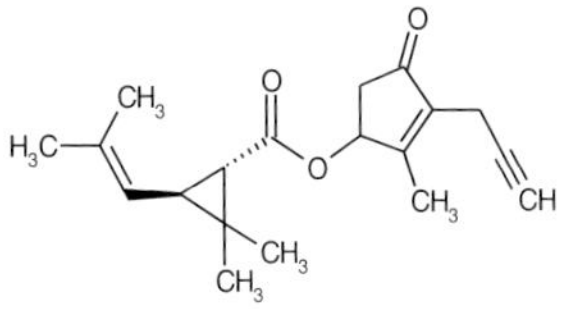	Prallethrin	23031-36-9	1171	3	–0.59
340	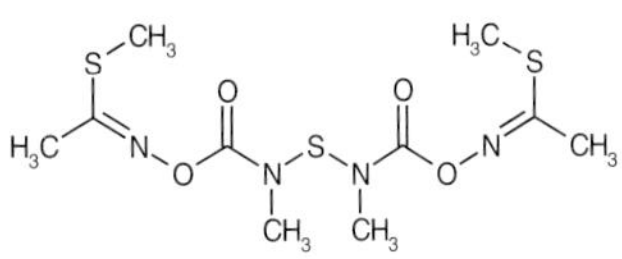	Thiodicarb	59669-26-0	2023	3	–0.76
346	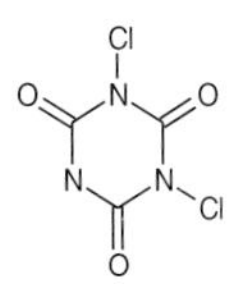	Dichloroiso cyanuric acid	2782-57-2	1590	3	–0.90

ID	Structure	Name	CAS NB	Toxicity LD_{50}		
				mg/kg	Class	–Log mmol/kg
347*	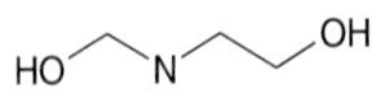	2-(Hydroxymethyl amino)ethanol	34375-28-5	1743	3	–1.28
351	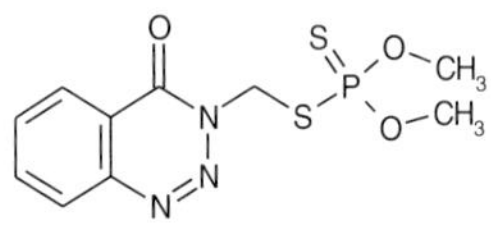	Azinphos-methyl	86-50-0	33	1	0.98
353	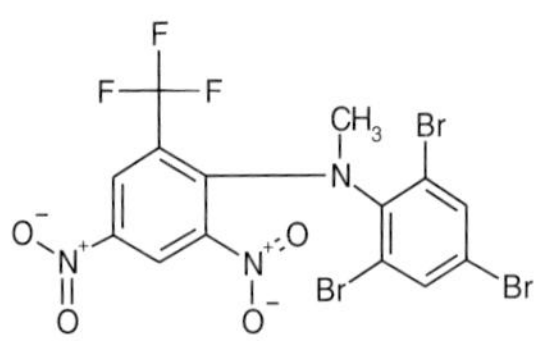	Bromethalin	63333-35-7	5	1	2.10
354	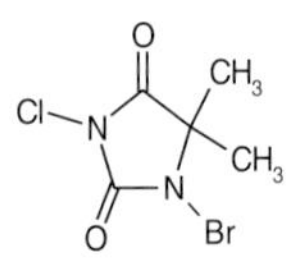	Bromo-3-chloro-5,5-dimethyl hydantoin (BCDMH)	16079-88-2	1070	3	–0.65
355	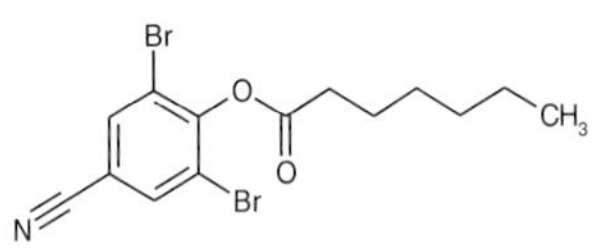	Bromoxynil heptanoate	56634-95-8	359	2	0.03
361*	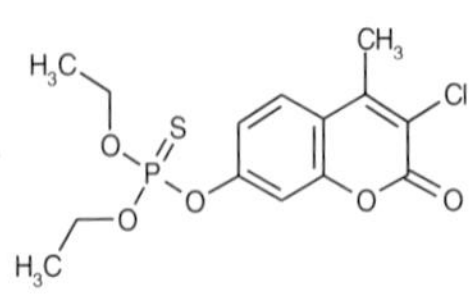	Coumaphos	56-72-4	2	1	2.19
364	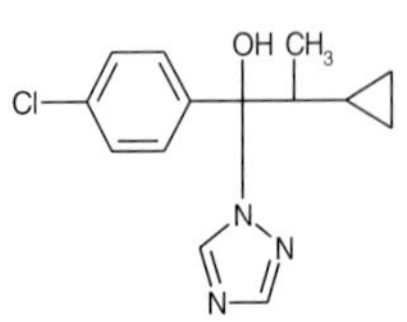	Cyproconazole	94361-06-5	150	2	0.29

ID	Structure	Name	CAS NB	Toxicity LD_{50}		
				mg/kg	Class	–Log mmol/kg
366	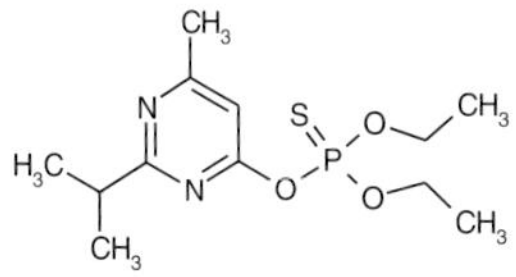	Diazinon	333-41-5	5	1	1.77
368	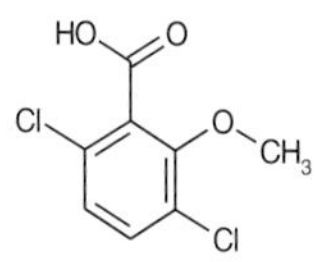	Dicamba	1918-00-9	216	2	0.01
369	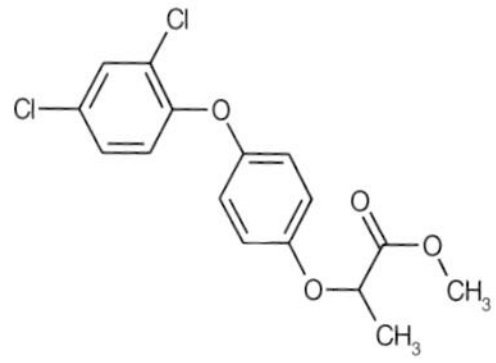	Diclofop-methyl	51338-27-3	4400	3	–1.11
371	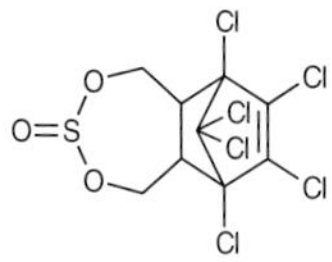	Endosulfan	115-29-7	42	1	0.99
372	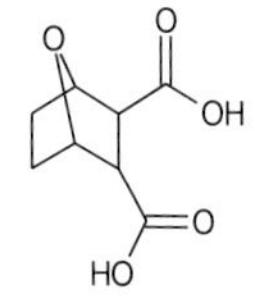	Endothall	145-73-3	494	2	–0.42
376	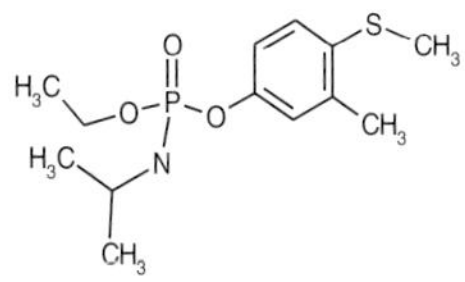	Fenamiphos	22224-92-6	2	1	2.28

ID	Structure	Name	CAS NB	Toxicity LD_{50}		
				mg/kg	Class	–Log mmol/kg
377	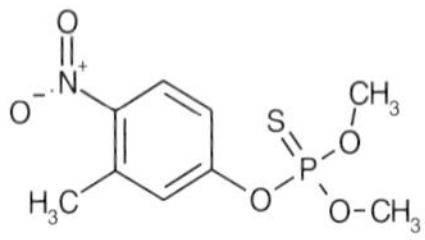	Fenitrothion	122-14-5	24	1	1.07
378	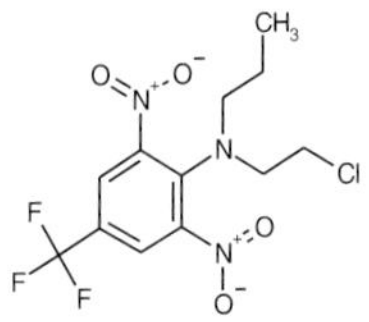	Fluchloralin	33245-39-5	7000	3	–1.29
385*	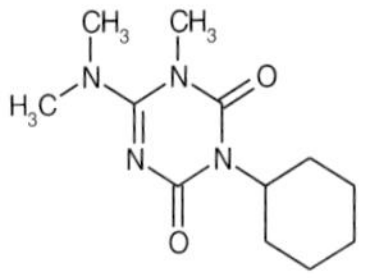	Hexazinone	51235-04-2	2251	3	–0.95
386	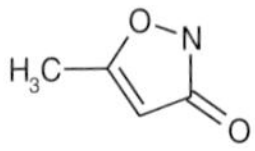	Hymexazol	10004-44-1	1479	3	–1.17
391*	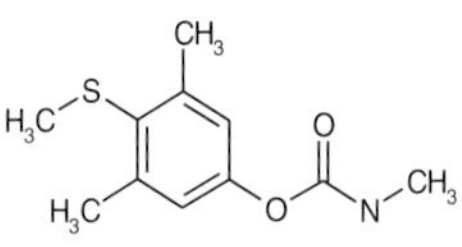	Methiocarb	2032-65-7	20	1	1.06
394	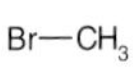	Methyl bromide	74-83-9	73	2	0.11
395	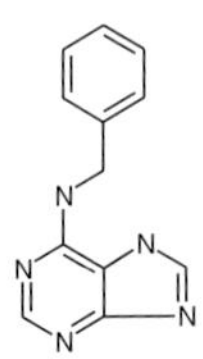	6-Benzyladenine	1214-39-7	1599	3	–0.85

ID	Structure	Name	CAS NB	Toxicity LD_{50}		
				mg/kg	Class	–Log mmol/kg
400	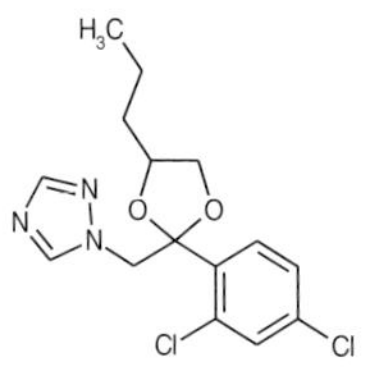	Propiconazole	60207-90-1	2825	3	–0.92
411	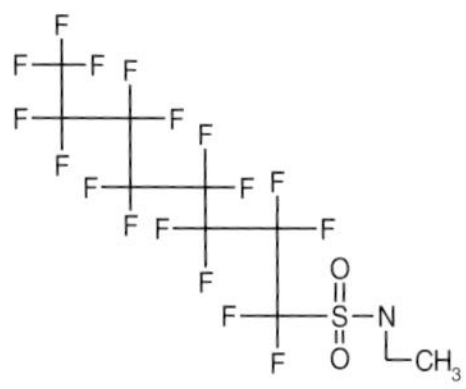	Sulfluramid	4151-50-2	474	2	0.05
412	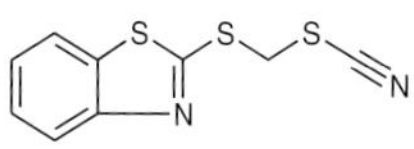	TCMTB	21564-17-0	661	3	–0.44
423*	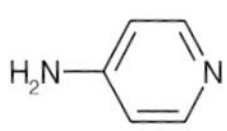	4-Aminopyridine	504-24-5	15	1	0.80
424	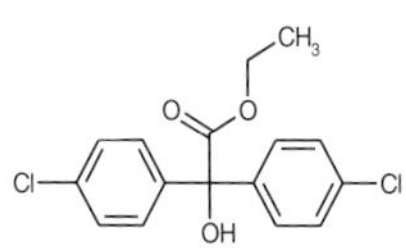	Chlorobenzilate	510-15-6	607	3	–0.27
425*	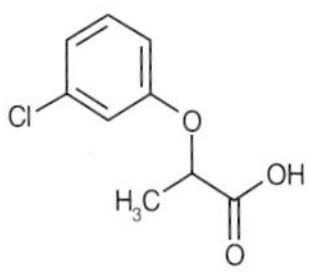	Chloroprop	101-10-0	1186	3	–0.77
426	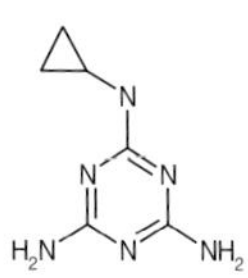	Cyromazine	66215-27-8	1785	3	–1.03

ID	Structure	Name	CAS NB	Toxicity LD_{50}		
				mg/kg	Class	–Log mmol/kg
427	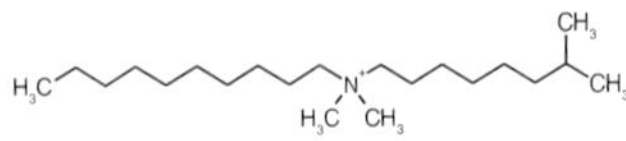	Decyl isonomyl dimethyl ammonium	138698-36-9	31	1	1.01
428	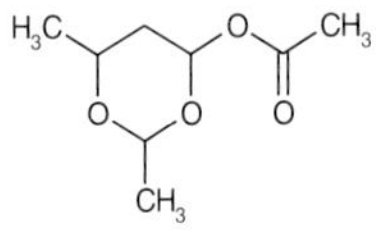	Dimethoxane	828-00-2	1585	3	–0.96
429	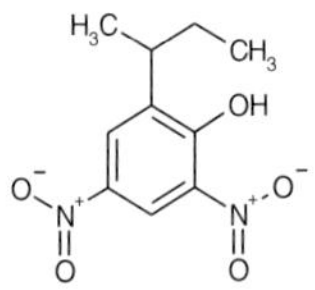	Dinoseb acid	88-85-7	40	1	0.78
430	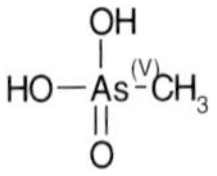	Methanearsonate	144-21-8	477	2	–0.53
431*	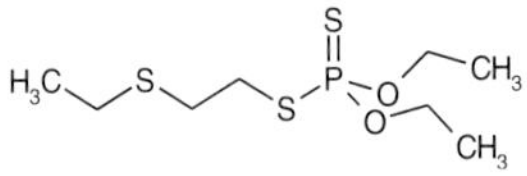	Disulfoton	298-04-4	12	1	1.36
432	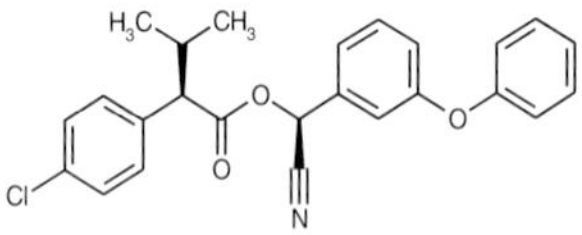	Esfenvalerate	66230-04-4	381	2	0.04
433	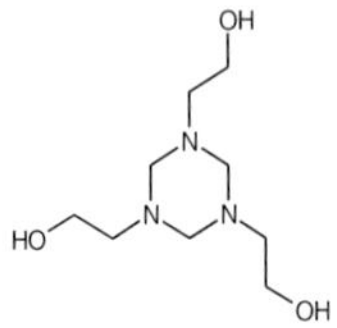	Grotan	4719-04-4	1520	3	–0.84

ID	Structure	Name	CAS NB	Toxicity LD_{50}		
				mg/kg	Class	–Log mmol/kg
434*	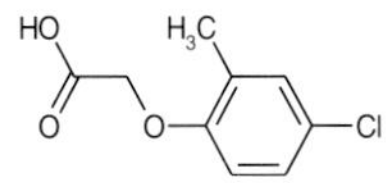	MCPA Acid	94-74-6	377	2	–0.27
435	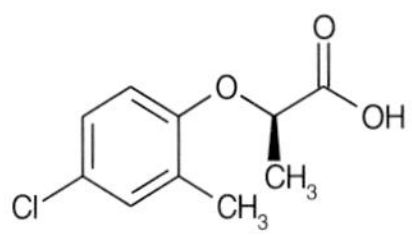	Mecoprop-P	16484-77-8	546	3	–0.41
436	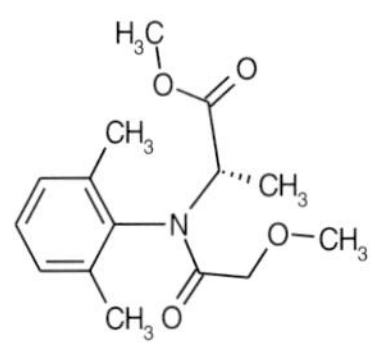	Mefenoxam	70630-17-0	981	3	–0.55
437	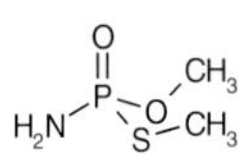	Methamidophos	10265-92-6	10	1	1.15
438	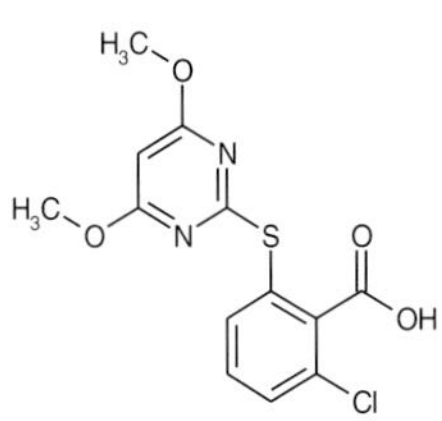	Pyrithiobac	123342-93-8	1498	3	–0.66
439*	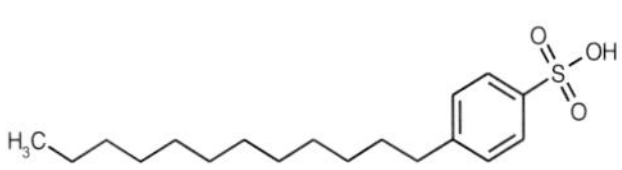	Dodecyl Benzene sulfonate	27176-87-0	1270	3	–0.59
440	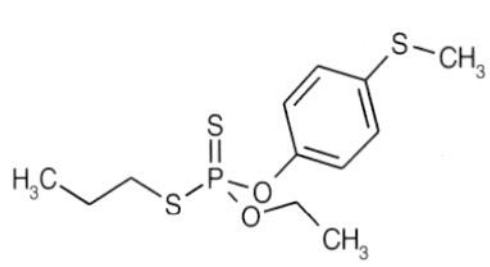	Sulprofos	35400-43-2	47	1	0.84

ID	Structure	Name	CAS NB	Toxicity LD_{50}		
				mg/kg	Class	–Log mmol/kg
441	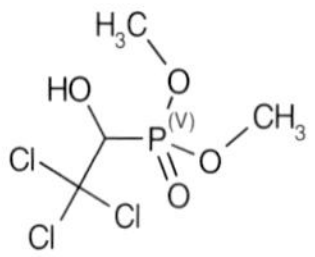	Trichlorfon	52-68-6	22	1	1.06
442	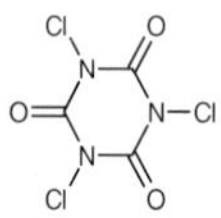	Trichloro-*s*-triazinetrione	87-90-1	1674	3	–0.86

LIST OF THE 105 COMPOUNDS INCLUDED IN THE BEE DATA SET

Three toxicity classes were generated according to the intervals established by the EC regulation, after regrouping the toxicity classes as too weakly represented:

LD_{50} (class 1) $< 1\,\mu g$/bee; $1\,\mu g$/bee $\leq LD_{50}$ (class 2) $< 50\,\mu g$/bee; LD_{50} (class 3) $\geq 50\,\mu g$/bee.

The symbol '*' in the ID column indicates the test set compounds.

ID	Structure	Name	CAS NB	Toxicity LD_{50}		
				μg/bee	Class	–Log μmol/bee
20	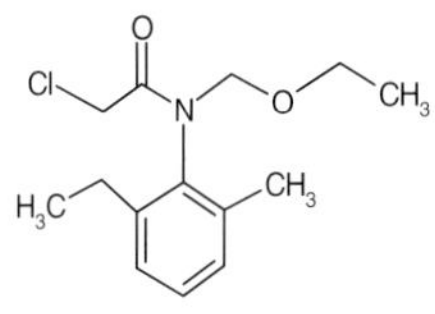	Acetochlor	34256-82-1	1715	3	–0.80
22	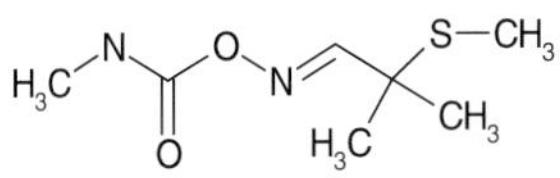	Aldicarb	116-06-3	0.285	1	2.82
25	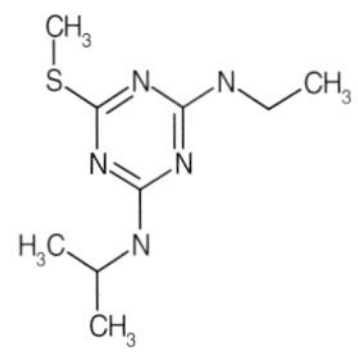	Ametryne	834-12-8	100	3	0.36
32	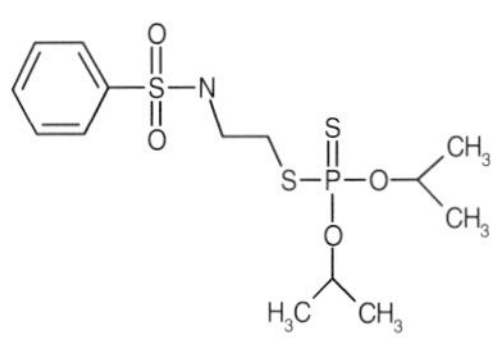	Bensulide	741-58-2	24	2	1.22
33	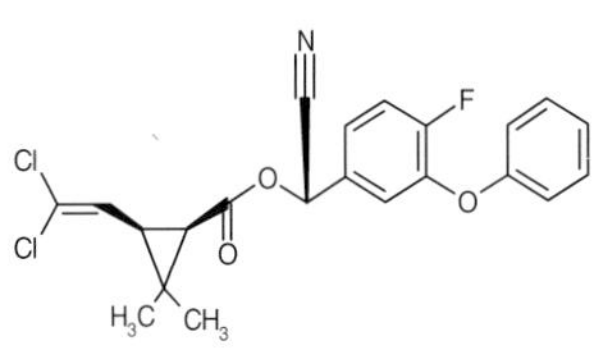	Beta-cyfluthrin	68359-37-5	0.037	1	4.07

ID	Structure	Name	CAS NB	Toxicity LD_{50}		
				µg/bee	Class	–Log µmol/bee
35*	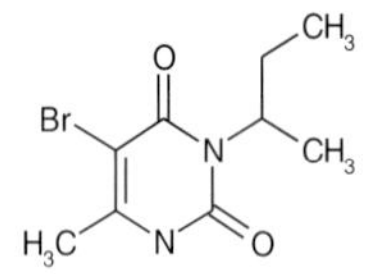	Bromacil	314-40-9	193.4	3	0.13
36	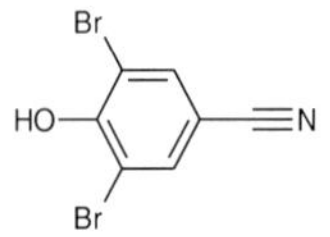	Bromoxynil	1689-84-5	14.5	2	1.28
37	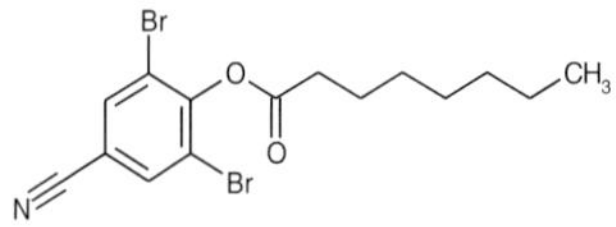	Bromoxynil octanoate	1689-99-2	2	2	2.30
44	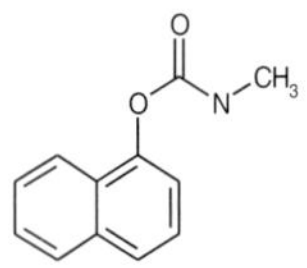	Carbaril	63-25-2	1.3	2	2.19
46	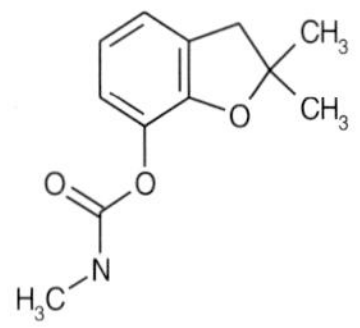	Carbofuran	1563-66-2	0.16	1	3.14
47	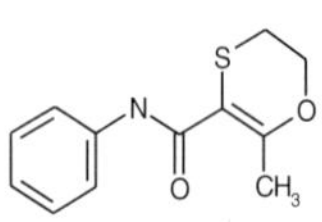	Carboxin	5234-68-4	181	3	0.11
48	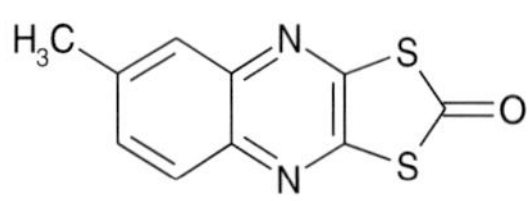	Chino methionat	2439-01-2	66.5	3	0.55

ID	Structure	Name	CAS NB	Toxicity LD_{50}		
				μg/bee	Class	–Log μmol/bee
50	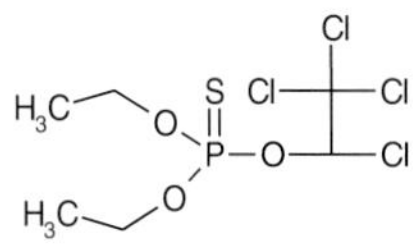	Chloro ethoxyfos	54593-83-8	0.09	1	3.57
52	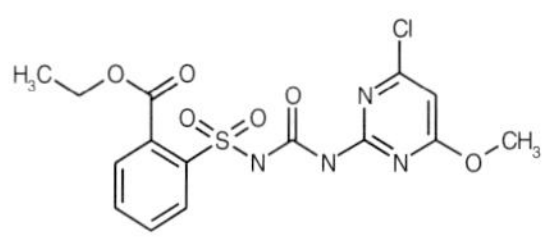	Chlorimuron ethyl	90982-32-4	12.5	2	1.52
68	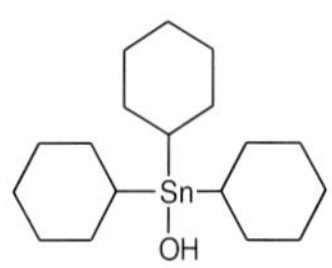	Cyhexatin	13121-70-5	35.9	2	1.03
73*	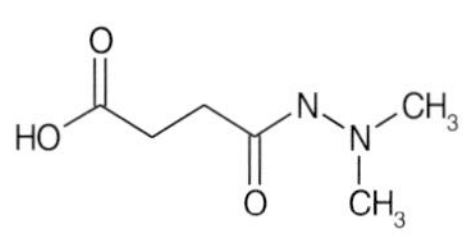	Daminozide	1596-84-5	205.5	3	–0.11
77	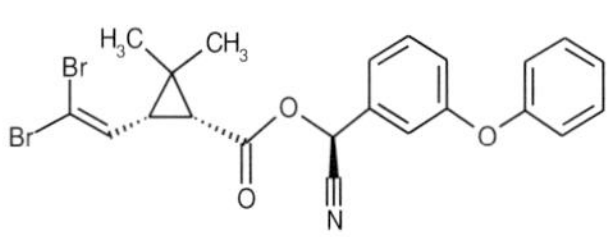	Deltamethrin	52918-63-5	0.0015	1	5.53
85	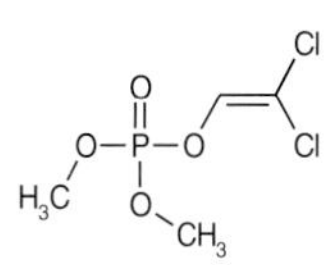	Dichlorvos	62-73-7	0.5	1	2.65
89	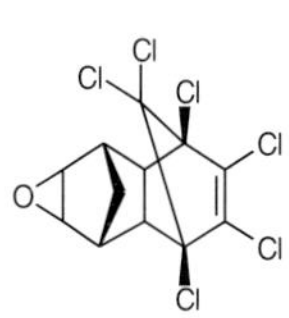	Dieldrin	60-57-1	0.14	1	3.44

ID	Structure	Name	CAS NB	Toxicity LD_{50}		
				μg/bee	Class	–Log μmol/bee
95	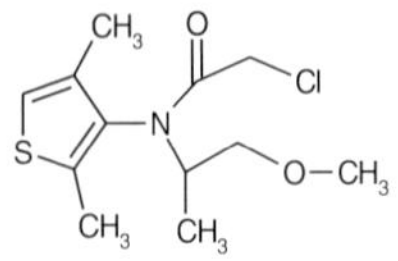	Dimethen amid	87674-68-8	94	3	0.47
96	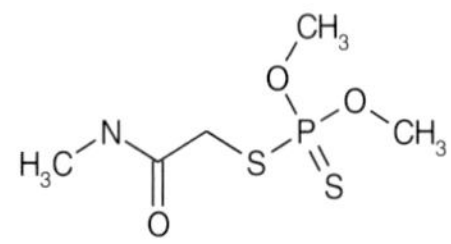	Dimethoate	60-51-5	0.16	1	3.16
101	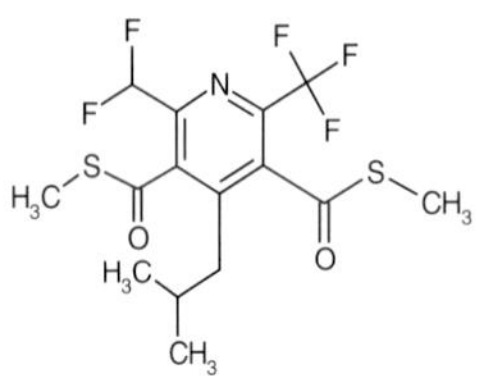	Dithiopyr	97886-45-8	81	3	0.70
102	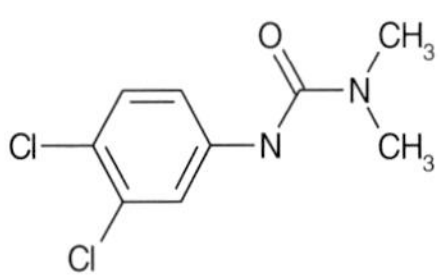	Diuron	330-54-1	145	3	0.21
103	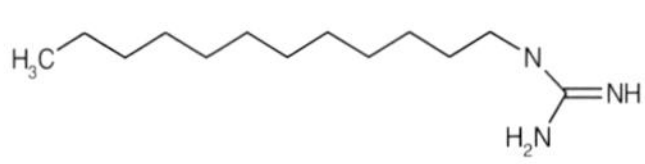	Dodine	2439-10-3	9.57	2	1.38
107	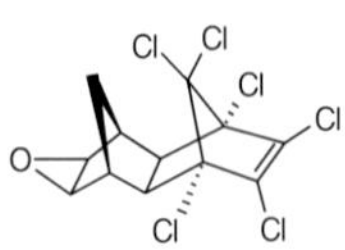	Endrin	72-20-8	2.02	2	2.28
108	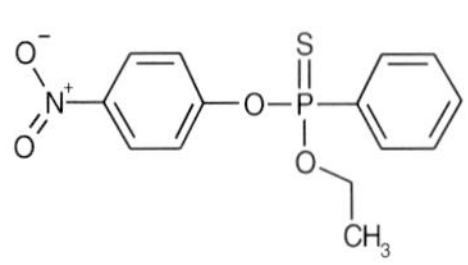	EPN	2104-64-5	0.245	1	3.12

ID	Structure	Name	CAS NB	Toxicity LD_{50}		
				µg/bee	Class	–Log µmol/bee
110	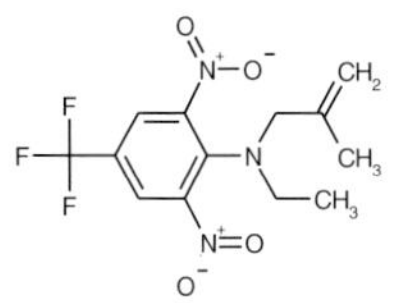	Ethalfluralin	55283-68-6	51	3	0.82
111	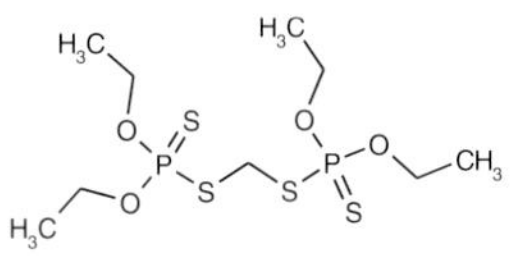	Ethion	563-12-2	20.55	2	1.27
120	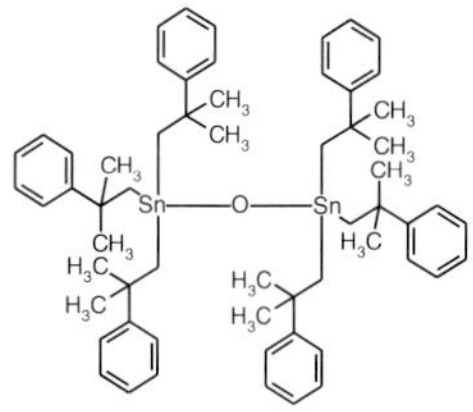	Fenbutatin oxide	13356-08-6	3982	3	-0.58
122*	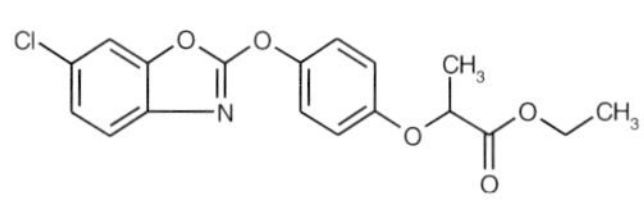	Fenoxa propethyl	66441-23-4	0.1	1	3.56
126	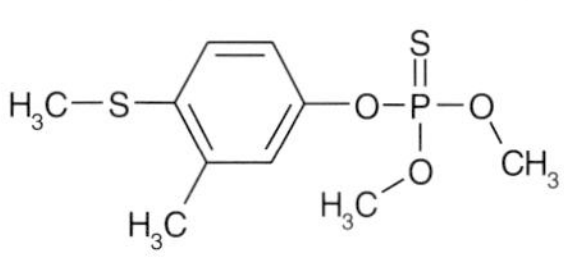	Fenthion	55-38-9	0.31	1	2.96
127	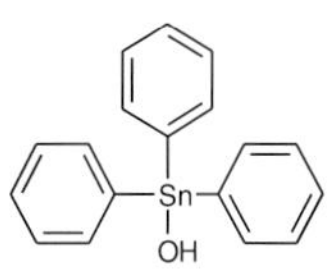	Fentin hydroxide	76-87-9	114.8	3	0.50

ID	Structure	Name	CAS NB	Toxicity LD_{50}		
				μg/bee	Class	–Log μmol/bee
137	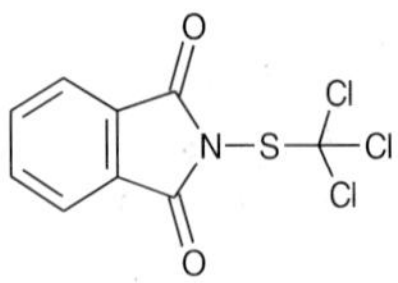	Folpet	133-07-3	12.1	2	1.39
139	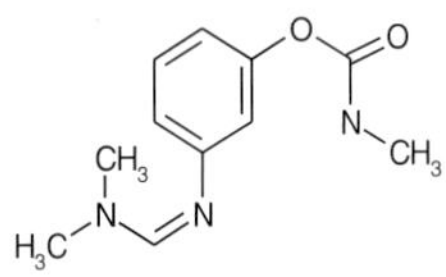	Formetanate	22259-30-9	12.25	2	1.26
142	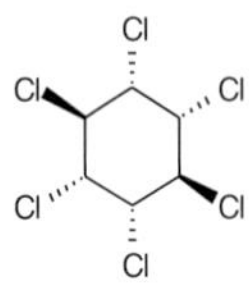	gamma-HCH	58-89-9	0.56	1	2.72
146	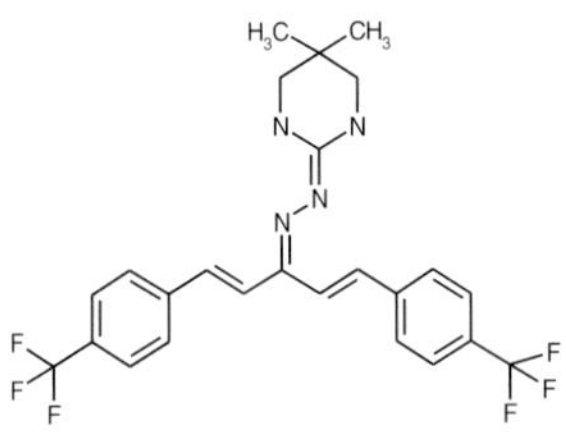	Hydra methylnon	67485-29-4	67	3	0.87
158*	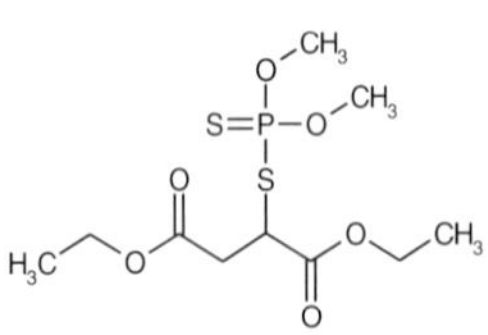	Malathion	121-75-5	0.2	1	3.22
165	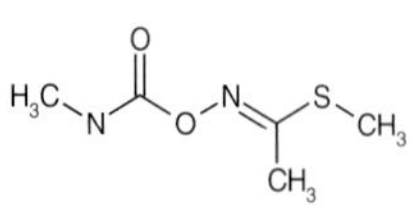	Methomyl	16752-77-5	0.16	1	3.01

ID	Structure	Name	CAS NB	Toxicity LD_{50}		
				μg/bee	Class	–Log μmol/bee
172	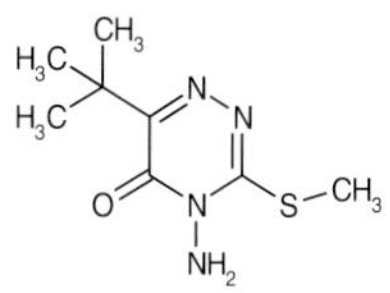	Metribuzin	21087-64-9	60.4	3	0.55
179	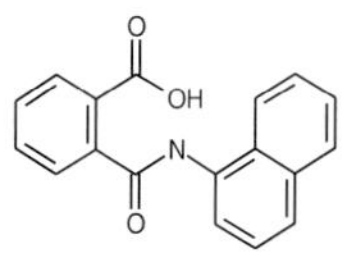	Naptalam	132-66-1	105.27	3	0.44
185	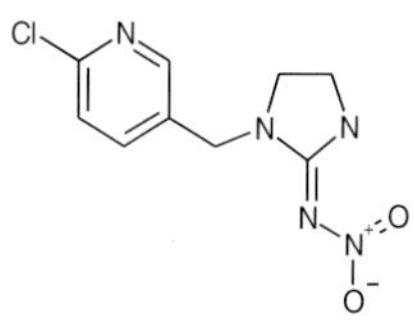	Imidacloprid	105827-78-9	0.08	1	3.52
190	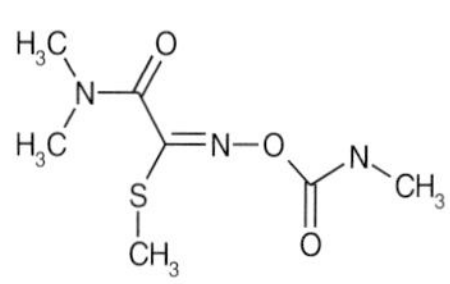	Oxamyl	23135-22-0	10.3	2	1.33
192	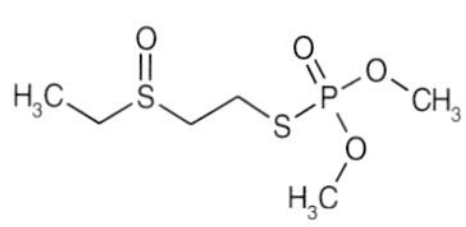	Oxydemeton-methyl	301-12-2	3	2	1.91
196	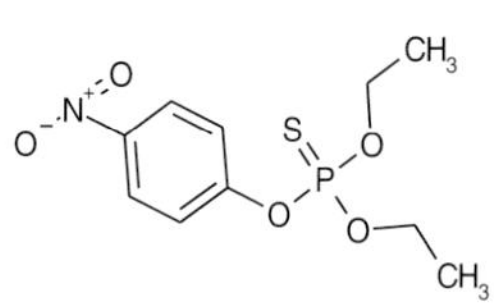	Parathion	56-38-2	0.175	1	3.22

ID	Structure	Name	CAS NB	Toxicity LD_{50}		
				μg/bee	Class	–Log μmol/bee
197	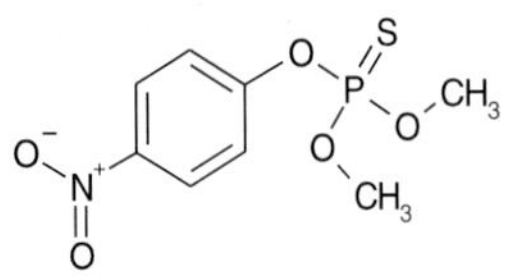	Parathion-methyl	298-00-0	0.111	1	3.38
200*	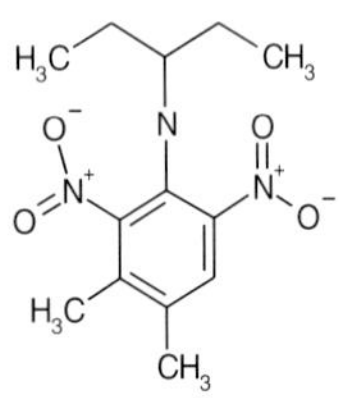	Pendi methalin	40487-42-1	49.8	2	0.75
202	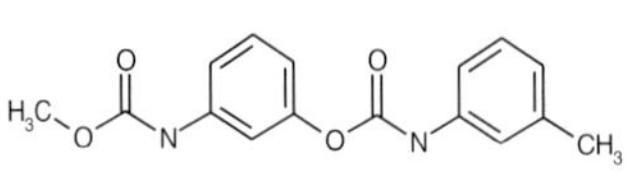	Phen medipham	13684-63-4	241.7	3	0.09
204	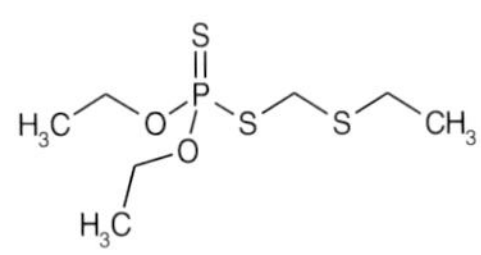	Phorate	298-02-2	10.1	2	1.41
205	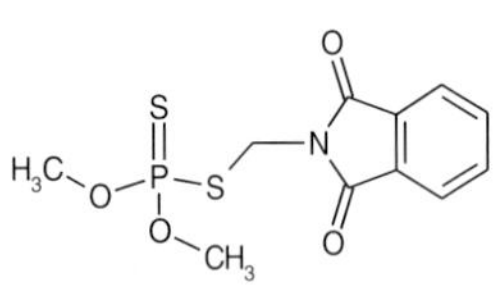	Phosmet	732-11-6	1.1	2	2.48
213	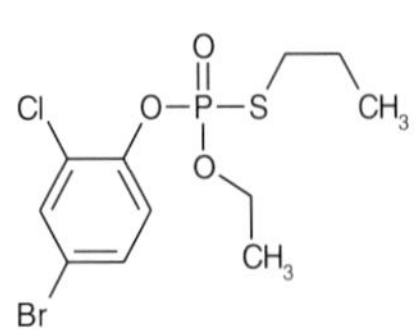	Profenofos	41198-08-7	0.09	1	3.59

ID	Structure	Name	CAS NB	Toxicity LD_{50}		
				μg/bee	Class	–Log μmol/bee
214	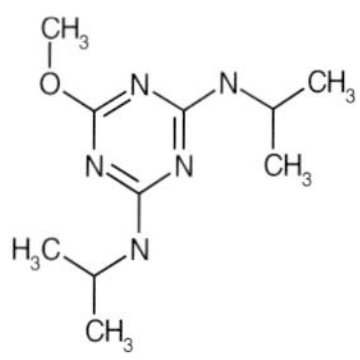	Prometon	1610-18-0	36	2	0.80
215	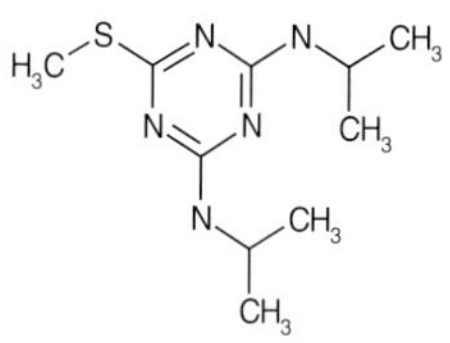	Prometryn	7287-19-6	96.7	3	0.40
218	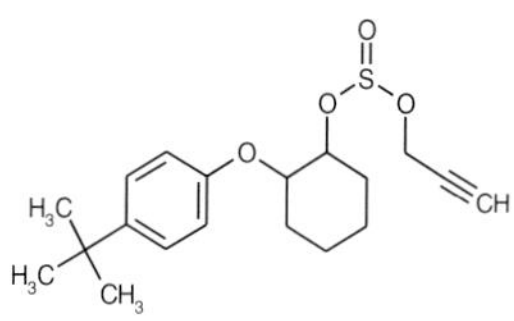	Propargite	2312-35-8	15	2	1.37
225	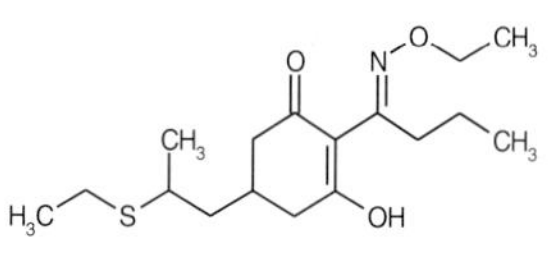	Sethoxydim	74051-80-2	10	2	1.52
231	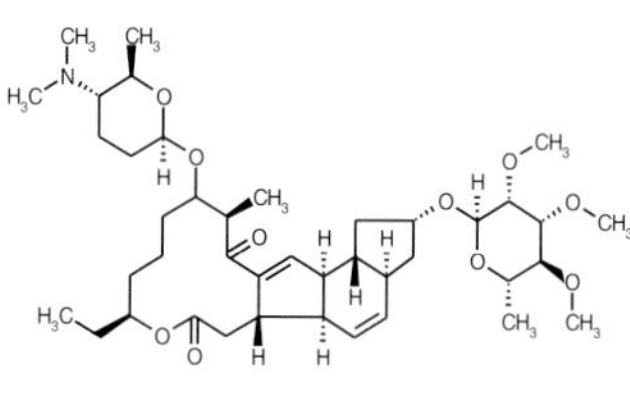	Spinosad	131929-60-7	0.003	1	5.40
247	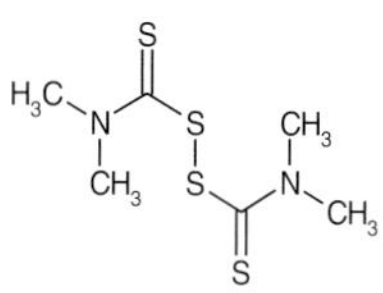	Thiram	137-26-8	74	3	0.51

ID	Structure	Name	CAS NB	Toxicity LD_{50}		
				µg/bee	Class	–Log µmol/bee
248*	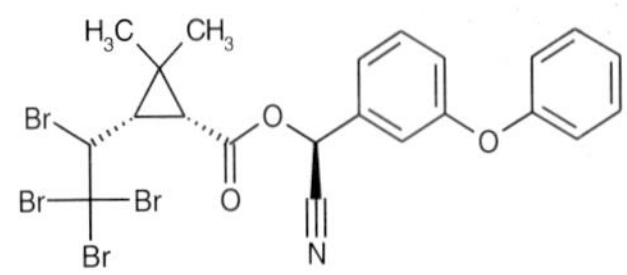	Tralomethrin	66841-25-6	0.13	1	3.71
260	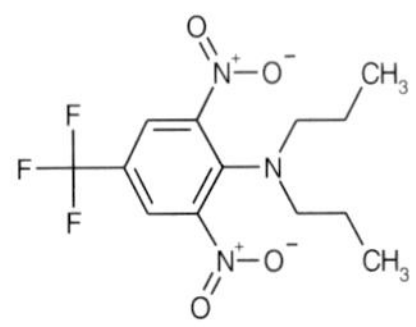	Trifluralin	1582-09-8	24.2	2	1.14
276	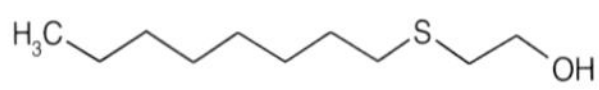	2-Hydroxyethyl octyl sulfide	3547-33-9	56.9	3	0.52
281*	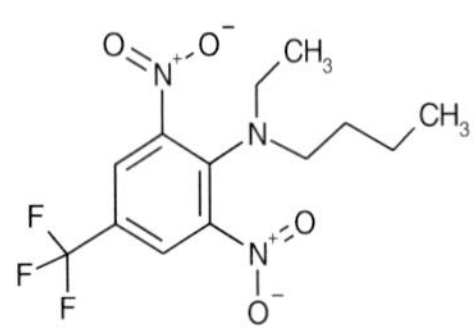	Benfluralin	1861-40-1	14.5	2	1.36
284	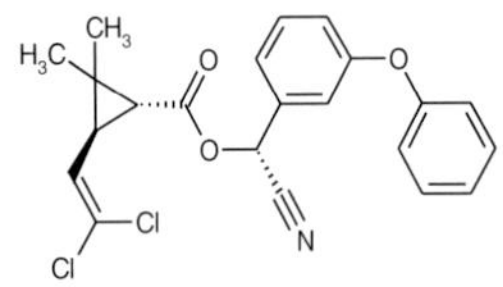	Beta cypermethrin	66841-24-5	0.02	1	4.26
285	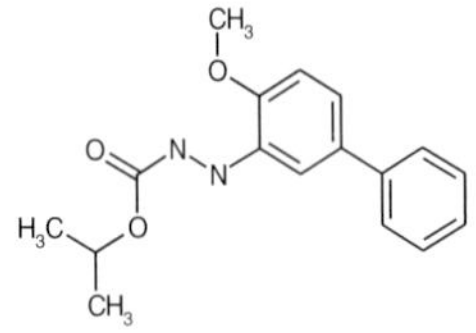	Bifenazate	149877-41-8	7.8	2	1.59

ID	Structure	Name	CAS NB	Toxicity LD_{50}		
				µg/bee	Class	–Log µmol/bee
293	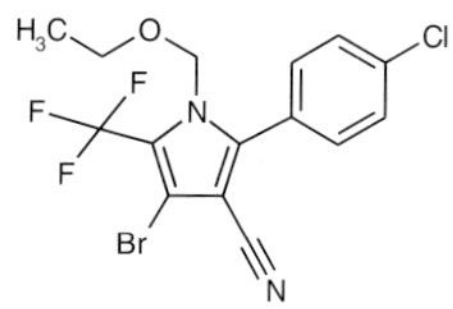	Chlorfenapyr (Pirate)	122453-73-0	0.12	1	3.53
295	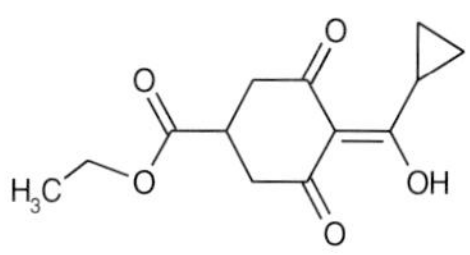	Cimecticarb	95266-40-3	47	2	0.73
319*	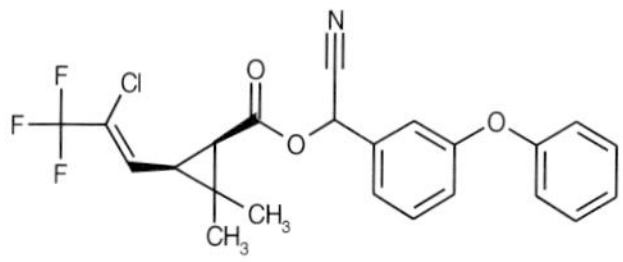	Lambda-cyhalothrin	91465-08-6	0.04	1	4.07
329	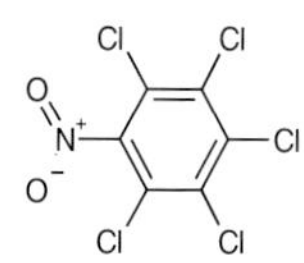	PCNB	82-68-8	0.1	1	3.47
330	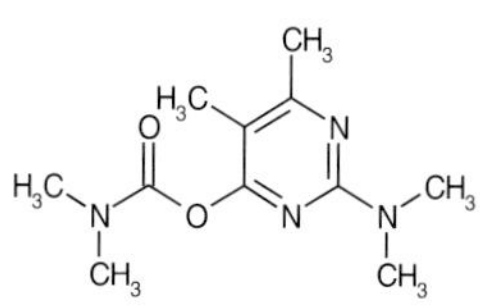	Pirimicarb	23103-98-2	18.7	2	1.10
331	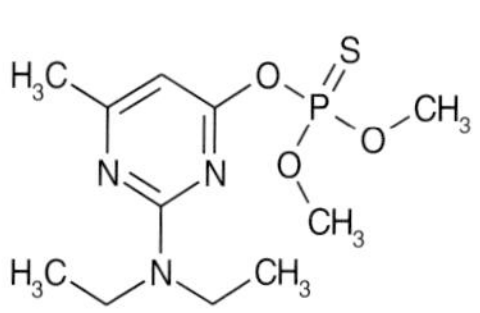	Pirimiphos methyl	29232-93-7	0.39	1	2.89

ID	Structure	Name	CAS NB	Toxicity LD_{50}		
				μg/bee	Class	–Log μmol/bee
333	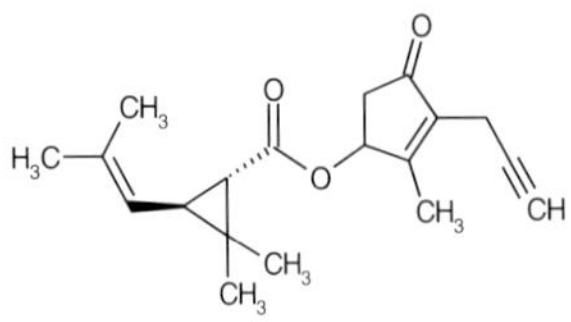	Prallethrin	23031-36-9	0.03	1	4.03
334	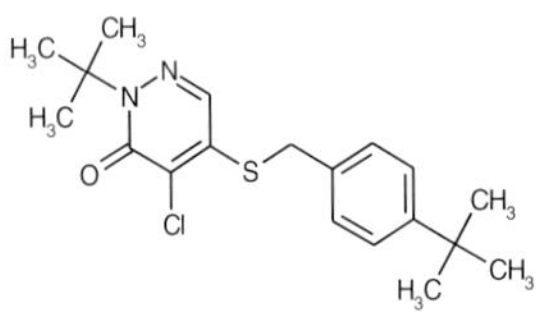	Pyridaben	96489-71-3	0.024	1	4.18
349	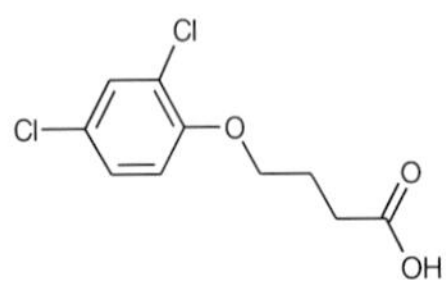	2,4-DB	94-82-6	14.5	2	1.24
351	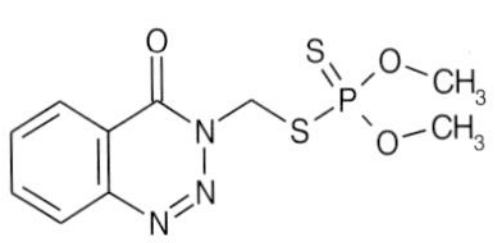	Azinphos-methyl	86-50-0	0.42	1	2.88
359	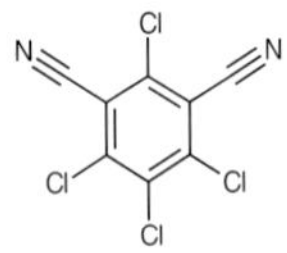	Chloro thalonil	1897-45-6	181.3	3	0.17
362	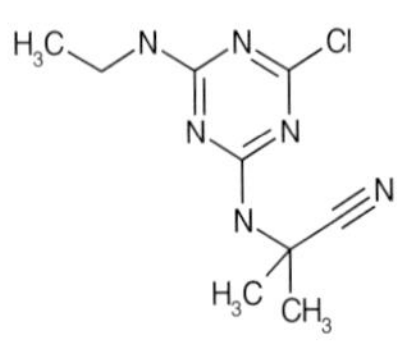	Cyanazine	21725-46-2	193.4	3	0.10

ID	Structure	Name	CAS NB	Toxicity LD_{50}		
				μg/bee	Class	–Log μmol/bee
366	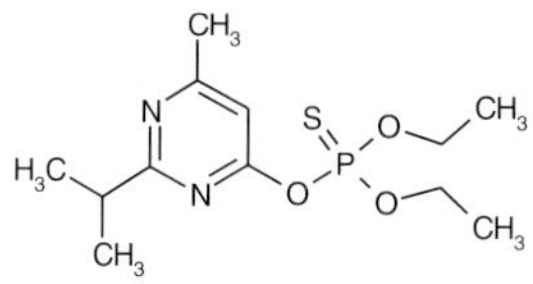	Diazinon	333-41-5	0.2	1	3.18
370	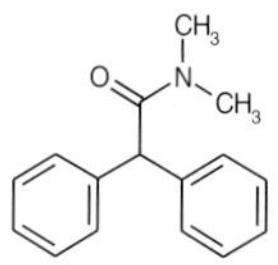	Diphenamid	957-51-7	2431.7	3	–1.01
371	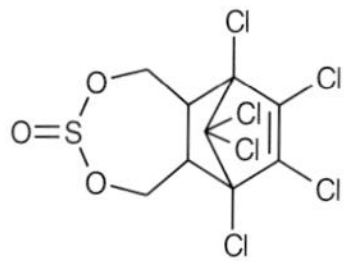	Endosulfan	115-29-7	4.5	2	1.96
373	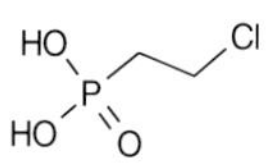	Ethephon	16672-87-0	12.1	2	1.08
376	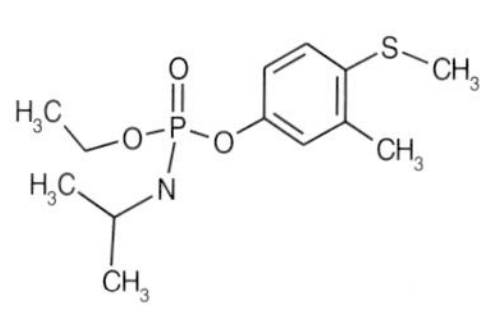	Fenamiphos	22224-92-6	1.87	2	2.21
382	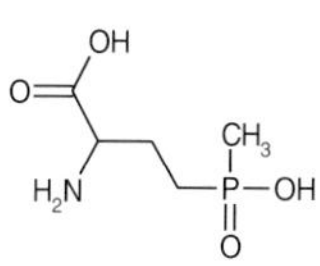	Glufosinate	51276-47-2	315.9	3	–0.24
391	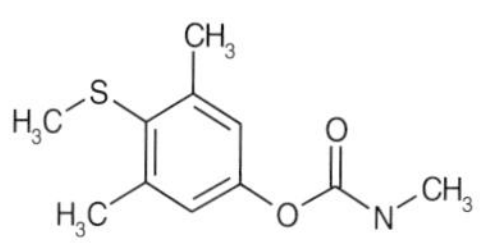	Methiocarb	2032-65-7	0.375	1	2.78

ID	Structure	Name	CAS NB	Toxicity LD50		
				μg/bee	Class	–Log μmol/bee
397	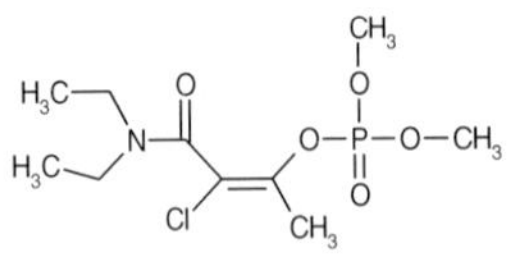	Phosph amidon	13171-21-6	1.46	2	2.31
398	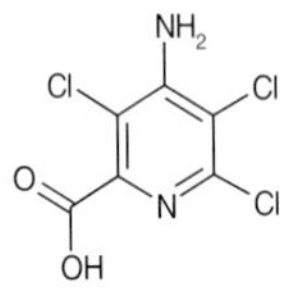	Picloram	1918-02-1	14.5	2	1.22
413	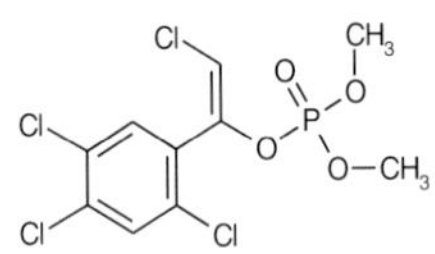	Tetra chlorvinphos	22248-79-9	1.37	2	2.43
414	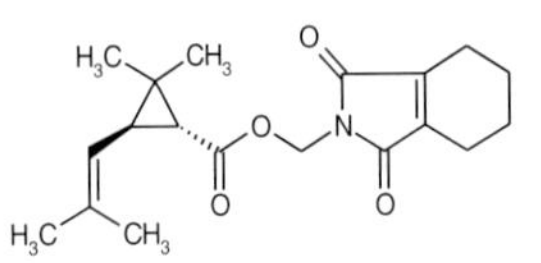	Tetramethrin	7696-12-0	0.155	1	3.33
432	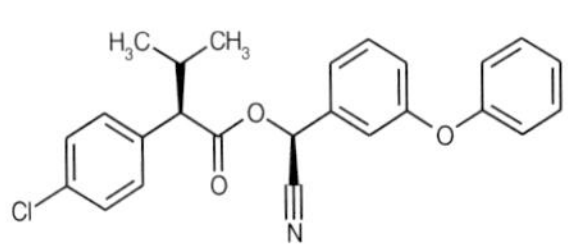	Esfen valerate	66230-04-4	0.41	1	3.01
437	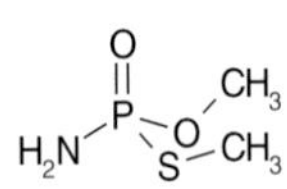	Metha midophos	10265-92-6	1.37	2	2.01
441	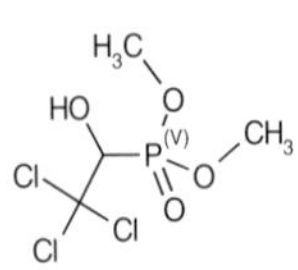	Trichlorfon	52-68-6	59.8	3	0.63

ID	Structure	Name	CAS NB	Toxicity LD_{50}		
				µg/bee	Class	–Log µmol/bee
444	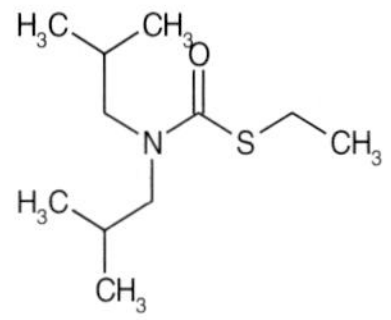	Butylate	2008-41-5	29	2	0.87
448	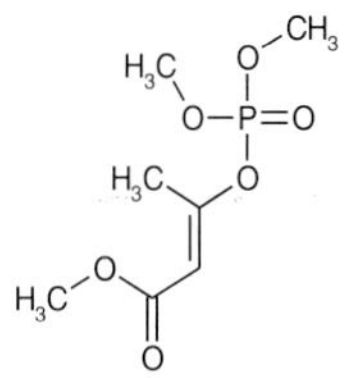	Mevinphos	7786-34-7	0.07	1	3.51
450	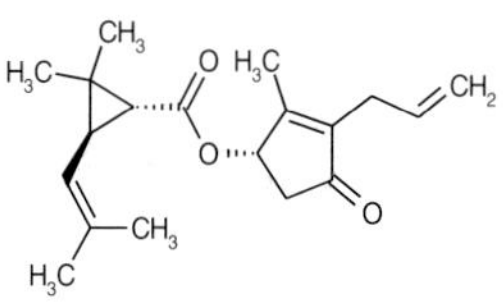	*s*-Bioallethrin	28434-00-6	3.4	2	1.95
451	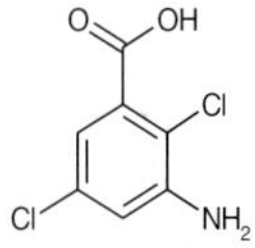	Chloramben	133-90-4	13.4	2	1.19
452	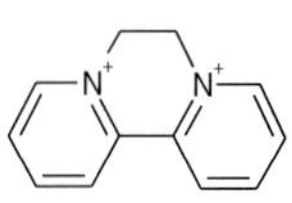	Diquat	2764-72-9	53.55	3	0.54
453	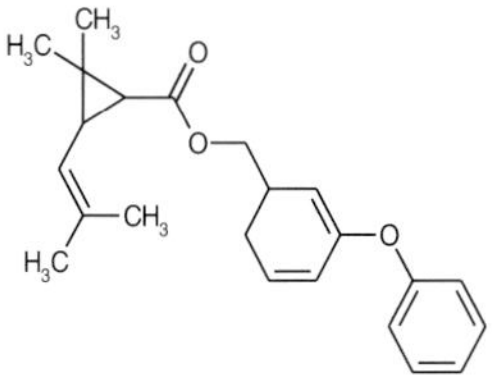	D-Phenothrin	26002-80-2	0.07	1	3.72

ID	Structure	Name	CAS NB	Toxicity LD_{50}		
				μg/bee	Class	–Log μmol/bee
454	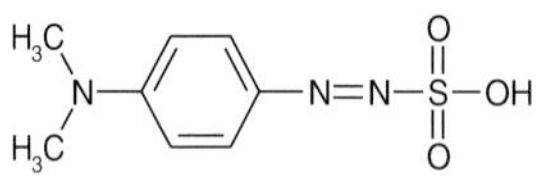	Fenamino sulf	140-56-7	93.19	3	0.39
455	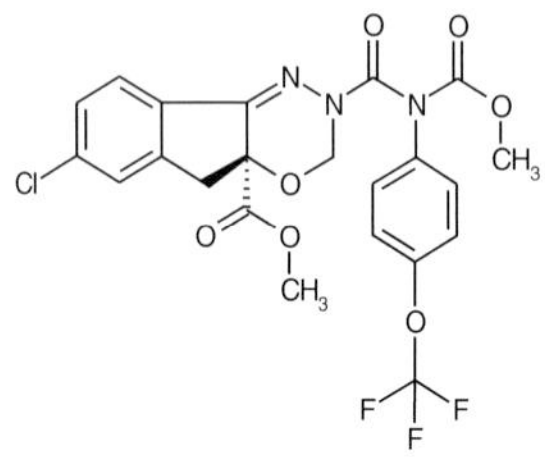	Indoxacarb (DPX-MP062)	173584-44-6	0.2	1	3.47
456	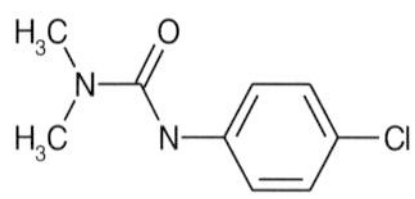	Monuron	150-68-5	110	3	0.26
457	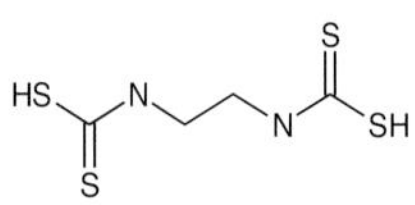	Nabam	142-59-6	10.02	2	1.33
458	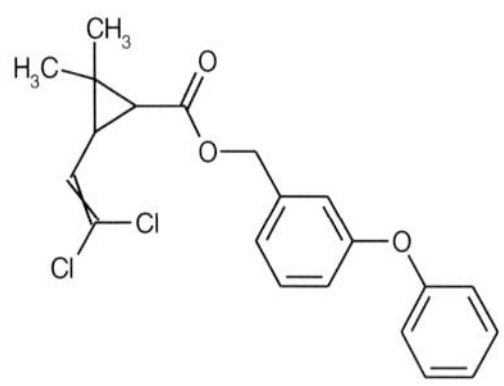	Permethrin	52645-53-1	0.02	1	4.21
459	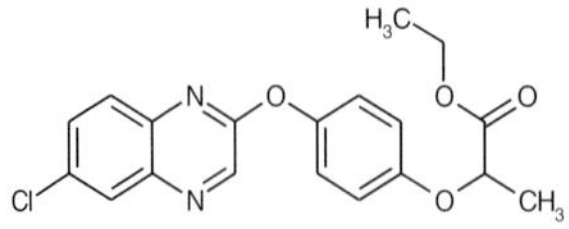	Quizalofop Ethyl	76578-14-8	50	3	0.87
460	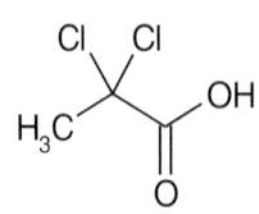	Dalapon	127-20-8	20.98	2	0.83

ID	Structure	Name	CAS NB	Toxicity LD_{50}		
				μg/bee	Class	–Log μmol/bee
461	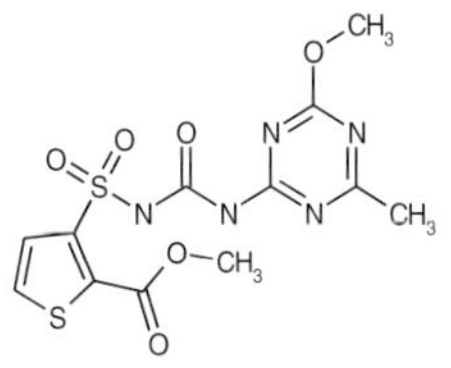	Thiameturon-methyl	79277-27-3	12.5	2	1.49
462	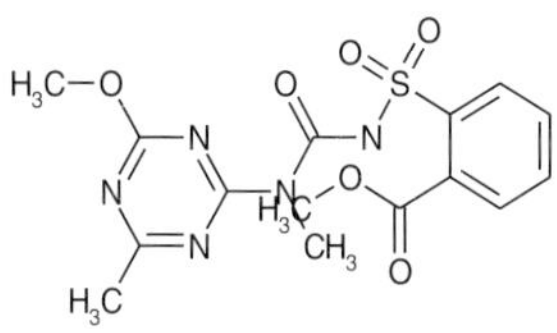	Tribenuron methyl	101200-48-0	100	3	0.60
464	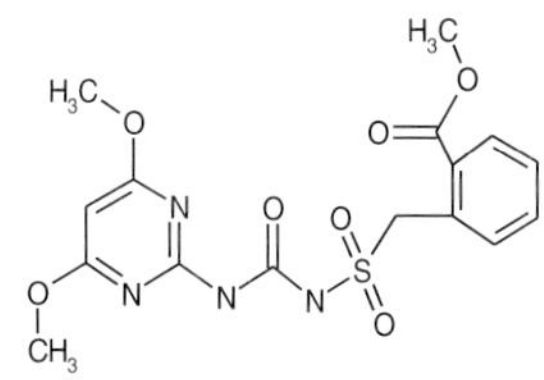	Bensulfuron Methyl	83055-99-6	12.5	2	1.52
465	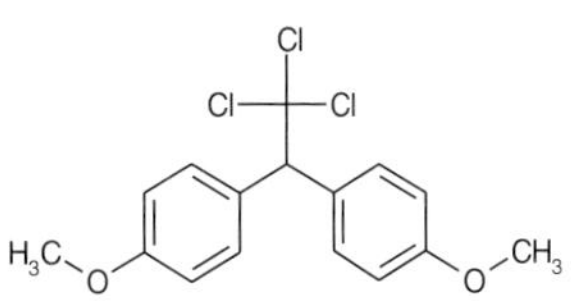	Methoxy chlor	72-43-5	23.6	2	1.17

Edited by Emilio Benfenati

Appendix C

Example procedures in molecular modelling

Marco Pintore and Christophe Wechman

BCX, Biochemics Consulting SAS, Orléans, France

MINIMIZATION PARAMETERS UNDER MACROMODEL

The 'conformational search' module of macromodel was used to find the best conformations of all compounds. More particularly, the parameters used for this conformational search are as follows.

FOR THE POTENTIAL PANEL

The force field used was MMF94s, without any solvent treatment. A constant dielectric of 1.0 was employed for the electrostatic treatment, whereas the charges were assigned by the force field (Fig. 1).

Moreover, an extended non-bonded cut-off was set, for taking into account the best maximum distances over which hydrogen bonding, Van der Waals and electrostatic contributions to the molecular potential energy were evaluated. The cut-off distances are 4 Å for hydrogen bonding, 7 Å for Van der Waals and 12 Å for electrostatics.

FOR THE MINIMIZATION PANEL

The method used for the minimization step is the PRCG method or Polak-Ribiere conjugate gradient. This is a conjugate gradient minimization scheme that uses the Polak-Ribiere first derivative algorithm that restarts every $3N$ iterations (Fig. 2).

A value of 500 max iterations was used to specify when the calculation should end if the specified convergence criterion has not been met. This criterion corresponds to a gradient setting with a convergence threshold of 0.05 (default value).

FOR THE CONFORMATIONAL SEARCH PANEL

A serial low-mode algorithm was used for exploring the better conformations. This method analyses the low-frequency eigen vectors of the system, which are expected to follow 'soft' degrees of freedom, such as torsions. Moreover, it explores the conformational space aggressively enough to switch the chirality of atoms within the structure provided. Finally, it performs a separate conformational search for

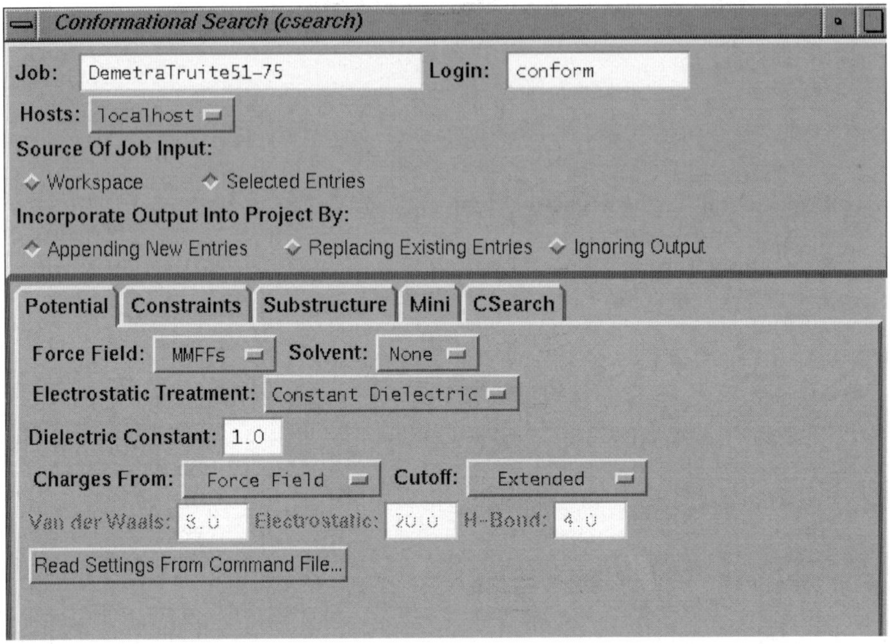

Figure 1: Potential panel parameters specifications in Macromodel. The image was generated using Schrödinger's Maestro program (version 7.0).

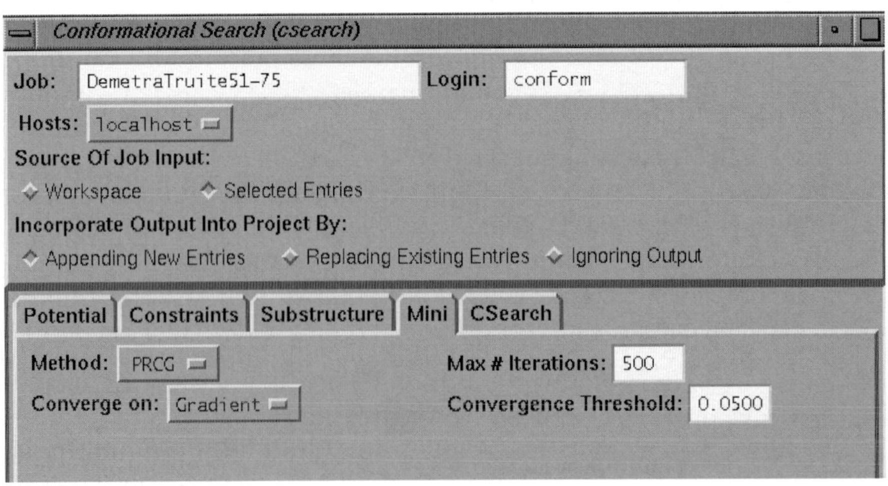

Figure 2: Minimization panel parameters specifications in Macromodel. The image was generated using Schrödinger's Maestro program (version 7.0).

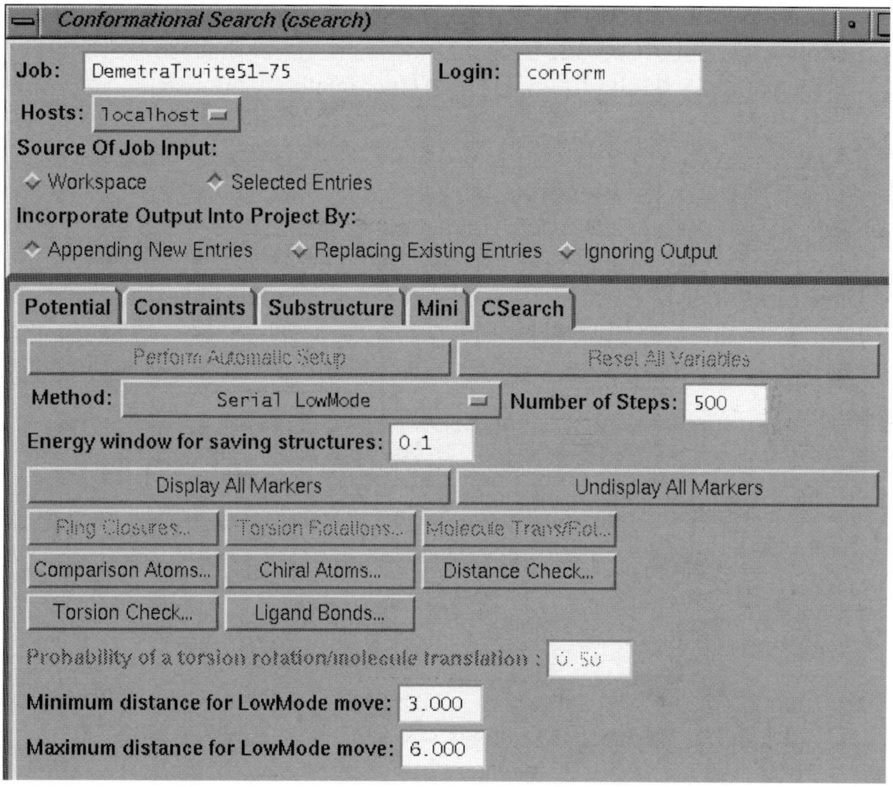

Figure 3: Conformational search panel parameters specifications in Macromodel. The image was generated using Schrödinger's Maestro program (version 7.0).

each structure in the input file. This takes advantage of its ability to define fruitful search directions without specifying variable torsions (Fig. 3).

Finally, minimum and maximum distance for low-mode move parameters were set to 3–6 Å, respectively. During a search analysis, the 'fastest moving' atom is displaced at random distances ranging between these minimum and maximum values.

Each analysis lasted 500 steps, and a 0.1-kJ/mol 'window' was used to compare trial structures. That is, only new minimized structures within 0.1 kJ/mol above the current global minimum were kept.

USE OF MOPAC UNDER SYBYL

MOPAC package under Sybyl is a collection of four semi-empirical molecular orbital (MO) methods: MINDO/3 (Modified Intermediate Neglect of Differential Overlap), MNDO (Modified Neglect of Differential Overlap), AM1 and PM3.

AM1 was used for these first data sets, according to previous BCX (Biochemics Consulting) experiences and suggestions by CSL (Central Science Laboratory).

More particularly, the following AM1 parameters were used:

MOPAC COMPUTE PANEL

AM1 with a singlet state was used. The net charge parameter is the only one that can vary in the OPC (Organa-Phosphorous Compounds) data set (0, +1, −1 . . .). The time limit was fixed at 3600 s. Normal and precise convergence, with a full optimization, was chosen to be able to apply a SCF (Self-Consistent Field) convergence procedure. All bonds, angles and dihedrals were optimized (Fig. 4).

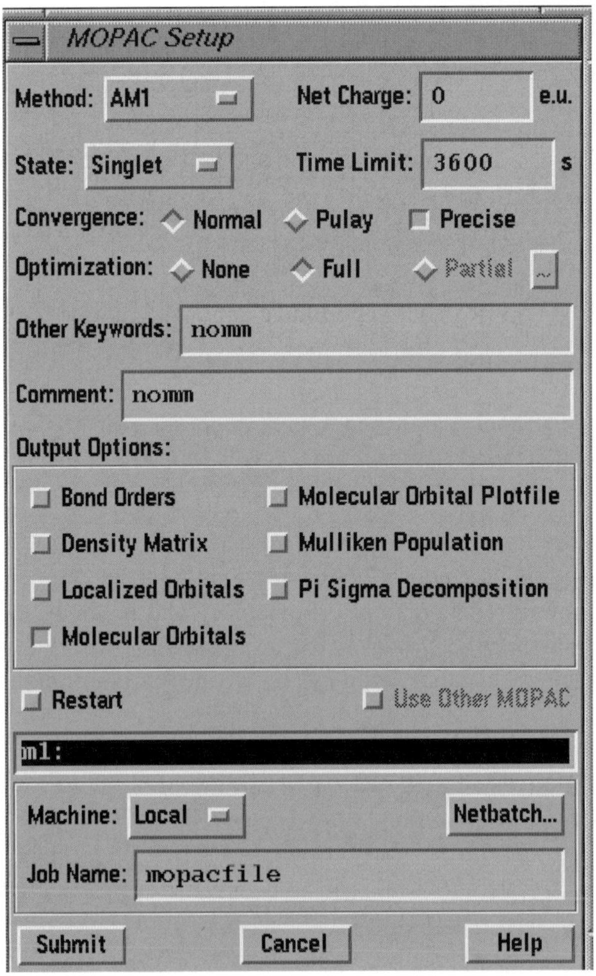

Figure 4: MOPAC compute panel. The image was generated using Tripos Sybyl program (version 7.1).

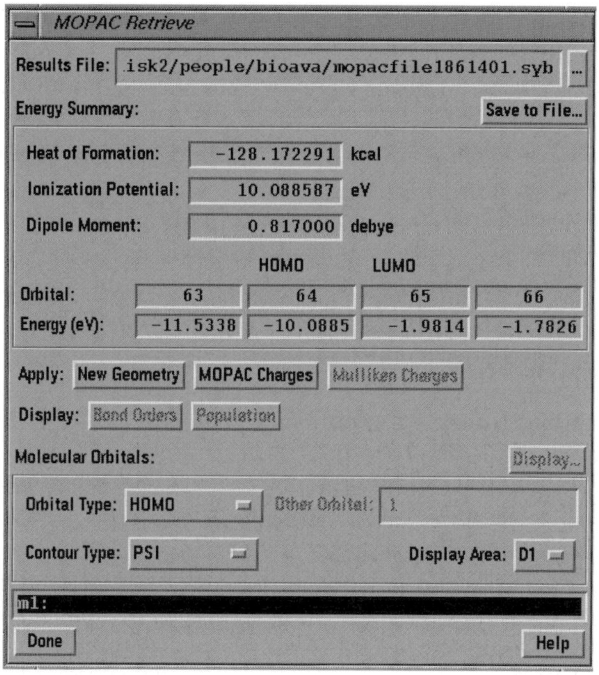

Figure 5: MOPAC result panel. The image was generated using Tripos Sybyl program (version 7.1).

MOPAC RESULTS PANEL

For completing the modelling work, only two options were exploited (Fig. 5):

1. New geometry for updating compound co-ordinates.
2. MOPAC charge for assigning MOPAC-determined atomic charge.

AB INITIO MODELLING PROTOCOL FOR DEMETRA STRUCTURES

Source structures: MOL2 files searched by conformational search with macromodel as above.

Conversion utility used to prepare Gaussian input files: Babel for DOS ver. 1.3.

Full Gaussian optimization with vibrational analysis, command line: No. AM1 FOPT Freq=NoRaman.

Optimized structures with Gaussian 03W are consistent with the optimized structures obtained with Hyperchem 7.0 for Windows but are different from ones obtained with MOPAC 6. Gaussian outputs were analysed with GaussView:

1. The majority of structures was optimized into stable geometry – number of imaginary frequencies was equal to 0.
2. Several structures finished in the transitional state (quasi 'ideal geometry' due to idealized starting geometry, for example, 100% planar amino group bound to phenyl ring), where number of imaginary frequencies was equal to 1. After inspection and small geometry change of unstable part of the molecule (slight adjustment of the torsion angle) and subsequent re-optimization, Nimag was equal to 0.
3. Four structures could not be optimized so that Nimag would be equal to 0. These include *cis*-unsaturated aliphatic chains (three cases) and a triazine heterocycle with a methyl group bound to it (one case). However, the structures seem to have reasonable geometries, and imaginary vibrations have small intensities. Source MOL2 files were updated with optimized co-ordinates by internal program utility. Source MOL2 files were updated with final AM1 charges by internal program utility.

Edited by Emilio Benfenati

Appendix D

The descriptors selected for each data set

1. TROUT

Reference number	Descriptor name	Description
1.1.	Number of three-membered rings	Count of all rings with three members
1.2.	Log *D* at pH 7	Measure of the lipophilicity calculated as an apparent partition coefficient on two different phases, one organic and the other aqueous, at the corresponding pH
1.3.	Log *P*	Measure of lipophilicity calculated as octanol–water partition coefficient
1.4.	Mean atomic polarizability (scaled on carbon atom) – constitutional descriptors	Measure of the molecular polarizability understood as the magnitude of the dipole provoked by an electric field over a molecule. This descriptor is calculated as the mean of all the atomic polarizabilities scaled over the Carbon atoms $$Mp = \frac{1}{NC}\sum_{i=1}^{atoms}\alpha_i$$
1.5.	Number of total hydroxyl groups – functional groups	Count of all hydroxyl groups present in the molecule R−OH
1.6.	Number of phenols – functional groups	Count of the number of hydroxyl groups bonded to benzene rings 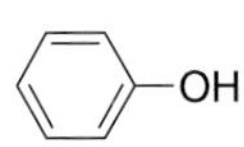
1.7.	Number of donor atoms for H-bonds (with N and O) – functional groups	Count of all hydrogen atoms linked to oxygen and nitrogen atoms 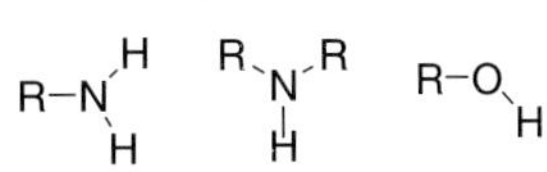

Reference number	Descriptor name	Description
1.8.	Number of acceptor atoms for H-bonds (N, O, and F) – functional groups	Count of all nitrogen, oxygen, and fluoride atoms present in the molecule able to form hydrogen bonds. That fact excludes N and O present as heteroatoms in aromatic rings 
1.9.	X–CR–X – atom-centered fragments	Count of all the sp^2 carbons in aromatic rings with two non-carbon substitutions bonded to any kind of carbon atom 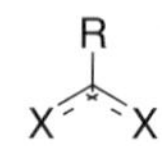
1.10.	Fragment-based polar surface area – properties	Measure of the molecular areas over the molecular surface susceptible to present charge deviation from the neutral charge. This descriptor is calculated by the sum of the areas present in some identified fragments. Values for every fragment defined in the calculating package are derived from experimental measures
1.11.	Eigenvalue sum from Z-weighted distance matrix (Barysz matrix) – topological descriptor	It is the sum of the eigenvalues of the weighted distance matrix accounting simultaneously for the presence of heteroatoms and multiple bonds. Barysz matrix is defined as $$[D^Z]_{ij} = \begin{cases} 1 - \dfrac{Z_C}{Z_i} & i = j \\ \sum_{b=1}^{d_{ij}} \left(\dfrac{1}{\pi_b^*} \times \dfrac{Z_C^2}{Z_{b(1)} \times Z_{b(2)}} \right) & i \neq j \end{cases},$$ where Z_C is the atomic number of the carbon atom, Z_i the atomic number of the ith atom, p^* is the conventional bond order, the sum runs over all d_{ji} atoms involved in the shortest path between vertices v_i and v_j, d_{ij} is the topological distance, and $b(1)$ and $b(2)$ represent vertices incident to the considered bond

Reference number	Descriptor name	Description
1.12.	Distance/detour ring index of order 3 – topological descriptor	It is a descriptor based on operations over the distance/detour matrix D/Δ. The detour matrix is a square symmetric matrix that contains the ratios of the lengths of the shortest to the longest path between any pair of vertices. It is calculated by $$D/\Delta = \frac{1}{2}\sum_{i=1}^{A}\sum_{j=1}^{A}(D/\Delta)_{ij}$$ where Δ is the detour distance
1.13.	Distance/detour ring index of order 9 – topological descriptor	Same that of the previous one, but in order 9s
1.14.	Self-returning walk count of order 5 – molecular walk counts	It is the total number of five-step walks returning again to the same atom in graph and is calculated by the 5th power of the adjacency matrix $$srw^5 = \sum_{i=1}^{A} srw_i^5 = tr(A^k)$$
1.15.	Number of [≥C–OH] – fragment descriptor	Number of hydroxyl groups bonded to a carbon. No matter whether the carbon atom is aromatic or not
1.16.	Moran autocorrelation – lag 3/weighted by atomic Sanderson electronegativities – 2D autocorrelations	It is an index of spatial autocorrelation defined over the Sanderson electronegativities (in Paulings) of every periodic element. It is calculated by $$I(3) = \frac{\frac{1}{\Delta}\sum_{i=1}^{A}\sum_{j=1}^{A}\delta_{ij}(\chi_i-\overline{\chi})\cdot(\chi_j-\overline{\chi})}{\frac{1}{A}\sum_{i=1}^{A}(\chi_i-\overline{\chi})},$$ where χ_i is the Sanderson electronegativity of every atom present in the molecule, $\overline{\chi}$ is the average electronegativity over the molecule, A is the number of atoms present in the molecule, 3 is the topological distance used in this case, δ_{ij} is a Kronecker delta ($\delta_{ij} = 1$ if $d_{ij} = 3$, 0 otherwise), and Δ is the sum of the obtained Kroneker deltas

Reference number	Descriptor name	Description
1.17.	Moran autocorrelation – lag 8/weighted by atomic Sanderson electronegativities – 2D autocorrelations	The same as the Moran autocorrelation – lag 3, already described in 1.16, but with $d = 8$.
1.18.	Geary autocorrelation – lag 2/weighted by atomic masses – 2D autocorrelations	It is another index of spatial autocorrelation but, in this case, defined over the atomic masses. It is calculated by $$c(2) = \frac{\frac{1}{2\Delta}\sum_{i=1}^{A}\sum_{j=1}^{A}\delta_{ij}.(AM_i - AM_j)^2}{\frac{1}{(A-1)}\sum_{i=1}^{A}(AM_i - \overline{AM})^2},$$ where 2 is the considered topological distance (d), A is the number of atoms present in the molecule, δ_{ij} is a Kronecker delta, Δ is the sum of the obtained Kroneker deltas, and AM_i is the atomic mass and $\overline{AM}$ the average of the atomic masses
1.19.	Geary autocorrelation – lag 3/weighted by atomic masses – 2D autocorrelations	The same as the Geary autocorrelation – lag 2 but with $d = 3$
1.20.	Number of esters (aliphatic) – functional groups	Count of all aliphatic esters present in the molecule 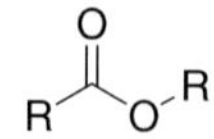

2. DAPHNIA

Reference number	Descriptor name	Description
2.1.	Molecular weight	It is the sum of all the atomic masses of all the atoms present in the molecule. It is calculated by $$MW = \sum_{i=1}^{A} m_i$$
2.2.	Mean atomic polarizability (scaled on carbon atom)	Already defined in 1.4.
2.3.	Number of phosphorus atoms	Count of all phosphorus atoms present in the molecule without taking into account the functionality they represent
2.4.	Number of tertiary amines (aromatic)	Count of all nitrogen atoms present in aromatic rings 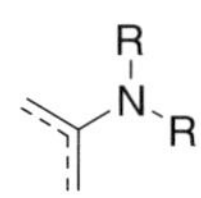
2.5.	Number of acceptor atoms for H-bonds (N, O, and F)	Already defined in 1.8.
2.6	Phenol/enol/carboxyl OH	Count of all hydroxyl groups bonded to benzene or to an sp^2 non-aromatic carbon 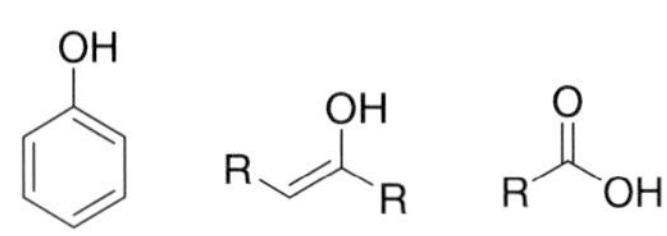
2.7.	AI–O–Ar/Ar–O–Ar/ R···O···O/R–O–C = X	Count of oxygen atoms that are actual esters of different acid functions 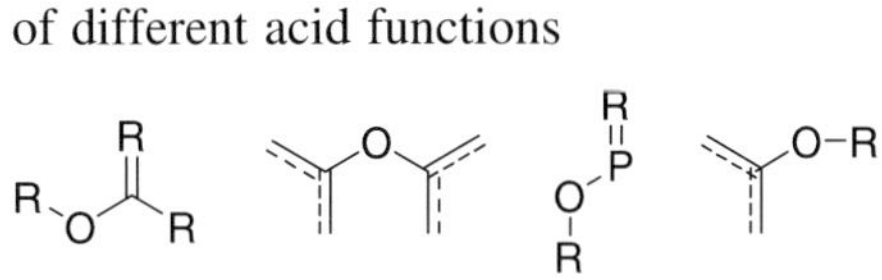
2.8.	R2S/RS–SR	Count of sulphur atoms involved thioether or disulphur bonds —S— —S–S—
2.9.	Moriguchi octanol–water partition coefficient (log *P*)	It is another measure of lipophilicity calculated as octanol–water partition coefficient. In this case, the calculation is made by a regression equation based on 13 structural parameters.

Reference number	Descriptor name	Description
		$\log P = -1.014 + 1.244\ (F_{CX})^{0.6} - 1.017\ (N_O + N_N)^{0.9} + 0.406\ F_{PRX} - 0.145\ N_{UNS}{}^{0.8} + 0.511\ I_{HB} + 0.268\ N_{POL}\ - 2.215\ F_{AMP} + 0.912\ I_{ALK} - 0.392\ I_{RNG} - 3.684\ F_{QN} + 0.474\ N_{NO2} + 1.582\ F_{NCS} + 0.773\ I_{\beta L}$, where F_{CX} is the weighted summation of number of carbon and halogen atoms, $N_O + N_N$ is the number of nitrogen and oxygen atoms, F_{PRX} is the proximity effect of N/O, N_{UNS} is the number of unsaturated bonds, I_{HB} is a variable taking into account the presence of intramolecular hydrogen bonds, N_{POL} is the number of polar substituents, F_{AMP} is the amphoteric property, I_{ALK} is a variable taking into account alkane, alkene, cycloalkane, and cycloalkenes, I_{RNG} is a variable counting the number of ring structures (not benzene and its condensed rings), F_{QN} is the count of quaternary nitrogens, N_{NO2} is the number of nitro groups, F_{NCS} is the count of isothiocyanates and thiocyanates, and $I_{\beta L}$ is the presence of β-lactam
2.10.	Mean Wiener index	It consists in an operation over the Wiener index. This index is defined as the sum over all bonds of the product of the number of vertices on each side of the bond. It is indeed the sum of all topological distances. Wiener index definition is $W = \frac{1}{2} \sum_{i=1}^{A} \sum_{j=1}^{A} d_{ij}.$ So the mean Wiener index is calculated by $\overline{W} = \frac{2 \cdot W}{A \cdot (A-1)}.$
2.11.	Mean information content on the distance equality	It is based on the equality of distances in the topological graph, and it is calculated by $\overline{I_D^E} = \frac{A \cdot (A-1)}{2} \cdot \log_2 \frac{A \cdot (A-1)}{2} - \sum_{g=1}^{G} {}^{g}f \cdot \log_2^{g} f,$

Reference number	Descriptor name	Description
		where A is the number of atoms, ${}^{g}f$ is the number of distances with equal g value in the triangular D submatrix, and G is the maximum distance value
2.12.	Information content index (neighbourhood symmetry of order 2)	It is a measure of the symmetry, displayed as equality between vertices and edges of a molecular graph, at a defined order/topological distance, 2 in this case $$I_c = \sum_{g=1}^{G} n_g \log_2 n_g$$ where G is the number of equivalent classes and n_g the number of elements in the 2nd (for this case) class
2.13.	Leading eingenvalue from polarizability weighted distance matrix	It is the largest eigenvalue of the symmetrical distance matrix weighted by the polarizability of the atoms. It is a measure of branching able to relate such property with the charge distribution
2.14.	Sum of topological distances between F···Cl	It is taken from values of the weighted distance matrix between fluorine and chlorine atoms. This matrix is built measuring the topological distance, the number of edges in the shortest path between atom pairs of a molecule. In this case, each distance is the minimum sum of edge weights referred to polarizability along the considered path
2.15.	Self-returning walk count of order 5	Already defined in 1.14.
2.16.	Highest eigenvalue number 1 of Burden matrix/ weighted by atomic masses	It is the highest of the eigenvalues obtained from the Burden matrix. Burden matrix is a modified connectivity matrix where the diagonal elements are substituted by the atomic masses (in this case) and every element out of the diagonal represents the conventional bond order

3. ORAL QUAIL

Reference number	Descriptor name	Description
3.1.	CH_3X	Count of the methyl groups within the molecule
3.2.	X–CX–X	Count of atom carbons with three substituents different from carbon in an aromatic ring 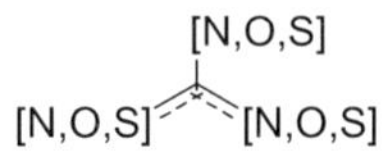
3.3.	Al–O–Ar/Ar–O–Ar/R···O···R/R–O–C = X	Already described in 2.7.
3.4.	E-state topological parameter	The electrotopological state index is calculated form the chemical graph. This index is based on the electronic effect of each atom on the other atoms in the molecule. Each atom has an assigned intrinsic state value I_i calculated as follows: $I_i = \frac{\left(\frac{2}{N}\right)^2 \cdot \delta^v + 1}{\delta}$, where N is the principal quantum number of the atom i, δ^v is the number of electrons in the skeleton $(Z^v - h)$, and d is the number of electron in the skeleton $(s - h)$. The E-state $S(A_i)$ for the atom is the modified intrinsic value $S(A_i) = I_i + \Delta I_i$, where ΔI_i quantifies the perturbing effect on the intrinsic atom value. This perturbation is assumed to be a function of the difference in the intrinsic values I_i and I_j $\Delta I_i = \sum_{j=1}^{N} \frac{I_i - I_j}{r_{ij}^2}$. And the total index is the sum of all $S(A_i)$ values

Reference number	Descriptor name	Description
3.5.	Average valence connectivity index χ^1	Is based on the Randic connectivity index defined as $$\chi_R \equiv {}^1\chi = \sum_b (\delta_i \cdot \delta_j)^{-1/2},$$ where b runs over the B bonds $i-j$ of the molecule; δ_i and δ_j are the vertex degrees of the atoms bonded. The mean connectivity index is defined as $$\overline{\chi_R} = \frac{\chi_R}{B},$$ where B is the number of edges in the molecular graph
3.6.	Mean information content on the distance degree equality	It consist in applying the mean information content equation (see 2.11) over the degree of distance equality, calculated as $$\sigma_i = \sum_{k=1}^{\eta_i} {}^k f_i \cdot k,$$ where ${}^k f_i$ is the vertex distance count of kth order and the sum runs over the different distance values
3.7.	Total information content index (neighborhood symmetry of 0 order)	It represents the residual information contained in the relationships defined among the elements of the molecule. It is calculated by $$I = n \log_2 n - \sum_{g=1}^{G} n_g \log_2 n_g,$$ where n is the number of elements and n_g is the number of elements of the gth class
3.8.	Sum of topological distances between S$\cdots$S	Already defined in 2.14, but applied to the presence of sulphur atoms
3.9.	Highest eigenvalue number 3 of Burden matrix/weighted by atomic mass	Already defined in 2.16.; in this case, it is the 3rd higher eigenvalue

Reference number	Descriptor name	Description
3.10.	Mean topological charge index of order 6	Topological charge indices were proposed to evaluate the charge transfer between pairs of atoms and therefore the charge transfer over the molecule. It is calculated by $$J_k = \frac{\frac{1}{2}\sum\sum\lvert CT_{ij}\rvert \cdot \delta(k, d_{ij})}{A-1},$$ where $\lvert CT_{ij}\rvert$ is the charge term matrix which is a square and unsymmetric matrix; every term, defined as pair of vertices, represents the charge transfer between pairs of atoms; $\delta(k; d_{ij})$ is a Kroneker delta which is 1 if $\delta_{ij}=$ order (6 in this case), 0 otherwise, and A is the number of atoms
3.11.	Moran autocorrelation – lag 2/weighted by atomic Sanderson electronegativities	Already defined in 1.16. but with lag 2 for this case
3.12.	Geary autocorrelation – lag 1/weighted by atomic Sanderson electronegativities	Already defined in 1.16. but with lag 1 for this case
3.13.	Weighted presence of local and global invariants	This is an in-house-built descriptors, defined as the presence of some local invariants (counting the presence of carbon atoms, aromatic carbon atoms, oxygen, nitrogen,. . . ,double bonds, triple, branching, etc.) and global invariants (counting the molecules with defined moieties like molecules which have nitrogen, oxygen, chlorine, double bonds, branching, etc.). Initially every invariant has a value of 1, but they are weighted by means of a Monte Carlo algorithm to obtain the best fitting between the toxicity and the total descriptor in the form of a linear equation expressed in the general way $$pTox = a \cdot D(SMILES) + b$$ and D(SMILES) is calculated by $$D(SMILES) = CW(CC)\prod_{k=1}^{N} CW(s_k),$$ where CC is the global invariant calculated, $CW(CC)$ is a correlation weight of the CC, and s_k is the SMILES fragment defined

4. DIETARY QUAIL

Reference number	Descriptor name	Description
4.1.	Number of nitrogen atoms	Count of the number of nitrogen atoms involved in any functionality 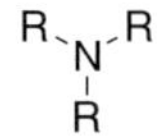
4.2.	Number of phosphorous atoms	Count of the number of phosphorous atoms involved in any functionality 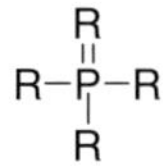
4.3.	Number of fluorine atoms	Count of the number of fluorine atoms. No matter the substitution R–F
4.4.	Number of ethers (aromatic)	Count of the number of ethers involved in the union of at least one aromatic ring 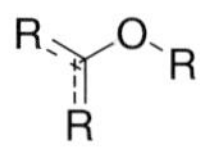
4.5.	Number of sulphurs	Count of the number of sulphurs present in the molecule in any functionality 
4.6.	Number of X–C on aromatic ring	Count of the number of halogen atoms substituting in aromatic rings 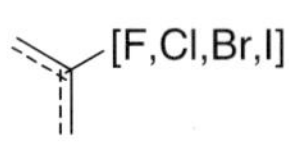
4.7.	H attached to C0 (sp^3) with 1 X attached to next C	It is the count of hydrogen bonded to an sp^3 carbon bonded to another carbon with halogen substitution 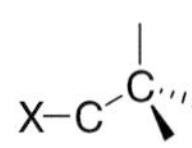
4.8.	Mean square distance index (Balaban)	It is the mean of the square of all distances calculated by means the Balaban distance connectivity index, defined as $ADDD = \frac{1}{A}\sum_{i=1}^{A}\sum_{j=2}^{A}\left(\frac{r_{ij}}{d_{ij}}\right)^2,$

Reference number	Descriptor name	Description
		where r_{ij} is the geometric distance and d_{ij} is the topological distance
4.9.	Second Mohar index TI2	This index is calculated over de Laplacian matrix which is a square symmetric matrix defined as $$[\mathbf{L}]_{ij} = \begin{cases} \delta_i & if \quad i=j \\ -1 & if \quad \{i,j\} \in E(G), \\ 0 & if \quad \{i,j\} \notin E(G) \end{cases}$$ whose diagonalization gives a real eigenvalue λ_i. The second Mohar index is calculated by $$(TI)_2 = \frac{4}{A \cdot \lambda_{A-1}},$$ where λ_{A-1} is the first non-zero eigenvalue
4.10.	Maximal electrotopological negative variation	It is the sum of the negative members of the E-state index, defined in 3.4.
4.11.	Average connectivity index χ^1	It is defined in the same way as the valence connectivity indices 3.5 replacing the valence vertex degree by the vertex degree. Order 4 in this case
4.12.	Average connectivity index χ^3	Already described in 4.11. but with order 3 for this case
4.13.	Average connectivity index χ^4	Already described in 4.11. but with order 4 for this case
4.14.	Valence connectivity index χ^4	Already described in 3.5. Order 4
4.15.	Path/walk 4 – Randic shape index	It is defined as the ratio between the atomic path count ${}^mP_i$ and the atomic walk count $awc_i^{(m)}$ of the same length m $$(p/w)_i^m = \frac{{}^mP_i}{awc_i^{(m)}}$$
4.16.	Eigenvector coefficient sum from adjacency matrix	It is the sum of the eigenvectors of the adjacency matrix. This matrix represents the set of connections present in the molecule, each term of the matrix has value 1 if vertices v_i and v_j are adjacent, 0 otherwise

Reference number	Descriptor name	Description
4.17.	Average eigenvector coefficient sum from adjacency matrix	It is the average of the eigenvectors coefficients from the adjacency matrix defined in 4.16.
4.18.	Average Randic-type eigenvector-based index from adjacency matrix	It is the average of the eigenvalues of a matrix derived from substitute Randic indices in the adjacency matrix $x^1 = \sum_b (\delta_i \cdot \delta_j)_b^{-1/2}$
4.19.	Sum of topological distances between O···S	Already defined in 2.14. but applied to the presence of oxygen and sulphur atoms
4.20.	Sum of topological distances between O···P	Already defined in 2.14. but applied to the presence of oxygen and phosphorous atoms
4.21.	Sum of topological distances between O···Cl	Already defined in 2.14. but applied to the presence of oxygen and chlorine atoms
4.22.	Self-returning walk count of order 5	Already defined in 1.14.
4.23.	Self-returning walk count of order 7	Already defined in 1.14. Order 7 in this case
4.24.	Self-returning walk count of order 9	Already defined in 1.14. Order 9 in this case
4.25.	Highest eigenvalue number 1 of Burden matrix/weighted by atomic masses	Already defined in 2.16.
4.26.	Highest eigenvalue number 8 of Burden matrix/weighted by atomic masses	Already defined in 2.16. This is the 8th higher eigenvalue
4.27.	Topological charge index of order 10	The topological charge has been defined in 3.10. In this case, the average is not necessary
4.28.	Mean topological charge index of order 4	Same as defined in 3.10. Order 4 in this case
4.29.	Broto-Moreau autocorrelation of a topological structure – lag 7/weighted by atomic Sanderson electronegativities	It is another measure of autocorrelation, defined over the molecular graph as $ATS_d = \sum_{i=1}^{A} \sum_{j=1}^{A} \delta_{ij} (\chi_i \cdot \chi_j)_d$ at the considered topological distance 7

Reference number	Descriptor name	Description
4.30.	Broto-Moreau autocorrelation of a topological structure – lag 1/weighted by atomic polarizabilities	Already defined in 4.29, at $d = 1$
4.31.	Moran autocorrelation – lag 2/weighted by atomic Sanderson electronegativities	Already defined in 1.16. Lag 2 in this case
4.32.	Moran autocorrelation – lag 5/weighted by atomic Sanderson electronegativities	Already defined in 1.16. Lag 5 in this case
4.33.	Moran autocorrelation – lag 3/weighted by atomic polarizabilities	Already defined in 1.16. Lag 3 in this case and substituting electronegativities by polarizabilities
4.34.	Geary autocorrelation – lag 2/weighted by atomic Sanderson electronegativities	Already defined in 1.18.
4.35.	Relative number of C atoms	It is the ratio between the number of carbon atoms and the total number of atoms of the whole molecule
4.36.	Relative number of F atoms	It is the ratio between the number of fluorine atoms and the total number of atoms of the whole molecule
4.37.	Relative number of P atoms	It is the ratio between the number of phosphorous atoms and the total number of atoms of the whole molecule
4.38.	Number of rings	It is the count of all the possible rings in the molecule
4.39.	Gravitation index (all bonds)	It is a measure of the mass distribution of the molecule. It is defined by $$G_1 = \sum_{i=1}^{A-1} \sum_{j=1}^{A} \frac{m_i \cdot m_j}{d_{ij}^2},$$ where m is the atomic mass and d_{ij} is the topological distance
4.40.	Randic index (order 0)	Already defined in 3.5. but with order 0, one bond topological distance (δ)

Reference number	Descriptor name	Description
4.41.	Kier&Hall index (order 1)	It is an upgrade of the Randic index to calculate higher order descriptors $$^{m}\chi_q = \sum_{k=1}^{K}\left(\prod_{a=1}^{n}\delta_a\right)_k,$$ where k runs over all the mth order subgraphs and K is the total number of mth order subgraphs. The product is over the simple vertex degrees δ of all the vertices involved in each subgraph. In this case, $m = 1$
4.42.	Kier&Hall index (order 3)	Already defined in 4.41. In this case, $m = 3$
4.43.	Kier shape index (order 2)	Already defined in 4.41. In this case, $m = 2$

5. BEE

Reference number	Descriptor name	Description
5.1.	Molecular weight	Already defined in 2.1.
5.2.	Number of multiple bonds	It is the count of all the double triple and aromatic bonds over the molecule
5.3.	Number of oxygen atoms	It is the count of all the oxygen atoms present in the molecule without considering the substitution
5.4.	Number of esters (aliphatic)	Already defined in 1.20.
5.5.	X–CX–X	Already defined in 3.2.
5.6.	H attached to C3 (sp^3)/C2 (sp^2)/C3 (sp^2)/C3 (sp)	It is the count of all the possible hydrogen atoms bonded to carbon atoms with the indicated hybridations

Reference number	Descriptor name	Description
5.7.	O =	It is the count of the carbonyl groups present in the molecule. No matter if they are also involved in higher functionalities
5.8.	Al–O–Ar/Ar–O–Ar/R ···O···R/R–O–C = X	Already described in 2.7.
5.9.	X3–P = X (phosphate)	It is the count of all the phosphate groups present in the molecule
5.10.	Balaban-type index from van der Waals weighted distance matrix	It is an index calculated by the Balaban formula over a distance matrix weighted by the van der Waals atomic radius
5.11.	Average valence connectivity index χ^0	Already described in 3.5. Order 0
5.12.	Average valence connectivity index χ^5	Already described in 3.5. Order 5
5.13.	Eccentric connectivity index	It is defined as the sum of the products between eccentricity η and vertex degree δ over the atoms of the graph. It is calculated by $$\xi^c = \sum_{i=1}^{A} \eta_i \cdot \delta_i$$
5.14.	Mean information content on the distance degree equality	Already defined in 3.6.
5.15.	Graph vertex complexity index	It is derived from the distance matrix and is calculated as the molecular average vertex complexity. It is calculated as $$H_V = \frac{1}{A}\sum_{i=1}^{A}\left(-\sum_{g=0}^{\eta_i}\frac{{}^g f_i}{A}\cdot\log_2\frac{{}^g f_i}{A}\right),$$ where ${}^g f_i$ is the number of distances from the vertex v_i equal to g and A is the number of atoms
5.16.	Lovasz-Pelikan index (leading eigenvalue)	It is a measure of molecular branching. It is the leading eigenvalue of the adjacency matrix
5.17.	Eigenvector coefficient sum from adjacency matrix	It is the sum of the coefficient eigenvectors of the adjacency matrix

Reference number	Descriptor name	Description
5.18.	Sum of topological distances between S···P	Already defined in 2.14. but applied to the presence of sulphur and phosphorous atoms
5.19.	Self-returning walk count of order 5	Already defined in 1.14.
5.20.	Relative molecular weight	It is the sum of all the atomic masses of all the atoms present in the molecule divided by the number of atoms. It is calculated by $MW = \frac{1}{A}\sum_{i=1}^{A} m_i$

Quantitative Structure-Activity Relationships (QSAR) for Pesticide Regulatory Purposes
Edited by Emilio Benfenati

Appendix E

List of abbreviations

Mosè Casalegno[1]

[1]Istituto di Ricerche Farmacologiche "Mario Negri", Milano, Italy

Term	Definition
3D	Three-dimensional (descriptors or structures)
2D	Bi-dimensional (descriptors or structures)
AP	Associated product
AS	Active substance
AC	Analog complexing
AFP	Adaptive fuzzy partition
AI	Artificial intelligence
ANN	Artificial neural networks
APC	Average probability-based combination
AQUIRE	Aquatic toxicity information retrieval database
AUC	Area under the ROC curve
BBA	German Federal Biological Research Centre for Agriculture and Forestry
BVL	German Federal Office for Consumer Protection and Food Safety
bw	Body weight
CAS	Chemical Abstract Service
CDS	Crystallographic Database Service
CEPA	Canadian Environmental Protection Act
CML	Chemical markup language
CoMFA	Comparative molecular field analysis
CPDBAS	Carcinogenic potency database
CPSA	Partial charge surface area
CSCL	Japanese chemical substance control law
CSD	Cambridge structural database
CSSR	Cambridge structure search and retrieval
CSTEE	Scientific Committee on Toxicity, Ecotoxicity and the Environment
CTB	Board for the Authorisation of Pesticides (Netherlands)
cw	Correlation weight

Term	Definition
DBPCAN	EPA Water Disinfection By-Products with Carcinogenicity Estimates
DCA	Department of Community Affairs (U.S.)
DDT	Para-dichloro-diphenyl-trichloroethene
DEFRA	Department for Environment, Food, and Rural Affairs (U.K.)
DN	Data normalization
DP	Descriptive power
DSSTox	Distribute structure-searchable toxicity
DT	Decision tree
$DT90_f$	Disappearance time 90%, the time takes a dissipation study until 90% of initial amount has disappeared
EAV	Entity attribute value
EbC_{50}	Effective concentration that results in a 50% reduction in algal biomass
EC_{10}	Effective concentration 10 – the concentration at which there is 10% effect of the test population
EC_{20}	Effective concentration 20 – the concentration at which there is 20% effect of the test population
EC_{25}	Effective concentration 25 – the concentration at which there is 25% effect of the test population
EC_{50}	Effective concentration 50 – the concentration at which there is 50% effect of the test population
ECB	European Chemical Bureau
ECCO	European Community COordination
EEC	European Economic Community
EK	Explicit knowledge
EPA	Environmental Protection Agency
EPACTP	EPA Computational Toxicology Program
EPAFHM	EPA Fathead Minnow Acute Toxicity
EPPO	European and Mediterranean Protection Plant Organization
ErC_{50}	Median effective concentration as determined by measuring the average specific growth rate
ESCORT 2	European Standard Characteristics of Non-Target Arthropod Regulatory Testing
ETE	Estimated theoretical exposure
EU	European Union
EXTOXNET	EXtension TOXicology NETwork database
FADINAP	Fertilizer Advisory, Development and Information Network for Asia and the Pacific
FDA	Food and Drug Administration (U.S.)
FDAMDD	FDA maximum daily dose database

Term	Definition
FEM	Fire each module
FEMF	Fuzzy inference of linguistic outputs
FEMS	Statistical combination of crisp outputs
FIS	Fuzzy inference system
FL	Fuzzy logic
FS	Features selection
FTP	Fast transfer protocol
GA	Genetic algorithm
GAO	Graph of atomic orbitals
GLP	Good laboratory practice
HECS	Healthy Environments & Consumer Safety Branch
HFG	Hydrogen-filled graph
HIS	Hybrid intelligent systems
HNN	Hybrid neural network
HQ	Hazard quotient
HSG	Hydrogen-suppressed graph
IBL	Instance-based learner
ICPS	International Centre for Pesticide and Health Risk Prevention
IK	Implicit knowledge
ILSI	International Life Science Institute
InChi	International chemical identifier
INRA	French National Institute for Agricultural Research
ISO	International Organization for Standardization
IUPAC	International Union of Pure and Applied Chemists
JRC	Joint Research Centre (Italy)
KDD	Knowledge discovery from databases
KNN	*K*-nearest neighbor
LC_{50}	Lethal concentration for 50% of a population
LD_{50}	Lethal dose for 50% of a population
LOEC	Lowest-observed-effect-concentration
log *P*	Logarithm (base 10) of the octanol water partition coefficient, often indicated also as log Kow
LOO	Leave one out
LR_{50}	Lethal residue that cause mortality to 50% of population
LRI	Long range initiative
LSO	Leave several out
LVQ	Learning vector quantization
MCS	Multiple classifier system
METI	Japanese Ministry of Economy, Trade and Industry
MHLW	Japanese Ministry of Health, Labor and Welfare

Term	Definition
MLP	Multi-layer perceptron
MM	Molecular mechanics
MO	Molecular orbital
MOE	Japanese Ministry of Environment
MPC	Maximum permitted concentration
MPC	Maximal probability-based combination
MVC	Majority voting-based combination
NCTRER	National Centre for Toxicological Research Estrogen Receptor Binding Database
NIEHS	US National Institute of Environmental Health Science
NITE	National Institute of Technology and Evaluation
NMR	Nuclear magnetic resonance
NN	Neural network
NOEC	No observable effect concentration
NOEL	No observable effects limit
NR	Noise removal
NTP	National Toxicology Program (U.S.)
OCWLGI	Optimization of Correlation Weights of Local Graph Invariants
OECD	Organization for Economic Co-operation and Development
OPP	U.S. Office of Pesticide Programs
Opt-3D	Optimized three-dimensional molecular structure
ORD	Office of Research and Development (U.S.)
PAN	Pesticide action network (North America)
PCA	Principal component analysis
PEC	Predicted environmental concentration
PHYTOTOX	Terrestrial plants toxicity database (EPA)
PLS	Partial least squares
PMN	Pre-manufacture notice
PNSA1	Partial negative surface area
PNSA2	Total charge weighted partial negative surface area
PNSA3	Atomic charge weighted partial negative surface area
ppm	part per million
PPP	Plant production product
PPSA	Partial positive surface area
PRESS	fraction of residual (or unexplained) variance
PSD	Pesticide safety directorate (U.K.)
QA	Quality assurance
QSAR	Quantitative structure-activity relationship
R^2	correlation coefficient
RBFNN	Radial basis function neural network

Term	Definition
REACH	Registration, Evaluation and Authorisation of CHemicals
REC	Regression error characteristic
RIPPER	Repeated incremental pruning to produce error reduction
RIVM	National Institute for Public Health and Environment (The Netherlands)
ROC	Receiver operating characteristic
(DG) SANCO	Directorate-General for Health and Consumer Protection
SAR	Structure-activity relationship
SDF	Structure data file
SEEM	Statistical evaluation of Available Ecotoxicology data on plant protection products and their Metabolites
SGN	Supervised-trained gating network
SMILES	Simplified Molecular Input Line Entry System
SOSLM	Self-organizing statistical learning methods
SOSLN	Self-organizing statistical learning networks
SSY	Total variance
TD_{50}	Toxic dose for 50% animals tested
TER	Toxicity-exposure ratio
TERRETOX	EPA terrestrial species toxicity database
TRI	Toxic chemical release inventory
TSCA	Toxic Substances Control Act (USA, 1976)
UGN	Unsupervised-trained gating network
UNICORE	UNiform Interface to Computer REsources
UnOpt-3D	Unoptimized three-dimensional molecular structure
USDA	U.S. Department of Agriculture
USEPA	U.S. Environmental Protection Agency
USFWS	U.S. Fish and Wildlife Service
WPC	Weighted probability-based combination
XML	eXtenisble markup language

Edited by Emilio Benfenati

Appendix F

Software tool for toxicity prediction of pesticides, candidate pesticides, and their derivatives (user guide)

Severin Bumbaru[1], Marian Viorel Crăciun[1], Viorel Mînzu[1], Daniel Neagu[2], Anna Maria Pezzotti[3]

[1]Department of Computer Science and Engineering, University "Dunarea de Jos", Galati, Romania

[2]Department of Computing, School of Informatics, University of Bradford, Bradford, UK

[3]Laboratory of Environmental Chemistry and Toxicology, Istituto di Ricerche Farmacologiche "Mario Negri", Milano, Italy

INTRODUCTION

The aim of this guide is the description of a software tool for toxicity prediction of molecules of pesticides and related compounds. The software is based on the integration of the knowledge acquired in an EU project in a homogeneous manner using the best algorithms obtained as the basis for hybrid combinative models to be used for predictive purposes.

The tool allows processing of chemical compounds one by one and in batch for the prediction of toxicity against five endpoints: acute toxicity for rainbow trout (*Oncorhynchus mykiss*): LC_{50} 96-h exposure; acute toxicity for water flea (*Daphnia magna*): LC_{50} 48-h exposure; acute oral toxicity for bobwhite quail (*Colinus virginianus*): LD_{50} 14-day exposure; dietary toxicity for bobwhite quail (*C. virginianus*): LD_{50} 8-day exposure; acute contact toxicity for honey bee (*Apis melifera*): LD_{50} 48-h exposure.

The input is the chemical structure of the molecule, characterized by a list of numerical chemical descriptors. The algorithms, as quantitative structure activity relationships (QSARs), output the estimated toxicity value. The values outside the expertise domain of the predictive models are also indicated.

INTENDED AUDIENCE

The list of the potential users of the software includes regulatory bodies, industries, non-governmental agencies, researchers from academics and non-academics areas, and general public.

The tool allows regulatory evaluators to assess data submitted in support of approval applications.

WHAT IS DEMETRA?

DEMETRA – Development of Environmental Modules for Evaluation of Toxicity of pesticide Residues in Agriculture (QLK5-CT-2002-00691) – is an EU FP5 project.

The major objective of this project is to produce software for toxicity prediction of pesticides and related compounds (such as metabolites), directly and immediately useful for the evaluation of pesticides and related compounds within the Dossier preparation for pesticide registration.

PREPARING THE INPUT: DESCRIPTORS CALCULATION

The user should refer to the manuals of the software packages for descriptor calculations. Below we provide a simplified description, which is valid for the software versions we used.

The following procedure serves to generate molecular descriptors for the compounds under investigation. This procedure could be skipped if the molecular descriptors have already been computed or made available. Two simple steps have to be carried out: drawing the molecular structure and computing the descriptors by submitting the molecular structure to the following software packages: Codessa, Dragon, and ACDLabs.

DRAWING THE COMPOUND STRUCTURES

If we have a structure drawn on a sheet, first of all, it is important to create the .mol file representative of this molecule. To generate this file, we can use the software ACDLabs 9.0 (ChemSketch).

In this phase, it is important to *add explicitly* all the *hydrogens* that are present in the molecule (to make this, we have to select 'add explicit hydrogens' in the 'tools' panel) within the versions we indicated in Chapter 2 (Benfenati *et al.*, this volume), and specified below.

DESCRIPTORS CALCULATION

The second step deals with the calculation of the descriptors of interest. We should have identified the descriptors representative of the modeling equation, listed for each endpoint at the Web site http://demetra-tox.net.

These descriptors have to be calculated through the following software packages:

1) Dragon (Web Version 3.0; http://www.talete.mi.it/)

   It is a well-known software for the calculation of more than 1600 molecular descriptors from several input formats (MDL, SYBYL, HyperChem, and SMILES).

   If we use the MDL file as input format, we have to add the '$$$$' characters at the end of the text file (if not present).

   If we prefer the file .smi, in this case, we should create a text file in this way: every row should be composed of the SMILES notation of the compound plus (after a 'tab' space) the CAS RN of the same compound. The advantage of the file .smi is that we have only one file that comprises all the molecules we want to screen.

   In spite of this, it is important to say that the file .smi in some cases has created problems, rejecting the calculation of descriptors for some compounds. The reason is that the descriptors that require the hydrogens explicitly added cannot be calculated. Due to the fact that it has been proved that explicit hydrogens are necessary, the SMILES notation cannot be used.

   Now, a new version of Dragon is available, and several of the above-described situations have been improved.

2) Codessa (Version 2.20)

   MDL .mol files format have been used as input files to calculate descriptors.

   For descriptors calculation, go to the Data Menu Commands in the program menus. This tool provides commands for entering and manipulating data: structures, properties, and descriptors.

   The command 'load' is aimed for loading input data from a text (ASCII) file.

   An input text file should be organized in columns separated by a blank, and each column should contain data of one type (e.g., structures names, property values, and files names).

   An example of an input text file for Codessa is the following:

```
MOLPATHS STRNAME:
C:\finBEE\mol001.mol mol001
C:\finBEE\mol002.mol mol002
C:\finBEE\mol003.mol mol003
C:\finBEE\mol004.mol mol004
C:\finBEE\mol005.mol mol005
C:\finBEE\mol006.mol mol006
```

C:\finBEE\mol007.mol mol007

C:\finBEE\mol008.mol mol008

C:\finBEE\mol009.mol mol009

The first column stands for the MDL's .mol file paths and the second column stands for the structure names. Indeed, this text file explains that we can find the .mol files of all our compounds of interest in the folder 'finBEE' present in the local disk 'C'.

A simpler solution for calculating few descriptors with Codessa is to use the commands: 'view descriptor' to select our descriptors and then use the command 'print descriptor'. We will obtain the list of values associated with the name of the molecule in a semicolon-separated text file. As for the second column, it repeats the name of every structure.

At this time, the program can scan the file and calculate the descriptor values.

The output of Codessa software is a text file, but it has to be rearranged in a way that can be manipulated with a standard spreadsheet.

3) ACDLabs 9.0 (http://www.acdlabs.com)

This software has been used for the calculation of only one descriptor, log D, pH 7, and exclusively for the trout endpoint.

The input file that has been chosen, for practicality, is the MDL SDfile (.SDF). Indeed, this file format can contain multiple compounds together, with properties and references.

Alternatively, other input files can be used, such as ACD/ChemFolder database (.CFD); SMILES text file (.TXT); and ISIS database (.DB).

How to open the input file of interest through the program: when we run ACD/log D, the program opens the ChemSketch window. Then, we have to click on the panel 'history' on the bottom of the spreadsheet. At this step, we have to go to the 'tools' option in the history panel and choose the import command: we can select our SDfile according to its location and we have the possibility to choose the properties to calculate (in this case, the log D).

We can see that all our molecules have been imported and we can also observe every structure of these.

Then, we click on View/Log D results/Custom pH and we select 7 as a single value.

Finally, we click History/Tools/Export and we can save all our results in a folder. At this step, it is also important to select the available fields that allow us exporting our data options, for example, the 'file name' and the 'internal id', to have the values of log D associated to the corresponding name of each compound.

Once all the descriptors of interest have been computed, one can submit (e.g., by copy/paste) the computed values in the applet for predicting the desired endpoint value.

GETTING THE SOFTWARE

The tool is provided as downloadable, standalone Java application and Java applet, running within a Web browser using the Java Plug-in (Microsoft® Internet Explorer, Netscape Navigator®, Mozilla, Opera, etc.).

The compiled file (*bytecode*) can be downloaded from http://www.demetra-tox.net.

The applet is available at http://demetra-tox.net.

PLATFORMS AND REQUIREMENTS

The software is written in Java™ and can run in any machine supporting Java or Java 2 runtime environment (J2RE) Version 1.4 or later.

The system requirements for the computers running DEMETRA software tool are the same as the recommended minimum necessities for the J2RE standard edition (http://java.sum.com):

- Software requirements (operating systems):
  - Microsoft Windows 98 (1st and 2nd editions), ME, NT 4.0 (Service Pack 6a), XP Home Edition, XP Professional Edition (Service Pack 1), 2000 Professional (Service Pack 3), Server 2003;
  - Linux; and
  - Unix (Solaris 7, 8, and 9).
- Hardware requirements:
  - x86 processors, Intel, and 100% compatible processors (minimum Pentium 166 MHz), at least 32 MB RAM to run graphically based applications and 48 MB RAM for applets running within a Web browser, minimum 75 MB of free disk space.

However, the project's technical annex recommends as minimum hardware configuration: 600 MHz processor, 64 MB RAM, 500 MB available on hard disk.

ADDITIONAL RECOMMENDATION

To allow the applet to access the clipboard (for copy and paste facilities), the user must set a policy file. If the file already exists, the following lines must be added at the end of it. Otherwise, such a file must be created and it must contain the following lines:

```
grant codeBase "http://www.demetra-tox.net/prototype/*" {
permission java.awt.AWTPermission "accessClipboard";
};
```

In Windows environments, the policy file is usually located here:

```
C:\Program Files\Java\j2re*\lib\security\java.policy
```

In Linux, the file is usually located in the Java home directory in the user's home directory:

```
~/.java.policy
```

USING THE SOFTWARE – THREE EASY STEPS

The standalone application could be started using the command 'java -jar DemetraTool.jar' in the directory where the jar file was saved or simply executing the jar file.

The applet starts when the Web page is loaded within the Web browser.

Both the applet and the Java application have the same graphical user interface (GUI). The user is guided in a three-step wizard to obtain the toxicity values against one of the five endpoints, giving the values of the chemical descriptors of its molecule.

The user can navigate through steps using either the links in the menu on the left side of the window or the two buttons in the bottom of the window (Fig. 1).

CHOOSE THE ENDPOINT

The first step is the selection of the endpoint. In the main part of the application window there is the list of the five endpoints. The endpoints are described as text, and a picture of the life form is provided on the right side.

The user can select an endpoint by a simple mouse click or pressing the *<TAB>* and then *<Enter>* key, Next button, or by clicking the link *2) Input values* in the menu. The selected endpoint is highlighted. The effect of the endpoint's selection is the advancement at second step.

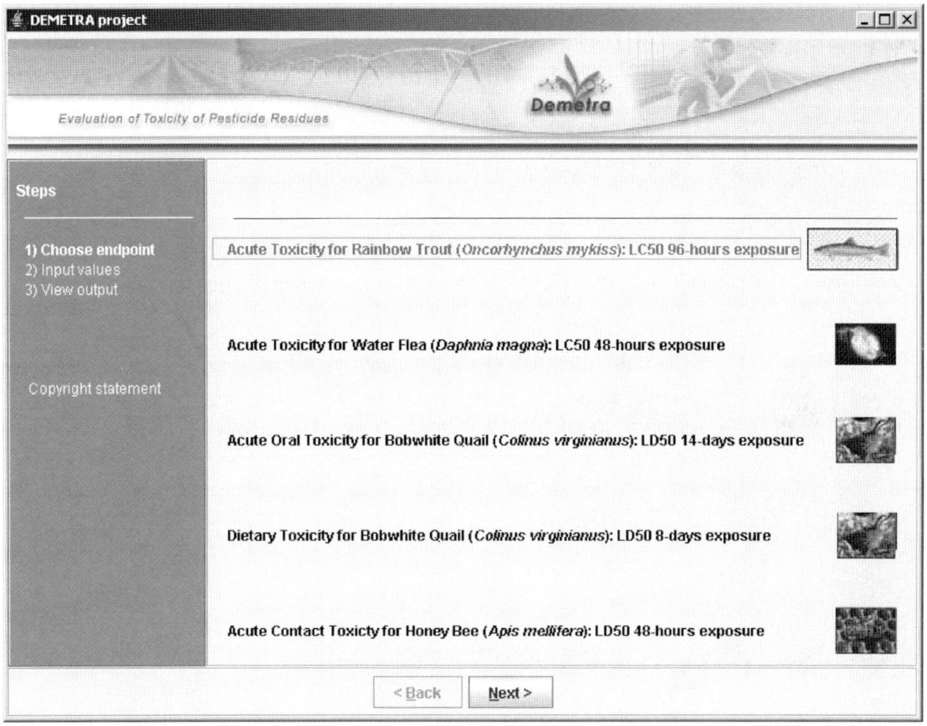

Figure 1: First step – choosing the endpoint.

INPUT THE VALUES OF THE DESCRIPTORS

In the second step, the user have to enter the values of the descriptors after, eventually, introducing the number of the chemical compounds he or she wants to evaluate. The values of the descriptors are entered into a table (Fig. 2).

First column in the table contains the name of the descriptors followed by the name of suggested software to compute their values (in parentheses). If the mouse is pointed over a row in the table, a tooltip gives few more details regarding the corresponding descriptor (Fig. 2).

If the user has already computed the values of the descriptors and those values are presented into an Excel worksheet (for example, see Fig. 3) he or she can copy and paste the numbers. The well-known key combinations CTRL + C and CTRL + V or a pop-up menu (mouse right click) can be used (Fig. 4).

To use this copy and paste facilities within the applet version of the tool, please refer to the *Platforms and requirements* section for details.

Before pasting data, the user must select the starting cell in the table, paying attention to the dimension of the data in the clipboard, and the number of rows and columns in the table. The results of this is presented in Fig. 5.

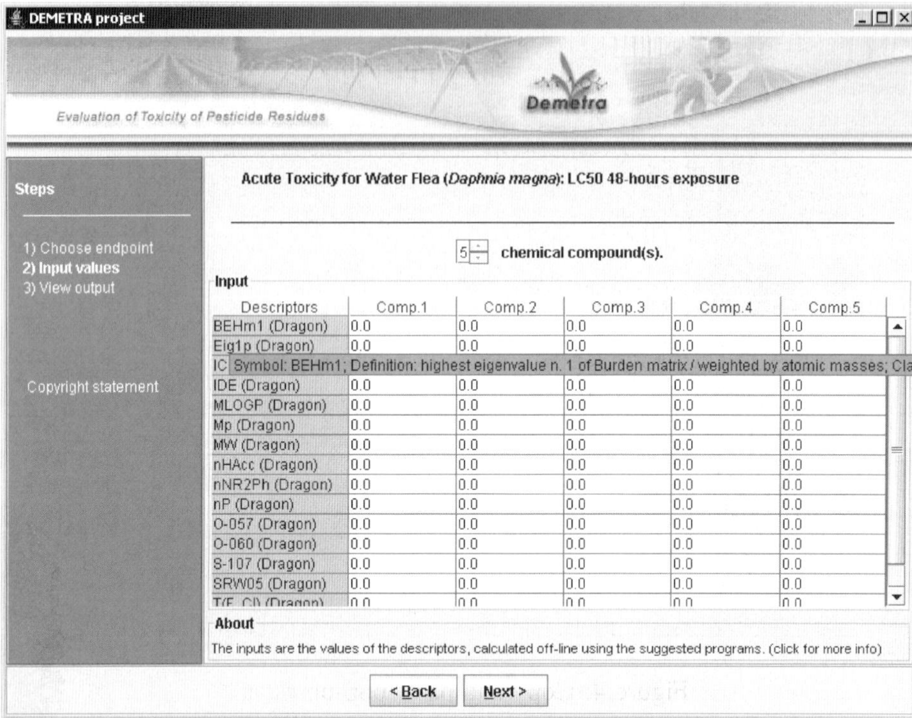

Figure 2: Descriptors details in a tooltip.

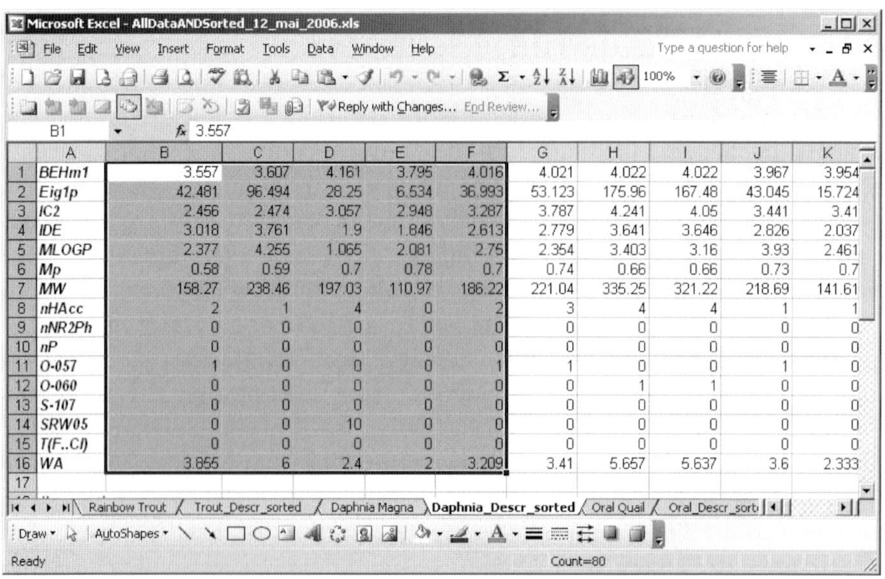

Figure 3: Excel worksheet screenshot.

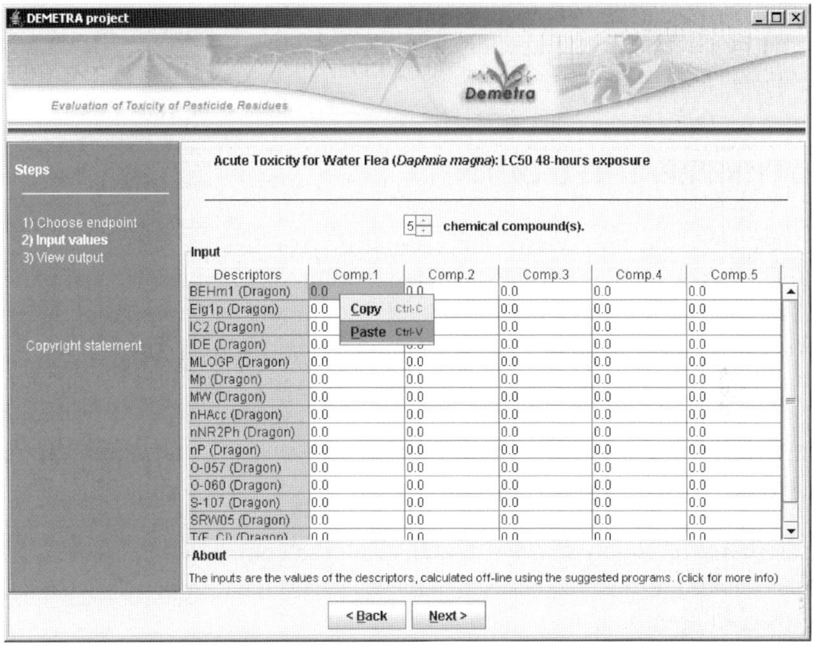

Figure 4: Copy and paste pop-up menu.

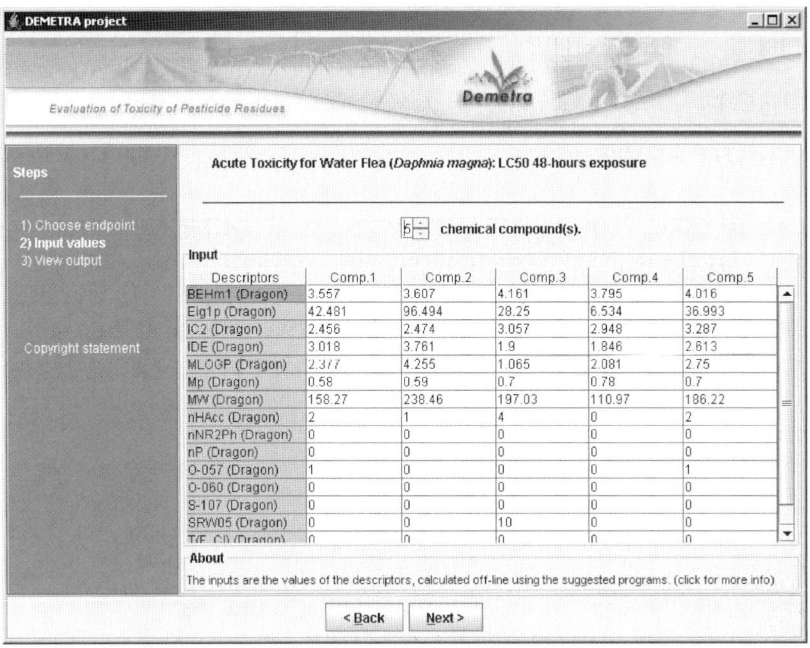

Figure 5: Second step – entering the descriptors.

The values of the descriptors could be introduced within the cells of the table, one by one, using the keyboard, too.

VIEW THE PREDICTED OUTPUTS

In the last step, the user views the estimated toxicities as negative logarithmic values. The expertise range is also depicted as minimum and maximum predictions.

All the predictive models implemented in this software tool are hybrid combination of two or more individual models trained on a special set each for each endpoint. The minimum and maximum values in the predictions are computed based on to the minimum and maximum of the individual models on the training set and the predicted values for the evaluated compound. These values do not refer to the range of the hybrid model.

The outputs having values outside this (minimum or maximum) domain are considered *unreliable predictions*, and instead, the computed values, < *minimum* or > *maximum*, are given. The user is the one who have to decide what value to use further, minimum or maximum.

The algorithm to compute the *domain of expertise* is proposed by the partners in DEMETRA project, and it can be obtained from the project Web site or from the coordinator of the project.

The predictive hybrid models, QSARs, used to predict the toxicity against those five endpoints are also available on the project Web site or can be provided by the coordinator of the project (Fig. 6).

COPYRIGHT STATEMENT

Copyright (c)2002–2006 DEMETRA Project. All DEMETRA programs and accompanying documentation are protected by copyleft.

The DEMETRA programs are free software applications; you can redistribute them under terms of the GNU General Public License as published by the Free Software Foundation; either version 2 of the License, or (at your opinion) any later version, but changing of the software and distributed models it is not allowed.

The DEMETRA programs are distributed in the hope that they will be useful, but WITHOUT ANY WARRANTY; without even the implied warranty of MERCHANTABILITY or FITNESS FOR A PARTICULAR PURPOSE. See the GNU General Public License for more details.

You should have received a copy of the GNU General Public License along with the DEMETRA programs; if not, look on the web at http://www.gnu.org/copyleft/gpl.html or write to the Free Software Foundation, Inc., 59 Temple place - Suite 330, Boston, MA 02111-1307, USA.

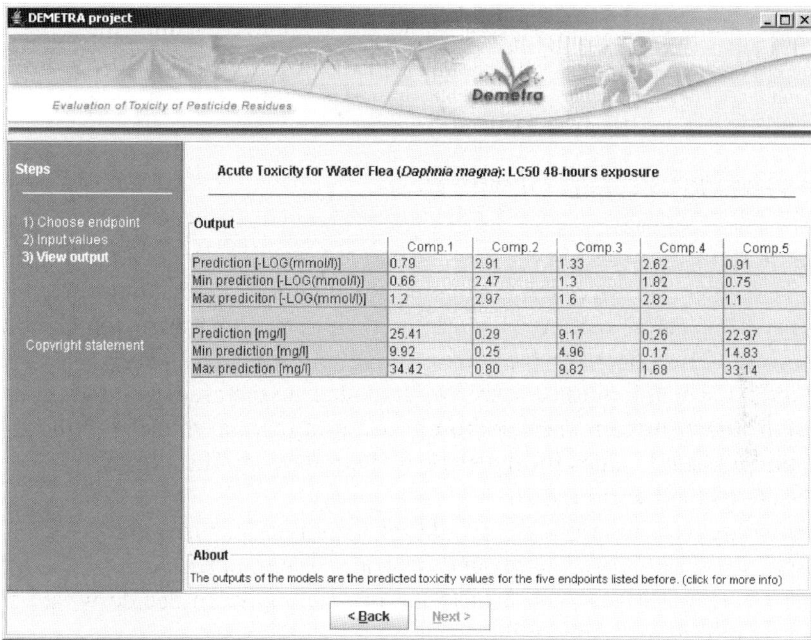

Figure 6: Last step – viewing the predicted toxicity.

To contact the director of the DEMETRA project, send email to Dr. Emilio BENFENATI, benfenati@marionegri.it or write him at Laboratory of Environmental Chemistry and Toxicology, Instituto di Ricerche Farmacologiche "Mario Negri", Via La Masa 19, 20157 Milano, Italy.

Index